Recent Advances in Polyphenol Research

Recent Advances in Polyphenol Research

Volume 3

Edited by

Véronique Cheynier
Research Director, Food Chemistry
Institut National de la Recherche Agronomique
UMR1083 Sciences pour l'Œnologie
Montpellier, France

Pascale Sarni-Manchado
Research Associate, Plant and Food Biochemistry
Institut National de la Recherche Agronomique
UMR1083 Sciences pour l'Œnologie
Montpellier, France

Stéphane Quideau
Professor, Organic and Bioorganic Chemistry
Institut des Sciences Moléculaires, CNRS-UMR 5255
Institut Européen de Chimie et Biologie
Université de Bordeaux, France

WILEY-BLACKWELL

A John Wiley & Sons, Ltd., Publication

Dedications

To Jean-Jacques Macheix—a board member of Groupe Polyphénols for many years and its President from 1986 to 1990—whose career has been devoted to phenolic compounds in plants.

To Ismaïl El-Hadrami—an active and enthusiastic member of the Groupe Polyphénols board for many years, and a member of the editorial board of the RAPR series—*in memoriam.*

Acknowledgments

The editors wish to thank all of the members of the Groupe Polyphénols Board Committee (2008–2010) for their guidance and assistance throughout this project.

Groupe Polyphénols Board 2008–2010

Dr. Catherine Chèze
Prof. Gilles Comte
Dr. Kevin Davies
Prof. Ismaïl El-Hadrami
Dr. Sylvain Guyot
Prof. Victor de Freitas
Dr. Hélène Fulcrand
Dr. Paul A. Kroon
Dr. Virginie Leplanquais
Dr. Stephan Martens
Prof. Stéphane Quideau
Prof. Anna-Lisa Romani
Dr. Pascale Sarni-Manchado
Prof. Celestino Santos-Buelga
Prof. Dieter Treutter
Prof. Kristiina Wähälä

Contents

5 Colouring up Plant Biotechnology 131

Cathie Martin, Yang Zhang, Laurence Tomlinson, Kalyani Kallam, Jie Luo, Jonathan D.G. Jones, Antonio Granell, Diego Orzaez and Eugenio Butelli

6 Anthocyanin Biosynthesis, Regulation, and Transport: New Insights from Model Species 143

Lucille Pourcel, Andrés Bohórquez-Restrepo, Niloufer G. Irani and Erich Grotewold

A color plate section is located between pages 12 and 13.

Contributors

Lorne R. Adam, Department of Plant Science, University of Manitoba, 66 Dafoe Road, Winnipeg, MB, R3T 2N2, Canada

Elena Bernardini, Biblioteca Centrale di Farmacia, Università degli Studi di Milano, Via Balzaretti 9, 20133 Milano, Italy

Andrés Bohórquez-Restrepo, Plant Biotechnology Center, 206 Rightmire Hal, 1060 Carmack Road, Columbus, OH 43210, USA

Alain-Michel Boudet, Laboratoire Surfaces Cellulaires et Signalisation chez les Végétaux—UMR-CNRS 5546, Università Paul Sabatier, Pôle de Biotechnologies Végétale, 24 chemin de Borde Rouge, B.P. 42617, Auzeville, 31326 Castanet-Tolosan, France

Eugenio Butelli, John Innes Centre, Norwich Research Park, Colney, Norwich NR4 7UH, UK

Angela Cardinali, Istituto di Scienze delle Produzioni Alimentari—CNR, Via Amendola 122/O, 70126-Bari, Italy

Véronique Cheynier, INRA, UMR1083 Sciences pour l'œnologie, 2 place Viala, F-34060 Montpellier Cedex 1, France

Fouad Daayf, Department of Plant Science, University of Manitoba, 66 Dafoe Road, Winnipeg, MB, R3T 2N2, Canada

Holly Derksen, Department of Plant Science, University of Manitoba, 66 Dafoe Road, Winnipeg, MB, R3T 2N2, Canada

Montserrat Dueñas, Grupo de Investigación de Polifenoles (GIP-USAL), Facultad de Farmacia, Universidad de Salamanca, Campus Miguel de Unamuno, 37007 Salamanca, Spain

Ahmed F. El-Bebany, Department of Plant Science, University of Manitoba, 66 Dafoe Road, Winnipeg, MB, R3T 2N2, Canada; and Department of Plant Pathology, Faculty of Agriculture, Alexandria University, Aflaton Street, El-Shatby, 21545, Alexandria, Egypt

Abdelbasset El Hadrami, Laboratoire de Biotechnologies, Protection et Valorisation des Ressources Végétales (Biotec-VRV), Faculté des Sciences Semlalia, Université Cadi Ayyad, B.P. 2390, 40 000 Marrakech, Morocco

Ismaïl El-Hadrami, Laboratoire de Biotechnologies, Protection et Valorisation des Ressources Végétales (Biotec-VRV), Faculté des Sciences Semlalia, Université Cadi Ayyad, B.P. 2390, 40 000 Marrakech, Morocco

Susana González-Manzano, Grupo de Investigación de Polifenoles (GIP-USAL), Facultad de Farmacia, Universidad de Salamanca, Campus Miguel de Unamuno, 37007 Salamanca, Spain

Ana M. González-Paramás, Grupo de Investigación de Polifenoles (GIP-USAL), Facultad de Farmacia, Universidad de Salamanca, Campus Miguel de Unamuno, 37007 Salamanca, Spain

Antonio Granell, IBMCP-CSIC-UPV, Universidad Politécnica de Valencia, Avda Los Naranjos SN, 46022 Valencia, Spain

Erich Grotewold, Department of Molecular Genetics and Plant Biotechnology Center, 206 Rightmire Hal, 1060 Carmack Road, Columbus, OH 43210, USA

Ann E. Hagerman, Department of Chemistry & Biochemistry, Miami University, 701 E. High Street, Hughes Laboratories, Oxford, OH 45056, USA

Maria A. Henriquez, Department of Plant Science, University of Manitoba, 66 Dafoe Road, Winnipeg, MB, R3T 2N2, Canada

Yung-Fen Huang, INRA, UMR1083 Sciences pour l'Œnologie, and INRA, UMR AGAP (Amélioration Génétique et Adaptation des Plantes), 2 place Viala, F-34060 Montpellier Cedex 1, France

Niloufer G. Irani, Department of Plant Systems Biology, Flanders Institute for Biotechnology, and Department of Plant Biotechnology and Genetics, Ghent University, 9052 Ghent, Belgium

Jonathan D.G. Jones, Sainsbury Laboratory, Norwich Research Park, Colney, Norwich NR4 7UH, UK

Kalyani Kallam, John Innes Centre, Norwich Research Park, Colney, Norwich NR4 7UH, UK

Tadao Kondo, Graduate School of Bioagricultural Science & Graduate School of Information Science, Nagoya University, Chikusa, Nagoya 464-8601, Japan

Vincenzo Lattanzio, Dipartimento di Scienze Agro-Ambientali Chimica e Difesa Vegetale, Facoltà di Agraria, Università degli Studi di Foggia, Via Napoli 25, 71100 Foggia, Italy

Vito Linsalata, Istituto di Scienze delle Produzioni Alimentari—CNR, Via Amendola 122/O, 70126-Bari, Italy

Jie Luo, Department of Metabolic Biology, John Innes Centre, Norwich Research Park, Colney, Norwich NR4 7UH, UK; and National Key Laboratory of Crop Genetic Improvement, National Center of Plant Gene Research, Huazhong Agricultural University, Wuhan 430070, People's Republic of China

Andrew Marston, Chemistry Department, University of the Free State, Nelson Mandela Drive, Bloemfontein 9300, South Africa

Cathie Martin, John Innes Centre, Norwich Research Park, Colney, Norwich NR4 7UH, UK

Diego Orzaez, IBMCP-CSIC-UPV, Universidad Politécnica de Valencia, Avda Los Naranjos SN, 46022 Valencia, Spain

Kin-ichi Oyama, Chemical Instrumentation Facility, Research Center for Materials Science, Nagoya University, Chikusa, Nagoya 464-8601, Japan

Lucille Pourcel, Département de Botanique et Biologie Végétale, Université de Genève, Sciences III, 30 quai Ernest-Ansemet, 1211 Genève, Switzerland

Celestino Santos-Buelga, Grupo de Investigación de Polifenoles (GIP-USAL), Facultad de Farmacia, Universidad de Salamanca, Campus Miguel de Unamuno, 37007 Salamanca, Spain

Scott A. Snyder, Department of Chemistry, Columbia University, 3000 Broadway, Havemeyer Hall, NY 10027, USA

Angelique Stalmach, College of Medical, Veterinary and Life Sciences, University of Glasgow, Glasgow G12 8QQ, UK

Nancy Terrier, INRA, UMR1083 Sciences pour l'œnologie, 2 place Viala, F-34060 Montpellier Cedex 1, France

Laurence Tomlinson, Sainsbury Laboratory, Norwich Research Park, Colney, Norwich, NR4 7UH, UK

Francesco Visioli, Laboratory of Functional Foods, IMDEA-Food, Campus de Cantoblanco, 28049 Madrid, Spain

Gary Williamson, School of Food Science and Nutrition, University of Leeds, Leeds LS2 9JT, UK

Zhen Yao, Department of Plant Science, University of Manitoba, 66 Dafoe Road, Winnipeg, MB, R3T 2N2, Canada

Kumi Yoshida, Graduate School of Information Science, Nagoya University, Chikusa, Nagoya 464-8601, Japan

Yang Zhang, John Innes Centre, Norwich Research Park, Colney, Norwich NR4 7UH, UK

UMR47—Diversité, Adaptation et Développement des Plantes (Université Montpellier II). Five topics were covered:

(1) *Chemistry and physicochemistry*: structure, reactivity, physicochemical properties, synthesis, . . .
(2) *Biosynthesis, genetics, and metabolomic engineering*: molecular biology, enzymology, gene expression and regulation, transport, biotechnology, . . .
(3) *Roles in plants and ecosystems*: plant growth and development, plant–insect relationships, biotic and abiotic stress, resistance, . . .
(4) *Health and nutrition*: medicinal properties, bioavailability and metabolism, mode of action, nutraceuticals, cosmetics, . . .
(5) *Analysis and metabolomics*: analytical methods, omics, . . .

Some 365 participants, from government institutional research and private business, representing 44 countries from all over the world, attended ICP2010, where 40 oral communications and 300 posters were presented. The present and third volume of *Recent Advances in Polyphenol Research* (RAPRIII), a series initiated by Groupe Polyphenols in 2008, includes chapters from the 11 guest speakers and some invited contributors. Essential complement to Polyphenols Communications 2010, the proceedings of ICP2010, RAPRIII offers in-depth knowledge on selected aspects of current polyphenol research, pursuing the role of ICP in being a base for debates and exchange on all research topics related to plant polyphenols.

In conclusion, we are pleased to observe that research advances in polyphenol science, enabling progress of our understanding of polyphenols at both the chemical and biological levels, are based on different approaches from different research areas and interactions between them. This would not be possible without the constant involvement of "Groupe Polyphénols" in maintaining ICP and coordinating this book series. So, we wish to thank deeply its Board and the scientific committee of ICP2010 for their contribution to the advancement of polyphenol research worldwide.

This 25th International Conference on Polyphenols would not have been possible without the generous support of public donors such as the French *Région Languedoc Roussillon, Montpellier Agglomération, INRA,* and *Université Montpellier II.* Grants from *Groupe Polyphénols* and from the *Phytochemical Society of Europe* for junior and senior attendees are also gratefully acknowledged. Other sponsors included Agilent Technology, GlaxoSmithKine, Indena, L'Oréal, PhenoFarm, Sanofi Aventis, and Waters.

Last, but not least, ICP2010 and RAPRIII would not be without the members of the local organizing committee, as well as many other "volunteers," whose dedicated effort and support ensured a smooth and eventless scientific and logistic organization. Our sincere thanks to all of them.

Véronique Cheynier
Pascale Sarni-Manchado
Stéphane Quideau

Preface

Plant polyphenolics are secondary metabolites that constitute one of the most common and widespread groups of substances in plants. They are structurally diverse, from rather simple compounds (e.g., anthocyanins, flavonols, isoflavones, catechins, and resveratrol) to highly complex polymeric species, and exhibit a large and diverse array of biological properties, for both plants and humans. Synthesis of polyphenolic compounds, which contribute to the pigmentation of flowers, fruits, leaves, or seeds, and play protective roles against biotic and abiotic stresses, is part of the adaptative strategies of plants. Polyphenolic compounds also contribute to the development of color and taste properties of plant-based foods and beverages, such as tea, wine, or chocolate, and they may play a part in the health protecting effects associated with the dietary consumption of such food products, although the actual benefit and mechanisms involved are yet to be proven. Finally, they are potentially helpful as therapeutic agents against various pathologies.

The list of plant (poly) phenolic compounds is constantly expanding, and, in spite of recent progress in the development of analytical methods, in particular for metabolomics, these molecules still present a considerable challenge to the analyst. Biological studies are aimed at understanding their role and status *in planta*, but also their fate *in vivo* after ingestion from food and beverages. Most of the work is sustained by the analysis of their chemical characteristics and physicochemical properties. There has been much effort over the last years to understand polyphenol biosynthesis and build the knowledge required to engineer or better harness their production in plants. Alternative strategies rely on organic synthesis to prepare polyphenolic target compounds in sufficient quantities to explore their properties and use them in various applications.

The diversity of structure and activity of (poly) phenolic compounds resulted in a multiplicity of research areas such as chemistry, biotechnology, ecology, physiology, nutrition, medicine, and cosmetics. The International Conference on Polyphenols, organized under the auspices of "Groupe Polyphénols," every other year, is a unique opportunity for scientists in these and other fields to get together and exchange their ideas and new findings.

The 25th edition of this conference (ICP2010) was held in Montpellier, France, from August 24 to 27, 2010, and organized by the Polyphenols and Interactions group of UMR1083—Sciences pour l'Oenologie (INRA Montpellier), in partnership with

Chapter 1
Plant Phenolics: A Biochemical and Physiological Perspective

Vincenzo Lattanzio, Angela Cardinali and Vito Linsalata

Abstract: The plant polyphenols are a very heterogeneous group, some universally and others widely distributed among plants, and often present in surprisingly high concentrations. During the evolutionary adaptation of plants to land, the biosynthesis of different phenolics classes in plants has evolved in response to changes in the external environment. Besides a bulk of phenolic substances having cell wall structural roles, a great diversity of non-structural constituents was also formed, having such various roles as defending plants, establishing flower colour and contributing substantially to certain flavours. The accumulation of phenolics in plant tissues is considered a common adaptive response of plants to adverse environmental conditions, therefore increasing evolutionary fitness. In addition, these secondary metabolites may still be physiologically important as a means of channelling and storing carbon compounds, accumulated from photosynthesis, during periods when nitrogen is limiting or whenever leaf growth is curtailed.

Keywords: phenolics; abiotic/biotic stress; primary/secondary metabolism relationships; metabolic costs of resistance

1.1 The general phenolic metabolism in plants

Phenolic compounds are found throughout the plant kingdom but the type of compound present varies considerably according to phylum. Phenolics are uncommon in bacteria, fungi and algae, and few classes of phenols are recorded: flavonoids are almost completely absent. Bryophytes are regular producers of polyphenols including flavonoids, but it is in the vascular plants that the full range of polyphenols is found (Swain, 1975; Harborne, 1980; Stafford, 1991). The plant polyphenols are a very heterogeneous group; some are universally and others widely distributed among plants, and they are often present in surprisingly high concentrations. They are not distributed evenly throughout the plant – either

Recent Advances in Polyphenol Research, Volume 3, First Edition. Edited by Véronique Cheynier, Pascale Sarni-Manchado and Stéphane Quideau.

quantitatively or qualitatively – in space and in time. The pattern of secondary metabolites in a given plant is complex because it changes in a tissue- and organ-specific way. Differences can regularly be seen between different developmental stages (e.g. organs important for survival and reproduction have the highest and most potent secondary metabolites), and between individuals and populations and these differences are subject to environmental as well as genetic control (Swain, 1977; Harborne, 1980; Wink, 1988; Osbourn *et al.*, 2003; Wink, 2003; Noel *et al.*, 2005; Singh & Bharate, 2006; Yu & Jez, 2008). Phenolic metabolism in plants is a complex process resulting from the interaction of at least five different pathways. The glycolytic pathway that produces phosphoenolpyruvate; the pentose phosphate pathway that produces erythrose-4-phosphate; the shikimate pathway that synthesises phenylalanine; the general phenylpropanoid metabolism that produces the activated cinnamic acid derivatives and the plant structural component lignin, and the diverse specific flavonoid pathways (Boudet *et al.*, 1985; Hrazdina, 1994; Schmid & Amrhein, 1995; Winkel-Shirley, 2001; Austin & Noel, 2003) (Fig. 1.1). Phenolic metabolism must be regarded as a dynamic system involving steady-state concentrations of the various phenolic compounds, which during certain phases of growth and development are subject to substantial qualitative and quantitative changes. This turnover may involve three types of reactions: (i) interconversions which are involved in biosynthetic sequences; (ii) catabolic reactions where the products are converted to primary metabolic constituents and (iii) oxidative polymerisation reactions leading to insoluble structures of high molecular weight (Barz & Hoesel, 1975, 1979).

Plants, as sessile organisms, evolve and exploit metabolic systems to produce a vast and diverse array of phenolic and polyphenolic compounds with a variety of ecological and physiological roles. The ability to synthesise phenolic compounds has been selected throughout the course of evolution in different plant lineages when such compounds addressed specific needs, thus permitting plants to cope with the constantly changing environmental challenges over evolutionary time (Pichersky & Gang, 2000; Noel *et al.*, 2005). For example, the successful adaptation to land by some higher members of the Charophyceae – which are regarded as prototypes of amphibious plants that presumably preceded true land plants when they emerged from an aquatic environment onto the land – was achieved largely by massive formation of 'phenolic UV light screens' (Swain, 1975; Lowry *et al.*, 1980; Stafford, 1991; Graham *et al.*, 2000). Regarding the structure of phenolic compounds involved in this photoprotective role of plant phenolics, there was an exciting discussion between Tony Swain and Brian Lowry. Lowry's speculative viewpoint was that 'when plants invaded the land habitat and were exposed to solar-ultraviolet radiation more intense than that found today, an early obvious protective adaptation strategy used by plants would be the accumulation of substituted cinnamic acids from the deamination of aromatic amino acids' (Lowry *et al.*, 1980). Swain's objection to this speculative hypothesis was that 'cinnamic acids absorbing at 310–325 nm do not have the right absorption characteristics to enable them to act efficiently in this way and thus prevent UV photodestruction of either nucleic acids or proteins (λ_{max} ca 260 and 280 nm, respectively)'. Swain's opinion was that flavonoids (λ_{max} ca 260 and 330 nm), cell wall polysaccharide acylation by cinnamic acids and suberin could all presumably have aided in the success of land plants (Swain, 1981). Lowry's reply was that, 'given the presence of even trace amounts of ozone in the atmosphere during the time

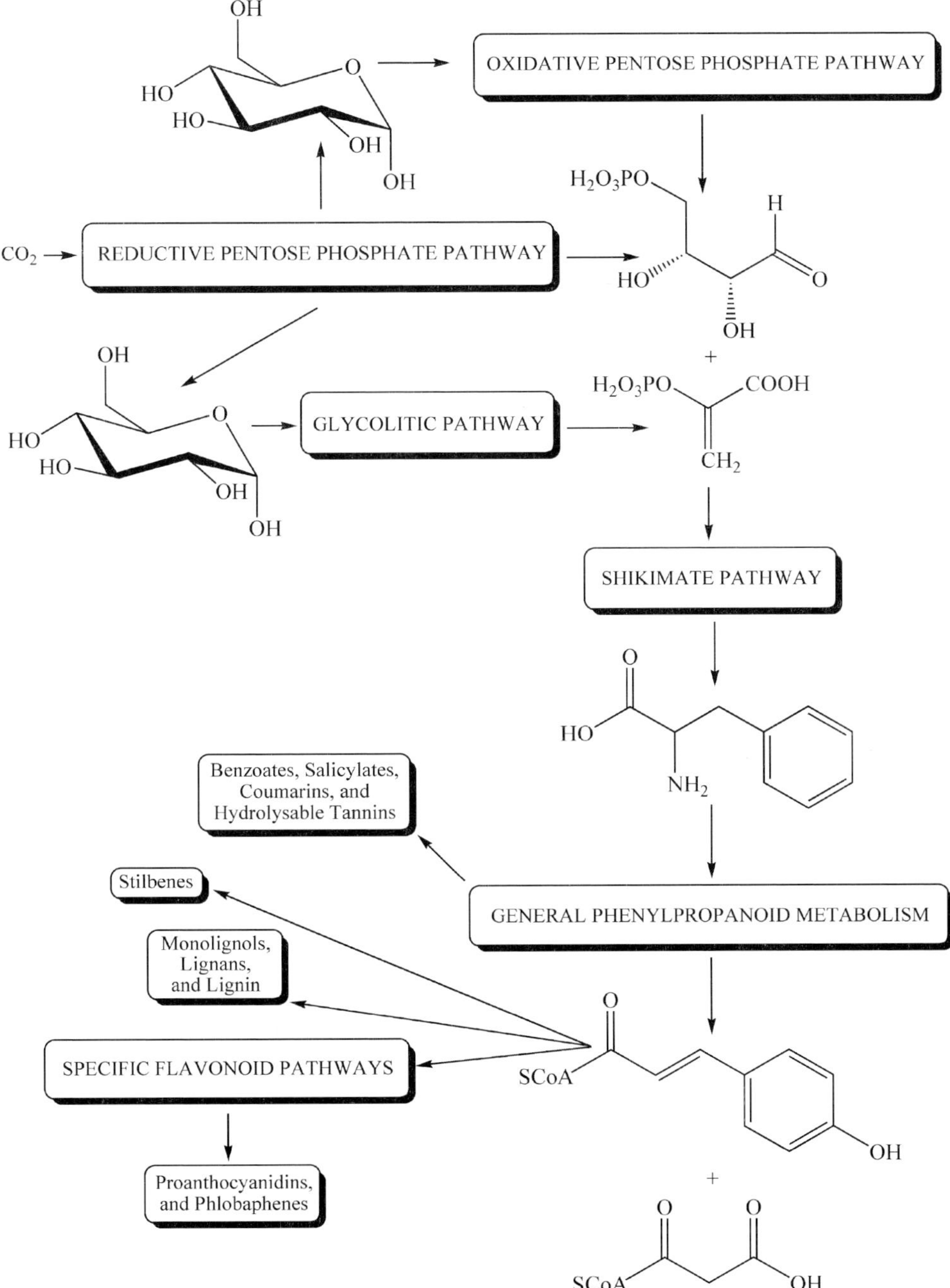

Fig. 1.1   Carbon fluxes towards the phenolic metabolism.

leading up to the Silurian and early Devonian (starting some 420 million years ago), it is extremely unlikely that terrestrial organisms would have been exposed to UV-C radiation (less than 280 nm)' and that DNA and proteins are both damaged by radiation in the UV-B region (280–315 nm) (Lowry *et al.*, 1983). A wide array of flavones have been reported for *Takakia lepidozioides*, believed to be amongst the most primitive of extant liverworts and the possible ancestor of modern bryophytes. This suggested that metabolic pathways leading to flavonoid synthesis appeared quite early in the evolutionary record of plants, perhaps even before the development of vascular tissues (Markham & Porter, 1979). Bryophyte lines that mainly synthesised flavones and flavonols, branched off within populations of pioneering land plants. Within other populations of early land plants, the evolution of the enzymes unique to the lignin pathway permitted the evolution of vascular plants, the tracheophytes. Proanthocyanidins and flavan-3-ols became widespread in some fern groups, while these and 3-hydroxyanthocyanidins became dominant flavonoids in gymnosperms and, especially, in angiosperms. Proanthocyanidins remained as major constitutive defence compounds in leaves of long-lived woody plants, but became relatively rare in short-lived herbaceous angiosperms, except in the seed coats of some of these plants. The pterocarpan pathways producing inducible phytoalexins for chemical defence purposes were evolved in a few angiosperm taxons (Stafford, 1991). Broadly, it is now well known that charophyte green algae can inhabit extreme habitats (highly saline and acidic waters with high levels of heavy metals) and that green algae are also common on land. Terrestrial algae grow in some of the most difficult habitats on earth, such as desert soils. Morphological and molecular analyses of some of these charophyte green algae have indicated multiple transitions to arid habitats from aquatic ancestors. During the evolutionary adaptation of plants to land, the biosynthesis of different phenolics classes in plants has evolved in response to changes in the external environment. In addition to a bulk of phenolic substances with cell wall structural roles, an amazing diversity of non-structural constituents was also formed, having such various roles as defending plants, determining the durability of different woods and barks, establishing flower colour and contributing substantially to certain flavours. In addition, phenolics – and ultimately flavonoids – were also selected for their protection against ultraviolet damage and autotoxicity. All these diverse functions performed by the different classes of phenolic compounds are essential for the continued survival of all types of vascular plants (Lowry *et al.*, 1980; Cooper-Driver & Bhattacharya, 1998; Flechtner *et al.*, 1998; Croteau *et al.* 2000; Bieza & Lois, 2001; Lewis & Mccourt, 2004; Teklemariam & Blake, 2004; Caldwell *et al.*, 2007; Lattanzio *et al.*, 2008).

However, it is not true that all plants lack mobility, although, plants are generally rooted and unable to move from place to place by themselves. Some plants are now known to be able to move in certain ways; some plants are known to open their leaves in the daytime and 'sleep' at night with their leaves folded. This circadian rhythmic leaf movement known as nyctinasty is widely observed in leguminous plants. It was thought that nyctinastic movement was controlled by Schildknecht's turgorins (chemical factors controlling the turgor changes in plants which induce turgor-controlled movements including nyctinasty), which induce leaf-closing movement of the plants (Schildknecht & Schumacher, 1982; Schildknecht, 1983). Ueda and his collaborators found that nyctinastic plants have a pair of endogenous bioactive substances that control nyctinastic leaf movement (Ueda & Yamamura

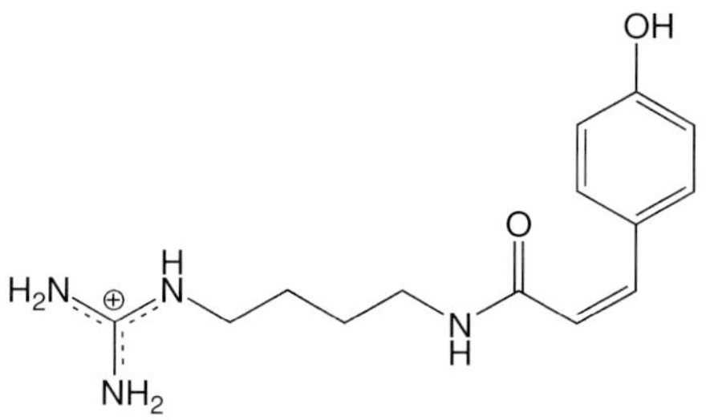

cis-p-Coumaroylagmatine
leaf-opening factor (*Albizzia julibrissin* Durazz)

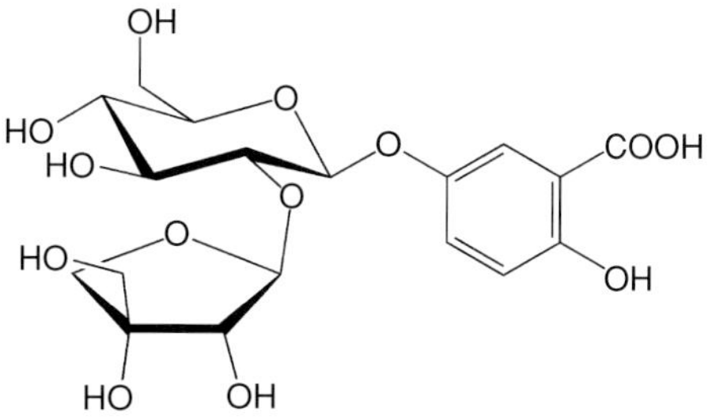

Gentisic acid 5-*O*-[β-D-apiofuranosyl-(1→2)-
β-D-glucopyranoside] leaf-closing factor
(*Mimosa pudica* L.)

Fig. 1.2 Leaf-movement factors from nyctinasic plants.

2000; Ueda & Nakamura 2006). One of these is a leaf-opening factor that 'awakens' plant leaves, and the other is a leaf-closing factor that reverses this process, so that the plant leaves 'sleep' (Fig. 1.2). All leaf-opening factors, which are effective under physiological pH and in a physiological concentration, have the common structural feature of *p*-coumaroyl moiety, and this result suggests that this structural feature is deeply involved in the common mechanism for leaf-opening (Ueda & Nakamura, 2010).

The highly ordered interactions between plants and their biotic and abiotic environments have been a major driving force behind the emergence of specific natural products. The accumulation of phenolics in plant tissues is considered a common adaptive response of plants to adverse environmental conditions, increasing evolutionary fitness. In addition, these secondary metabolites may still be physiologically important as a means of channelling and storing carbon compounds, accumulated from photosynthesis, during periods when nitrogen is limiting or whenever leaf growth is curtailed. Large increases in the amount of phenolic compounds can occur in stressed plants and those undergoing mechanical damage. Plant phenolics are considered to have a key role as defence compounds when environmental stresses such as bright light, low temperatures, pathogen infection, herbivores and nutrient deficiency can lead to increased production of free radicals and other oxidative species in plants. A growing body of evidence suggests that plants respond to these biotic and abiotic stress factors by increasing their capacity to scavenge reactive oxygen species. In addition, in order to establish a protective role for a given metabolite, it is necessary to monitor concentrations over the life cycle of the plant, to survey plant populations, to determine specific localisation within tissues and to carry out bioassays against insects and microorganisms. Finally, changes in secondary chemistry may also occur during ontogeny and protection may be restricted to the most vulnerable plant organs (Robbins *et al.*, 1985; Harborne, 1990; Lattanzio *et al.*, 1994; Dixon & Paiva, 1995; Facchini, 1999; Winkel-Shirley, 2002, Blokhina *et al.*, 2003).

The bewildering array of phenolic compounds produced by plant tissues (several thousand different chemical structures have been characterised) belong to various classes, such as esters, amides and glycosides of hydroxycinnamic acids, glycosylated flavonoids, especially flavonols, proanthocyanidins and their relatives and the polymeric lignin and suberin. Some soluble phenolics, for example chlorogenic acid, are widely distributed, but the distribution of many other structures is restricted to specific genera or families making them convenient biomarkers for taxonomic studies. Even if the potential value of plant secondary metabolites to taxonomy has been recognised for nearly 200 years, their practical application has been restricted to the twentieth century and predominantly to the last 40 years. The use of secondary compounds has clear advantages over the use of primary compounds in establishing phylogenetic relationships because differences in the complement of secondary compounds are qualitative differences whereas differences in the concentrations of primary compounds are quantitative differences, and these are subject to both environmental and genetic control. Phenolic compounds are often similar within members of a clade and therefore the existence of a common pattern of secondary compounds may indeed provide much clearer evidence of common ancestry than morphological similarities attributable either to common ancestry or to convergent evolution (Bell, 1980; Lattanzio *et al.*, 1996; Wink, 2003).

1.2 Effect of non-freezing low temperature stress on phenolic metabolism in crop plants

Of the various environmental stresses, exposure to non-freezing low temperatures is one of the most important abiotic stress factors for plants. The precise way in which plants adapt to low temperature is obviously of scientific interest, but there are also practical and economic aspects. Many important crop plants of tropical and subtropical origin are, in general, sensitive to low non-freezing temperatures less than $10°C$ to $12°C$. Several studies have suggested that exposure to low temperatures usually triggers a variety of biochemical, physiological and molecular changes that allow the plants to adjust to stress conditions and this response is characterised by a greater ability to resist injury or survive an otherwise lethal low temperature stress. This process is known as cold acclimation (Lyons, 1973; Graham & Patterson, 1982; Janas *et al.*, 2000; Sharma *et al.*, 2005). Lowering temperatures will thermodynamically reduce the kinetics of metabolic reactions. Exposure to low temperatures will shift the thermodynamic equilibrium so that there is an increased likelihood of non-polar side chains of proteins becoming exposed to the aqueous medium of the cell. This leads to a disturbance in the stability of proteins, or protein complexes and also to a disturbance of the metabolic regulations. Lower temperatures induce rigidification of membranes, leading to a disturbance of all membrane properties (permeability, electric field, cation concentration and water ordering, and this leads to disturbance of the conformation and thus the activity, of membrane-bound enzymes). Chilling is also associated with the accumulation of reactive oxygen species (ROS). The activities of the scavenging enzymes will be lowered by low temperatures, and the scavenging systems will then be unable to counterbalance the ROS formation that is always associated with mitochondrial and chloroplastic electron transfer reactions. The accumulation of ROS has deleterious effects, especially on membranes. Some plants are able to adapt through mechanisms based on protein synthesis, membrane composition changes, and activation of active oxygen scavenging systems. There is an increasing body of evidence that many of these biochemical and physiological changes are regulated by low temperature through changes in gene expression. In recent years, a number of low temperature-responsive genes have been cloned from a range of both dicotyledon and monocotyledon species (Wolfe, 1978; Howarth & Ougham, 1993; Hughes & Dunn, 1996; Thomashow, 1998; Siddiqui & Cavicchioli, 2006; Ruelland *et al.*, 2009).

Low temperature stress induces accumulation of phenolic compounds that protect chilled tissues from damage by free radical-induced oxidative stress. It has also been observed that cold stress increases the amount of water-soluble phenolics and their subsequent incorporation into the cell wall either as suberin or lignin (Chalker-Scott & Fuchigami, 1989; Ippolito *et al.*, 1997). Many papers report the effects of low temperature on phenolic metabolism, and these have shown that phenolic metabolism is enhanced under chill stress and that the behaviour of the same metabolism is further dependent on the storage temperature. There is a low critical temperature below which an increase of phenylpropanoid metabolism is stimulated during the storage of plant tissues and this temperature varies from commodity to commodity. The threshold temperature for increasing phenolic metabolism is related to the threshold temperature at which chilling injury is also induced and it has been shown that low temperature treatments stimulate phenylpropanoid metabolism as well as flavonoid metabolism in various plant tissues, including artichoke, carrot, gherkin, maize, olive, pea, pear, potato, tomato and watermelon (Rhodes & Wooltorton, 1977, 1978; Rhodes *et al.*, 1981; Blankenship & Richardson, 1985; Lattanzio & Van Sumere, 1987; Lattanzio *et al.*, 1989; Christie *et al.*, 1994; Leyva *et al.*, 1995; Chalker-Scott, 1999; Solecka *et al.*, 1999; Gil-Izquierdo *et al.*, 2001; Golding *et al.*, 2001; Rivero *et al.*, 2001; Ortega-García & Peragón, 2009). Figure 1.3a shows changes in the total flavonoid (quercetin and phloretin glycosides) content in Golden Delicious apple skin during storage at 2°C. During the first 60 days of cold storage, there is a relevant increase in flavonoid content, but flavonoid content gradually decreases in fruits stored for a longer period. Similar changes have been observed in the levels of phenolic compounds, mono- and di-caffeoylquinic acids, in artichoke heads stored at 4°C (Fig. 1.3b). The timing of the observed peak in the phenol level during cold storage depends on the species or cultivar, the harvesting time and the storage conditions (Lattanzio *et al.*, 1989, 2001; Lattanzio, 2003a, 2003b).

In connection with the increased synthesis of phenolic compounds at low temperatures, some studies have been carried out on some enzymes of phenolic metabolism,

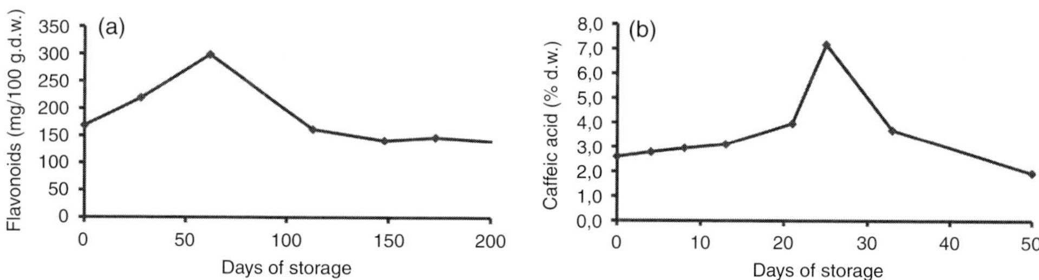

Fig. 1.3   Changes in the total flavonoid content in apple skin during storage at 2°C (a) and in the levels of mono- and di-caffeoylquinic acids (as % of caffeic acid on dry weight) in artichoke heads stored at 4°C (b).

phenylalanine ammonia lyase (PAL, EC 4.3.1.5), cinnamic acid 4-hydroxylase (CA4H) (1.14.13.11), *p*-coumarate CoA ligase (4CL, EC 6.2.1.12), hydroxycinnamoyl CoA quinate hydroxycinnamoyl transferase (HQT, EC 2.3.1.133) and chalcone synthase (CHS, EC 2.3.1.74). Generally, this low temperature effect on the phenol level involves a cold-induced stimulation of PAL, the branch point enzyme between primary (shikimate pathway) and secondary (phenolic) metabolism. It is well known that activity of this key enzyme of phenolic biosynthesis is induced in response to different external stimuli including low temperature stress (Engelsma 1970; Camm & Towers, 1973; Engelsma, 1974; Jones, 1984; Shaw *et al.*, 1990; Orr *et al.* 1993; Leyva *et al.*, 1995; Liu & McClure, 1995; Sarma & Sharma, 1999; Campos-Vargas and Saltveit, 2002; Gomez-Vasquez *et al.*, 2004; Tattini *et al.*, 2005). An enhanced PAL activity has been observed during cold storage of tomato and potato (Rhodes & Wooltorton, 1977; Rhodes *et al.*, 1981), citrus fruits (Sanchez-Ballesta *et al.*, 2000a; Lafuente *et al.*, 2001), olive (Ortega-García & Peragón, 2009) and onion (Benkeblia, 2000). PAL activity increased about fivefold in stored artichoke heads during the first days of storage at 4°C, and thereafter this activity decreased again to a low level (Lattanzio *et al.*, 1989).

The observed increases in PAL activity induced by low temperature might involve both enzyme *de novo* synthesis and release of PAL from a pre-existing but inactive enzyme–inhibitor complex. In any case, stimulation of PAL activity and, in turn, of phenylpropanoid pathway has been considered as a part of the response mechanism of fruits and vegetables to cold stress (Siriphanich & Kader, 1985a; Lattanzio & Van Sumere, 1987; Christie *et al.*, 1994; Dixon & Paiva, 1995; Leyva *et al.*, 1995; Janas *et al.*, 2000; Sanchez-Ballesta *et al.*, 2000a, 2000b; Lattanzio *et al.*, 2001; Hannah *et al.*, 2006; Olsen *et al.*, 2009; Ortega-García & Peragón, 2009). It is likely that endogenous ethylene, produced in plant tissue exposed to low temperature stress, promotes the induction of PAL activity and this is consistent with data showing that cold-induced PAL activity is reduced by inhibitors of ethylene production or by inhibitors of the action of ethylene. The onset of ethylene production in stressed plant tissues occurs at approximately the same time as an increase in PAL activity. Moreover, the effect of exogenously-added ethylene on most tissues is to cause increased production of PAL. The concentration of ethylene that affects PAL levels varies in different plants (Hyodo & Yang, 1971; Rhodes & Wooltorton, 1971; Chalutz, 1973; Hyodo *et al.*, 1978; Blankenship & Richardson, 1985; Blankenship & Unrath, 1988; Ke & Saltveit, 1989; Nigro *et al.*, 2000; Lafuente *et al.*, 2001).

Low temperature induction of PAL activity alone in plant tissues does not produce a corresponding increase in phenol production. At low temperatures, it is possible that the subsequent steps in the biosynthesis of phenolic compounds may limit their formation. In this connection, reference must be made to some excellent papers showing that other enzymes important in the phenolic biosynthetic pathway (e.g. CA4H, CQT, 4CL and CHS) can be stimulated by low temperature treatments. This phenomenon is largely dependent on the plant material studied, the storage temperature and the controlled or modified atmosphere used. In tomatoes stored at 2°C, besides PAL activity, during the first days of storage, a sizeable increase was observed in the activity of CQT, an enzyme involved in chlorogenic acid metabolism. A similar pattern of changes was observed in the enzymes CQT and *p*-coumarate CoA ligase in potato tubers stored at 0°C (Rhodes & Wooltorton, 1977, 1978;

Rhodes *et al.* 1981). Siriphanich and Kader (1985b) recorded an increase in CA4H activity in lettuce tissues stored at 0°C and potato disks kept at 5°C. Low temperature stress, besides affecting enzymes involved in the general phenylpropanoid pathway, also affects CHS the key enzyme of the flavonoid pathway. An increase in the CHS mRNA level after low temperature treatment has also been observed in soybean, maize and parsley (Christie *et al.*, 1994; Hasegawa *et al.*, 2001; Kasai *et al.*, 2009). The increase in these enzymes of phenolic metabolism presumably contributes to the increased production of phenols at low temperature.

An increase in the activity of the enzymes, as well as in the level of phenolic compounds, could combine with the temperature-dependent phase changes in the cellular membrane, to affect the shelf life of stored fruit and vegetables by providing an adequate substrate to the browning reactions. Browning in plant tissues during handling and storage of fresh fruit and vegetables commonly result from either non-enzymatic or enzymatic reactions involving plant phenols, oxygen and environmental contaminants such as metal ions. Enzymatic browning in fruit and some vegetables starts with the enzymatic oxidation of phenols by polyphenol oxidases (PPOs, EC 1.14.18.1 and EC 1.10.3.1), which are Cu enzymes almost ubiquitous in plants and catalyse the conversion of monophenols to *o*-diphenols and *o*-dihydroxyphenols to *o*-quinones. The quinone products can then polymerise and react with amino acid groups of cellular proteins, resulting in black or brown pigment deposits (melanins). Such damage causes considerable economic and nutritional loss in the commercial production of fruit and vegetables. PPOs are located in plastids, and they are not integral membrane proteins, although they are membrane associated. *In vivo*, the phenolic substrates of PPOs are localised in the vacuole and browning only occurs as a result of tissue damage leading to a loss of this sub-cellular compartmentalisation (Mathew & Parpia, 1971; Pollard & Timberlake, 1971; Mayer & Harel, 1981; Vaughn *et al.*, 1988; Martinez & Whitaker, 1995; Friedman, 1996; Guyot *et al.*, 1996; Amiot *et al.*, 1997; Lattanzio, 2003a, 2003b; Pourcel *et al.*, 2007; Guyot *et al.*, 2008). Non-enzymatic causes of browning in plant tissues may be attributable to the interactions between phenols and heavy metals – especially iron – which yield coloured complexes. It is generally accepted that a dark coloured complex of ferric iron and an orthodihydric phenol is responsible for discolouration. It has been suggested that a phenolic compound involved may be chlorogenic acid (5-*O*-caffeoylquinic acid) and that subcellular decompartmentalisation of plant cells during senescence allows the organic ligand to chelate the iron. Since the metal is originally present in the reduced state, a colourless complex is first formed and when exposed to oxygen, oxidises to yield a coloured compound. Therefore, while enzymatic oxidations of phenolics generally promote brown discolouration in mechanically damaged plant tissues, iron-phenol complexes are relevant during processing and/or storage of some fruits and vegetables such as potatoes, cauliflowers, asparagus and olives (Tinkler, 1931; Bate-Smith *et al.*, 1958; Hughes *et al.*, 1962; Hughes & Swain, 1962a, 1962b; Lattanzio *et al.*, 1994; Brenes *et al.*, 1995; Cheng & Crisosto, 1997; Coetzer *et al.*, 2001; Marsilio *et al.*, 2001; Lattanzio, 2003a, 2003b).

Plate 1.1 shows non-enzymatic browning reactions, caused by iron-polyphenol complexing, in cold stored non-mechanically damaged plant tissues. Figure 1.4 shows the total phenol content (mono- and dicaffeoylquinic acids) in browned tissues of artichoke (*Cynara*

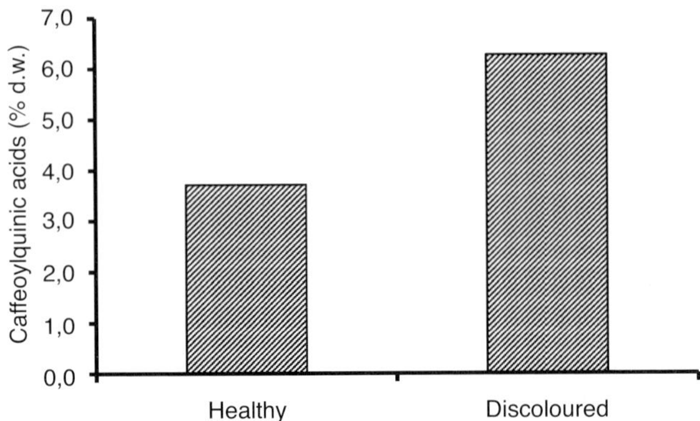

Fig. 1.4   Total caffeoylquinic acids in healthy and discoloured tissues of cold (at 4°C) stored artichoke heads.

cardunculus L. var. *scolymus* (L.) Fiori) heads stored at 4°C. It is noticeable that in discoloured tissues, phenol content is higher than in the healthy tissues of the same artichoke bract, which does not agree with the hypothesis of enzymatic browning. In this case we would expect a remarkable lowering of phenol content, due to the enzymatic oxidative phenomena. Furthermore, when artichoke tissues suffered enzymatic browning after mechanical damages and brief exposure to air, the phenolic content found was much lower than that of intact tissues. When iron complexes of chlorogenic acid (the most representative phenolic compound of artichoke heads) are considered, it has been observed that, at the physiological pH of artichoke tissues, chlorogenic acid forms coloured complexes with Fe^{3+}; the colour of these complexes changes from green to green-blue, grey-blue and brown as the pH of the medium increases from 5.5 to 8.0 or increasing the molar ratio chlorogenic acid/iron. It has been suggested that at pH 6.5 the main complex iron/chlorogenic acid is the 1:2 form, but there may also be some 1:1 complex, the stable form at lower pH and 1:3 complex, which is the stable form at higher pH. *In vivo* the 'discolouration' may vary if the pH in plant tissues increases due to physiological factors such as senescence. In the absence of oxygen, the same substrate forms colourless complexes with Fe^{2+}. After exposure to air, the complexed Fe^{2+} is quickly oxidised to Fe^{3+} and gives coloured compounds. Citric acid produces a 100% reduction in colour when an iron/citric acid ratio of 1:10 is used and the solution pH is kept unchanged. Citrate must be considered a strong sequestering agent for iron and this sequestering action is mainly responsible for its beneficial effects on artichoke head discolouration by preventing and/or reversing the formation of iron–phenolic complexes. Finally, HPLC analyses have also demonstrated that more than 85% of chlorogenic acid is released from the complex when the solution pH decreases from 6 to 3 (Hughes & Swain, 1962b; Lattanzio *et al.*, 1989, 1994; Cheng & Crisosto, 1997).

As far as the localisation of iron is concerned, different studies all agree that plastids contain the bulk of the cell's iron and that most of this iron is present in the ferric state: this is consistent with the fact that ferritin is the major eukaryotic iron-storage protein. In plants, ferritin is known to be present in chloroplasts, and, especially, in the plastids of non-photosynthesising tissues. Ferritin serves to solubilise and sequester iron: good evidence exists that iron is delivered to ferritin as Fe^{2+}, deposited as Fe^{3+}, and released upon reduction back to Fe^{2+} (Price, 1968; Jones *et al.*, 1978; Bienfait & Van der Briel, 1980). Several chelating agents are able to promote the release of ferritin iron in the presence of a reducing agent. It has been shown that plant phenols, including caffeic acid and chlorogenic acid, can promote the reductive release of ferritin iron: a direct correlation exists between oxidation–reduction potential and the rate of iron release. It has also been suggested that electrons are carried to the centre of the protein by ferrous ions produced in the entrance to a channel by interaction of labile Fe^{3+} with the reducing agents. In addition, reductant access to the ferritin iron core is also likely, when molecules are relatively small, (Price, 1968; Boyer *et al.*, 1988a, 1988b, 1989; Jacobs *et al.*, 1989, 1990).

From these data, and the results concerning the phenolic metabolism and changes in PPO and PAL activities during the cold storage of artichoke heads (Lattanzio & Van sumere, 1987; Lattanzio *et al.*, 1989, 1994) a non-enzymatic browning mechanism has been suggested in non-mechanically damaged tissues (Plate 1.1). During storage of artichoke heads at 4°C, low-temperature induction of PAL activity caused a biosynthetic increase of phenolics, especially chlorogenic acid. On the other hand, PPO activity did not change significantly during the cold storage period. The increased content of phenolics provided an adequate substrate for the browning. These reactions started from the chloroplasts, considered to be the site of chlorogenic acid biosynthesis (Ranjeva *et al.*, 1977a, 1977b; Alibert & Boudet 1982; Mondolot *et al.*, 2006), and where the iron is stored as ferritin. A release of ferritin iron, as Fe^{2+}, was induced by the chlorogenic acid, thus creating a colourless complex with the excess of chlorogenic acid. Afterwards, oxidising conditions from the senescence process, leading to membrane modification and progressive cell decompartmentalisation and/or low temperature-induced toxic oxygen forms caused the formation of a grey-blue chlorogenic acid/Fe^{3+} complex followed by browning. This complexed phenolic substrate, removed from the regular post-harvest metabolism occurring during cold storage of artichoke, was released in the free form when acidic pH conditions of the medium during HPLC analyses of artichoke caffeoylquinic acids caused the complex to break down.

1.3 Plant phenolics as defence compounds

The role of plant phenolics in chemoecology, especially on the feeding behaviour of herbivores, has been recognised since 1959 when Fraenkel described phenolic compounds as 'trigger' substances which induce or prevent the uptake of nutrients by animal herbivores. Ehrlich and Raven (1964) were among the first to propose a defined ecological role for plant secondary metabolites as defence agents against herbivorous insects. These substances are repellent to most insects and may often be decisive in patterns of food plant selection. Through occasional mutations and recombination, angiosperms have produced a series of chemical compounds not directly related to their basic metabolic pathways, but not inimical to normal growth and development. By chance some of these compounds reduce or destroy the palatability of the plant in which they are produced (Fraenkel, 1959; Ehrlich & Raven, 1964). Most research concerning insect anti-feeding agents has shown the involvement of phenylpropanoids, flavonoids and lignans in the plant resistance mechanism against insects.

Tannins are protein complexing compounds and enzyme inhibitors, and may also affect the growth of insects. The concentration of the phenolic compounds in the plant is a key factor in deterrence and it is the accumulation of phenols in particular parts of the plant that represents a feeding barrier. The effectiveness of phenolics as a resistance factor to animal feeding is enhanced, as aforesaid, by oxidation to polymers, which reduces digestibility, palatability and nutritional value (Ananthakrishnan, 1997; Lattanzio *et al.*, 2000, 2005; Harborne, 2001; Simmonds, 2001, 2003; Harmatha & Dinan, 2003). In addition, plants may be unsuitable as hosts for fungal pathogens because of pre-formed antifungal phenolics and/or induced defence phenolics synthesised in response to biotic stress, as part of an active defence response, when a pathogen manages to overcome constitutive defence barriers (Nicholson & Hammerschmidt, 1992; Lattanzio *et al.*, 2006; Treutter, 2006).

Plants encounter numerous pests and pathogens in the natural environment. An appropriate response to attack by such organisms can lead to tolerance or resistance mechanisms that enable the plant to survive (Paul *et al.*, 2000; Roy & Kirchner, 2000; Taylor *et al.*, 2004). Most plants produce a broad range of secondary metabolites that are toxic to pathogens and herbivores, either as part of their normal programme of growth and development or in response to biotic stress (Treutter, 2005; Agati *et al.*, 2008; Witzell & Martin, 2008; Lattanzio *et al.*, 2008; Abdel-Farid *et al.*, 2009; Eyles *et al.*, 2009). Both tolerance and resistance traits require the reallocation of host resources, therefore defensive chemicals are considered to be costly for plants, reducing the fitness of the host in the absence of disease, because resistance genes might impose metabolic costs on plants (e.g. lower growth rates than their sensitive counterparts). One way for a plant to reduce these costs is to synthesise defence compounds only after there has been some degree of initial damage by a pathogen or insect: this strategy is inherently risky because the initial attack may be too rapid or too severe for an effective defence response. Therefore, plants that are likely to suffer frequent and/or serious damage may benefit from investing mainly in constitutive defences, whereas plants that are attacked rarely may rely predominantly on induced defences (Morrissey & Osbourn, 1999; Purrington, 2000; Brown, 2002; Wittstock & Gershenzon, 2002; Brown, 2003; Koricheva *et al.*, 2004; Dietrich *et al.*, 2005).

1.3.1 Phenolic-mediated induced resistance of apples against fungal pathogens

It is estimated that there are about 250,000 species of higher plants, but six times as many (1.5 million) species of fungi. Fortunately for plants, their relationship with fungi is usually a mutually beneficial one (saprophytic fungi, mycorrhizae and endophytes). A small minority of fungal species has developed further and broken the fine balance of mutual benefit to become plant pathogens. This is because (i) the plant is unable to support the nutrient requirements of a potential pathogen and is thus a non-host; or (ii) the plant possesses pre-formed physical or chemical barriers that confine successful infections to specialised pathogen species; or (iii) when the attacking pathogen is recognised, defence mechanisms are elaborated and the invasion remains localised (Hammond-Kosack & Jones, 1996; Grayer & Kokubun, 2001). Indeed, survival of the plant in the plant–fungus encounter is controlled by the quality, the timing, the coordination, and the local extent of

Plate 1.1   Browning phenomena in artichoke heads that are not mechanically damaged and stored at 4°C.

Plate 1.2 Rotting of stored apples by *Phlyctaena vagabunda*.

Plate 1.3 *Vigna unguiculata* leaves infested by aphids.

Plate 1.4 Image of cowpea weevil infestation of two cultivated accessions of *Vigna*.

Plate 1.5 Response of oregano shoot growth to nutritional stress.

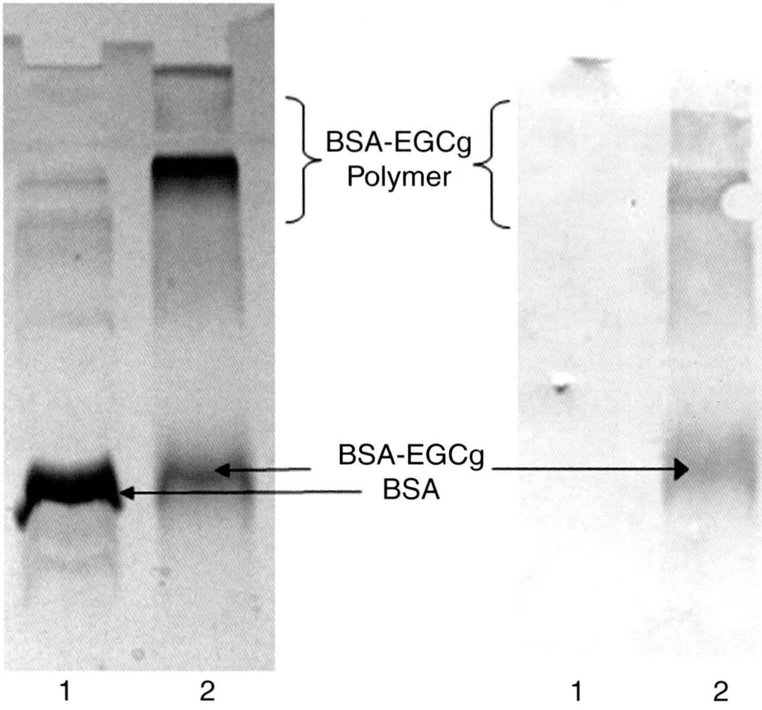

Plate 3.1 Coomassie-stained SDS–PAGE gel (left) and corresponding NBT-stained blot (right) for unreacted bovine serum albumin (lane 1) and for the reaction product with a 67fold molar excess of EGCg (lane 2). The arrows indicate the mobility of unmodified BSA, the slightly lower mobility of the soluble BSA–EGCg product, and very low mobility BSA–EGCg polymers.

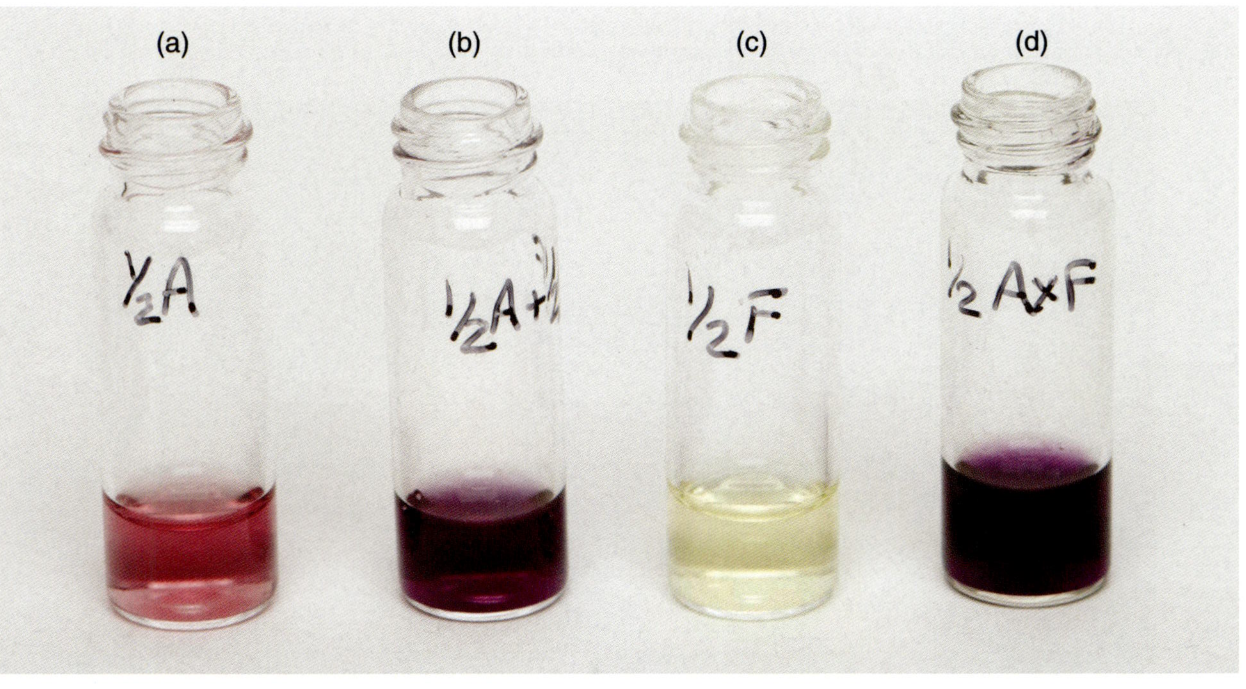

Plate 5.1   Effects of co-pigmentation on colour conferred by anthocyanins: (a) Plant extract enriched in anthocyanin (delphinidin 3-*O*-(coumaroyl) rutinoside 5-*O*-glucoside), (b) Mixture of plant extract enriched in anthocyanins with extract enriched in flavonols, (c) Plant extract enriched in flavonols (kaempferol and quercetin 3-*O*-rutinosides), (d) Extract of hybrid plant producing both high anthocyanins and high flavonols.

Plate 5.2   Pattern of purple anthocyanin production driven by the promoter of the PNH gene (Estornell *et al.*, 2009) expressed in tomato fruit, as visualised by the promoter driving expression of a MYB transcription factor (Rosea1; Schwinn *et al.*, 2006) and the CaMV 35S promoter driving constitutive expression of a bHLH protein (Delila; Goodrich *et al.*, 1992).

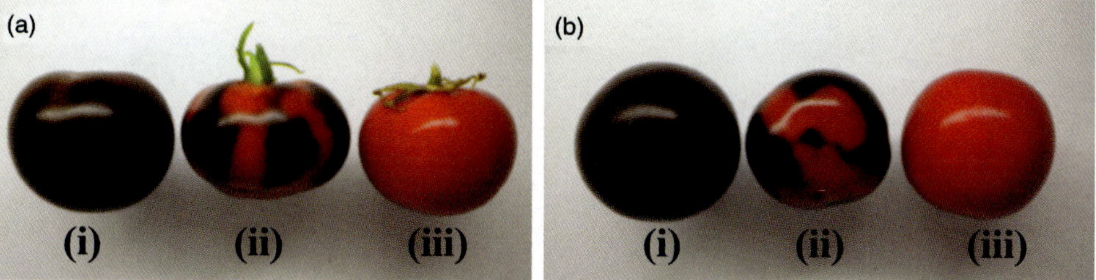

Plate 5.3 Use of *Del/Ros1* purple tomato in a visually traceable system to monitor VIGS following Agro-inoculation of fruit: (i) *Del/Ros1* purple fruit with high anthocyanin content (ii) *Del/Ros1* fruit inoculated with Agrobacterium carrying *Del* and *Ros1* sequences in TRV2. The red areas represent regions of silencing of *Del* and *Ros1*. Inclusion of other gene sequences in TRV2 causes co-incident silencing with *Del* and *Ros1*. The red sectors go through the fruit and can be dissected for analysis. (iii) Red control fruit. (a) Side view of fruit (b) Fruit viewed from beneath.

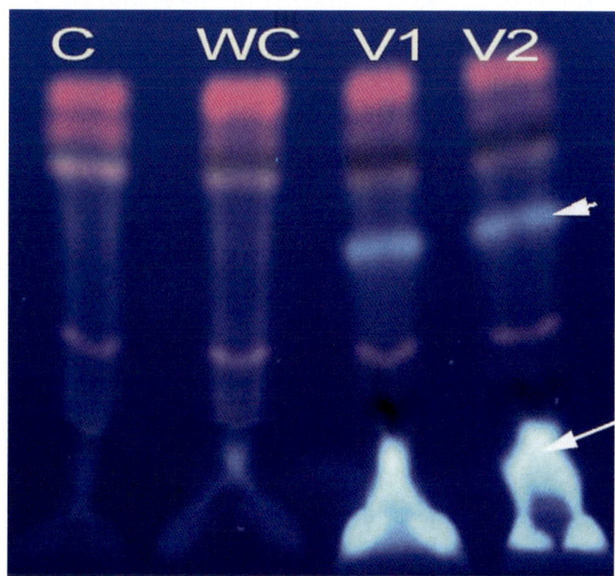

Plate 8.1 Thin layer chromatography observed under UV light showing the accumulation of scopoletin (top arrow) and other phytoalexins (bottom arrow). C, control tissue; WC, wounded control; V1 and V2, tissue inoculated with *Verticillium dahliae*.

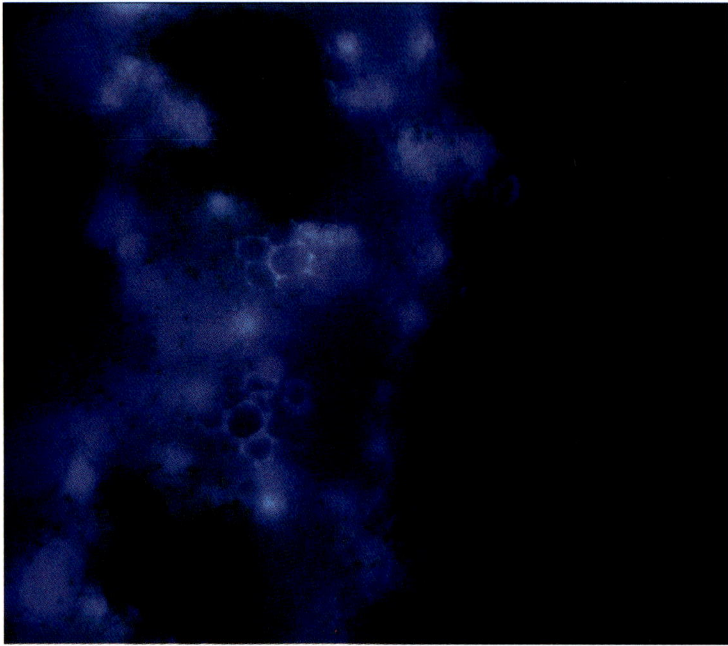

Plate 8.2 Fluorescent compounds accumulating in cotton tissues around the vessels in response to inoculation with *Verticillium dahliae*.

activation of biosynthetic pathways that may then lead to a resistant phenotype. Although only some plant families – notably belonging to the *Fabaceae* and *Apiaceae* – produce polyphenolic phytoalexins, the very early activation of the phenolic metabolism appears to play a pivotal role in the expression of disease resistance in all plants (Matern & Grimmig, 1993; Harborne, 1995). The first demonstrated example from the early plant pathology literature of phenolic compounds providing disease resistance was the case of coloured onion scales accumulating sufficient quantities of catechol and protocatechuic acid to prevent the germination of *Colletotrichum circinans* spores (Link *et al.*, 1929; Angell *et al.*, 1930; Link & Walker, 1933; Walker & Stahmann, 1955). Pre-formed antibiotic phenolics (phytoanticipins) are stored in plant cells mainly as inactive bound forms but are readily converted into biologically active antibiotics by plant hydrolysing enzymes (glycosidases) in response to pathogen attack. These compounds are considered as pre-formed antibiotics because the plant enzymes that activate them are already present but are separated from their substrates by compartmentalisation, enabling rapid activation without a requirement for the transcription of new gene products (Osbourn, 1996; Lattanzio *et al.*, 2008). When a pathogen manages to overcome constitutive defence barriers, it may be recognised at the plasma membrane of plant cells. Activation of inducible plant defence responses is probably brought about by the recognition of invariant pathogen-associated molecular patterns (PAMP) that are characteristic of whole classes of microbial organisms. PAMP perception systems trigger signalling cascades whose recognition is very likely to activate defence responses in natural plant–pathogen encounters (Nürnberger & Lipka, 2005). Plants respond to pathogens by activating broad-spectrum innate immune responses that can be expressed locally at the site of pathogen invasion as well as systemically in the uninfected tissue.

Rotting of stored apples (*Malus domestica* Borkh) by *Phlyctaena vagabunda* Desm. (syn. *Gloeosporium album* Osterw) (Plate 1.2) is an important cause of wastage. An important characteristic of the fungus is that spores of *P. vagabunda* are produced by small infections on the wood of the tree throughout the year and are spread by rain and dew on the fruit, which is thus exposed to infection during the entire growing season. There are conditions depending on the fungus and the nature of vegetable tissue, in which infections, which take place in lenticels, can develop during storage to produce lesions. The available evidence (Lattanzio *et al.*, 2001) does not support the hypothesis that pre-formed phenolic compounds (chlorogenic acid, (+)-catechin, (−)-epicatechin, phloretin glycosides and quercetin glycosides) may be involved in the constitutive resistance of apple to *P. vagabunda*. *In vitro* bioassays have shown that none of these naturally-occurring phenolics in concentrations like those encountered in fresh fruit exhibit inhibitory activity against spore germination or mycelial growth of *P. Vagabunda*. If pre-existing antifungal phenolics are not sufficient to stop the development of the infectious process, plant cells usually respond (hypersensitive reaction) by blocking or delaying the microbial invasion. Reactive oxygen species are often generated as warning signals within the cell or neighbouring cells, triggering off various reactions. These include the rapid increase of pre-existing antifungal phenols at the infection site, after an elicited increased activity of the key enzymes (PAL and chalcone synthase) of the biosynthetic pathway; this functions to slow or even halt the growth of the pathogen and to allow for the activation of secondary strategies that would restrict the pathogen more thoroughly. This initial defence response must occur so rapidly that it is unlikely to

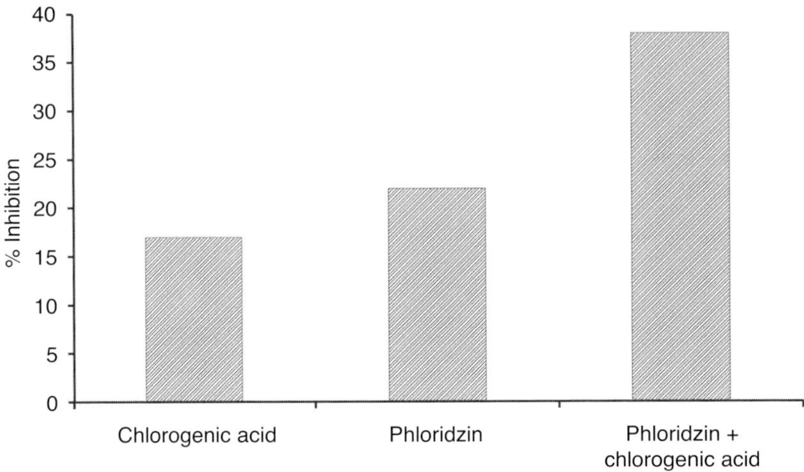

Fig. 1.5   *In vitro* inhibition of fungal spore germination of *Phlyctaena vagabunda* by apple oxidised phenolics.

involve a *de novo* transcription and translation of genes, which would be characteristic of the second level of defence (Ingham, 1973; Nicholson & Hammerschmidt, 1992; Lamb & Dixon, 1997). PPO activity also increased in these tissues, to 2–3 times that in healthy tissues. Post-infection accumulation of pre-existing phenolics, especially phloridzin and chlorogenic acid which are good substrates of apple PPO, provides an adequate substrate to the increased PPO activity. Thus, it cannot be excluded that after oxidative transformation phenolics are involved in induced resistance. The enzyme consumes oxygen and produces quinones or semiquinones, highly reactive compounds with potential toxic properties, and this makes the medium unfavourable to further development of pathogens (Byrde *et al.*, 1960; Friend, 1979; Butt, 1985; Friend, 1985; Cowan, 1999; Pontais *et al.*, 2008). *In vitro* bioassays showed that, when a crude extract of apple PPO was added to a spore suspension of *P. vagabunda* containing 10^{-3} M of each apple phenolics, an inhibition of fungal spore germination was observed. These bioassays also showed a potential synergistic effect of phloridzin and chlorogenic acid (Fig. 1.5). Phloridzin alone oxidised slowly and formed the light yellowish reaction products. However, the simultaneous presence of chlorogenic acid in a model system increases the oxidation rate of phloridzin in the presence of PPO by decreasing the lag period of the enzymatic reaction. This synergistic effect should probably be considered in the overall defensive strategy of apple against fungal attack (Oszmianski & Lee, 1991; Lattanzio *et al.*, 2001). From these data, it appears that infection of apple tissue elicited an active glycosidase and PPO capable of converting phloridzin to phloretin, which was subsequently oxidised. Simultaneously with hydrolysis to phloretin, phloridzin is oxidised *via* 3-hydroxyphloridzin to the corresponding *o*-quinone. The formed *o*-quinones are transient intermediates that may rapidly undergo oxidative condensation reactions (Fig. 1.6). These transformation reactions of phloridzin in the presence of apple PPO indicate that oxidation products may be involved in the defence mechanism of apple against the fungus *P. Vagabunda*. This metabolism of phloridzin, initiated by cell decompartmentalisation at the site of fungal infection, and the synergistic effect of chlorogenic acid, that accelerates

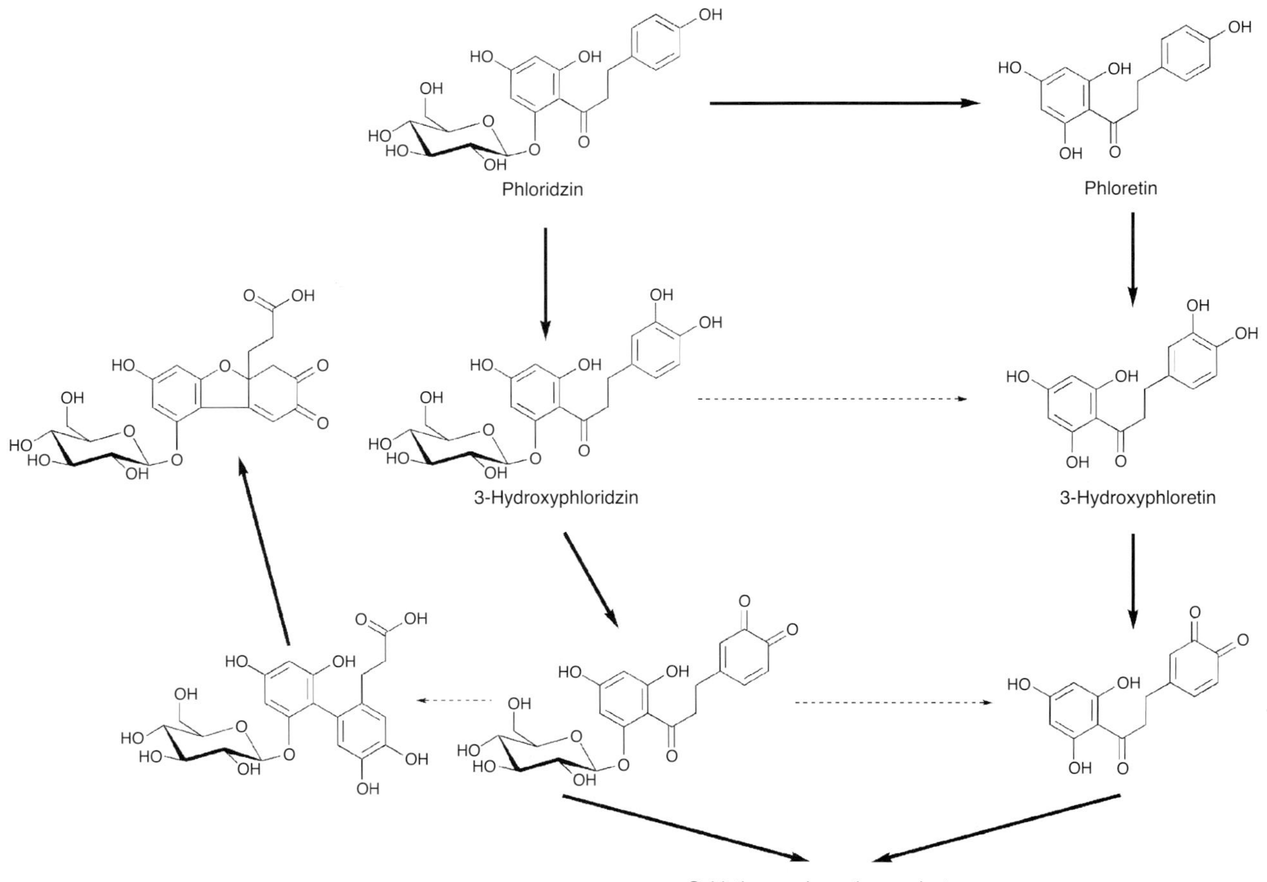

Fig. 1.6 Transformation reactions of phloridzin in the presence of apple enzymes.

the oxidation of phloridzin in the presence of PPO could play a prominent role in host resistance to further pathogen development (Noveroske *et al.*, 1964; Pierpoint, 1966; Raa, 1968; Raa & Overeem, 1968; Pierpoint, 1969; Synge, 1975; Pierpoint, 1983; Le Guernevé *et al.*, 2004; Guyot *et al.*, 2007).

1.3.2 Contribution of vigna phenolics to plant protection against insects

Whether a plant is accepted or rejected as food by insects depends largely on its chemical composition in addition, of course, to physical factors such as toughness, thickness and hairiness. Chemical inhibitors also play an important role in the inhibition of oviposition on the host-plant, and, in turn, on insect larval growth and the survival of progeny. Studies on the role of inhibitors in host plant selection indicate that many different chemicals may be expected to have an inhibitory effect on feeding by different insects. It is now generally accepted that plant phenolics play a role in protecting plants from insects (Painter, 1941; Thorsteinson, 1960; Dethier, 1970; Chapman, 1974; Joerdens-Roettger, 1979; Ferguson *et al.*, 1983; Pereyra & Bowers, 1988; Roessingh *et al.*, 1997; Constabel, 1999; Stotz *et al.*, 1999; Bernays & Chapman, 2000; Harborne & Williams, 2000; Harborne, 2001; Lattanzio *et al.*, 2008). Plant flavonoids affect the behaviour, development and growth of a number of insects (Hedin & Waage, 1986; Simmonds & Stevenson, 2001; Simmonds, 2001, 2003). Some cotton flavonoids are feeding stimulants for the boll weevil, *Anthonomus grandis* (Hedin *et al.*, 1988), or oviposition stimulants of a *Citrus*-feeding swallowtail butterfly, *Papilio xuthus* L. (Nishida *et al.*, 1987) or, finally, antibiotics effective against phytophagous insects (Todd *et al.*, 1971; Elliger *et al.*, 1980; Hanny, 1980; Hedin *et al.*, 1983; Harborne, 1997, 1999, 2001).

Cowpea (*Vigna unguiculata* (L.) Walp.) is an important food legume in many countries in sub-Saharan Africa and Latin America. The major constraints to cowpea production are insect pests, plant diseases, plant parasitic weeds, drought and heat (Murdock, 1992; Singh *et al.*, 1992; Thottappilly *et al.*, 1992). Aphids are one of the world's major insect pest groups on crop plants. Aphids feed by sucking plant sap, directly resulting in plant damage due to a reduction in the plant's resources. Additional plant damage can also be caused by plant viruses that some aphid species transmit. It has been estimated that 60% of all plant viruses are spread by aphids (Dreyer & Campbell, 1987). There are two *Aphis* spp. (Homoptera: Aphididae) reported as pests of cowpeas: *Aphis craccivora* Koch (cowpea aphid), which is the main aphid infesting cowpeas throughout Africa and Asia, and *Aphis fabae* Scopoli (black bean aphid), which has been reported as a minor pest in Africa and whose biology appears to be similar to that of *A. craccivora*. Cowpea aphids primarily infest seedlings, but large populations also infest flowers and green pods of older plants (Plate 1.3) (Singh & Jackai, 1985; Annan *et al.*, 1996). Flavonoid HPLC analyses (Lattanzio *et al.*, 2000) have shown that cultivated lines of *V. unguiculata* (L.) Walp. are qualitatively very similar, always containing three flavonoid aglycones: quercetin, kaempferol and isorhamnetin. In addition, a positive relationship was found between resistance/susceptibility characteristics to aphids and total flavonoid glycoside content of cowpea lines. The resistant lines have a higher total flavonoid content than susceptible lines. This relationship was further confirmed when the flavonoid aglycone content of two near-isogenic lines of *V. unguiculata* was considered: the

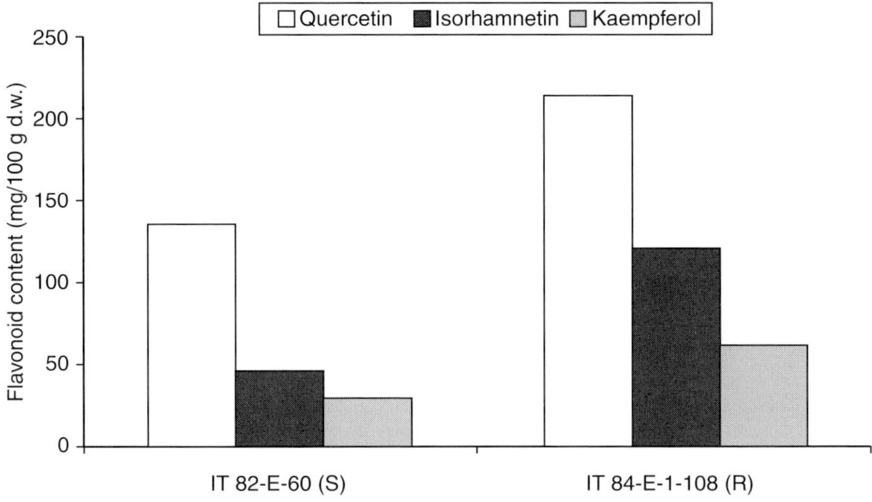

Fig. 1.7   Flavonoid content in near-isogenic lines of *V. unguiculata* (S, susceptible; R, resistant).

level of flavonoids in IT 84-E-1-108 (resistant) is twice as high as in IT 82-E-60 (susceptible) (Fig. 1.7). Figure 1.8 shows the *in vitro* inhibitory effect of *Vigna* endogenous flavonoids (0.1 mM), relative to the control, upon nymph deposition by *A. Fabae*: quercetin is the most active whereas kaempferol has little effect on the reproduction rate. Many flavonoids can act as feeding deterrents to phytophagous insects at relatively low concentrations. Therefore, the concentrations of flavonoids in plants are normally far higher than those needed for a deterrent effect on aphid feeding. However, aphids tend to feed on tissues such as the phloem, which are generally low in flavonoids, and thus they will normally only encounter high levels while probing the plant tissues for phloem sap, and not while feeding (Harborne & Grayer, 1993).

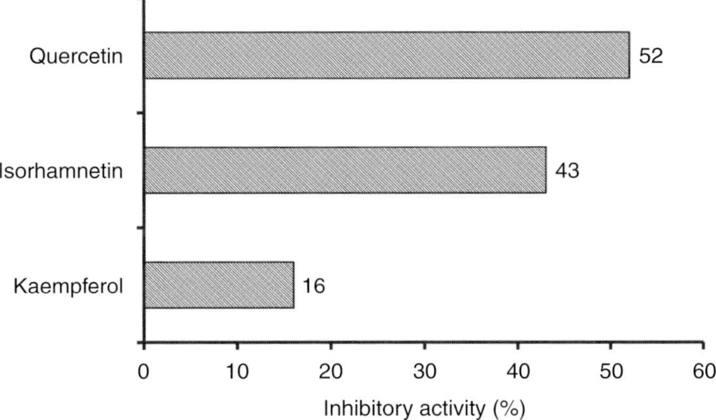

Fig. 1.8   Inhibitory activity, relative to the control (10% MeOH), of *Vigna* endogenous flavonoids (0.1 mM in 10% MeOH) on *Aphis fabae* progeny deposition.

Table 1.1   Daily production of *Aphis fabae* nymphs on *Vigna* accessions.

Accession	Daily larval deposition
Kempferol chemotype:	
Vigna luteola (Jacq.) Bentham TVnu 172	7.73
Vigna marina (Burm.) Merrill var. *marina* TVnu 717	7.25
Quercetin chemotype:	
V. luteola (Jacq.) Bentham TVnu 475	1.50
Isorhamnetin chemotype:	
V. marina (Burm.) Merrill var. *oblonga* TVnu 174	0.67

Flavonoid HPLC fingerprints of wild *Vigna* species support evidence for the existence of different flavonoid chemotypes in some *Vigna* species (Lattanzio *et al.*, 1997, 2000). There are kaempferol chemotypes, in which kaempferol is the only or the main aglycone detected, quercetin chemotypes, containing only quercetin glycosides, and isorhamnetin chemotypes. From an ecological point of view, the most interesting chemotypes are some accessions, belonging to the same species, which make it possible to study, *ceteris paribus*, the role of endogenous flavonoids in plant resistance to aphids. Two chemotypes were found amongst *Vigna marina* accessions: *V. marina* var. *oblonga* TVnu 1174 (isorhamnetin chemotype) and *V. marina* var. *marina* TVnu 717 (kaempferol chemotype). *V. luteola* accessions also showed two different chemotypes: TVnu 475 contains only quercetin glycosides, while the other accession, the kaempferol chemotype TVnu 172, contains robinin (kaempferol-3-robinoside-7-rhamnoside). When the resistance characteristics to aphids in different chemotypes of the same species were tested (Table 1.1), it became evident that quercetin or isorhamnetin chemotypes show a higher level of resistance compared to the kaempferol chemotypes of the same species, thus demonstrating a direct involvement of quercetin or isorhamnetin in the resistance mechanism.

The cowpea seed beetle, *Callosobruchus maculatus* Fabricius (Coleoptera: Bruchidae) is a major pest of stored cowpeas, but actually infests the green pods while they are still in the field. The adult beetles lay eggs on drying cowpea pods in the field and/or seeds in storage. Larvae hatch from eggs and penetrate the pod wall or the seed testa with their mouthparts. Larval feeding in the cotyledons causes significant losses in seed weight, germination viability and seed marketability (Singh & Jackai, 1985; Singh *et al.*, 1990; Murdock *et al.*, 1997; Appleby & Credland, 2003; Zannou *et al.*, 2003; Chi *et al.*, 2009). Regarding resistance/susceptibility characteristics of stored cowpeas to *C. maculatus* legume seeds do not generally rely on one type of chemical defence and may accumulate several chemicals of one class or compounds of several classes to increase their defence levels. Therefore, the strong resistance of some cultivated or wild *Vigna* species to *C. maculatus* may be due to the presence of multiple chemical factors with additive or synergistic action to protect seeds from predation.

A moderate level of resistance to cowpea bruchid was identified in accession TVu 2027 by screening a world germoplasm collection, and the resistance mechanism was found to be antibiosis that caused larval mortality (Singh & Rachie, 1985). The resistance of Tvu 2027 to bruchids was investigated by Gatehouse *et al.* (1979) who concluded that resistance

Table 1.2   Antinutritional factors of cowpea seeds.

Cowpea accession	Proanthocyanidin content (mg/g dry seed coat)	α-Amylase inhibitory activity $(I_{50})^a$
Vita 7	32.0	nd
IT 84E-1-108	2.4	26.0

$^a I_{50} = \mu$g of protein that gives 50% inhibition of insect α-amylase.
nd, not detectable.

derived from an elevated level of trypsin inhibitor within the cowpea seeds. However, some researchers suggest that the trypsin inhibitor alone does not account for bruchid resistance in cowpea, thus indicating the need for further investigations. Plant α-amylase inhibitors are particularly abundant in cereals and leguminosae, and some wheat α-amylase inhibitors inhibit insect α-amylases strongly. When added in low concentrations (1%) to an artificial diet, bean α-amylase inhibitors proved toxic to the cowpea weevil and adzuki bean weevil larvae (Ishimoto & Kitamura; 1989; Shade *et al.*, 1994; Schroeder *et al.*, 1995; Franco *et al.*, 2000; Pedra *et al.*, 2003).

In addition, seed coat tannins are present at high levels in most plant seeds and grains, and are generally considered to be harmful to phytophagous insects. Tannins may affect the growth of insects in three main ways: they have an astringent taste which affects palatability and decreases feed consumption; they combine with proteins to form complexes of reduced digestibility; and they act as enzyme inactivators (Winkel-Shirley, 1998). Recent work by Raymond Barbehenn and coworkers about tannin oxidation in insects suggests that tannin activity cannot be explained quite this simply, as tannin oxidation should also be taken into account as a defence mechanism for plants (Constabel & Barbehenn, 2008; Barbehenn *et al.*, 2008, 2009a, 2009b, 2010). In stored cowpea, seed coat proanthocyanidins contribute to resistance against cowpea weevil (*C. maculatus*) infestation (Lattanzio *et al.*, 2005). Plate 1.4 shows two accessions of stored cowpea seeds presenting different degrees of bruchid damage during storage: IT 84E-1-108 exhibit an high level of infestation (about 30%), while Vita 7 does not show damage caused by cowpea weevil larvae. No α-amylase inhibitory activity has been found in cotyledons of Vita 7 seeds, while IT 84E-1-108 exhibited a moderate level of α-amylase inhibitory activity (Table 1.2). On the contrary, the seed coat tannin content was found to be 13 times higher in undamaged Vita 7 seeds than in IT 84E-1-108 infested seeds. These results support the hypothesis that, if bruchids infest cowpea when the grain is stored after harvest, seed coat tannins are effectively involved in the biochemical defence mechanisms, which can deter, poison or starve the bruchid larvae that feed on cowpea seeds.

1.4 Diversion of carbon skeletons from primary to phenolic-related secondary metabolism

The accumulation of phenolics in plant tissues is a distinctive characteristic of plant stress: phenolic compound may be increased or *de novo* synthesised in plants as a response to various biotic stresses, such as herbivores, pests and fungal pathogens, and to abiotic stresses, including visible and UVB radiation, cold temperatures, water stress, and nutrient deficiency. This means that plant phenolics confer various physiological functions for plants to survive and to adapt to environmental disturbances (Bennett & Wallsgrove, 1994; Leyva *et al.*, 1995; Bachereau *et al.*, 1998; Cooper-Driver & Bhattacharya, 1998; Chalker-Scott, 1999; Logemann *et al.*, 2000; Kidd *et al.*, 2001; Stewart *et al.*, 2001; Casati & Walbot, 2003; Treutter, 2005; 2006; Lattanzio *et al.*, 2006; Caldwell *et al.*, 2007; Lillo *et al.*, 2008; Olsen *et al.*, 2008; Adams-Phillip *et al.*, 2010). This chemical response to changing environments has led to the enormous structural variation in the major groups of phenolic compounds, which are evident in plants today. More detailed knowledge of these effects should enable prediction and selection of growth conditions in order to achieve a desirable content of these secondary metabolites. Manipulation of environmental factors should – at least to some degree – represent an alternative to genetic engineering for achieving special effects on the level of plant components. Furthermore, understanding of the regulatory and biochemical mechanisms that control the types and amounts of phenolic compounds synthesised under different conditions continues to be a high priority for research, with a view to possible engineering of crop plants to overproduce antioxidant phenolics.

Broadly speaking, plant growth and productivity are greatly affected by environmental stresses. Both abiotic and biotic stresses divert substantial amounts of substrates from primary metabolism into secondary defensive product formation and this could lead to constraints on growth. Plants have limited resources to support their physiological processes, so that all requirements cannot be met simultaneously, and trade-offs occur between growth and defence (Coley *et al.*, 1985; Herms & Mattson, 1992; Van der Plas *et al.*, 1995). Therefore, a principal feature of plant metabolism is the flexibility to accommodate developmental changes and respond to the environment. The cellular and molecular responses of plants to environmental stress include mechanisms by which plants perceive environmental signals and transmit the signals to cellular machinery to activate adaptive responses, and this is of fundamental importance to biology. Knowledge about stress signal transduction is also vital for the continued development of strategies to improve stress tolerance in crops (Xiong *et al.*, 2002; Yamaguchi-Shinozaki & Shinozaki, 2006; Weigelt *et al.*, 2009). In addition, plant responses to both biotic and abiotic stresses require the reallocation of resources, therefore these responses are considered to be costly for plants because of the energy consumed in the biosynthesis of defensive phenolics and the ecological consequences of their accumulation. Costs can be described as resource-based trade-offs between resistance and fitness, as ecological costs, or as allocation costs (Heil *et al.*, 2000; Heil & Baldwin, 2002; Strauss *et al.*, 2002). In order to quantify these costs in plants, researchers have attempted to link a measure of plant success (usually, growth rate) with levels of defensive compounds. Zangerl *et al.* (1997) examined the effects of damage-induced synthesis of furanocoumarins, known defence compounds, on the growth of wild parsnip. Plants that had 2% of their leaf area removed accumulated 8.6% less total biomass and 14% less root biomass than intact plants over a 4-week period. Pavia *et al.* (1999) investigated the potential cost of polyphenolic (phlorotannin) production in brown seaweed *Ascophyllum nodosum* by testing for phenotypic trade-offs between phlorotannin content and annual growth. Data showed that there was a significant negative relationship between phlorotannins and growth. Shoots with a relatively high phlorotannin content (>9% dry weight) presented a mean growth reduction that varied from 25% to 54%, compared to shoots with relatively low concentrations (<6% dry weight) of phlorotannins. Resource-based allocation theory predicts a trade-off mechanism between plant reproduction, growth and defence functions that regulates carbon fluxes between primary and secondary metabolism, and that is specifically required for protective adaptation to environmental stresses (Coley *et al.*, 1985; Bazzaz *et al.*, 1987; Chapin *et al.*, 1987; Herms & Mattson, 1992; Purrington, 2000; Brown, 2003; Burdon & Thrall, 2003; Siemens *et al.*, 2003; Dietrich *et al.*, 2005).

In many plants, free proline also accumulates as a common physiological response to a wide range of biotic and abiotic stresses. Furthermore, proline accumulation is considered to be one of the stress signal influencing adaptive multiple responses that are part of the adaptation process. Transgenic approaches have confirmed the beneficial effect of proline overproduction during stress. Accumulation of proline could be due to *de novo* synthesis, to decreased degradation, or to both of these. Most attempts to account for the phenomenon have focused on the ability of proline to mediate osmotic adjustment, to scavenge free radicals, and to act as a source of reducing power and as a source of carbon and/or nitrogen. Accumulated proline has been proposed to protect enzymes, membranes and polyribosomes during environmental disturbances, and to protect cellular functions by scavenging reactive oxygen species (Kushad & Yelenosky, 1987; Saradhi *et al.*, 1995; Kiyosue *et al.*, 1996; Hare & Cress, 1997; Hare *et al.*, 1998, 1999; Maggio *et al.*, 2002; Parida *et al.*, 2002; Deuschle *et al.*, 2004; Kavi Kishor *et al.*, 2005; Sharma & Dietz, 2006 Verbruggen & Hermans, 2008). Is there a link between increased phenolic levels and increased proline levels in plant tissues under stress? In this connection, it must be stressed that the oxidative pentose phosphate pathway (OPPP) is the source of reducing equivalents (NADPH) for phenylpropanoid biosynthesis, and that this pathway also provides the erythrose 4-phosphate that, along with phosphoenolpyruvate formed from glycolysis, serves as a precursor for phenylalanine biosynthesis via the shikimic acid pathway (Fahrendorf *et al.*, 1995). In addition, the increased $NADP^+$/NADPH ratio, mediated by stress-induced proline biosynthesis, is likely to enhance the activity of the OPPP. The two dehydrogenases responsible for transforming glucose-6-phosphate into ribose-5-phosphate are primarily regulated by the $NADP^+$/NADPH ratio, with both enzymes strongly inhibited by NADPH. Dehydrogenase reactions that consume NADPH and produce $NADP^+$ would positively interfere with OPPP activity: the alternating oxidation of NADPH by proline synthesis and reduction of $NADP^+$ by the two oxidative steps of the OPPP would link these two pathways (Fig. 1.9) (Hare & Cress, 1997; Kavi Kishor *et al.*, 2005).

1.4.1 Metabolic costs of adaptive responses to adverse environmental conditions

Primary metabolism is an important source of precursors for the synthesis of secondary phenolic metabolites. On the other hand, central metabolism requires high levels of limited plant resources and during intense growth the synthesis of phenolic metabolites may be substrate- and/or energy limited. A fixed amount of resources is usually assumed to be divided among fixed maintenance cost, growth and reproduction. This suggests that an organism's growth rate should be at its physiological maximum whenever it is not

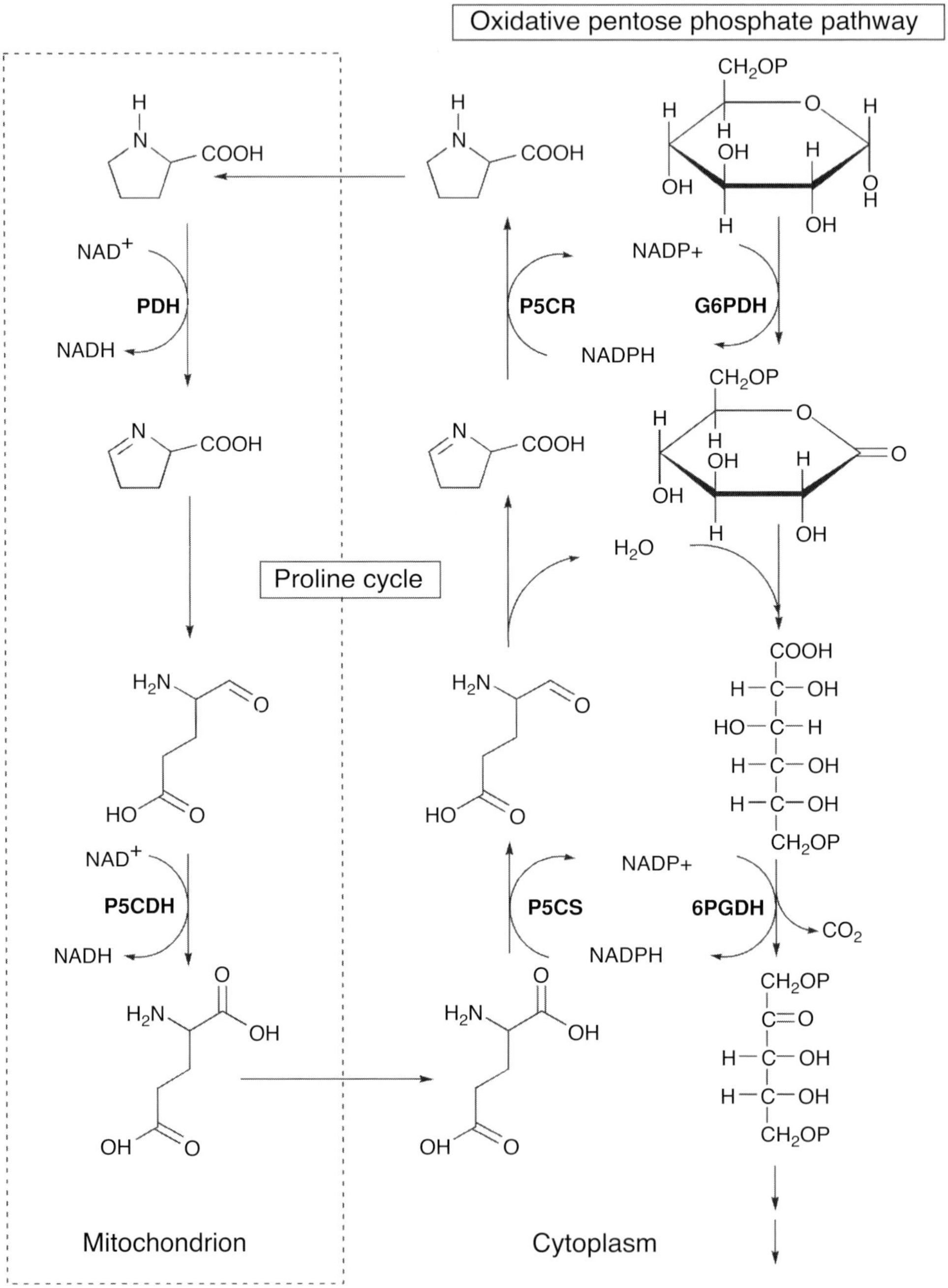

Fig. 1.9   Relationship between proline cycle and oxidative steps of cytosolic pentose phosphate pathway. The enzymes are: G6PDH: Glucose-6-phosphate dehydrogenase (EC 1.1.1.49); 6PGDH: 6-Phosphogluconate dehydrogenase (EC 1.1.1.44); P5CS: Δ^1-pyrroline-5-carboxylate synthetase (EC 2.7.2.11 + EC 1.2.1.41); P5CR: Δ^1-pyrroline-5-carboxylate reductase (EC 1.5.1.2); PDH: Proline dehydrogenase (EC 1.4.3); P5CDH: Δ^1-pyrroline-5-carboxylate dehydrogenase (EC 1.5.1.12).

reproducing. Growth rates are correlated with the ecological conditions in which each species is living in nature; slow growth is adaptive for dealing with environmental stresses. Growing plants, therefore, continuously face a dilemma regarding the partitioning of their available carbon resources. If priority is given to the plant growth processes, the availability of carbon resources (and other nutrients) may become limiting for plant defence-related processes, and vice versa. So far, four main plant defence hypotheses have been put forward to explain patterns and variations in the concentration of carbon-based secondary compounds in plant tissues, according to availability of resources. These theories hinge on the presence of resistance costs, because, in the absence of costs, selection is expected to favour the best-defended genotypes. So, the problem is to explain the costs and trade-offs that cause organisms to grow below their physiological maximum (Bergelson & Purrington, 1996; Strauss *et al.*, 2002).

The carbon–nutrient balance hypothesis (CNBH; Bryant *et al.*, 1983) predicts how resources affect phenotypic expression of plant defence, often with studies concerned about the allocation cost of defence. This hypothesis also suggests that carbon-based secondary metabolites tend to accumulate when growth is limited by low levels of mineral nutrients. The optimal defence theory (ODT; McKey, 1974) has served as the main framework for investigation of genotypic expression of plant defence, with the emphasis on the allocation cost of defence. This theory addresses how the defensive needs of a plant contribute to the evolution of secondary metabolites, with defence costs paid to maximise plant fitness. In essence, this hypothesis states that any defensive pattern is possible if it is adaptive. The protein competition model (PCM; Margna, 1977; Margna *et al.*, 1989; Jones & Hartley, 1999) predicts total phenol allocations in higher plants suggesting that a location trade-off for carbon and nitrogen probably occur among metabolic pathways and even within pathways. PCM states that protein and phenol synthesis compete for the common, limiting resource phenylalanine, so that protein and phenolic allocation are inversely correlated. Phenol allocation can be predicted from the effects of development, inherent growth rate and environment on leaf functions that create competing demands for proteins or phenolics. PCM is considered an alternative to the CNBH, and a hypothesis that complements the growth-differentiation balance hypothesis (GDBH). The GDBH (Herms & Mattson, 1992) predicts how plants allocate resources between growth-related processes (any process that requires substantial cell division and elongation) and differentiation-related processes (enhancement of the structure or function of existing cells, such as secondary metabolism) in different environmental conditions. Allocation to differentiation includes the cost of enzymes, transport and storage structures involved in defence. Growth and secondary metabolism can compete for available photosynthates and so there is a trade-off for carbon allocation. GDBH states that there is a physiological trade-off between growth and secondary metabolism imposed by developmental constraints in growing cells, and competition between primary and secondary metabolic pathways in mature cells. This hypothesis also predicts that any factor that slows growth more than it slows photosynthesis can increase the internal resources available for allocation to differentiation. For instance, growth is slowed by the limitation of nutrients, whereas photosynthesis is less sensitive to it. Consequently, carbohydrates accumulate beyond growth demands, and may thus be converted to secondary metabolites. These four hypotheses suggest that plants continuously make effective use of costly versus beneficial investments towards defence versus growth processes, with the trade-off mainly conditioned by resource availability. A clearer understanding of these hypotheses – and what we have learned from investigations that use them – can facilitate development of well-designed experiments that address the gaps in knowledge of plant defence (Lorio & Sommers, 1986; Arendt, 1997; Stamp, 2003, 2004; Glynn *et al.*, 2007; Le Bot *et al.*, 2009).

1.4.2 Transduction pathway between nutrient depletion and enhanced polyphenol content

An intriguing question linked to the role of plant phenolics in plant responses to environmental stress is the identification of a signal transduction; this is an ordered sequence of biochemical reactions inside the cell, resulting in a signal transduction pathway, which transfer an environmental signal from the outside of the cell into the plant cell, thus producing a physiological response. In nature, cellular functions are propagated by cascades of molecules, which interact with one another. Generally speaking, one reaction depends on a previous step.

It has recently been proposed that there is a link between primary and secondary metabolism that couples the accumulation of the stress metabolite proline with the energy transfer towards phenylpropanoid biosynthesis *via* the oxidative pentose phosphate pathway (Hare & Cress, 1997; Lattanzio *et al.*, 2009). Following the imposition of a nutritional stress, the growth of oregano shoots is reduced (–40%) in comparison to the control shoots (Plate 1.5). In contrast with this reduced growth of oregano shoots, the total phenolic content (carbon-based secondary metabolites) is greatly enhanced (+120%); this increase runs parallel to an increase (moderate) in intracellular free proline, enhancing the tolerance of cellular components to reactive oxygen species synthesised by plants experiencing stress conditions (Smirnoff, 1993). These data are consistent with the scheme proposed in Fig. 1.10, which involves a continuous cycling of proline, and is based on the fact that the plant cell is a highly integrated system, ensuring a tight regulation of interacting pathways by their coupling through common intermediates, including pyridine nucleotides. Plant tissues are forced to accumulate free proline under stress conditions. In these conditions the increased proline synthesis maintains $NAD(P)^+/NAD(P)H$ ratios at values compatible with metabolism under normal conditions, because proline synthesis is accompanied by the oxidation of NADPH; this may constitute a form of metabolic response within the plant cell, triggered in the signal transduction pathway between perception of nutritional stress and physiological response. The increased $NADP^+/NADPH$ ratio, mediated by proline biosynthesis, is likely to enhance the activity of the oxidative pentose phosphate pathway providing precursors for phenolic biosynthesis via the shikimic acid pathway (Chandler & Thorpe, 1987; Chen & Kao, 1995; Hare & Cress, 1997). The alternating oxidation of NADPH by cytosolic proline synthesis and reduction of $NADP^+$ by the two oxidative steps of the oxidative pentose phosphate pathway serve to link both pathways and thereby facilitate the continuation of high rates of proline synthesis during stress and lead to a simultaneous accumulation of phenolic compounds. Figure 1.10 shows that mitochondrial proline oxidation could drive the oxidative pentose phosphate pathway

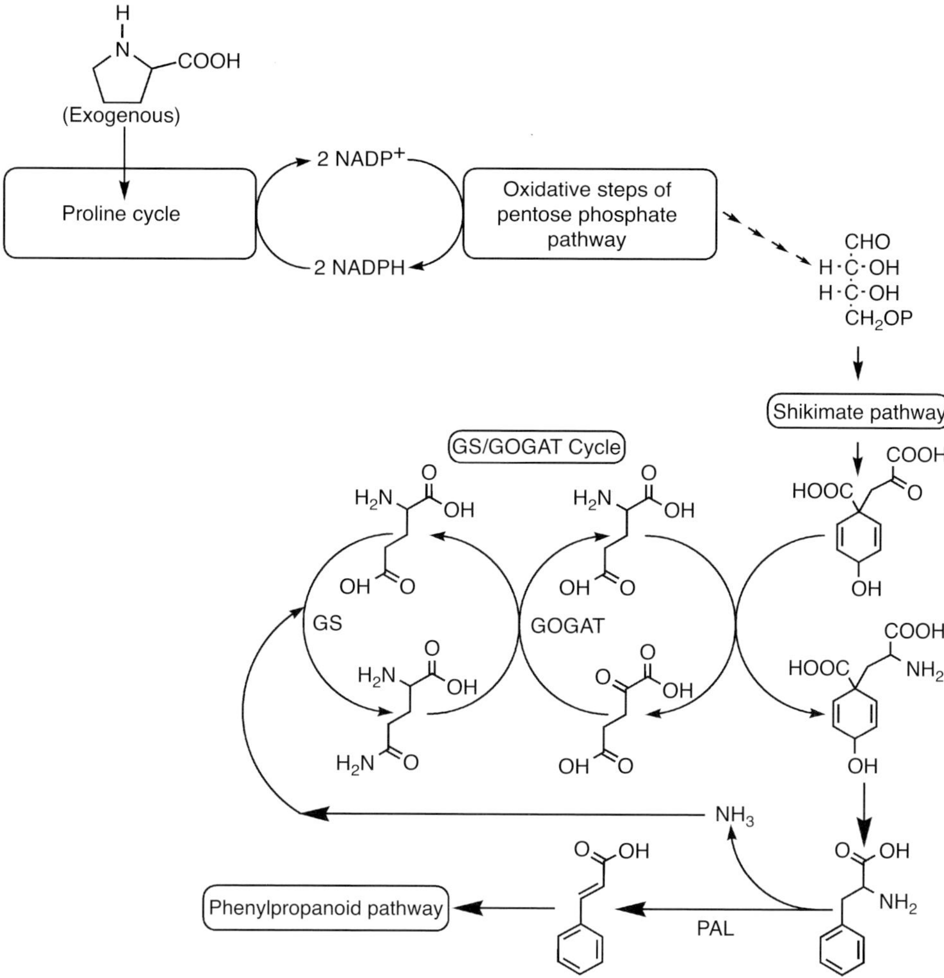

Fig. 1.10 Scheme showing the relationships between primary and secondary metabolism and the role of endogenous and exogenous proline in stimulating phenylpropanoid pathway. The enzymes are: PAL, Phenylalanine ammonia lyase (EC 4.3.1.5); GS, Glutamine synthetase (EC 6.1.1.3); GOGAT, Glutamate synthase (EC 1.4.1.14).

by recycling glutamic acid into the cytosol to generate a proline redox cycle (Zheng *et al.*, 2001). Figure 1.10 also shows that cytosolic glutamic acid may be also utilised for recycling ammonium ions, produced in the first step of the phenylpropanoid biosynthesis, by means of the glutamine synthetase (EC 6.1.1.3; GS) and glutamate synthase (syn. glutamine ox-oglutarate aminotransferase) (EC 1.4.1.14; GOGAT) cycle. It has been suggested that the ammonium ion released during active phenylpropanoid metabolism is not made available for general amino acid/protein synthesis. Rather, it is rapidly recycled back to regenerate phenylalanine, thereby providing an effective means of maintaining active phenylpropanoid metabolism with no additional nitrogen requirement. The ammonium ion released during lysis is metabolised via the GS/GOGAT cycle to generate glutamate thereby permitting arogenate synthesis, via prephenate transamination, which, in turn, regenerates phenylalanine (van Heerden *et al.*, 1996).

Finally, in good agreement with the scheme proposed in Fig. 1.10, it has been also observed that an application of 0.5mM proline to the nutrient medium of *in vitro* grown oregano seedlings elicit the accumulation of rosmarinic acid and other phenolic compounds in that plant (Lattanzio *et al.*, 2009). Therefore, it can be concluded that the effect of both, exogenously added proline and stress elicited proline, on phenolic metabolism is linked to replenishment of the $NADP^+$ supply to oxidative pentose phosphate pathway which, in turn, is a source of NADPH and carbon skeletons for phenylpropanoid biosynthesis.

References

Abdel-Farid, I.B., Jahangir, M., van den Hondel, C.A.M.J., Kim, H.K., Choi, Y.H. & Verpoorte, R. (2009) Fungal infection-induced metabolites in *Brassica rapa*. *Plant Science*, **176**, 608–615.

Adams-Phillip, L., Briggs, A.G. & Bent, A.F. (2010) Disruption of poly(ADP-ribosyl)ation mechanisms alters responses of arabidopsis to biotic stress. *Plant Physiology*, **152**, 267–280.

Alibert, G. & Boudet, A.M. (1982) Compartimentation tissulaire et intracellulaire des polyphenols. *C. R. Groupe Polyphenols* **10**, 129–150.

Agati, G., Cerovic, Z.G., Dalla, Marta, A., *et al.* (2008) Optically-assessed preformed flavonoids and susceptibility of grapevine to Plasmopara viticola under different light regimes. *Functional Plant Biology*, **35**, 77–84.

Amiot, M.J., Fleuriet, A., Cheynier, V. & Nicolas, J. (1997) Phenolic compounds and oxidative mechanisms in fruits and vegetables. In: *Phytochemistry of Fruit and Vegetables* (eds. F.A. Tomás-Barberán & J. Robins), pp. 51–85. Clarendon Press, Oxford.

Ananthakrishnan, T.N. (1997) Gallic and salicylic acids: sentinels of plant defence against insects. *Current Science*, **73**, 576–579.

Angell, H.R., Walker, J.C. & Link, K.P. (1930) The relation of protocatechuic acid to disease resistance in the onion. *Phytopathology*, **20**, 431–438.

Annan, I.B., Schaefers, G.A. & Tingey, W.M. (1996) Impact of density of *Aphis craccivora* (Aphididae) on growth and yield of susceptible and resistant cowpea cultivars. *Annals of Applied Biology*, **128**, 185–193.

Austin, M.B. & Noel, J.P. (2003) The chalcone synthase superfamily of type III polyketide synthases. *Natural Product Reports*, **20**, 79–110.

Appleby, J.H. & Credland, P.F. (2003) Variation in responses to susceptible and resistant cowpeas among West African populations of *Callosobruchus maculatus* (Coleoptera: Bruchidae). *Journal of Economic Entomology*, **96**, 489–502.

Arendt, J.D. (1997) Adaptive intrinsic growth rates: an integration across taxa. *The Quarterly Review of Biology*, **72**, 149–177.

Bachereau, F., Marigo, G. & Asta, J. (1998) Effect of solar radiation (UV and visible) at high altitude on CAM-cycling and phenolic compound biosynthesis in *Sedum album*. *Physiologia Plantarum*, **104**, 203–210.

Barbehenn, R., Dukatz, C., Holt, C., *et al.* (2010) Feeding on poplar leaves by caterpillars potentiates foliar peroxidase action in their guts and increases plant resistance. *Oecologia*, **164**, 993–1004.

Barbehenn, R., Weir, Q. & Salminen, J.-P. (2008) Oxidation of ingested phenolics in the tree feeding caterpillar *Orgyia leucostigma* depends on foliar chemical composition. *Journal of Chemical Ecology*, **34**, 748–756.

Barbehenn, R.V., Jaros, A., Lee, G., Mozola, C., Weir, Q. & Salminen, J.-P. (2009a) Tree resistance to *Lymantria dispar* caterpillars: importance and limitations of foliar tannin composition. *Oecologia*, **159**, 777–788.

Barbehenn, R.V., Jaros, A., Lee, G., Mozola, C., Weir, Q. & Salminen, J.-P. (2009b) Hydrolyzable tannins as "quantitative defenses": limited impact against *Lymantria dispar* caterpillars on hybrid poplar. *Journal of Insect Physiology*, **55**, 297–304.

Barz, W. & Hoesel, W. (1975) Metabolism of flavonoids. In: *The Flavonoids* (eds. J.B. Harborne, T.J. Mabry & H. Mabry), pp. 916–969. Chapman and Hall, London, UK.

Barz, W. & Hoesel, W. (1979) Metabolism and degradation of phenolic compounds in plants. In: *Biochemistry of Plant Phenolics, Recent Advance in Phytochemistry* (eds. T. Swain, J.B. Harborne & C.F. Van Sumere), vol. 12, pp. 339–369. Plenum Press, New York.

Bate-Smith, E.C., Hughes, J.C. & Swain, T. (1958) After- cooking discoloration in potatoes. *Chemistry & Industry (London)*, **21**, 627–630.

Bazzaz, F.A., Chiariello, N.R., Coley, P.D. & Pitelka, L.F. (1987) Allocating resources to reproduction and defense. *BioScience*, **37**, 58–67.

Bell, E.A. (1980) The possible significance of secondary compounds in plants. In: *Secondary Plant Products, Enciclopedia of Plant Physiology New Series* (eds. E.A. Bell & B.W. Charlwood), Vol. 8, pp. 11–21. Springer-Verlag, Berlin.

Benkeblia, N. (2000) Phenylalanine ammonia-lyase, peroxidase, pyruvic acid and total phenolics variations in onion bulbs during long-term storage. *Lebensmittel-Wissenschaft und-Technologie*, **33**, 112–116.

Bennett, R.N. & Wallsgrove, R.M. (1994) Secondary metabolites in plant defence mechanisms. *New Phytologist*, **127**, 617–633.

Bergelson, J. & Purrington, C.B. (1996) Surveying patterns in the cost of resistance in plants. *American Naturalist*, **148**, 536–558.

Bernays, E.A. & Chapman, R.F. (2000) Plant secondary compounds and grasshoppers: beyond plant defenses. *Journal of Chemical Ecology*, **26**, 1773–1794.

Bienfait, H.F. & Van der Briel, M.L. (1980) Rapid mobilization of ferritin iron by ascorbate in the presence of oxygen. *Biochimica et Biophysica Acta*, **631**, 507–510.

Bieza, K. & Lois, R. (2001) An *Arabidopsis* mutant tolerant to lethal ultraviolet-B levels shows constitutively elevated accumulation of flavonoids and other phenolics. *Plant Physiology*, **126**, 1105–1115.

Blankenship, S.M. & Richardson, D.G. (1985) Changes in phenolic acids and internal ethylene during long-term cold storage of pear. *Journal of American Society for Horticultural Science*, **110**, 336–339.

Blankenship, S.M. & Unrath, C.R. (1988) PAL and ethylene content during maturation of Red and Golden Delicious apples. *Phytochemistry*, **27**, 1001–1003.

Blokhina, O., Virolainen, E. & Fagerstedt, K.V. (2003) Antioxidants, oxidative damage and oxygen deprivation stress: A review. *Annals of Botany*, **91**, 179–194.

Boudet, A.M., Graziana, A & Ranjeva, R. (1985) Recent advances in the regulation of the prearomatic pathway. In: *The Biochemistry of Plant Phenolics* (eds. C. F. Van Sumere & P. J. Lea), pp. 135–160. Clarendon Press., London.

Boyer, R.F., Clark, H.M. & La Roche, A.P. (1988a) Reduction and release of ferritin iron by plant phenolics. *Journal of Inorganic Biochemistry*, **32**, 171–181.

Boyer, R.F., Clark, H.M. & Sanchez, S. (1989) Solubilisation of ferrihydride iron by plant phenolics: a model for rhizosphere processes. *Journal of Plant Nutrition*, **12**, 581–592.

Boyer, R.F., Grabill, T.W. & Petrovich, R.M. (1988b) Reductive release of ferritin iron: a kinetic assay. *Analytical Biochemistry*, **174**, 17–22.

Boyer, R.F., McArthur, J.S. & Cary, T.M. (1990) Plant phenolics as reductant for ferritin iron release. *Phytochemistry*, **29**, 3717–3719.

Brenes, M., Romero, C., García, P. & Garrido, A. (1995) Effect of pH on the color formed by Fe-phenolic complexes in ripe olives. *Journal of the Science of Food and Agriculture*, **67**, 35–41.

Brown, J.K.M. (2002) Yield penalties of disease resistance in crops. *Current Opinion in Plant Biology*, **5**, 339–344.

Brown, J.K.M. (2003) A cost of disease resistance: paradigm or peculiarity? *Trends in Genetics*, **19**, 667–671.

Bryant, J.P., Chapin, F.S. III & Klein. D.R. (1983) Carbon/nutrient balance of boreal plants in relation to vertebrate herbivory. *Oikos*, **40**, 357–368.

Burdon, J.J. & Thrall, P.H. (2003) The fitness costs to plants of resistance to pathogens. *Genome Biology*, **4**, 227.1–227.3.

Butt, V.S. (1985) Oxygenation and oxidation in the metabolism of aromatic compounds. In: *The Biochemistry of Plant Phenolics* (eds. C.F. Van Sumere & P.J. Lea), pp. 349–365. Clarendon Press, Oxford.

Byrde, R.J.W., Fielding, A.H. & Williams, A.H. (1960) The role of oxidized polyphenols in the varietal resistance of apple to brown rot. In: *Phenolics in Plants and Disease* (ed. J.B. Pridham), pp. 95–99. Pergamon Press, Oxford.

Caldwell, M.M., Bornman., J.F., Ballaré, C.L., Flint, S.D. & Kulandaivelu, G. (2007) Terrestrial ecosystems, increased solar ultraviolet radiation, and interactions with other climate change factors. *Photochemical & Photobiological Science*, **6**, 252–266.

Camm, E.J. & Towers, G.H.N. (1973) Phenylalanine ammonia-lyase. *Phytochemistry* **12**, 961–973.

Campos-Vargas, R. & Saltveit, M.E. (2002) Involvement of putative chemical wound signals in the induction of phenolic metabolism in wounded lettuce. *Physiologia Plantarum*, **114**, 73–84.

Casati, P. & Walbot, V, (2003) Gene expression profiling in response to ultraviolet radiation in *Zea mays* genotypes with varying flavonoid content. *Plant Physiology*, **132**, 1739–1754.

Chalker-Scott, L. (1999) Environmental significance of anthocyanins in plant stress responses. *Photochemistry and Photobiology*, **70**, 1–9.

Chalker-Scott, L. & Fuchigami, L.H. (1989) The role of phenolic compounds in plants stress responses. In: *Low Temperature Stress Physiology in Crops* (ed. H.L. Paul), pp. 67–79. CRC Press, Boca Raton, Florida.

Chalutz, E. (1973) Ethylene-induce phenylalanine ammonia-lyase activity in carrot roots. *Plant Physiology*, **51**, 1033–1036.

Chandler, S.F. & Thorpe, T.A. (1987) Characterisation of growth, water relations, and proline accumulation in sodium sulphate tolerant callus of *Brassica napus* L. cv Westar (Canola). *Plant Physiology*, **84**, 106–111.

Chapin, F.S. III, Bloom, A.J., Field, C.B. & Waring, R.H. (1987) Plant responses to multiple environmental factors. *BioScience*, **37**, 49–57.

Chapman, R.F. (1974) The chemical inhibition of feeding by phytophagous insects: A review. *Bulletin of Entomological Research*, **64**, 339–363.

Chen, S.L. & Kao, C.H. (1995) Cd induced changes in proline level and peroxidase activity in roots of rice seedlings. *Plant Growth Regulation*, **17**, 67–71.

Cheng, G.W. & Crisosto, C.H. (1997) Iron-polyphenol complex formation and skin discoloration in peaches and nectarines. *Journal of the American Society for Horticultural Science*, **122**, 95–99.

Chi, Y. H., Salzman, R. A., Balfe, S., *et al.* (2009) Cowpea bruchid midgut transcriptome response to a soybean cystatin – costs and benefits of counter-defence. *Insect Molecular Biology*, **18**, 97–110.

Christie, R.J., Alfenito, M.R. & Walbot, V. (1994) Impact of low temperature stress on general phenylpropanoid and anthocyanin pathways: enhancement of transcript abundance and anthocyanin pigmentation in maize seedlings. *Planta*, **194**, 541–549.

Coetzer, C., Corsini, D., Love, S., Pavek, J. & Tumer, N. (2001) Control of enzymatic browning in potato (*Solanum tuberosum* L.) by sense and antisense RNA from tomato polyphenol oxidase. *Journal of Agricultural and Food Chemistry*, **49**, 652–657.

Coley, P.D., Bryant, J.P. & Chapin, F.S. III (1985) Resource availability and plant antiherbivore defense. *Science*, **230**, 895–899.

Constabel, C.P. (1999) A survey of herbivore-inducible defensive proteins and phytochemicals. In: *Induced Plant Defenses Against Herbivores and Pathogens: Biochemistry, Ecology, and Agriculture* (eds. A.A. Agrawaal, S. Tuzun & E. Bent), pp. 137–166. APS Press, St. Paul, Minnesota.

Constabel, C.lP. & Barbehenn, R. (2008) Defensive roles of polyphenol oxidase in plants. In: *Induced Plant Resistance to Herbivory* (ed. A. Schaller), pp. 253–269. Springer Science+Business Media B.V., Netherlands.

Cooper-Driver, G.A. & Bhattacharya, M. (1998) Role of phenolics in plant evolution. *Phytochemistry*, **49**, 1165–1174.

Cowan, M.M. (1999) Plant products as antimicrobial agents. *Clinical Microbiology Reviews*, **12**, 564–582.

Croteau, R., Kutchan, T.M. & Lewis, N.G. (2000) Natural products (secondary metabolites). In: *Biochemistry & Molecular Biology of Plants* (eds. B. Buchanan, W. Gruissem & R. Jones), pp. 1250–1318. American Society of Plant Physiologist, Rockville, Maryland.

Dethier, V.G. (1970) Chemical interactions between plants and insects. In: *Chemical Ecology* (eds. E. Sondheimer, & J.B Simeone), pp. 83–102. Academic Press, New York.

Deuschle, K., Funck, D., Forlani, G., *et al.* (2004) The role of Δ^1-pyrroline-5-carboxylate dehydrogenase in proline degradation. *Plant Cell*, **16**, 3413–3425.

Dietrich, R., Ploss, K. & Heil, M. (2005) Growth responses and fitness costs after induction of pathogen resistance depend on environmental conditions. *Plant, Cell & Environment*, **28**, 211–222.

Dixon, R.A. & Paiva, N.L. (1995) Stress-induced Phenylpropanoid metabolism. *Plant Cell*, **7**, 1085–1097.

Dreyer, D.L. & Campbell, B.C. (1987) Chemical basis of host-plant resistance to aphids. *Plant, Cell & Environment*, **10**, 353–361.

Ehrlich, P.R. & Raven, P.H. (1964) Butterflies and plants: a study in coevolution. *Evolution* **18**, 586–608.

Elliger, C.A., Chan, B.C. & Waiss, A.C. (1980) Flavonoids as larval growth inhibitors. *Naturwissenschaften*, **67**, 358–359.

Engelsma, G. (1970) Low-temperature effects on phenylalanine ammonia-lyase activity in gherkin seedlings. *Planta*, **91**, 246–254.

Engelsma, G. (1974) On the mechanism of the changes in phenylalanine ammonia-lyase activity induced by ultraviolet and blue light in gherkin hypocotyls. *Plant Physiology*, **54**, 702–705.

Eyles, A., Bonello, P., Ganley, R. & Mohammed, C. (2009) Induced resistance to pests and pathogens in trees. *New Phytologist*, **185**, 893–908.

Facchini, P.J. (1999) Plant secondary metabolism: out of the evolutionary abyss. *Trends in Plant Science*, **4**, 382–384.

Fahrendorf, T., Ni, W., Shorrosh, B.S. & Dixon, R.A. (1995) Stress responses in alfalfa (*Medicago sativa* L.) XIX. Transcriptional activation of oxidative pentose phosphate pathway genes at the onset of the isoflavonoid phytoalexin response. *Plant Molecular Biology*, **28**, 885–900.

Ferguson, J.E., Metcalf, E.R., Metcalf, R.L. & Rhodes, A.M. (1983) Influence of cucurbitacin content in cotyledons of Cucurbitaceae cultivars upon feeding behavior of Diabroticine beetles (Coleoptera: Chrysomelidae). *Journal of Economic Entomology*, **76**, 47–51.

Flechtner, V.R., Johansen, J.R. & Clark, W.H. (1998) Algal composition of microbiotic crusts from the central desert of Baja California, Mexico. *Great Basin Naturalist*, **58**, 295–311.

Fraenkel, G. (1959) The *raison d'être* of secondary plant substances. *Science*, **129**, 1466–1470.

Franco, O.L., Rigden, D.J., Melo, F.R., Bloch, Jr C., Silva, C.P. & Grossi-de-Sá, M.F. (2000) Activity of wheat α-amylase inhibitors towards bruchid α-amylases and structural explanation of observed specificities. *European Journal of Biochemistry*, **267**, 2166–2173.

Friedman, M. (1996) Food browning and its prevention: An overview. *Journal of Agricultural and Food Chemistry*, **44**, 631–653.

Friend, J. (1979) Phenolic substances and plant disease. In: *Biochemistry of Plant Phenolics* (eds. T. Swain, J.B. Harborne & C.F. Van Sumere), pp. 557–588. Plenum Press, London.

Friend, J. 1985. Phenolic substances and plant disease. In: *The Biochemistry of Plant Phenolics, Annual Proceedings of the Phytochemical Society of Europe* (eds. C.F. Van Sumere & P.J. Lea), vol. 25, pp. 367–392. Clarendon Press, Oxford.

Gatehouse, A.M.R., Gatehouse, J.A., Dobie, P., Kilminster, A.M. & Boulter, D. (1979) Biochemical basis of insect resistance in *Vigna unguiculata*. *Journal of The Science of Food and Agriculture*, **30**, 948–958.

Gil-Izquierdo, A., Gil, M.I., Conesa, M.A. & Ferreres, F. 2001: The effect of storage temperature on vitamin C and phenolics content of artichoke (*Cynara scolymus* L.) heads. *Innovative Food Science & Emerging Technologies*, **2**, 199–202.

Glynn, C., Herms, D.A., Orians, C.M., Hansen, R.C. & Larsson, S. (2007) Testing the growth-differentiation balance hypothesis: dynamic responses of willows to nutrient availability. *New Phytologist*, **176**, 623–634.

Golding, J.B., McGlasson, W.B., Wyllie, W.B. & Leach, D-M. (2001) Fate of apple peel phenolics during cool storage. *Journal of Agricultural and Food Chemistry*, **49**, 2283–2289.

Gomez-Vasquez, R., Day, R., Buschmann, H., Randles, S., Beeching, J.R. & Cooper, R.M. (2004) Phenylpropanoids, phenylalanine ammonia lyase and peroxidases in elicitor challenged Cassava (*Manihot esculenta*) suspension cells and leaves. *Annals of Botany*, **94**, 87–97.

Graham, D. & Patterson, B.D. (1982) Response of plants to low, non-freezing temperatures: proteins, metabolism, and acclimation. *Annual Review of Plant Physiology*, **33**, 347–372.

Graham, L.E., Cook, M.E. & Busse, J.S. (2000) The origin of plants: Body plan changes contributing to a major evolutionary radiation. *Proceedings of the National Academy of Sciences USA*, **97**, 4535–4540.

Grayer, R.J. & Kokubun, T. (2001) Plant-fungal interactions: the search for phytoalexins and other antifungal compounds from higher plants. *Phytochemistry*, **56**, 253–263.

Guyot, S., Bernillon, S., Poupard, P. & Renard, C.M.G.C. (2008) Multiplicity of phenolic oxidation products in apple juices and ciders, from synthetic medium to commercial products. In: *Recent Advances in Polyphenols Research* (eds. F. Daayf & V. Lattanzio), vol. 1, pp. 278–292. Wiley-Blackwell, Oxford.

Guyot, S., Serrand, S., Le Quéré, J. M., Sanoner, P. & Renard, C.M.G.C. (2007) Enzymatic synthesis and physicochemical characterisation of Phloridzin Oxidation Products (POP), a new water-soluble yellow dye deriving from apple. *Innovative Food Science and Emerging Technologies*, **8**, 443–450.

Guyot, S., Vercauteren, J. & Cheynier, V. (1996) Structural determination of colourless and yellow dimers resulting from (+)-catechin coupling catalysed by grape polyphenoloxidase. *Phytochemistry*, **42**, 1279–1288.

Hammond-Kosack, K.E., Jones, J.D.G. (1996) Resistance gene-dependent plant defense responses. *Plant Cell*, **8**, 1773–1791.

Hannah, M.A., Wiese, D., Freund, S., Fiehn, O., Heyer, A.G. & Hincha, D.K. (2006) Natural genetic variation of freezing tolerance in *Arabidopsis*. *Plant Physiology*, **142**, 98–112.

Hanny, B.W. (1980) Gossypol, flavonoid and condensed tannin content of cream and yellow anthers of five cotton (*Gossypium hirsutum* L.) cultivars. *Journal of Agricultural and Food Chemistry*, **28**, 504–506.

Harborne, J.B. (1980) Plant Phenolics. In: *Encyclopedia of Plant Physiology, New Series, Vol. 8, Secondary Plant Products* (eds. E.A. Bell & B.W Charlwood), pp. 329–402. Springer-Verlag, Berlin.

Harborne, J.B. (1990) Role of secondary metabolites in chemical defence mechanisms in plants. In: *Bioactive Compounds from Plants, Ciba Foundation Symposium 154*, pp. 126–139. John Wiley & Sons, New York.

Harborne, J.B. (1995) Plant polyphenols and their role in plant defence mechanisms. In: *Polyphenols 94* (eds. R. Brouillard, M. Jay & A. Scalbert), pp. 19–26. INRA Editions, Paris.

Harborne, J.B. (1997) Recent advances in chemical ecology. *Natural Product Reports*, **14**, 83–97.

Harborne, J.B. (1999) Recent advances in chemical ecology. *Natural Product Reports*, **16**, 509–523.

Harborne, J.B. (2001) Twenty-five years of chemical ecology. *Natural Product Reports*, **18**, 361–379.

Harborne, J.B. & Grayer, R.J. (1993) Flavonoids and insects. In: *Flavonoids – Advances in Research since 1986* (ed. J.B. Harborne), pp. 589–618. Chapman & Hall, London.

Harborne, J.B. & Williams, C.A. (2000) Advances in flavonoid research since 1992. *Phytochemistry*, **55**, 481–504.

Hare, P.D. & Cress, W.A. (1997) Metabolic implications of stress-induced proline accumulation in plants. *Plant Growth Regulation*, **21**, 79–102.

Hare, P.D., Cress, W.A. & van Staden, J. (1998) Dissecting the roles of osmolyte accumulation in plants. *Plant, Cell & Environment*, **21**, 535–553.

Hare, P.D., Cress, W.A. & van Staden, J. (1999). Proline synthesis and degradation: a model system for elucidating stress-related signal transduction. *Journal of Experimental Botany*, **50**, 413–434.

Harmatha, J. & Dinan, L. (2003) Biological activities of lignans and stilbenoids associated with plant-insect chemical interactions. *Phytochemistry Reviews*, **2**, 321–330.

Hasegawa, H., Fukasawa-Akada, T., Okuno, T., Niizeki, M. & Suzuki, M. (2001) Anthocyanin accumulation and related gene expression in Japanese parsley (*Oenanthe stolonifera*, DC.) induced by low temperature. *Journal of Plant Physiology*, **158**, 71–78.

Hedin, P.A., Jenkins, J.N., Collum, D.H., White, W.H., Parrot, W.L. & MacGown, M.W. (1983) Cyanidin-3-glucoside, a newly recognized basis for resistance in cotton to the tobacco budworm *Heliothis virescens* (Fab.) (lepidoptera: Noctuidae). *Experientia*, **39**, 799–801.

Hedin, P.A., Jenkins, J.N., Thompson, A.C., *et al.* (1988) Effect of bioregulators on flavonoids, insect resistance and yield of seed cotton. *Journal of Agricultural and Food Chemistry*, **36**, 1055–1061.

Hedin, P.A. & Waage, S.K. (1986) Roles of flavonoids in plant resistance to insects. In: *Plant Flavonoids in Biology and Medicine: Biochemical, Pharmacological, and Structure-Activity Relationships* (eds. V. Cody, E. Middleton & J.B. Harborne), pp. 87–100. Alan R. Liss Inc., New York.

Heil, M. & Baldwin, I.T. (2002) Fitness costs of induced resistance: emerging experimental support for a slippery concept. *Trends in Plant Science*, **7**, 61–67

Heil, M., Hilpert, A., Kaiser, W. & Linsenmair, E. (2000) Reduced growth and seed set following chemical induction of pathogen defence: does systemic acquired resistance (SAR) incur allocation costs? *Journal of Ecology*, **88**, 645–654.

Herms, D.A. & Mattson, W.J. (1992) The dilemma of plants: to grow or defend. *The Quarterly Review of Biology*, **67**, 283–335.

Howarth, C.J. & Ougham, H.J. (1993) Gene expression under temperature stress, *New Phytologist*, **125**, 1–26.

Hrazdina, G. (1994) Compartmentation in phenolic metabolism. *Acta Horticulture*, **381**, 86–96.

Hughes, J.C., Ayers, J.E. & Swain, T. (1962) After-cooking blackening in potatoes. I. Introduction and analytical methods. *Journal of The Science of Food and Agriculture*, **13**, 224–229.

Hughes, J.C. & Swain, T. (1962a) After-cooking blackening in potatoes. II. Core experiments. *Journal of The Science of Food and Agriculture*, **13**, 229–236.

Hughes, J.C, & Swain, T. (1962b) After-cooking blackening in potatoes. III. Examination of the interaction of factors by *in vitro* experiments. *Journal of The Science of Food and Agriculture*, **13**, 358–363.

Hughes., M.A. & Dunn, M.A. (1996) The molecular biology of plant acclimation to temperature. *Journal of Experimental Botany*, **47**, 291–305.

Hyodo, H., Kuroda, H. & Yang, S.F. (1978) Induction of phenylalanine ammonialyase and increase in phenolics in lettuce leaves in relation to the development of russet spotting caused by ethylene. *Plant Physiology*, **62**, 31–35.

Hyodo. H. & Yang, S.F. (1971) Ethylene-enhanced synthesis of phenylalanine ammonia-lyase in pea seedlings. *Plant Physiology*, **47**, 765–770.

Ippolito, A, Nigro, F., Lima, G., *et al.* (1997) Mechanism of resistance to *Botrytis cinerea* in wound of cured kiwufruits. *Acta Horticulture*, **444**, 719–724.

Ingham, J.L. (1973) Disease resistance in higher plants. The concept of pre-infectional and post-infectional resistance. *Phytopathologische Zeitschrift*, **78**, 314–335.

Ishimoto, M. & Kitamura, K. (1989) Growth inhibitory effects of an α-amylase inhibitor from the kidney bean. *Phaseolus vulgaris* (L.) on three species of bruchids (Coleoptera: Bruchidae). *Applied Entomology and Zoology*, **24**, 281–2867.

Jacobs, D.L., Watt, G.D., Frankel, R.B., & Papaefthymiou, G.C. (1989) Redox reactions associated with iron release from mammalian ferritin. *Biochemistry*, **28**, 1650–1655.

Janas, K.M., Cvikrovà, M., Palagiewicz, A. & Eder, J. (2000) Alterations in phenylpropanoid content in soybean roots during low temperature acclimation. *Plant Physiology and Biochemistry*, **38**, 587–593.

Jones, C.G. & Hartley, S.E. (1999) A protein competition model of phenolic allocation. *Oikos*, **86**, 27–44.

Jones, D.H. (1984) Phenylalanine ammonia-liase: regulation of its induction and its role in plant development. *Phytochemistry*, **23**, 1349–1359.

Jones. T., Spencer, R. & Walsh, C. (1978) Mechanism and kinetics of iron release from ferritin by dihydroflavins and dihydroflavin analogues. *Biochemistry*, **17**, 4011–4017.

Joerdens-Roettger, D. (1979) The role of phenolic substances for host-selection behaviour of the black bean aphid, *Aphis fabae*. *Entomologia Experimentalis et Applicata*, **26**, 49–54.

Kasai, A., S. Ohnishi, H. Yamazaki, H., *et al.* (2009) Molecular mechanism of seed coat discoloration induced by low temperature in yellow soybean. *Plant and Cell Physiology*, **50**, 1090–1098.

Kavi Kishor, P.B., Sangam, S., Amrutha, R.N., *et al.* (2005) Regulation of proline biosynthesis, degradation, uptake and transport in higher plants: Its implications in plant growth and abiotic stress tolerance. *Current Science*, **88**, 424–438.

Ke, D. & Saltveit, M.E. (1989) Wound-induced ethylene production, phenolic metabolism and susceptibility to russet spotting in iceberg lettuce. *Physiologia Plantarum*, **76**, 412–418.

Kidd, P.S., Llugany, M., Poschenrieder, C., Gunsé, B. & Barceló, J. (2001) The role of root exudates in aluminium resistance and silicon-induced amelioration of aluminium toxicity in three variety of maize (*Zea mays* L.). *Journal of Experimental Botany*, **52**, 1339–1352.

Kiyosue, T., Yoshiba, Y., Yamaguchi-Shinozaki, K. & Shinozaki, K. (1996) A nuclear gene encoding mitochondrial proline dehydrogenase, an enzyme involved in proline metabolism, is upregulated by proline but downregulated by dehydration in Arabidopsis. *Plant Cell*, **8**, 1323–1335.

Koricheva, J., Nykänen, H. & Gianoli, E. (2004) Meta-analysis of trade-offs among plant antiherbivore defenses: are plants jacks-of-all-trades, masters of all? *American Naturalist*, **163**, E64–E75.

Kushad, M.M. & Yelenosky, G. (1987) Evaluation of polyamine and proline levels during low temperature acclimation of citrus. *Plant Physiology*, **84**, 692–695.

Lafuente, M.T., Zacarias, L., Martínez-Téllez, M.A., Sanchez-Ballesta, M.T., & Dupille, E. (2001) Phenylalanine ammonia-lyase as related to ethylene in the development of chilling symptoms during cold storage of citrus fruits. *Journal of Agricultural and Food Chemistry*, **49**, 6020–6025.

Lamb, C.J. & Dixon, R. (1997) The oxidative burst in plant disease resistance. *Annual Review of Plant Physiology and Molecular Biology*, **48**, 251–275.

Lattanzio, V. (2003a) The role of plant phenolics in the postharvest physiology and quality of fruit and vegetables. In: *Advances in Phytochemistry* (ed. F. Imperato), pp. 49–83. Research Signpost, Trivandrum, Kerala.

Lattanzio, V. (2003b) Bioactive polyphenols: their role in quality and storability of fruit and vegetables. *Journal of Applied Botany*, **77**, 128–146.

Lattanzio, V., Arpaia, S., Cardinali, A., Di Venere, D. & Linsalata, V. (2000) Role of endogenous flavonoids in resistance mechanism of *Vigna* to aphids. *Journal of Agricultural and Food Chemistry*, **48**, 5316–5320.

Lattanzio, V., Cardinali, A., Di Venere, D., Linsalata, V. & Palmieri, S. (1994) Browning phenomena in stored artichoke (*Cynara scolymus* L.) heads: enzymic or chemical reactions? *Food Chemistry*, **50**, 1–7.

Lattanzio, V., Cardinali, A., Linsalata, V., Perrino, P. & Ng, N.Q. (1996) A chemosystematic study of the flavonoids of *Vigna. Genetic Resources and Crop Evolution*, **43**, 493–504.

Lattanzio, V., Cardinali, A., Linsalata, V., Perrino, P. & Ng, N.Q. (1997) Flavonoid HPLC fingerprints of wild *Vigna* species. In: *Advances in Cowpea Research* (eds. B.B. Singh, D.R. Mohan Raj, K.E. Dashiell & L.E.N. Jackai), pp. 66–74. IITA/JIRCAS Publishers, Ibadan.

Lattanzio, V., Cardinali, A., Ruta, C., *et al.* (2009) Relationship of secondary metabolism to growth in oregano (*Origanum vulgare* L.) shoot cultures under nutritional stress. *Environmental and Experimental Botany*, **65**, 54–62.

Lattanzio, V., Di Venere, D., Linsalata, V., Bertolini, P., Ippolito, A. & Salerno, M. (2001) Low temperature metabolism of apple phenolics and quiescence of *Phlyctaena vagabunda. Journal of Agricultural and Food Chemistry*, **49**, 5817–5821.

Lattanzio, V., Kroon, P.A., Quideau, S. & Treutter, D. (2008) Plant phenolics – Secondary metabolites with diverse functions. In: *Recent Advances in Polyphenols Research* (eds. F. Daayf & V. Lattanzio), vol. 1, pp. 1–35. Wiley-Blackwell, Oxford.

Lattanzio, V., Lattanzio, V.M.T. & Cardinali, A. (2006) Role of phenolics in the resistance mechanisms of plants against fungal pathogens and insects. In: *Phytochemistry: Advances in Research* (ed. F. Imperato), pp. 23–67. Research Signpost, Trivandrum, Kerala.

Lattanzio, V., Linsalata, V., Palmieri, S. & Van Sumere, C.F. (1989) The beneficial effect of citric and ascorbic acid on the phenolic browning reaction in stored artichoke (*Cynara scolymus* L.) heads. *Food Chemistry*, **33**, 93–106.

Lattanzio, V., Terzano, R., Cicco, N., Cardinali, A., Di Venere, D. & Linsalata, V. (2005) Seed coat tannins and bruchid resistance in stored cowpea seeds. *Journal of the Science of Food and Agriculture*, **85**, 839–846.

Lattanzio, V. & Van Sumere, C.F. (1987) Changes in phenolic compounds during the development and cold storage of artichoke (*Cynara scolymus* L.) heads. *Food Chemistry*, **24**, 37–50.

Le, Bot, J., Bénard, C., Robin, C., Bourgaud, F. & Adamowicz, S. (2009) The 'trade-off' between synthesis of primary and secondary compounds in young tomato leaves is altered by nitrate nutrition: experimental evidence and model consistency. *Journal of Experimental Botany*, **60**, 4301–4314.

Le Guernevé, C., Sanoner, P., Drilleau, J.-F. & Guyot, S. (2004) New compounds obtained by enzymatic oxidation of phloridzin. *Tetrahedron Letters*, **45**, 6673–6677.

Lewis, L.A. & Mccourt, R. M. (2004) Green algae and the origin of land plants. *American Journal of Botany*, **91**, 1535–1556.

Leyva, A., Jarillo, J.A., Salinas, J. & Martinez-Zapater, J.M. (1995) Low temperature induces the accumulation of phenylalanine ammonia-lyase and chalcone synthase mRNAs of *Arabidopsis thaliana* in a light-dependent manner. *Plant Phvsiology*, **108**, 39–46.

Lillo, C., Lea, U.S. & Ruoff, P. (2008) Nutrient depletion as a key factor for manipulating gene expression and product formation in different branches of the flavonoid pathway. *Plant Cell & Environment*, **31**, 587–601.

Link, K.P., Dickson, A.D. & Walker, J.C. (1929) Further observations on the occurrence of protocatechuic acid in pigmented onion scales and its relation to disease resistance in the onion. *Journal of Biological Chemistry*, **84**, 719–725.

Link, K.P. & Walker, J.C. (1933) The isolation of catechol from pigmented onion scales and its significance in relation to disease resistance in onions. *Journal of Biological Chemistry*, **100**, 379–383.

Liu, L. & McClure, J.W. (1995) Effects of UV-B on activities of enzymes of secondary phenolic metabolism in barley primary leaves. *Physiologia Plantarum*, **93**, 734–739.

Logemann, E., Tavernaro, A., Schulz, W., Somssich, I.E. & Hahlbrock, K. (2000) UV light selectively coinduces supply pathways from primary metabolism and flavonoid secondary product formation in parsley. *Proceedings of the National Academy of Sciences USA*, **97**, 1903–1907.

Lorio, P.L. Jr & Sommers, R.A. (1986) Evidence of competition for photosynthates between growth processes and oleoresin synthesis in *Pinus taeda* L. *Tree Physiology*, **2**, 301–306.

Lowry, B., Lee, D. & Hébant, C. (1980) The origin of land plants: a new look at an old problem. *Taxon*, **29**, 183–197

Lowry, J.B., Lee, D.W. & Hébant, C. (1983) The origin of land plants: a reply to Swain. *Taxon*, **32**, 101–103.

Lyons, J.M. 1973. Chilling injury in plants. *Annual Review of Plant Physiology*, **24**, 445–466.

Maggio, A., Miyazakim, S., Veronesem, P., *et al.* (2002) Does proline accumulation play an active role in stress-induced growth reduction? *Plant Journal*, **31**, 699–712.

Margna, U. (1977) Control at the level of substrate supply. An alternative in the regulation of phenylpropanoid accumulation in plant cells. *Phytochemistry*, **16**, 419–426.

Margna, U., Vainjärv, T. & Laanest, L. (1989) Different L-phenylalanine pools available for the biosynthesis of phenolics in buckwheat seedling tissues. *Phytochemistry*, **28**, 469–475.

Markham, K.R. & Porter, L. (1979) Flavonoids of the primitive liverwort *Takakia* and their taxonomic and phylogenetic significance. *Phytochemistry*, **18**, 611–615.

Marsilio, V., Campestre, C. & Lanza, B. (2001) Phenolic compounds change during California-style ripe olive processing. *Food Chemistry*, **74**, 55–60.

Martinez, M.V. & Whitaker, J.R. (1995) The biochemistry and control of enzymatic browning. *Trends in Food Science & Technology*, **6**, 195–200.

Matern, U. & Grimmig, B. (1993) Polyphenols in plant pathology. In: *Polyphenolic Phenomena* (ed. A. Scalbert), pp. 143–147. INRA Edition, Paris.

Mathew, A.C. & Parpia, H.A.B. (1971) Food browning as a polyphenol reaction. *Advances in Food Research*, **19**, 75–145.

Mayer, A.M. & Harel, E. (1981) Polyphenol oxidases in fruits. Changes during ripening. In: *Recent Advances in the Biochemistry of Fruits and Vegetables* (eds. J. Friend & M.J.C. Rhodes), pp. 161–180. Academic Press, London.

McKey, D. (1974) Adaptive patterns in alkaloid physiology. *American Naturalist*, **108**, 305–320.

Mondolot, L., La Fisca, P., Buatois, B., Talansier, E., De Kochko, A. & Campa, C. (2006) Evolution in caffeoylquinic acid content and histolocalization during *Coffea canephora* leaf development, *Annals of Botany*, **98**, 33–40.

Morrissey, J.P. & Osbourn, A.E. (1999) Fungal Resistance to Plant Antibiotics as a Mechanism of Pathogenesis. *Microbiology and Molecular Biology Reviews*, **63**, 708–724.

Murdock, L.L. (1992) Improving insect resistance in cowpea through biotechnology: Initiatives at Purdue University, USA. In: *Biotechnology: Enhancing Research on Tropical Crops in Africa* (eds. G. Thottappilly, L.M. Monti, D.R. Mohan Raj & A.W. Moore), pp. 313–320. CTA/IITA Publishers, Ibadan.

Murdock, L.L., Shade, R.E., Kitch, L.W., *et al.* (1997) Postharvest storage of cowpea in sub-Saharan Africa. In: *Advances in Cowpea Research* (eds. B.B. Singh, D.R. Mohan Raj, K.E. Dashiell & L.E.N. Jackai), pp. 302–312. IITA/JIRCAS, Ibadan.

Nicholson, R.L. & Hammerschmidt, R. (1992) Phenolic compounds and their role in disease resistance. *Annual Review of Phytopathology*, **30**, 369–389.

Nigro, F., Ippolito, A., Lattanzio, V., Di Venere, D. & Salerno, M. (2000) Effect of ultraviolet-C light on postharvest decay of strawberry. *Journal of Plant Pathology*, **82**, 29–37.

Nishida, R., Ohsugi, T., Kokubo, S. & Fukami, H. (1987) Oviposition stimulants of a citrus-feeding swallowtail butterfly, *Papilio xuthus* L. *Experientia*, **43**, 342–344.

Noel, J.P., Austin, M.B. & Bomati, E.K. (2005) Structure–function relationships in plant phenylpropanoid biosynthesis. *Current Opinion in Plant Biology*, **8**, 249–253.

Noveroske, R.L., Kuc, J. & Williams, E.B. (1964) Oxidation of phioridzin and phloretin related to resistance of *Malus* to *Venturia inaequalis*. *Phytopathology*, **54**, 92–98.

Nürnberger, T. & Lipka, V. (2005) Non-host resistance in plants: new insights into an old phenomenon. *Molecular Plant Pathology*, **6**, 335–345.

Olsen, K.M., Lea, U.S., Slimestad, R., Verheul, M. & Lillo, C. (2008) Differential expression of the four *Arabidopsis PAL* genes –*PAL1* and *PAL2* have functional specialization in abiotic environmental triggered flavonoid synthesis. *Journal of Plant Physiology*, **165**, 1491–1499.

Olsen, K.M., Slimestad, R., Lea, U.S., *et al.* (2009) Temperature and nitrogen effects on regulators and products of the flavonoid pathway: Experimental and kinetic model studies. *Plant, Cell and Environment*, **32**, 286–299.

Orr, J.D., Edwards, R. & Dixon, R.A. (1993) Stress responses in alfalfa (*Medicago sativa* L.) XIV. Changes in the levels of phenylpropanoid pathway intermediates in relation to regulation of L-phenylalanine ammonia-lyase in elicitor-treated cell-suspension cultures. *Plant Physiology*, **101**, 847–856.

Ortega-García, F. & Peragón, J. (2009) The response of phenylalanine ammonia-lyase, polyphenol oxidase and phenols to cold stress in the olive tree (*Olea europaea* L. cv. Picual). *Journal of the Science of Food and Agriculture*, **89**, 1565–1573.

Osbourn, A.E. (1996) Preformed antimicrobial compounds and plant defense against fungal attack. *Plant Cell*, **8**, 1821–1831.

Osbourn, A.E., Qi, X., Townsend, B. & Qin, B. (2003) Dissecting plant secondary metabolism – constitutive chemical defences in cereals. *New Phytologist*, **159**, 101–108.

Oszmianski, J. & Lee, C.Y. (1991) Enzymatic oxidation of phloretin glucoside in model system. *Journal of Agricultural and Food Chemistry*, **39**, 1050–1052.

Painter, R.H. (1941) The economic value and biologic significance of insect resistance in plants. *Journal of Economic Entomology*, **34**, 358–367.

Parida, A., Das, A.B. & Das, P. (2002) NaCl stress causes changes in photo-synthetic pigments, proteins and other metabolic components in the leaves of a true mangrove, *Bruguiera parviflora*, in hydroponic cultures. *Journal of Plant Biology*, **45**, 28–36.

Paul, N.D., Hatcher, P.E. & Taylor, J.E. (2000) Coping with multiple enemies: an integration of molecular and ecological perspectives. *Trends in Plant Science*, **5**, 220–225.

Pavia, H., Toth, G. & Aberg, P. (1999) Trade-offs between phlorotannin production and annual growth in natural populations of the brown seaweed *Ascophyllum nodosum*. *Journal of Ecology*, **87**, 761–771.

Pedra, J.H.F., Brandt, A., Westerman, R., *et al.* (2003) Transcriptome analysis of the cowpea weevil bruchid: identification of putative proteinases and α-amylases associated with food breakdown. *Insect Molecular Biology*, **12**: 405–412.

Pereyra, P.C. & Bowers, M.D. (1988) Iridoid glycosides as oviposition stimulants for the buckeye butterfly, *Junonia coenia* (Nymphalidae). *Journal of Chemical Ecology*, **14**, 917–928.

Pichersky, E. & Gang, D. R. (2000) Genetics and biochemistry of secondary metabolites in plants: an evolutionary perspective. *Trends in Plant Science*, **5**, 439–445.

Pierpoint, W.S. (1966) The enzymatic oxidation of chlorogenic acid and some reactions of the quinone produced. *Biochemical Journal*, **98**, 567–580.

Pierpoint, W.S. (1969) o-Quinones formed in plant extracts. Their reaction with amino acids and peptides. *Biochemical Journal*, **112**, 609–616.

Pierpoint, W.S. (1983) Reactions of phenolic compounds with protein and their relevance to the production of leaf protein. In: *Leaf Protein Concentrates* (eds. L. Telek & H.D. Graham), pp. 235–267. AVI Publishing Co., Westport, Connecticut.

Pollard, A. & Timberlake, C.F. (1971) Fruit juices. In: *Biochemistry of Fruits and Their Products* (ed. A.C. Hulme), Vol. 2, pp. 573–621. Academic Press, London.

Pontais, I., Treutter, D., Paulin, J.P. & Brisset, M.N. (2008) *Erwinia amylovora* modifies phenolic profiles of susceptible and resistant apple through its type III secretion system. *Physiologia Plantarum*, **132**, 262–271.

Pourcel, L., Routaboul, J.M. & Cheynier, V. (2007) Flavonoid oxidation in plants: from biochemical properties to physiological functions. *Trends in Plant Science*, **12**, 29–36.

Price, C.A. (1968) Iron compounds and plant nutrition. *Annual Review of Plant Physiology*, **19**, 239–248.

Purrington, C.B. (2000) Cost of resistance. *Current Opinion in Plant Biology*, **3**, 305–308.

Raa, J. (1968) Polyphenols and natural resistance of apple leaves against *Venturia inaequalis*. *European Journal of Plant Pathology*, **74**, 37–45.

Raa, J. & Overeem, J.C. (1968) Transformation reactions of phloridzin in the presence of apple leaf enzymes. *Phytochemistry*, **7**, 721–731.

Ranjeva, R., Alibert, G. & Boudet, A.M. (1977a) Metabolisme des composes phenoliques chez le petunia V. utilisation de la phenylalanine par des chloroplastes isoles. *Plant Science Letters*, **10**, 225–234.

Ranjeva, R., Alibert, G. & Boudet, A.M. (1977b) Metabolisme des composes phenoliques chez le petunia VI. Intervention des chloroplastes dans la biosynthese de la naringenine et de l'acide chlorogenique. *Plant Science Letters*, **10**, 235–242.

Rhodes, M.J.C. & Wooltorton, L.S.C. (1971) The effect of ethylene on the respiration and the activity of phenylalanine ammonia-lyase in swede and parsnip root tissue. *Phytochemistry*, **10**, 1989–1997.

Rhodes, M.J.C. & Wooltorton, L.S.C. (1977) Changes in the activity of enzymes of phenylpropanoid metabolism in tomatoes stored at low temperatures. *Phytochemistry*, **16**, 655–659.

Rhodes, M.J.C. & Wooltorton, L.S.C. (1978) Changes in the activity of hydroxycinnamyl CoA: Quinate hydrocinnamyl transferase and in the level of chlorogenic acid in potatoes and sweet potatoes at various temperatures. *Phytochemistry*, **17**, 1225–1229.

Rhodes, M.J.C., Wooltorton, L.S.C. & Hill, A.C. (1981) Changes in phenolic metabolism in fruit and vegetable tissues under stress. In: *Recent advances in the biochemistry of fruit and vegetables* (eds. J. Friend & M.J.C. Rhodes), pp. 191–220. Academic Press, London.

Rivero, R.M., Ruiz, J.M., Garcia, P.C., Lopez-Lefebre, L.R., Sanchez, E. & Romero, L. (2001) Resistance to cold and heat stress: accumulation of phenolic compounds in tomato and watermelon plants. *Plant Science*, **160**, 315–321.

Robbins, M.P., Bolwell, G.P. & Dixon, R.A. (1985) Metabolic changes in elicitor-treated bean cells Selectivity of enzyme induction in relation to phytoalexin accumulation. European Journal of *Biochemistry*, **148**, 563–569.

Roessingh, P., Städler, E., Baur, R., Hurter, J. & Ramp, T. (1997) Tarsal chemoreceptors and oviposition behaviour of the cabbage root fly sensitive to fractions and new compounds of host-leaf surface extracts. *Physiological Entomology*, **22**, 140–148.

Roy, B.A. & Kirchner, J.W. (2000) Evolutionary dynamics of pathogen resistance and tolerance. *Evolution*, **54**, 51–63.

Ruelland, E., Vaultier, M.N., Zachowski, A. & Hurry, V. (2009) Cold signalling and cold acclimation in plants. In: Advances in Botanical Research *49* (eds. J-C. Kader & M. Delsney), pp. 35–150. Academic Press, London.

Sanchez-Ballesta, M.T, Lafuente, M.T., Zacarias, L. & Granell, A. (2000a) Involvement of phenylalanine ammonia-lyase in the response of Fortune mandarin fruits to cold temperature. *Physiologia Plantarum*, **108**, 382–389.

Sanchez-Ballesta, M.T., Zacarias, L., Granell, A. & Lafuente, M.T. (2000b) Accumulation of PAL trascript and PAL activity as affected by heat-conditioning and low temperature storage and its relation to chilling sensitivity in mandarin fruits. *Journal of Agricultural and Food Chemistry*, **48**, 2726–2731.

Saradhi, P.P., Alia, Arora, S. & Prasad, K.V.S.K. (1995) Proline accumulates in plants exposed to UV radiation and protects them against induced peroxidation. *Biochemical and Biophysical Research Communications*, **290**, 1–5.

Sarma, A.D. & Sharma, R. (1999) Purification and characterization of UV-B induced phenylalanine ammonia-lyase from rice seedlings. *Phytochemistry*, **50**, 729–737.

Schildknecht, H. & Schumacher, K. (1982) Nyctinastenes – An approach to new phytohormones. *Pure and Applied Chemistry*, **54**, 2501–2514.

Schildknecht, H. (1983) Turgorins, hormones of the endogenous daily rhythms of higher organized plants – detection, isolation, structure, synthesis, and activity. *Angewandte Chemie International Edition English*, **22**, 695–710.

Schmid, J. & Amrhein, N. (1995) Molecular organization of the shikimate pathway in higher plants. *Phytochemistry*, **39**, 737–749.

Schroeder, H.E., Gollasch, S., Moore, A., *et al.* (1995) Bean α-amylase inhibitor confers resistance to the pea weevil (*Bruchus pisorum*) in transgenic peas (*Pisum sativum* L.). *Plant Physiology*, **107**, 1233–1239.

Shade, R.E., Schroeder, H.E., Pueyo, J.J., *et al.* (1994) Transgenic pea seeds expressing the α-amylase inhibitor of the common bean are resistant to bruchid beatles. *Bio/Technology*, **12**, 793–796.

Sharma, P., Sharma, N. & Deswal, R. (2005) The molecular biology of the low-temperature response in plants. *Bioessays*, **27**, 1048–1059.

Sharma, S.S. & Dietz, K.J. (2006) The significance of amino acids and amino acid-derived molecules in plant responses and adaptation to heavy metal stress. *Journal of Experimental Botany*, **57**, 711–726.

Shaw, N.M., Bolwell, G.P. & Smith, C. (1990) Wound-induced phenylalanine ammonia-lyase in potato (*Solanum tuberosum*) tuber discs. *Biochemical Journal*, **267**, 163–170.

Siddiqui, K. S. & Cavicchioli, R. (2006) Cold-adapted enzymes. *Annual Review of Biochemistry*, **75**, 403–433.

Siemens, D.H., Lischke, H., Maggiulli, N., Schürch, S. & Roy, B.A. (2003) Cost of resistance and tolerance under competition: the defense-stress benefit hypothesis. *Evolutionary Ecology*, **17**, 247–263.

Simmonds, M.S.J. (2001) Importance of flavonoids in insect-plant interactions: feeding and oviposition. *Phytochemistry*, **56**, 245–252.

Simmonds, M.S.J. (2003) Flavonoid-insect interactions: recent advances in our knowledge. *Phytochemistry*, **64**, 21–30.

Simmonds, M.S.J. & Stevenson, P.C. (2001) Effects of isoflavonoids from *Cicer* on larvae of *Heliocoverpa armigera*. *Journal of Chemical Ecology*, **27**, 965–977.

Synge, R.L.M. (1975) Interactions of polyphenols with proteins in plants and plant products. *Qualitas Plantarum-Plant Foods For Human Nutrition*, **24**, 337–350.

Singh, I.P. & Bharate, S.B. (2006) Phloroglucinol compounds of natural origin. *Natural Product Reports*, **23**, 558–591.

Singh, S.R. & Jackai, L.E.N. (1985) Insect pests of cowpeas in Africa: Their life cycle, economic importance and potential for control. In *Cowpea Research, Production and Utilization* (eds. S.R. Singh & K.O. Rachie), pp. 217–231. John Wiley & Sons, Chichester.

Singh, S.R., Jackai, L.E.N., Dos Santos, J.H.R. & Adalla, C.B. (1990) Insect pests of cowpea. In: *Insect Pests of Tropical Food Legumes* (ed S.R. Singh), pp. 43–89. John Wiley & Sons, Chichester.

Singh, S.R., Jackai, L.E.N., Thottappilly, G., Cardwell, K.F. & Myers, G.O. (1992) Status of research on constraints to cowpea production. In: *Biotechnology: Enhancing Research on Tropical Crops in Africa* (eds. G. Thottappilly, L.M. Monti, D.R. Mohan Raj & A.W. Moore), pp. 21–26. CTA/IITA Publishers, Ibadan.

Singh, S.R. & Rachie, K.O. (1985) *Cowpea Research, Production and Utilization*. John Wiley & Sons, Chichester.

Siriphanich, J. & Kader, A.A. (1985a) Effect of CO_2 on total phenolics, phenylalanine ammonia lyase and polyphenol oxidase in lettuce tissues. *Journal of the American Society for Horticultural Science*, **110**, 249–253.

Siriphanich, J. & Kader, A.A. (1985b) Effect of CO_2 on cinnamic acid-4-hydroxylase in relation to phenolic metabolism in lettuce tissues. *Journal of the American Society for Horticultural Science*, **110**, 333–335.

Smirnoff, N. (1993) The role of active oxygen in the response of plants to water deficit and desiccation. *New Phytologist*, **125**, 27–58.

Solecka, D., Boudet, A.M. & Kacperska, A. (1999) Phenylpropanoid and anthocyanin changes in low-temperature treated winter oilseed rape leaves. *Plant Physiology and Biochemistry*, **37**, 491–496.

Stafford, H.A. (1991) Flavonoid evolution: an enzymic approach. *Plant Physiology*, **96**, 680–685.

Stamp, N. (2003) Out of the quagmire of plant defense hypotheses. *The Quarterly Review of Biology*, **78**, 23–55.

Stamp, N. (2004) Can the growth–differentiation balance hypothesis be tested rigorously? *Oikos*, **107**, 439–448.

Stewart, A.J., Chapman, W., Jenkins, G.I., Graham, I., Martin, T. & Crozier, A. (2001) The effect of nitrogen and phosphorus deficiency on flavonol accumulation in plant tissues. *Plant, Cell and Environment*, **24**, 1189–1197.

Stotz, H.U., Kroymann, J. & Mitchell-Olds, T. (1999) Plant-insect interactions. *Current Opinion in Plant Biology*, **2**, 268–272.

Strauss, S.Y., Rudgers, J.A., Lau, J.A. & Irwin, R.E. (2002) Direct and ecological costs of resistance to herbivory. *Trends in Ecology and Evolution*, **17**, 278–284.

Swain, T. (1975) Evolution of flavonoid compounds. In: *The Flavonoids* (eds. J.B. Harborne, T.J. Mabry & H. Mabry), pp. 1096–1138. Chapman & Hall, London.

Swain, T. (1977) Secondary compounds as protective agents. *Annual Review of Plant Physiology*, **28**, 479–501.

Swain, T. (1981) Point of view. The origin of land plants: a new look at an old problem. *Taxon*, **30**, 471.

Tattini, M., Guidi, L., Morassi-Bonzi, L., *et al.* (2005) On the role of flavonoids in the integrated mechanisms of response of *Ligustrum vulgare* and *Phillyrea latifolia* to high solar radiation. *New Phytologist*, **167**, 457–470.

Taylor, J.E., Hatcher, P.E. & Paul, N.D. (2004) Crosstalk between plant responses to pathogens and herbivores: a view from the outside in. *Journal of Experimental Botany*, **55**, 159–168.

Teklemariam, T. & Blake, T.J. 2004 Phenylalanine ammonia-lyase induced freezing tolerance in jack pine (*Pinus banksiana*) seedlings treated with low, ambient levels of ultraviolet-B radiation, *Physiologia Plantarum*, **122**, 244–253.

Thomashow, M.F. 1998 Role of cold-responsive genes in plant freezing tolerance. *Plant Physiology*, **118**, 1–7.

Thorsteinson, A.J. (1960) Host selection in phytophagous insects. *Annual Review of Entomology*, **5**, 193–218.

Tinkler, C.K. (1931) LXXXVII. The blackening of potatoes after cooking. *Biochemical Journal*, **25**, 773–776.

Thottappilly, G., Monti, L.M., Mohan Raj, D.R. & Moore, A.W. (1992) *Biotechnology: Enhancing Research on Tropical Crops in Africa*. CTA/IITA Publishers, Ibadan.

Todd, G.W., Gethahun, A. & Cress, D.C. (1971) Resistance in barley to the greenbug, *Schizaphis graminum*.1. Toxicity of phenolic and flavonoid compounds and related substances. *Annals of the Entomological Society of America*, **64**, 718–721.

Treutter, D. (2005) Significance of flavonoids in plant resistance and enhancement of their biosynthesis. *Plant Biology*, **7**, 581–591.

Treutter, D. (2006) Significance of flavonoids in plant resistance: a review. *Environmental Chemistry Letters*, **4**, 147–157.

Ueda, M. & Nakamura, Y. (2006) Metabolites involved in plant movement and 'memory': nyctinasty of legumes and trap movement in the Venus flytrap. *Natural Product Reports*, **23**, 548–557.

Ueda, M. & Nakamura, Y. (2010) Plant phenolic compounds controlling leaf-movement. In: *Recent Advances in Polyphenol Research* (eds. C. Santos-Buelga, M.T. Escribano-Bailon & V. Lattanzio), vol. 2, pp. 226–237. Wiley-Blackwell, Oxford, UK.

Ueda, M. & Yamamura, S. (2000) Chemistry and biology of plant leaf movements. *Angewandte Chemie International Edition*, **39**, 1400–1414.

Van der Plas, L.H.W., Eijkelboom, C. & Hagendoorn, M.J.M. (1995) Relation between primary and secondary metabolism in plant cell suspensions. *Plant Cell, Tissue and Organ Culture*, **43**, 111–116.

van Heerden, P.S., Towers, G.H.N. & Lewis, N.G. (1996) Nitrogen metabolism in lignifying *Pinus taeda* cell cultures. *Journal of Biological Chemistry*, **271**, 12350–12355.

Vaughn, K.C., Lax, A.R. & Duke, S.O. (1988) Polyphenol oxidase: the chloroplast oxidase with no established function. *Physiologia Plantarum*, **72**, 659–665.

Verbruggen, N. & Hermans, C. (2008) Proline accumulation in plants: a review. *Amino Acids*, **35**, 753–759.

Walker, J.C. & Stahmann, M.A. (1955) Chemical nature of disease resistance. *Annual Review of Plant Physiology*, **6**, 351–366.

Weigelt, K., Kuster, H., Rutten, T., *et al.* (2009) ADP-glucose pyrophosphorylase-deficient pea embryos reveal specific transcriptional and metabolic changes of carbon-nitrogen metabolism and stress responses. *Plant Physiology*, **149**, 395–411.

Wink, M. (1988) Plant breeding: importance of plant secondary metabolites for protection against pathogens and herbivores. *Theoretical and Applied Genetics*, **75**, 225–233.

Wink, M. (2003) Evolution of secondary metabolites from an ecological and molecular phylogenetic perspective. *Phytochemistry*, **64**, 3–19.

Winkel-Shirley, B. (1998) Flavonoids in seeds and grains: physiological function, agronomic importance and the genetics of biosynthesis. *Seed Science Research*, **8**, 415–422.

Winkel-Shirley, B. (2001) Flavonoid biosynthesis: a colorful model for genetics, biochemistry, cell biology and biotechnology. *Plant Physiology*, **126**, 485–493.

Winkel-Shirley, B. (2002) Biosynthesis of flavonoids and effects of stress. *Current Opinion in Plant Biology*, **5**, 218–223.

Wittstock, U. & Gershenzon, J. (2002) Constitutive plant toxins and their role in defense against herbivores and pathogens. *Current Opinion in Plant Biology*, **5**, 1–8.

Witzell, J. & Martin, J.A. (2008) Phenolic metabolites in the resistance of northern forest trees to pathogens – past experiences and future prospects. *Canadian Journal of Forest Research*, **38**, 2711–2727.

Wolfe, J. (1978) Chilling injury in plants: the role of membrane. *Plant, Cell and Environment*, **1**, 241–247

Xiong, L., Schumaker, K.S. & Zhu, J.K. (2002) Cell signaling during cold, drought, and salt stress. *Plant Cell*, **14** (Suppl.), S165–S183

Yamaguchi-Shinozaki, K. & Shinozaki, K. (2006) Transcriptional regulatory networks in cellular responses and tolerance to dehydration and cold stresses. *Annual Review of Plant Biology*, **57**, 781–803.

Yu, O. & Jez, J.M. (2008) Nature's assembly line: biosynthesis of simple phenylpropanoids and polyketides, *Plant Journal*, **54**, 750–762.

Zangerl, A.R., Arntz, A.M. & Berenbaum, M.R. (1997) Physiological price of an induced chemical defense: photosynthesis, respiration, biosynthesis, and growth. *Oecologia*, **109**, 433–441.

Zannou, E.T., Glitho, I.A., Huignard, J. & Monge, J.P. (2003) Life history of flight morph females of *Callosobruchus maculatus*: evidence of a reproductive diapause. *Journal of Insect Physiology*, **4**, 575–582.

Zheng, Z., Sheth, U., Nadiga, M., Pinkham, J.L. & Shetty, K. (2001) A model for the role of the proline-linked pentose phosphate pathway in polymeric dye tolerance in oregano. *Process Biochemistry*, **36**, 941–946.

Chapter 2

Polyphenols: From Plant Adaptation to Useful Chemical Resources

Alain-Michel Boudet

Abstract: Plants produce a large range of natural substances called secondary metabolites. Although these compounds do not appear to be directly involved in the basic activities of plant cells, they play an essential role in plant development and plant/environment interactions. Among these, an immense variety of phenolic compounds has been progressively synthesized by plants during the course of evolution. All of them are derived from primary metabolites that are fed via the shikimate pathway. Many of these compounds are of medical or socio-economic value and there is growing interest in research and industry in polyphenols. In this chapter, we will consider essentially three different aspects:

1) The emergence of phenolic metabolism during evolution and the adaptation of plants to a terrestrial environment.
2) The role and the specificities of the shikimate pathway in higher plants.
3) The diversified uses of phenolics in different human activities.

Keywords: plant evolution; vessels and conducting éléments; lignins, shikimate pathway; phenylpropanoid pathway; arogenate pathway; quinic acid; comparative genomics; polyphenols and health; lignocellulosics bioconversion

2.1 The emergence of phenolic metabolism and the adaptation of plants to a terrestrial environment

Fossil data indicate that the first land plants appeared around 500 million years ago (Kenrick & Crane, 1997). As emphasized by these authors "the origin and early diversification of land plants mark an interval of unparalleled innovation in the history of plant life with an extraordinary array of complex organs and specialized tissue systems." Independent evidence from morphological, ultrastructural, biochemical and molecular data

Recent Advances in Polyphenol Research, Volume 3, First Edition. Edited by Véronique Cheynier, Pascale Sarni-Manchado and Stéphane Quideau.

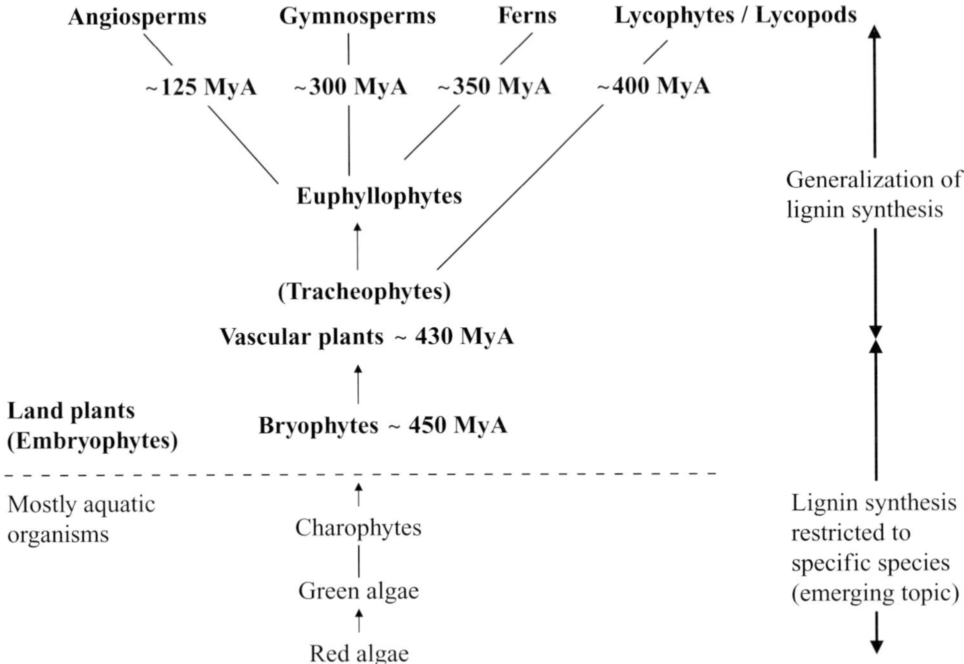

Fig. 2.1   Simplified phylogenetic tree showing the origin of land plants and the occurrence of lignins.

have shown that land plants (embryophytes) consisting of liverworts, hornworts, mosses and tracheophytes originated from charophycean green algae. This small group of predominantly freshwater green algae possesses several biosynthetic attributes that are expressed more fully among land plants including the capacity to produce sporopollenin, cutin, and some phenolic compounds (Kroken *et al.*, 1996). However, they lack other biochemical and morphological characteristics that likely evolved during the complete transition to land.

The fossil record of spores combined with phylogenetic studies indicates that groups related to living bryophytes were early colonizers of the land. Vascular plants arose later in the late Silurian period (~400 million years ago). They diversified rapidly, when an early split in the history of land plant evolution gave rise to two major lineages: the lycophytes and euphyllophytes (Kenrick & Crane, 1997) (Fig. 2.1). These lineages contain specialized water-conducting tracheary elements reinforced by lignins. Together, these changes resulted in more highly differentiated plants with different tissue systems including impermeable exterior surfaces and water conducting cells.

The appearance of phenolic compounds in large amount in plants is clearly related to their adaptation to terrestrial environments. The first plants that moved from an aquatic environment to emerged land had to face important stresses, including desiccation, varying temperatures, UV radiation and the attack of microorganisms. They had also to compete between each other for light and nutrients. The necessary biochemical adaptations to these new challenges were in large part issued from the phenyl propanoid pathway. This pathway provided different precursors at the origin of lignins for stem rigidity and vascularization, flavonoids for reproductive biology (fruit and flowers colors). Pathway products also provided protection against UV, microbial defense and also symbiotic plant-microbe interactions (e.g., leguminosae/rhyzobium). Other simple phenolics were also involved as attractants or deterrents in different adaptative strategies. Among these adaptations, the deposition of lignins in plant cell walls was likely a crucial mechanism that allowed the development of upright plants of large size adapted to a terrestrial habitat. Lignin provided structural rigidity for tracheophytes to stand upright and strengthened the cell wall of their water-conducting tracheary elements allowing them to withstand the negative pressure generated during transpiration.

The problem is to identify at which stage of evolution these different compounds appeared and how the enzymes/genes responsible for their synthesis emerged. As far as phenolic compounds are concerned, qualitative and quantitative differences occur in the biochemistry of embryophytes relative to charophytes, and between bryophytes and tracheophytes in the embryophytes. Classically, flavonoids seem to be restricted among extant plants to embryophytes, while true lignins are limited to (eu)tracheophytes.

Nevertheless, lignin-like polymers have been identified in the cell walls of the charalean alga *Nitella* and in primitive green algae (Delwiche *et al.*, 1989). Lignans (Raven, 2000; Suzuki & Umezawa, 2007) and other phenolic cell wall material different from lignins (Erickson & Miksche, 1974; Miksche & Yasuda, 1978) have been characterized in mosses and liverworts that do not harbor a vascular system. Interestingly, lignans, which are dimers derived from hydroxycinnamyl alcohols, allyl phenols or hydroxycinnamic acids, are widely distributed within embryophytes, but whereas hydroxycinnamyl alcohols are involved in their synthesis in tracheophytes, the lignans from bryophytes are derived from hydroxycinnamic acids (Scher *et al.*, 2003; Umezawa, 2003). All these different phenolic compounds have been associated to defense mechanisms against microorganisms or UV radiation (Ligrone *et al.*, 2008).

From the point of view of the biosynthetic pathways, the shikimate pathway, which provides phenylalanine as a protein amino acid, is present in procaryotes and has been conserved in unicellular and pluricellular eucaryotes including higher plants. However, the phenylpropanoid nucleus, from which most of phenolic compounds are derived, is also a product from phenylalanine via phenylalanine ammonia-lyase (PAL) and the presence of this enzyme is a prerequisite for an active phenolic synthesis. Although PAL enzymes have been extensively characterized in all land plant lineages, including bryophytes, their distribution in lower organisms is limited. PAL is present in some fungi where it participates in the catabolism of phenylalanine (MacDonald & D'Cunha, 2007) and has also been characterized from a few bacteria (Xiang and Moore, 2005; Moffit *et al.*, 2007). Emiliani and coworkers (Emiliani *et al.*, 2009) have performed an extensive analysis of the taxonomic distribution and phylogeny of PAL. The 160 representative sequences chosen for final tree construction were obtained from different prokaryotes and eukaryotes (plants and fungi). The authors conclude that PAL emerged in bacteria, likely with an antimicrobial role, and that a member of an early fungal lineage obtained a PAL via horizontal gene transfer from a bacterium. This fungal PAL was transferred to an ancestor of land plants via an ancient arbuscular mycorrhizal symbiosis where it paved the way for the development of the phenylpropanoid pathway. It is also possible that the ancestor of land plants and the ancestor of fungi independently acquired their PAL from two different but related bacteria. Even though they are quite speculative, these hypotheses are supported by phylogenetic analyses and can explain the emergence of PAL in land plants.

Despite the absence of data about the other genes involved in the phenyl propanoid pathway, it can be proposed that the assembly of the whole pathway likely occurred stepwise by the recruitment of pre-existing enzymes from other metabolic routes (Dixon & Steele, 1999; Lehfeldt *et al.*, 2000). As underlined by Weng and Chapple (2010), the enzymes of phenolic metabolism were gradually acquired from ancestors of primary metabolism through mutations and selection. As an example, PAL is homologous to histidine ammonia-lyase (HAL), an enzyme involved in histidine degradation in intermediary metabolism. Horizontal gene transfer from the few bacteria and fungi that harbor homologues of some enzymes of the pathway or mutations of some "enzyme precursors" of phenolic metabolism could have represented additional possibilities. As an example, we have found interesting homologies at the protein sequence level but also at the level of the structure of the genes (identical relative locations of introns and exons) between cinnamoyl-CoA reductase (CCR) and dihydroflavonol 4-reductase (DFR) (Lacombe *et al.*, 1997). These observations suggest that either these closely related genes, with different functions, derive from a common ancestor gene or alternatively that the DFR that is involved in the synthesis of anthocyanins (assumed to be anterior or concomitant to lignins during evolution) was a precursor of CCR.

As stressed by Emiliani *et al.* (2009), it is generally assumed that there is no evidence for the presence of a full phenylpropanoid metabolism in organisms other than land plants, although some bacteria and fungi harbor homologues of a few enzymes of the pathway (Moore *et al.*, 2002; Seshime *et al.*, 2005). However, recent results lead to reconsider this general dogma. Secondary walls and lignins have been recently characterized within cells of the red alga *Calliarthron cheilosporoides* (Martone *et al.*, 2009). A clear chemical characterization of *p*-hydroxyphenyl, guaiacyl and syringyl units demonstrated the presence of complex lignins in this organism. This result was confirmed by the detection in *Calliarthron* cell walls of lignin epitopes by polyclonal antibodies designed to localize lignins within terrestrial plant tissues. These data are particularly intriguing since they show that lignins exist within a red alga's calcified cells that lack hydraulic vasculature and have little need for additional support. The authors speculate that lignin biosynthetic pathways may have functioned in the common unicellular ancestor of red and green algae, thereby protecting cells from microbial infection or UV radiation. The enzymes and genes responsible for lignin synthesis in *Calliarthron* could have evolved convergently in *Calliarthron* and land plants. Alternatively, they could have been conserved having evolved prior to the divergence of red and green algae more than 1 billion years ago. In the latter case, future studies using the very sensitive methods available for lignin detection and comparative genomic approaches should reveal unexpected and exciting results allowing a revisit of our present views of phenolic metabolism. Taken together, the recent results of Martone *et al.* (2009) already provide a new vision of the phylogenetic distribution of phenolic compounds and lignins. It is interesting to observe that, independently of lignin synthesis, intermediates in the pathway could have been diverted to play other functions. Thus, monolignol derivatives such as dihydroconiferyl alcohol and dehydrodiconiferyl alcohol glucosides have been reported to play a hormonal role (Lynn *et al.*, 1987).

In addition to these results indicating a wider distribution of lignins than previously anticipated, recent data modify our views on lignin composition. Generally, ferns and gymnosperms deposit lignins that are derived primarily from guaiacyl monomers together with a small proportion of *p*-hydroxyphenyl units, whereas angiosperms lignins are guaiacyl/syringyl copolymers that can also contain some *p*-hydroxyphenyl monomers. These classical observations have led to the conclusion that syringyl lignins are recent adaptations restricted to angiosperms. Lycophytes, which arose in the early Silurian (380–400 Mya), represent a major lineage of vascular plants that has evolved in parallel with the ferns, gymnosperms and angiosperms. Syringyl monomers have been detected in lignins from Lycophytes, including species of *Selaginella*, (Logan & Thomas, 1985) and recently Weng and coworkers (Weng *et al.*, 2008) have identified the enzymatic system responsible of the synthesis of these monomers. They have characterized in the lycophyte *Selaginella moellendorffii* genome (fully sequenced), a P450-dependent monooxygenase—a ferulic acid/coniferaldehyde/coniferyl alcohol 5-hydroxylase (F5 H)—capable of diverting guaiacyl-substituted intermediates into syringyl lignin biosynthesis. Complementary data suggest this F5 H from *Selaginella* is functionally equivalent to, but phylogenetically independent of, angiosperm F5Hs. In a more recent study, the same group (Weng *et al.*, 2010a) has demonstrated that lignins are elaborated utilizing two distinct biosynthetic routes in *Selaginella* and angiosperms. *Selaginella* is part of one of the oldest divisions of vascular plants, resulting from an ancient split between the lycophytes and euphyllophytes that include all modern seed plants. In contrast to angiosperms, in which syringyl lignin biosynthesis requires two distinct meta-hydroxylases acting respectively on *p*-coumaroyl shikimate (C3′H) and ferulate (F5 H), an alternative pathway is used in the lycophyte *Selaginella*. This pathway involves a unique SmF5 H phenylpropanoid dual meta-hydroxylase acting both on *p*-coumarylaldehyde/alcohol and coniferylaldehyde/coniferylalcohol (Weng *et al.*, 2010a) that bypasses several steps of the canonical lignin biosynthetic pathway. This SmF5 H shares only 37% amino acid sequence identity with its angiosperm counterparts in agreement with the convergence evolution hypothesis. All together, these results support the chemical characterization of S lignins in *Selaginella* through molecular evidence at the gene and protein levels. They confirm the large distribution of S lignins among plants and suggest that different pathways at different evolution levels have been recruited for the synthesis of the same phenolic products. They represent a case of convergence evolution in which two distinct lineages that diverged from one to another more than 400 million years ago have evolved specific and independent biochemical strategies to synthesize syringyl lignins. This particular polymer likely provided evolutionary advantages in adaptation of *Selaginella* and angiosperms to their environment.

During the last years, the increasing number of sequenced genomes of lower and higher plants has allowed comparative genomic studies. If we consider the full set of genes involved in lignin biosynthesis, starting from PAL and including the peroxidases and laccases potentially involved in lignification, it appears as indicated by Delaux and coworkers (Delaux *et al.*, 2011, *in press*) that:

- some specific genes occur in green algae (CAD, CCR in *Chlamydomonas reinhardtii*);
- the complete set of genes is present in mosses (*Physcomitrella patens*), which are presumably not lignified organisms;
- an increase in the number of multigene families occurs with plant evolution.

The data obtained for mosses show that, assuming that pathways are functional on the basis of genomic data, they may not be valid without careful chemical determination of the corresponding metabolites. Considering the enzymes involved in the last step of lignin synthesis (i.e., the polymerization of monolignols), interesting hypotheses have been presented to relate the increase of atmospheric O_2 to the correlative production of reactive oxygen species (ROS) in living systems. It has been suggested (Lowry *et al.*, 1980) that the oxidative coupling of monolignols was initially permitted by the increase of oxygen concentration in the atmosphere during the Ordovician and Silurian periods. Thus, laccases, multicopper oxidases widely distributed in plants, fungi, bacteria, and animals would have been initially involved in the synthesis of primitive lignins. However, these enzymes were later replaced by the emergence of class III peroxidases (Weng & Chapple, 2010), a group of heme-containing oxidases that catalyse one-electron oxidation using H_2O_2. It has been speculated that the initial role of the class III peroxidases in early land plants was to fight against oxidative stress by exploiting phenylpropanoids as reducing agents (Kawaoka *et al.*, 2003). The resulting phenylpropanoid radicals would then have been able to polymerize into lignins. Whatever the probability of these hypotheses, different arguments strongly suggest that class III peroxidases are the dominant form of lignin polymerization enzymes (Weng & Chapple, 2010).

Another interesting biochemical innovation involving phenolics is related to the deposition of suberin and sporopollenin in specialized cells. When plants conquered land, they had to deal with major problems such as resistance to different stresses and uptake of water and nutrients from soil. To face these problems, they developed specialized complex polymers containing phenolics: suberin mainly located in the aerial periderm tissues, the root endoderm and the seed coat, and sporopollenin on the surface of pollen grains. Suberin is located between the cell wall and the plasma membrane and results from the association of a polyaliphatic domain with a phenolic domain (ferulate, *p*-coumarate or/and sinapate) linked to the cell wall (Bernards, 2002). Cell wall suberization of endoderm is particularly important to regulate the apoplastic transfer of water and solutes, but suberization of external tissues has also a crucial role in protecting the plant from the invasion of pathogens. However, knowledge of the synthesis of aromatic domains has been limited to pioneering works of the groups of Negrel (Lotfy *et al.*, 1994) and Kolattukudy (Cottle & Kolattukudy, 1982), who characterized an enzyme responsible of the synthesis of feruloylpalmitate. However, it was not until recently that a hydroxycinnamoyltransferase responsible for synthesizing suberin aromatics in *Arabidopsis* has been characterized (Gou *et al.*, 2009). The corresponding gene At5g41040, a member of the acyl-CoA dependent acyltransferase gene superfamily, exhibits an ω-hydroxyacid hydroxycinnamoyl-transferase activity with an *in vitro* preference for feruloyl-CoA and 16-hydroxypalmitic acid. The knockout of this gene in *Arabidopsis* reduces the quantity of ferulate and alters the permeability and sensitivity of roots to salt stress confirming the functional importance of suberin. In a parallel work, Molina and coworkers (Molina *et al.*, 2009) have shown that the promoter of the At5g41040 gene is specifically expressed in cell layers undergoing suberization and that the reduction of ferulate in At5g41040 knockout seeds is associated with an approximate stoichiometric decrease in aliphatic monomers containing ω-hydroxy groups.

Sporopollenin, the major part of the pollen exine, is a complex polymer highly resistant to chemical degradation. Similarly to suberin, it is formed both of hydroxylated fatty acids and phenylpropanoid residues (Dominguez *et al.*, 1999). Undoubtedly, the deposition of this very efficient protective layer was a crucial adaptation allowing pollen conservation and viability. In studying the biosynthetic phenolic pathways for pollen development, Matsuno and coworkers (Matsuno *et al.*, 2009) have incidentally illustrated the role of the diversification of plant cytochrome P450 enzymes in the emergence of novel phenolic pathways. They have shown that two new cytochrome P450 genes, CYP98A8 and CYP98A9, resulting from an ancestor, CYP98A3, which is known to meta-hydroxylate the *p*-coumarate esters of shikimic(SK)/quinic acids to form lignin monomers, are the main components of a new pathway. This pathway involves a cascade of six successive hydroxylations by two partially redundant cytochrome P450, leading from spermidine and *p*-coumaroyl-CoA to the formation of *N*1,*N*5-di(hydroxyferuloyl)-*N*10-sinapoyl spermidine, a major pollen constituent. The intermediates of this phenolamide route provide a putative new source of guaiacyl and syringyl precursors that are available for the synthesis of lignins and phenolic esters. CYP98A8 and CYP98A9 provide an example of adaptative gene evolution via retroposition, positive Darwinian selection, and subsequent duplication that led to novel enzymes, and plant metabolic pathways (Matsuno *et al.*, 2009). Even though it is no entirely clear whether these two CYPs diverged before angiosperm evolution or appeared recently, cytochrome P450 genes appear to be significant contributors to the remarkable phenolic plasticity of plants.

Overall, we are just beginning to identify the patterns of molecular evolution associated with the emergence of phenolic metabolism. This metabolic area represents a major example of important biochemical plant innovations for the adaptation of land plants in addition to other biochemical and morphological processes (Delaux *et al.* 2011, in press). As more plant genomes from primitive phylogenetic lineages are becoming available, we are likely on the way to bridge some of the evolutionary gaps that remain obscure in the field of phenolic metabolism (Bowman *et al.*, 2007). Future comparative genomic studies exploiting the increasing genomic resources and selected convenient taxa representatives of important stages of plant evolution will make these studies an exciting and rewarding domain.

2.2 The shikimate pathway: a complex and subtle interface between primary metabolism and phenolic metabolism

The shikimate pathway, initially characterized in microorganisms, is a fascinating pathway due to the number of reactions involved, the multiple regulation mechanisms and the occurrences in plants of alternative branches (quinic acid (QA) metabolism, arogenate

(ARO) or phenylpyruvate (PPY) as intermediates). We will concentrate in this review on results highlighting the specificities of this pathway in higher plants where it provides, as well as in microorganisms, the three aromatic amino acids; phenylalanine, tyrosine, and tryptophan, involved in protein synthesis. However, specifically in plants, phenylalanine and, secondarily, tyrosine are the precursors of a wide range of phenolic compounds.

The shikimate pathway includes a long chain of reactions classically subdivided into two segments. The prechorismate pathway starts with erythrose 4-phosphate and phosphoenolpyruvate and lead to chorismate in seven enzymatic steps. The second segment of the pathway leads from chorismate to the three aromatic amino acids and particularly to phenylalanine, the precursor of phenolics. Despite a basic similarity in the sequence of reactions, the molecular organization of the enzymes and their regulation mechanisms greatly differ between plants, fungi and bacteria. In microorganisms, a regulated carbon flux, through the different branches of the pathway is ensured through allosteric controls. These involve feedback regulation of 3-deoxy-D-arabino-heptulosonate-7-phosphate synthase (DAHP synthase), the first enzyme of the pathway, by the endproducts and allosteric activation/inhibition of the enzymes at the origin of phenylalanine/tyrosine (chorismate mutase) on one hand and tryptophan (anthranilate synthase) on the other (Herrmann & Weaver, 1999). As we will see later, the situation is more complex in plants.

The enzymes of the prechorismate pathway are differently organized according to the organisms. Fungi exhibit a specific association of five enzymes encoded by the pentafunctional arom gene. The corresponding enzymatic complex catalyses steps 2–6 in the prechorismate pathway (Hawkins *et al.*, 1993). In contrast, the same enzymes are independent and monofunctional in bacteria. In plants, two enzymes catalyzing the third and the fourth steps, dehydroquinate dehydratase (DHQase) and shikimate NADP oxidoreductase (Shorase), are associated in a bifunctional complex (Boudet & Lecussan, 1974; Singh & Christendat, 2006). Singh and Christendat have reported in 2006 the first crystal structure of *Arabidopsis* DHQase-Shorase. Their data support the role of the complex in increased flux efficiency (since the turnover rate for Shorase is 9 times faster than for DHQase) due to a maximized substrate transfer from one site to the next through the proximity effect. In higher plants, full-length cDNAs corresponding to the seven steps of the shikimate (prechorismate) pathway, including the cDNA encoding the bifunctional complex (DHQase-Shorase), have been characterized (Bischoff *et al.*, 1996; Bischoff *et al.*, 2001). All these genes contain sequences encoding *N*-terminal plastid-specific transit peptides (Schmid & Amrhein, 1999; Bischoff *et al.*, 2001) in agreement with the postulated plastidic location of the pathway. These different genes are transcriptionally induced after pathogen attack (Görlach *et al.*, 1995; Bischoff *et al.*, 1996; Bischoff *et al.*, 2001) likely to allow a convenient supply of precursors for the synthesis of phenolic compounds involved in defense. This coordinated enhanced transcription suggests that a general increase in transcripts/enzymes drive the increased flux in chorismate used for enhanced Phe/phenolic synthesis in these particular stress conditions.

From a more practical point of view, as the shikimate pathway is restricted to plants, fungi and bacteria, the related enzymes are good targets for the design of herbicides (such as glyphosate/round up) and antibiotics. Recently, a growing interest for SK was observed since this compound is used as a key starting material for the synthesis of the neuramidase inhibitor GS4104, which was developed under the name Tamiflu for treatment

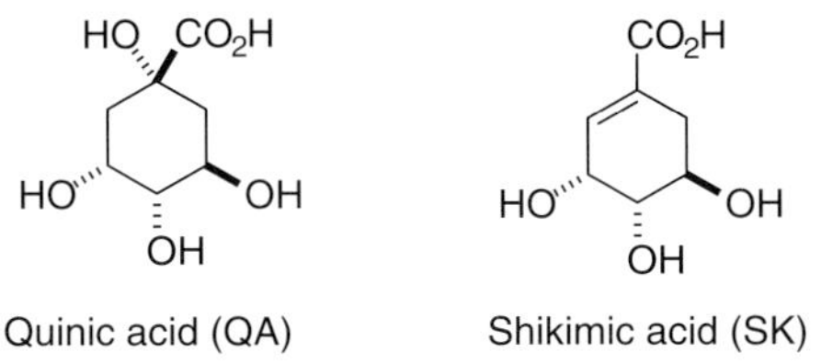

HO CO$_2$H
HO'' OH
OH
Quinic acid (QA)

CO$_2$H
HO'' OH
OH
Shikimic acid (SK)

Fig. 2.2 Quinic and shikimic acids, two alicylic acids precursors of phenolic compounds, highly accumulated in higher plants.

of antiviral infections. As an alternative to the isolation of SK from plant organs, its fermentative production by engineered microorganisms has been extensively investigated (Krämer *et al.*, 2003) and convenient strains of *Escherichia coli* have been designed for industrial production.

2.2.1 *Quinic acid, a specific component of higher plants*

Closely related to SK, without being an intermediate of the shikimate pathway, QA (Fig. 2.2) is a widely distributed and abundant metabolite in higher plants, particularly in woody species where its concentration can reach up to 8% of the dry mass in the leaves (Boudet, 1973). Interestingly, most of the species which accumulate QA do not contain a significant amount of SK. The converse is also true. SK is preferentially accumulated in gymnosperms (Yoshida *et al.*, 1975) and QA in angiosperms (Boudet, 1973). However, the two alicyclic acids exhibit a similar pattern of accumulation during an annual cycle, with a peak in spring during the period of intense growth, and a decrease in summer. This biphasic pattern suggests that these compounds could be first accumulated as a result of active photosynthetic activity in the leaves and then used as carbon sources for the synthesis of a wide range of phenolic compounds and particularly lignins, which are highly accumulated in woody plants. A redistribution of these alicyclic acids from the leaves (needles) to the trunk and the roots can be envisaged since both QA as well as SK have been characterized in the phloem of spruce and larch (Rohde *et al.*, 1996).

QA is also a building block for the synthesis of plant depsides including coumaroyl and caffeoyl quinic derivatives such as the widely distributed chlorogenic acid frequently accumulated in large amounts. These compounds are chemically diversified and different isomers of dicaffeoylquinic acid and of caffeoylquinic acid have been characterized in different plants (Moglia *et al.*, 2008). Quinate (and shikimate) *p*-coumaroyl and caffeoyl esters have also been identified as intermediates in lignin biosynthesis (Hoffmann *et al.*, 2004). However, the metabolic fate of QA, particularly when it is accumulated in large amount such as in the leaves of woody angiosperms, cannot be restricted to the synthesis of depsides. As previously suggested, a significant proportion of the carbon flux arising from QA is converted into phenylpropanoids and likely utilized for lignin synthesis.

Feeding experiments of *Quercus pedunculata* plantlets with labelled QA have been performed by Boudet (unpublished). The analysis of the derived compounds revealed that SK and also phenylalanine and tyrosine, as well as different phenolic acids, were strongly

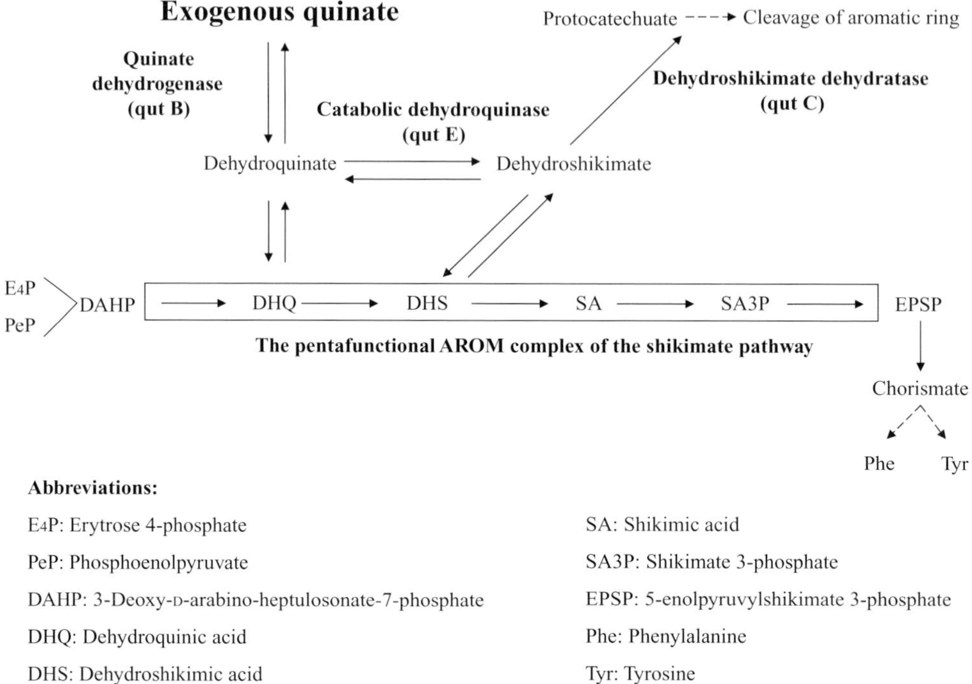

Abbreviations:

E4P: Erytrose 4-phosphate

PeP: Phosphoenolpyruvate

DAHP: 3-Deoxy-D-arabino-heptulosonate-7-phosphate

DHQ: Dehydroquinic acid

DHS: Dehydroshikimic acid

SA: Shikimic acid

SA3P: Shikimate 3-phosphate

EPSP: 5-enolpyruvylshikimate 3-phosphate

Phe: Phenylalanine

Tyr: Tyrosine

Fig. 2.3   The inducible quinate pathway in fungi.

labelled in agreement with previous results from Weinstein and coworkers (Weinstein *et al.*, 1959; Weinstein *et al.*, 1961) on the genus *Rosa*. These data support the involvement of QA in the synthesis of the aromatic amino acids, at least in part, through a conversion into SK. Interestingly, QA was also labelled when ^{14}C-SK was provided to the oak plantlets. The herbicide glyphosate, which inhibits the biosynthesis of aromatic amino acids by blocking the shikimate pathway downstream from SK, induces, as expected, an increase in shikimate content but also in quinate (Ulanov *et al.*, 2009; Orcaray *et al.*, 2010). All these data demonstrate potential biochemical connections between SK and QAs.

Quinate catabolism is well characterized in microorganisms. This compound, which accounts for up to 10% (w/w) of decaying plant matter, can be used as an important carbon source by soil-borne microbes, which do not produce QA but exhibit a sufficient plasticity to induce a specific pathway for its catabolism. This so-called quinate pathway (Lamb *et al.*, 1996) is encoded by three genes (the QA utilization gene cluster) that are co-ordinately regulated and encode quinate dehydrogenase or alternatively quinate: NAD(P) (+) oxidoreductase (Qorase), dehydroquinase (also named dehydroquinate hydrolase or hydrolyase) and dehydroshikimate dehydratase (Fig. 2.3). The Qorase is able to reversibly convert quinate intro dehydroquinate being also able in some species to use shikimate as a substrate (Hawkins *et al.*, 1993). The dehydroquinase activity is distinct from the biosynthetic dehydroquinase involved in the canonic shikimate pathway. In *Aspergillus nidulans*, the genes encoding the two dehydroquinases showed no similarity potentially identifying a case of convergent evolution. Subsequent work has introduced a new classification dividing

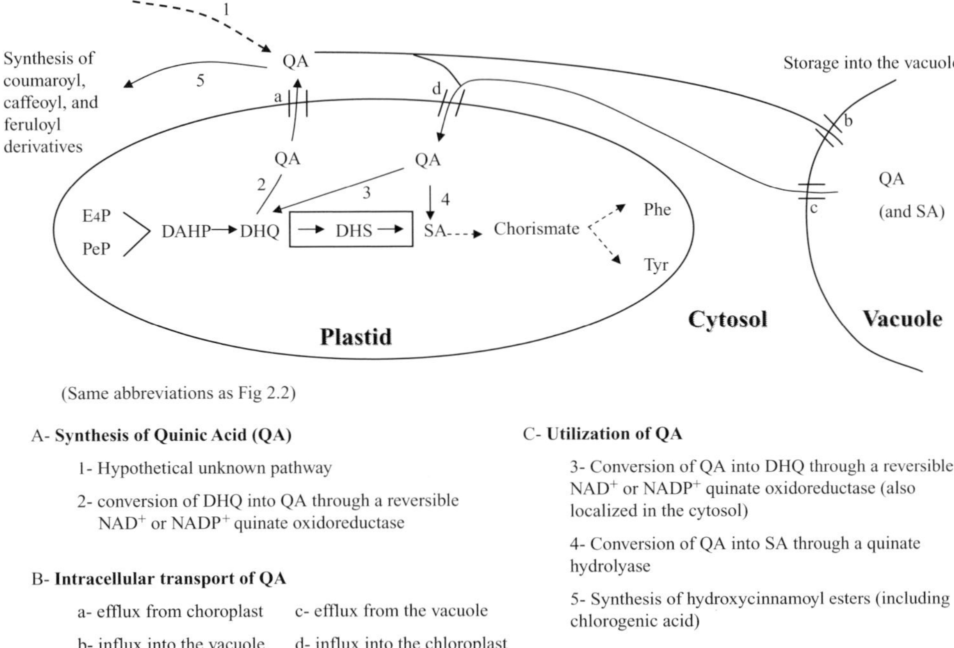

(Same abbreviations as Fig 2.2)

A- Synthesis of Quinic Acid (QA)

1- Hypothetical unknown pathway

2- conversion of DHQ into QA through a reversible NAD^+ or $NADP^+$ quinate oxidoreductase

B- Intracellular transport of QA

a- efflux from choroplast   c- efflux from the vacuole

b- influx into the vacuole   d- influx into the chloroplast

C- Utilization of QA

3- Conversion of QA into DHQ through a reversible NAD^+ or $NADP^+$ quinate oxidoreductase (also localized in the cytosol)

4- Conversion of QA into SA through a quinate hydrolyase

5- Synthesis of hydroxycinnamoyl esters (including chlorogenic acid)

Fig. 2.4   Potential routes for quinic acid synthesis and utilization in plants and subcellular compartments.

DHQases into Type 1 exclusively biosynthetic in nature and Type 2 involved in the catabolic quinate pathway. Finally, a dehydroshikimate dehydratase converts dehydroshikimate into protocatechuate further degraded by constitutive pathways (Lakshman *et al.*, 2006) (Fig. 2.3). The channeling function of this multienzyme complex is rather limited despite proposals suggesting that the AROM protein has a channeling function *in vivo* to keep separate the pools of DHQ and DHS in the quinate and shikimate pathways. Indeed, the two intermediates of the shikimate pathway DHQ and DHS can leak from the AROM complex, which has a low-level channeling function significantly only under nonoptimal conditions of nutrient supply and oxygenation (Lamb *et al.*, 1992). In this context, it is likely that exogenous DHQ and DHS can, at least partially, enter the complex.

Despite the marked accumulation of QA in higher plants, its metabolism is poorly understood in these organisms. Fragmentary data obtained from different species has allowed some potential enzymatic routes for its synthesis and its utilization to be proposed (Fig. 2.4), but a convincing general scheme is not yet available. QA could derive from an intermediate of the shikimate pathway dehydroquinic acid through a reversible quinate: NAD(P)+ oxidoreductase. This process would introduce a bypass allowing the storage of alicyclic precursors under the form of QA when the carbon flux through the shikimate pathway is high. This enzyme has been characterized in plants either with a strict (Minamikawa, 1977) or a partial (Kang & Scheibe, 1993) specificity towards NAD or with a strong preference for the use of NADP as a cofactor (Ossipov *et al.*, 1995). In *Pinus taeda* needles, Ossipov and coworkers (Ossipov *et al.*, 2000) have characterized two forms of a bifunctional enzyme having both quinate: NAD(P)+

oxidoreductase EC1.1.1.24, and shikimate: NADP+ oxidoreductase, EC1.1.125 activities. Being capable of accepting both quinate and shikimate as substrates, the enzyme resembles the bifunctional quinate/shikimate dehydrogenase of the inducible quinate pathway in microorganisms.

Recently, Marsh and coworkers (Marsh *et al.*, 2009) have studied QA storage and metabolism in three kiwifruit species that contain high levels of QA (2% w/w). QA accumulation occurred principally in the early stages of fruit development. A Qorase distinct from shikimate dehydrogenase (Shorase) was at a maximum around the time of greatest QA accumulation and markedly declined in late fruit development. It was also at high level in the species that accumulated the largest amount of QA. In contrast, Shorase activity declined slightly at later stages of fruit development. The authors suggested that Qorase could control the accumulation of QA in kiwi fruit.

As far as quinate utilization is concerned, a quinate hydrolyase catalyzing the irreversible conversion of QA into SK could be responsible for the entry of QA into the shikimate pathway. This enzyme activity has been initially characterized in pea (Tazaki *et al.*, 1974), in which the conversion of QA into SK is irreversible. More recently, the purification and subcellular compartmentation of QA hydrolyase of pea was reported by Lenschner and coworkers (Lenschner *et al.*, 1995). The quinate hydrolyase is localized in plastids. Interestingly, a quinate oxidoreductase activity was detectable in the cytosol of pea roots and could account for the conversion of DHQ into QA. The different localizations of these enzymes suggest a potential traffic of the substrates between cellular compartments. In corn, we have obtained evidence for the occurrence of an enzymatic complex able to convert QA into dehydroshikimic acid through the sequential involvement of a quinate: NAD+ oxidoreductase coupled to an isoform of DHQase II (Graziana *et al.*, 1980), but the intracellular location of this complex has not been studied.

The accumulation of QA and SKs in plant tissues on the one hand and the reactions involved in QA metabolism on the other represent specificities of higher plants likely related to the high demand of these alicyclic precursors for phenolic synthesis. However, additional studies are necessary using the tools of genetics, molecular biology, genomics, and cell biology to better understand the nature, location, and properties of the enzymes and transporters involved in the complex biochemical interactions between QA and the shikimate pathway.

The problems of compartmentation of substrates and related enzymes are undoubtedly crucial in this area of metabolism. Quinate hydrolyase in the same way as the enzymes of the classical shikimate pathway is localized in plastids. As QA is taken up by intact leucoplasts from pea roots, it could be synthesized in an extra plastidic compartment in agreement with the cellular location of pea Qorase and then move to the plastids for its conversion into SK. Nothing is known about the possible export of dehydroquinic from plastids, but this efflux could occur when the intraplastidial content of this intermediate becomes too high in order to store the alicyclic skeleton under the form of QA outside the plastids. QA and SK are also able to cross the tonoplast in order to be accumulated into the vacuole (Fig. 2.4). The release of the vacuolar quinic and SKs and their entry into plastids are thus potential regulatory steps in combination with the activity of the corresponding conversion enzymes when a high demand of phenylpropanoids is occurring.

From a general point of view, the transport mechanisms of phenolics, their conjugates and their precursors are poorly known. The exploitation of flavonoid mutants indicates the involvement of different transporters for the vacuolar transfer of flavonoids. In *Arabidopsis* and *Medicago*, glycosylated flavonoids are handled by a vacuolar membrane-localized multidrug and toxic compound extrusion (MATE) transporter (Marinova *et al.*, 2007; Zhao & Dixon, 2009). On the other hand, malonylated anthocyanins of maize are the substrates of typical MRP-type ABC transporters (Goodman *et al.*, 2004). Similar studies for a better understanding of the transport of shikimate pathway intermediates and simple phenolic compounds are urgently needed.

2.2.2 *The postchorismate branch of the shikimate pathway leading to phenylalanine: one or two metabolic routes in plants?*

Depending on the species, the conversion of chorismate into Phe involves in microorganisms either the PPY or the ARO pathways. The PPY route involves conversion of prephenate into PPY by the enzyme prephenate dehydratase (PDT) and the subsequent transformation of PPY into Phe or Tyr by the enzyme aromatic amino acid aminotransferase. The ARO route comprises the conversion of prephenate into ARO by the enzyme prephenate aminotransferase (PAT) and the subsequent conversion of ARO to Phe or Tyr, respectively, by the enzyme arogenate dehydratase (ADT) or arogenate dehydrogenase (ADS). In contrast to the experimental evidence supporting the use of ARO in plant Phe biosynthesis (De-Eknambul & Ellis, 1988; Cho *et al.*, 2007), it was not clear, until recently, whether PPY also serves as an intermediate in Phe biosynthesis (Fig. 2.5). Tzin *et al.* (2009) have provided new results showing of the conversion of PPY into Phe by plants. They have expressed in *Arabidopsis*

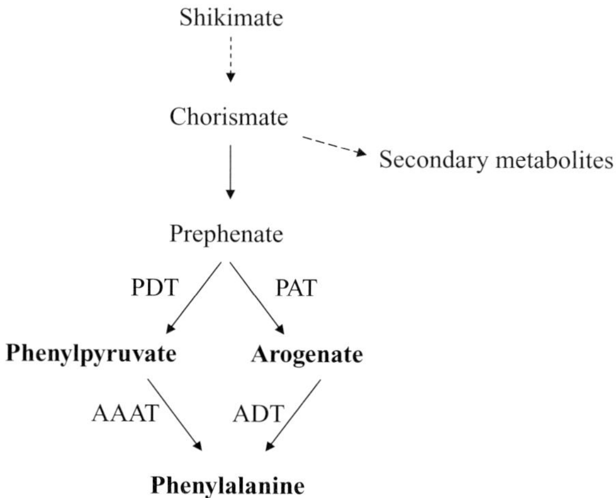

PDT: Prephenate dehydratase   AAAT: Aromatic amino acid aminotransferase
PAT: Prephenate aminotransferase   ADT: Arogenate dehydratase

Fig. 2.5 The biosynthetic routes for phenylalanine in higher plants.

thaliana a bacterial bifunctional Phe A (chorismate mutase/prephenate dehydratase) gene that converts chorismate via prephenate into phenylpyruvate. The pheA-expressing plants showed a large increase in the level of Phe, implying that they can convert phenylpyruvate, resulting from the over expression of Phe A, into Phe through an endogenous activity. These results support the hypothesis that plants can synthesize Phe via both the ARO and PPY routes. However, the relative importance of each individual route remains to be determined.

The enzymatic system converting ARO into phenylalanine through dehydration/decarboxylation (arogenate dehydratase ADT; EC4-2-1-51) is both complex and rather unusual. Indeed, recent biochemical characterization of six ADT genes from *Arabidopsis* revealed that three dehydratases have strict substrate specificity toward ARO while the remaining three can also accept prephenate but with much lower catalytic efficiencies (Yamada *et al.*, 2008). This lack of absolute substrate specificity and the inability to detect in plant tissues the unstable intermediates of the two parallel concurrent pathways (e.g., ARO/phenylpyruvate) makes it difficult to define the relative importance of the two biosynthetic pathways in Phe biosynthesis in plants.

Recently, Maeda and coworkers (Maeda *et al.*, 2010) using an RNA interference approach to suppress ADT with strict substrate specificity toward arogenate (ADT1) in *Petunia hybrida* petals demonstrated that Phe is predominantly produced via ARO in these tissues. Three ADT genes/proteins were occurring in Petunia (ADT1, ADT2, ADT3), but only the latter two exhibited PDT activity. ADT1 suppression significantly reduced ADT activity, levels of Phe and downstream phenylpropanoid/benzoid volatiles. Unexpectedly, shikimate levels were also decreased. An exogenous supply of shikimate led to prephenate and PPY accumulation and to a partial recovery of the reduced Phe levels in transgenic petals. This suggests that the PPY route is also functional in these tissues despite the fact that PPY was not detectable in control plants. These data are in line with the results of Tzin and coworkers (Tzin *et al.*, 2009) and Orlova and coworkers (Orlova *et al.*, 2006), all of which suggest that the PPY pathway can operate *in planta*. However, its relative contribution to Phe biosynthesis under specific physiological conditions requires further investigation.

2.2.2.1 Intracellular location of enzymes

The presence of a complete shikimate pathway, within plant plastids, leading to the three aromatic amino acids, has been clearly demonstrated (Schulze-Siebert *et al.*, 1984). However, a cytosolic form of chorismate mutase has also been characterized in several plant species. In *Arabidopsis* (Eberhard *et al.*, 1996), the isoform (AtCM2) occurred in cytosol in addition to the two well-characterized plastid isoforms (AtCM1 and AtCM3) and thereby raising the question of a potential duplication of the postchorismate branch of the shikimate pathway both in plastids and in cytosol. Several genes have been characterized for ADT and ADS in the *Arabidopsis* genome (Cho *et al.*, 2007). In a detailed and comprehensive study, Rippert and coworkers (Rippert *et al.*, 2009) have clearly excluded a subcellular localization of ADSs and ADTs in the cytosol. Transient expression analyses of ADS and ADT-green fluorescent protein fusions reveal that the two *Arabidopsis* ADS proteins and the six ADT proteins are all targeted within the plastids. The corresponding proteins were immunodetected in the chloroplasts of *Arabidopsis*. These data exclude any cytosolic postchorismate synthesis of Tyr and/or Phe at least in *Arabidopsis* green leaves and cultured cells. The physiological significance of a cytosolic chorismate mutase remains unexplained at the present time. Potential unknown conversions of prephenate in other products than Tyr and Phe could provide a rationale for the observed situation.

2.2.2.2 Complex and new regulatory mechanisms in the shikimate pathway

Yamada and coworkers (Yamada *et al.*, 2008) have identified a rice mutant (mtr1) that accumulates Phe, Trp and several phenylpropanoids. They have shown that the mutant gene encodes a form of rice bifunctional prephenate dehydratase/arogenate dehydratase (with a preference for ARO as substrate) with a point mutation in the putative allosteric region of the protein. The wild-type enzyme is feedback regulated by Phe, whereas the mutant enzyme exhibited reduced feedback sensitivity. These results show that, in addition to the well-known retroinhibition of chorismate mutase by Phe and Tyr occurring both in microorganisms and in plants (Eberhard *et al.* 1996), PDT is also feedback regulated by Phe. Similarly, arogenate dehydrogenase (ADS), which catalyzes the last step of Tyr biosynthesis, has been shown to be feedback regulated by Tyr in *Arabidopsis* (Rippert & Matringe, 2002). The accumulation of Phe in the mtr 1 mutant resulted in Trp overproduction and consequently in resistance to the toxic analog of Trp (5-methyltryptophan). These data suggest a connection between Phe accumulation and Trp pool size likely through an activation of a key enzyme of Trp biosynthesis by Phe. These different results show that tight allosteric regulations are occurring in the postchorismate pathway to keep, as in microorganisms, the pools of different amino acids relatively stable. However, fluxes of synthesis can increase for a given and constant pool size, and in addition, compartmentation processes may induce the formation of storage pools of Phe that can be mobilized when an increased synthesis of phenyl propanoids is necessary.

At the transcriptional level, Maeda and coworkers (Maeda *et al.*, 2010) have shown that together with ADT1 suppression the transcript levels of different genes involved in the shikimate pathway up to chorismate mutase were upregulated presumably to compensate for low levels of Phe. However, posttranscriptional regulation overrides this transcriptional compensation and down regulates the flux through the shikimate pathway. As exogenously supplied, shikimate bypasses this negative regulation, the posttranscriptional control should occur upstream of shikimate. These results suggest that a factor (or factors) downstream of chorismate likely serves as regulator(s) of the flux through the shikimate pathway. They are in agreement with a previous hypothesis proposing ARO as a DAHP synthase feedback regulator, thereby acting as a sensor of the overall Phe level (Jensen, 1986; Doong *et al.*, 1993).

Interestingly, in the aforementioned work of Tzin *et al.* (2009), Phe A overexpression enhanced the production of a number of lignin associated metabolites, as well as benzyl derivatives, but decreased the synthesis of anthocyanins that are located quite far downstream in the phenylpropanoid pathway. These data suggest an alteration of the competition between the lignin branch and the anthocyanin branch induced by Phe A overexpression. They confirm previous studies in which alterations of fluxes through various branches in the phenylpropanoid network were observed in response to other metabolic perturbations of Phe metabolism (Li *et al.*, 1993; Howles *et al.*, 1996).

These different results highlight the complexity and the specificity of regulatory mechanisms operating on the shikimate pathway in higher plants. The relatively well-characterized feedback regulation and activation mechanisms involved in microorganisms are likely inadequate to control efficiently the carbon fluxes and carbon distribution along a pathway tightly interconnected with the multibranched plant phenolic metabolism. As stressed by Vogt (Vogt, 2010) after considering the results of Lillo (Lillo *et al.*, 2008), "the response of individual genes of the shikimate pathway to variations in light or nutrient content appears tightly regulated and more complex than the transcriptional responses of the entry genes of phenylpropanoid and flavonoid biosynthesis, encoding phenylalanine ammonia-lyase or chalcone synthase".

2.3 Plant (poly)phenols: a diversified reservoir of useful chemicals

Endogenous phenolics impact human activities in the agro-industrial and food sectors by exerting either positive or negative influence on the processes or quality of the products. In addition, (poly)phenols have been widely used from very ancient time by mankind for diverse applications after their extraction from plant organs.

Many plants used in folk medicine contain a wide variety of phenolic compounds that likely act in the scavenging of free radicals and in the inhibition of lipid peroxidation. A typical example is given by salicylic acid (2-hydroxybenzoic acid), initially isolated from the bark of willow trees, known to ease aches and pains and reduce fevers. Its industrial substitute, aspirin, is likely the most widely used medication in the world. In agriculture, rotenone, a potent insecticide, is a complex chemical containing a benzene ring with two methoxy substituents. It is extracted from different species such as those belonging to the genus *Derris*.

Tannins, a group of polyphenols with molecular weights between 500 and 30,000 Da fall into two major groups, condensed tannins (proanthocyanidins) and hydrolysable tannins. These polymers are widely distributed in plants and usually found in large quantities in the bark of trees, where they act as a barrier for microorganisms. Tannins exert multiple effects in industrial and biological applications (Haslam, 2007; Serrano *et al.*, 2009). Besides in tanning (their initial exploitation to convert animal hides to leather), they are also used in dyeing, photography, refining beer and wine, as well as astringent in medicines. They can also be exploited as green chemicals. For example, phenol formaldehyde thermo-hardening resins can be replaced by cornstarch/quebracho tannin adhesives in the production of plywood panels (Moubarik *et al.*, 2010).

Natural colorants used as food additives are progressively replacing synthetic colorants that correspond to a huge industrial activity (several billions of dollars for more than 5000 artificial dyes). Among the polyphenols, anthocyanins (from various fruits and vegetables) and curcumin (yellow pigment extracted from the rhizome of *Curcuma longa*) are the most widely used sources of natural colorants. However, these compounds can lose their initial coloration depending on the pH and temperature and defining the conditions for improving their stability is an important challenge.

The interest in these natural colorants is continuously increasing because of new regulations that limit the use of artificial dyes and also because of their potential health benefits (Malien-Aubert & Amiot-Carlin, 2006). In the same way, plant extracts containing polyphenols (e.g., rosmarinic acid) are used as antioxidants.

If we consider the quality of plant products from the point of view of their organoleptic properties, polyphenols are responsible for the color, taste (bitterness, astringency), and scent of many fruits and vegetables. As we will see later, they are also supposed to exert health benefits. In addition, many polyphenols are involved in natural plant defense and the enhancement of these protecting compounds against pests is one the main objectives of a new ecologically intensive agriculture. The modification (suppression, enhancement) of these active molecules for potential benefits is now envisaged and the improvement of phenolic profile of useful plants and crops through classical and modern breeding strategies or through genetic engineering is an emerging research area. In the following section, we do not plan to present an exhaustive review of these various uses and strategies, but will focus on two areas in which there is an intense research activity, namely:

1) The health-promoting properties of polyphenols.
2) The biotechnological exploitation of lignocellulosics to provide biofuels and new chemicals.

2.3.1 *The health-promoting properties of polyphenols*

Considering that the role of dietary polyphenols in disease prevention has only attracted real scientific interest for about two decades, this field has seen a remarkable rate of progress. There has been an explosion of studies aiming to confirm the health benefits of these compounds. Polyphenols are becoming strong marketing arguments for consumers spontaneously attracted by natural substances and for the food industry, interest in phenolics has led to product innovation to increase the supply of polyphenol-containing food products (Boudet, 2007).

Simple (poly)phenolics such as hydroxycinnamic acid conjugates and flavonoids are important constituents of fruits, vegetables and beverages. These compounds show a wide range of antioxidant activities *in vitro* (Rice-Evans *et al.*, 1995) and are thought to exert protective effects against major diseases such as cancer and cardiovascular diseases (Scalbert *et al.*, 2005). Indeed, oxidative stress imposed by ROS plays a crucial role in the pathophysiology associated with neoplasia, atherosclerosis and neurodegenerative diseases. In this context, many interesting results indicate positive correlations between the dietary intake of polyphenol-containing food and the prevention of major diseases. Consequently, a wide range of molecular and *in vitro* studies have been undertaken to confirm these postulated effects of phenolic compounds. Epidemiological studies have also analyzed the health implications of dietary phenolic intake on various pathological situations. These studies, in addition to their own complexity, must take into account the biochemical composition of the components of the diet and the bioavailability of the individual compounds under study. Indeed, since the exact mechanism of absorption differs greatly from one polyphenol to another, the most abundant polyphenols in our diet do not necessarily achieve the highest concentration of active metabolites in target tissues. As an example of positive effects, a recent meta-analysis of the effects of polyphenols in humans showed that polyphenols provided protective effects on cardiovascular disease risk through the reduction of low-density lipoprotein (LDL) cholesterol and blood pressure, as well as through the improvement of flow-mediated dilation, an indication of endothelial function and a marker of cardiovascular risk (Hooper *et al.*, 2008). On the other hand, a large number of epidemiological studies have suggested that a high consumption of antioxidant-rich fruits and vegetables is inversely correlated with the incidence of cancer (Levi, 1999; Knekt *et al.*, 2002) and some mechanisms for the anticancer activity of polyphenols have been proposed (Guo *et al.*, 2009). However, one of the most crucial conclusions at the present time is that a low risk of cancer is more closely related to a diet rich in multiple antioxidants than one supplemented with one individual antioxidant. The combination of antioxidative agents with different modes of action is suggested to increase efficacy and minimize toxicity (Lee & Lee, 2006). Additive and synergistic effects of dietary phytochemicals obtained from fruits and vegetables are likely responsible for the anticancer activities in a more efficient way than expensive dietary supplements (Liu, 2004).

Future studies performed with standardized and structurally characterized mixtures of compounds or with single compounds will undoubtedly provide new opportunities to more clearly define the role of these food components. More studies in human (clinical and epidemiological) are needed to provide definitive proofs of the protective role of polyphenols. Such studies are also required to better evaluate the risks possibly resulting from a too high polyphenol consumption (Scalbert *et al.*, 2005). At this stage, dietary supplements cannot yet be promoted, but nutritional recommendations for an equilibrated diet in line with classical dietary guidelines are justified.

In conclusion, the positive effects of polyphenols on health are still a matter of intense debate and few health claims for such phytochemicals have so far been approved by regulatory authorities. It is clear that an increasing awareness among the public regarding the role of polyphenols in reducing the risk of chronic diseases has greatly stimulated the research in this area. However, science is not supposed to confirm the speculations of public opinion, its role is to provide facts and logical proposals deduced from experimental evidence. At the moment, it can be suggested that additive and synergistic effects of polyphenols with each other and with other protective compounds are likely to be the key of many of the observed effects.

2.3.2 A new time for lignocellulosics utilization through biotechnology

Fossil fuels were formed through the long transformation of phytomass accumulated during ancient periods. This progressive and complex process has resulted in the accumulation of huge reservoirs of coal, oil and gas, whose utilization has shaped industrial societies and the modern life style of most people on the planet. The main problem is that these resources are not renewable and are limited to supplies that are so intensively consumed that the end of their exploitation now appears very close. Mankind is thus faced with one of its biggest challenges, namely how to maintain the same level of activity, comfort and security with the expected shortage of the energy sources that are driving the system. Rising environmental and economic concerns linked to man's dependence on fossil fuels are thus currently driving research initiatives aimed at the development of viable energy alternatives. The use of biomass as a lignocellulosic feedstock to produce second-generation biofuels and new chemicals is one of these alternatives.

A rough estimate suggests that more than 1 billion hectares would be necessary to produce the biomass covering the global demand for biofuels for transportation. These figures show that the contribution of the biomass to the global energy demand is obligatorily limited. However, the conversion of lignocellulosic biomass into fuels and chemicals represents one of the society's important challenges. Basically, the core process involves the degradation of cell wall polymers into simple molecules and their subsequent conversion into end products. This process includes complex steps that are not at the present time globally economically viable. However, interesting possibilities are clearly open with the discovery of new enzymes and new microorganisms able to convert efficiently the components of the biomass into simple carbon sources that are potentially exploitable for the production of biofuels and green chemicals.

2.3.2.1 Biomass pretreatment and enzymatic conversion of polysaccharides

The chemical and physical complexity of the plant cell wall poses significant challenges to its enzymic deconstruction. The main components of the wall, cellulose, hemicelluloses and lignins, are tightly associated and their degradation by different biocatalysts implies a facilitated contact between the substrates and the enzymes. The presence of lignins in plant cell walls especially impedes the breakdown of wall polysaccharides to simple sugars. An improved knowledge of plant cell wall composition and molecular organization is still needed for efficient pretreatments of the plant biomass. For example, ester linked hydroxycinnamic derivatives such as ferulic and *p*-coumaric acids can link polysaccharide chains, as well as polysaccharides and lignin polymers (Hatfield *et al.*, 2009). Such bonds represent an additional target for lignocellulosic biomass deconstruction. Pretreatments necessary to facilitate the access of carbohydrates degrading enzymes to their corresponding substrates include procedures used in the pulp industry (kraft pulping), hydrothermolysis, ball milling, organosolvolysis or enzymatic action of ligninolytic enzymes.

In order to generate hydrolases with enhanced activities, detailed studies of the relationship between the structure and function of these enzymes are being conducted. Recent data reveal the complexity of and the intricacies of substrate recognition and catalysis by carbohydrate degrading enzymes. Carbohydrate-binding modules play a central role in the optimization of catalysis through their ability to bind specific plant structural polysaccharides and/or their possible disruptive effects on the plant cell wall network (Shoseyov *et al.*, 2006; Gilbert *et al.*, 2008). New combinations of catalytic sites and carbohydrate-binding modules are thus explored in order to improve the efficiency of the biocatalysts. In order to discover novel enzymes, scientists are also investigating a range of fungal and bacterial strains involved in wood degradation including white-rot and brown-rot fungi. Data collected on these diverse organisms suggest that optimizing biomass conversion in the future will require the activities expressed by microbial communities rather than single isolates. This more complex approach would closer represent natural conditions. Finally, enzymes are not suited for industrial applications due to their natural optimal pH and temperature, as well as stability. They must work in viscous media with high solid concentrations and in the presence of the products of the reaction and of various inhibitors. The improvement of enzymes along these different characteristics is vital to identify high-performance biocatalysts adapted to the multiple demands of industrial processes including a reduction of cost in order to set economically favorable strategies.

2.3.2.2 Lignins: degradation, bioconversion

Lignin is highly recalcitrant towards degradation. Peroxidases and laccases are the major enzymes involved in lignin degradation. However, here again the natural enzymes are far from working optimally under industrial conditions. An increased knowledge of the sequences of basidiomycetes peroxidases (including lignin peroxidase, manganese peroxidase, and versatile peroxidase) has contributed to a better evaluation of the structure-function relationships of these biocatalysts. In this context, peroxidase engineering is being investigated in order to develop tailor-made biocatalysts for industrial applications (Martinez *et al.*, 2009). Similar strategies are exploited for fungal laccases capable of degrading lignins. These enzymes have been successfully applied to paper manufacturing, enhancement of fibre properties, production of improved forages and pretreatment of lignocellulosic biomasses for fuel production (Hoegger *et al.*, 2006; Arora and Sharma, 2010). They are also exploited for the functionalization of lignocelluloses in order to obtain, for example, antimicrobial paper by coupling low-molecular mass phenols to unbleached pulp. In both cases, the high-throughput screening of large collections of fungal strains from different origins is being pursued for the selection of new strains with high enzymatic activities.

Ligninolytic enzymes are also considered to have potential for industrial applications such as biodegradation of environmental pollutants (e.g., polycyclic aromatic hydrocarbons, textile dyes) stain bleaching, biobleaching and biopulping of wood chips. Bioconversion of lignins in high value-added products is also an important objective. Indeed, the chemical nature of lignins (including aromatic and aliphatic moieties) makes it an interesting alternative source for aromatic chemicals. Progress has been made on the use of lignin as a feedstock for novel chemicals (reviewed in Boudet *et al.*, 2003). Some bacteria can produce polyester and polyamide plastics from a wide variety of dimeric and monomeric lignin compounds. The production of lignin-derived end products is still in its infancy at present, but the progress of metagenomics and of engineering microorganisms should allow the biotechnological production of novel chemicals from lignins.

For these different reasons, peroxidases and laccases represent important commercial economical markets. General strategies dealing with improvement of the performances of these biocatalysts are being developed with the final goal of producing recombinant optimized proteins (Ruiz-Dueñas *et al.*, 2007; Smith *et al.*, 2009).

2.3.2.3 The fermentation step towards the production of bioalcohols

Diverse strains of microorganisms that are able to convert simple sugars into alcohols are already available, but a general trend is to envisage their engineering in order to increase the yield of ethanol, as well as a tolerance to ethanol, acetate and inhibitors. Several studies at the laboratory scale are rapidly approaching a commercial realization. In addition to *Saccharomyces cerevisiae*, organisms such as *Thermoanaerobacterium saccharolyticum* and *Clostridium thermocellum* have been also exploited and significantly improved for practical applications.

Major investments have focused on the improvement of yeast strains for the overall conversion of lignocellulosic hydrolysates that contain hexoses and pentoses. The microbial conversion of pentoses, which represent up to 40% of total sugars in lignocellulosic materials, is a major challenge. Despite the fact that fermentation pathways of xylose and arabinose are complex, some groups have engineered industrial yeast strains, normally unable to transform pentoses, to ferment xylose and arabinose, as well as glucose in plant biomass hydrolysates. Other studies have characterized natural xylose-fermenting yeasts such as *Pichia stipitis* or *Candida shehatae* at the laboratory scale level (Bettiga *et al.*, 2009; Fromanger *et al.*, 2010).

Current research also deals with the engineering of *S. cerevisiae* for the production of butanol instead of ethanol, since the four-carbon alcohol butanol shows superior properties as a biofuel. However, some results have already been obtained on a laboratory strain and should be extended to commercial strains, which are more difficult to handle (Weber *et al.*, 2010).

2.3.2.4 Biorefinery pilot plants

A general trend in the exploitation of different sources of plant biomass is based on the concept of biorefining, i.e., a more complete utilization of lignocellulosic resources, involving the sequential or complementary use of different sub-fractions of the plant material for producing fuels and chemicals (Fig. 2.6). Some pilot plants for production of lignocellulosic bioethanol are already functioning. One in Denmark uses the "Inbicon" process that converts wheat straw and other soft lignocellulosic biomasses (e.g., corn stovers, rice straw) into three final products: ethanol for transportation, C5 molasses for animal feed and lignin, as a solid biofuel replacing coal in power and heat regeneration. The yield of final products for 1 ton of pretreated wheat straws is, at the moment, 435 kg of solid biofuel, 371 kg of C5 molasses and 144 kg ethanol. No chemical is involved in the process that converts 4 tons of wheat straw per hour and uses only steam, enzymes and yeast (Larsen *et al.*, 2008).

2.3.2.5 Quality and availability of the upstream resource

A key factor for success in the biotechnological conversion of lignocellulosics will require the ready availability of the plant resource. Improvement of that resource in turn will require more suitable plant materials (with low lignin content, for example) through breeding or genetic engineering strategies. In the latter case, targeting different key enzymes in the lignin biosynthetic pathway has been successfully explored, but it is clear that a trade-off

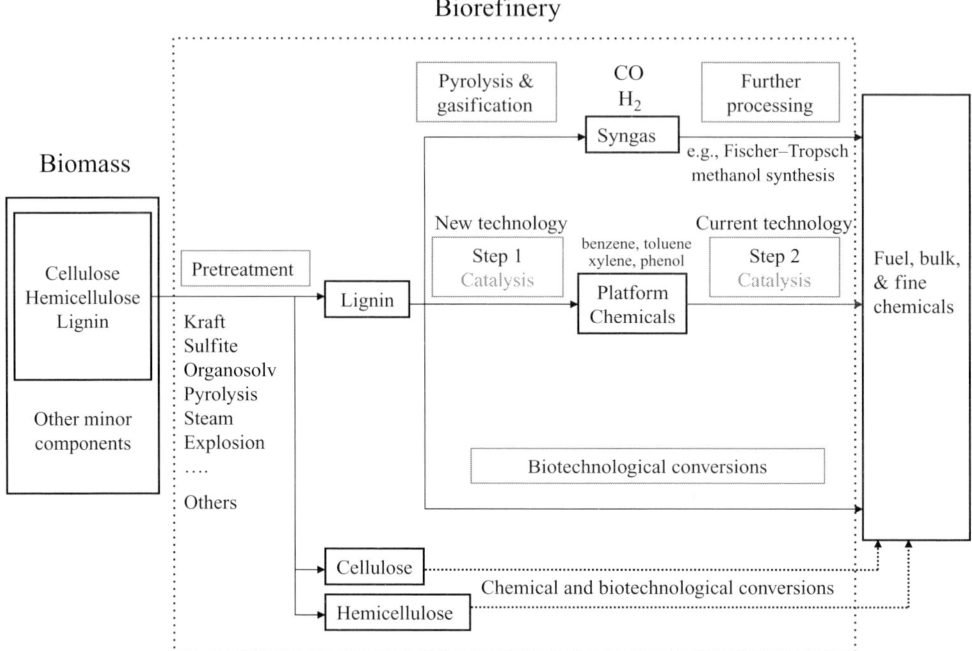

Fig. 2.6 An illustration of the concept of biorefining lignocellulosics with a special emphasis on lignin valorization (modified from Zakzeski *et al.*, 2010).

is needed between a low lignin content and a viable plant. Plants in which the cinnamyl alcohol dehydrogenase has been down regulated contain a modified lignin (aldehyde-rich) that is more easily extracted. Although the amount of lignin is not decreased, this approach seems to represent a good compromise at this time (Pilate *et al.*, 2002). In the same way, ferulate 5-hydroxylase overexpression gives extremely high-syringyl unit material (with no change in the amount of lignin) that would make delignification easier (Marita *et al.*, 1999; Weng *et al.*, 2010b).

2.3.2.6 Future prospects

Microorganisms engineered with optimized biosynthetic pathways and high-performance enzymes represent prerequisites for new alternatives for renewable energy and chemicals from lignocellulosic biomass. However, these strategies will succeed only if their production costs can compete or even outcompete with the production costs for current fuels and petrochemicals. Nevertheless, if costs of petroleum products (including the costs of wars, subsidies and the cost of global climate change) were factored into the real costs, competition will be much less of an issue!

Increased efficiency of the technology will be in any case very welcome and, in addition to classical approaches including site directed mutagenesis, *in vitro* enzyme evolution and random engineering strategies, recent advances in metagenomics offer new opportunities for the identification of high-performance enzymes (Simmons *et al.*, 2008). For example, new genes/enzymes are investigated through metagenomics in the digestive flora of termites, which are able to efficiently perform lignocellulose digestion. Other solutions can be provided in the future by synthetic biology, which allows one to design and build novel networks and pathways for re-engineering organisms for practical applications. After the discovery of more efficient biocatalysts, the next step will be their adaptation to the constraints of the industrial processes.

2.3.3 Chemical and catalytic valorization of polyphenols

In addition to biotechnological conversions, the catalytic valorization of lignins for the production of renewable chemicals is an alternative strategy. Despite the annual production of about 50 million tons of extracted lignins by the pulp and paper industry, only 2% are used to produce low-value products such as dispersing or binding agents with the remainder being burned as low-value fuel. However, a wide variety of bulk and fine chemicals, particularly aromatic compounds, are potentially obtainable from lignin.

A recent review (Zakzeski *et al.*, 2010) focuses on the different strategies for catalytic lignin conversion. These include among others (hydro)cracking lignin reduction reactions used to make fuels or bulk aromatic and phenolic compounds. In addition, lignin oxidation reactions are used to make functionalized aromatics for the production of fine chemicals. The authors have concentrated their attention on those chemical routes that employ a heterogeneous or homogeneous catalyst for the production of value-added low-molecular mass chemicals from lignin. They underline the limitations of the present approaches, particularly the lack of detailed information regarding the performance of catalysts. These catalysts, initially tested on pure lignin model compounds, are indeed used for the valorization of industrial lignins that contain different potential poisons for catalysts. However, they are confident in the future of biorefineries adopting the appropriate catalytic technology to produce renewable aromatic compounds for bulk and specialty chemicals.

Mäki-Arvela and coworkers (Mäki-Arvela *et al.*, 2007) have more widely reviewed the range of fine and specialty chemicals that can be obtained from phenolics extracted from wood. For example, eugenol can be catalytically isomerized to isoeugenol used in pharmaceuticals and in fragrance; pinosylvin, a stilbene present in the heartwood pines, is a precursor of dihydropinosylvin applied for skin lightening and for treatment of acne; propyl gallate produced from gallic acid, a hydrolytic product of tannins, is used as a food additive.

2.4 Concluding remarks

(Poly)phenols constitute the most abundant and the most widely distributed class of plant natural products. As clearly demonstrated by experimental evidence, they are crucial for the survival of higher plants because of their direct involvement in different adaptation strategies. This crucial importance does not fit well with the traditional term of secondary metabolites used to describe polyphenols and other natural substances. Hartmann (2007) has suggested that the term "ecochemicals" would be more appropriate.

In addition to the major classes of polyphenols such as lignins, flavonoids and tannins, higher plants contain a variety of phenolic structures, sometimes restricted to a few families or species, which also play important adaptive roles in response to environmental constraints. Among these are, for example, volatile compounds such as methylbenzoate and methylsalicylate, which are involved as floral attractants to pollinators, defense compounds against microorganisms, insects and mammalian predators or can display allelopathic effects on other plants (Dudareva *et al.*, 2004). This diversity and plasticity of phenolic metabolism are illustrated by the continuous discovery of new structures, new enzymes and new genes (Vogt, 2010). Plants adapt themselves their phenolic patterns to a changing environment through the emergence of new genes, such as cytochrome P450 oxidases, brought about by gene duplication and mutation and subsequent recruitment for adaptation to specific functions. It is reasonable to assume that these phenolic potential "sensors" or "adaptive devices" of plants have increased in number and sophistication with the increasing complexity of plant organization and plant–environment interactions. However, recent intriguing results, such as the occurrence of lignins in red algae, may lead to a revision of classical views. Newly available technical tools such as genomic resources and ultrasensitive analytical techniques will undoubtedly provide a renewed picture of phenolics emergence and distribution.

Despite a basic knowledge of the main biosynthetic pathways, some black boxes or open questions, such as the intracellular transport of phenolics, need additional study. In addition to the classical approaches ("feeding experiments," enzymology, cell biology), the efficient exploitation of mutants and transgenic plants will continue to provide an increased knowledge of the biosynthesis and functions of plant phenolics.

Until now, it appears that the phenolic patterns of higher plants have been shaped by a dialog between plants and their environment for the benefits of plants and their better adaptation to external conditions. At the moment, we shift to a new paradigm based on the extensive and rationale exploitation of plant chemical resources by mankind. Several examples have been presented in this chapter and the problems to be solved require the design of plants with specific phenolic profiles, more integrated approaches and more interactions with society selecting the more acceptable strategies. In the future, research programs that exploit emerging disciplines such as metagenomics, synthetic biology and some aspects of nanotechnology will undoubtedly provide an extended use of these compounds for human activities.

Acknowledgments

Professor Bolwell and Professor Conn are gratefully acknowledged for critically reading the manuscript. Sincere thanks to Sandrine Abadie for careful and efficient editorial assistance.

References

Arora, D.S. & Sharma, R.K. (2010) Ligninolytic fungal laccases and their biotechnological applications. *Applied Biochemistry and Biotechnology*, **160**, 1760–1788.

Bernards, M. (2002) Demystifying suberin. *Canadian Journal of Botany*, **80**, 227–240.

Bettiga, M., Bengtsson, O., Hahn-Hägerdal, B. & Gorwa-Grauslund, M.F. (2009) Arabinose and xylose fermentation by recombinant *Saccharomyces cerevisiae* expressing a fungal pentose utilization pathway. *Microbial Cell Factories*, **8**, 40.

Bischoff, M., Rösler, J., Raesecke, H-R., Görlach, J., Amrhein, N. & Schmid, J. (1996) Cloning of a cDNA encoding a 3-dehydroquinate synthase from a higher plant, and analysis of the organ-specific and elicitor-induced expression of the corresponding gene. *Plant Molecular Biology*, **31**, 69–76.

Bischoff, M., Schaller, A., Bieri, F., Kessler, F., Amrhein, N. & Schmid, J. (2001) Molecular characterization of tomato 3-dehydroquinate dehydratase-shikimate: NADP oxidoreductase 1. *Plant Physiology*, **125**, 1891–1900.

Boudet, A.-M. (1973) Les acides quinique et shikimique chez les angiospermes arborescentes. *Phytochemistry*, **12**, 363–370.

Boudet, A.-M. (2007) Evolution and current status of research in phenolic compounds. *Phytochemistry*, **68**, 2722–2735.

Boudet, A.-M., Kajita, S., Grima-Pettenati, J. & Goffner, D. (2003) Lignins and lignocellulosics: a better control of synthesis for new and improved uses. *Trends in Plant Science*, **8**, 576–581.

Boudet, A.-M. & Lecussan, R. (1974) Généralité de l'association (5-deshydroquinate hydrolyase, shikimate: NADP oxydoréductase) chez les végétaux supérieurs. *Planta*, **119**, 71–79.

Bowman, J.L., Floyd, S.K. & Sakakibara, K. (2007) Green genes-comparative genomics of the green branch of life. *Cell*, **129**, 229–234.

Cho, M.H., Corea, O.R., Yang, H., *et al.* (2007) Phenylalanine biosynthesis in *Arabidopsis thaliana*: identification and characterization of arogenate dehydratases. *Journal of Biology Chemistry*, **282**, 30827–30835.

Cottle, W. & Kolattukudy, P.E. (1982) Abscisic acid stimulation of suberization: induction of enzymes and deposition of polymeric components and associated waxes in tissue cultures of potato tuber. *Plant Physiology*, **70**, 775–780.

De-Eknambul, W. & Ellis, B.E. (1988) Purification and characterization of prephenate aminotransferase from *Anchusa officinalis* cell cultures. *Archives of Biochemistry and Biophysics*, **267**, 87–94.

Delaux, P.M., Nanda, A.K., Mathé, C., *et al.* (2011) Molecular and biochemical aspects of plant terrestrialization. *Perspectives in Plant Ecology, Evolution and Systematics* (in press). doi:10.1016/j.ppees.2011.09.001

Delwiche, C.F., Graham, L.E. & Thomson, N. (1989) Lignin-like compounds and sporopollenin in *Coleochaete*, an algal model for land plant ancestry. *Science*, **245**, 399–401.

Dixon, R.A. & Steele, L. (1999) Flavonoids and isoflavonoids—a gold mine for metabolic engineering. *Trends in Plant Science*, **4**, 394–400.

Dominguez, E., Mercado, J.A., Quesada, M. & Heredia, A. (1999) Pollen sporopollenin: degradation and structural elucidation. *Sexual Plant Reproduction*, **12**, 171–178.

Doong, R.L., Ganson, R.J. & Jensen, R.A. (1993) Plastid-localized 3-deoxy-D-*arabino*-heptulosonate 7-phosphate synthase (DS Mn): The early-pathway target of sequential feedback inhibition in higher plants. *Plant, Cell & Environment*, **16**, 393–402.

Dudareva, N., Pichersky, E. & Gershenzon, J. (2004) Biochemistry of plant volatiles. *Plant Physiology*, **135**, 1893–1902.

Eberhard, J., Ehrler, T.T., Epple, P., *et al.* (1996) Cytosolic and plastidic chorismate mutase isozymes from Arabidopsis thaliana: molecular characterization and enzymatic properties. *The Plant Journal*, **10**, 815–821.

Emiliani, G., Fondi, M., Fani, R. & Gribaldo, S. (2009) A horizontal gene transfer at the origin of phenylpropanoid metabolism: a key adaptation of plants to land. *Biology Direct*, **4**, 7.

Erickson, M. & Miksche, G.E. (1974) On the occurrence of lignin or polyphenols in some mosses and liverworts. *Phytochemistry*, **13**, 2295–2299.

Fromanger, R., Guillouet, S.E., Uribelarrea, J.L., Molina-Jouve, C. & Cameleyre, X. (2010) Effect of controlled oxygen limitation on Candida shehatae physiology for ethanol production from xylose and glucose. *Journal of Industrial Microbiology and Biotechnology*, **37**, 437–445.

Gilbert, H.J., Stålbrand, H. & Brumer, H. (2008) How the walls come crumbling down: recent structural biochemistry of plant polysaccharide degradation. *Current Opinion in Plant Biology*, **11**, 338–348.

Goodman, C.D., Casati, P. & Walbot, V. (2004) A multidrug resistance-associated protein involved in anthocyanin transport in Zea mays. *The Plant Cell*, **16**, 1812–1826.

Görlach, J., Raesecke, H-R., Rentsch, D., *et al.* (1995) Temporally distinct accumulation of transcripts encoding enzymes of the prechorismate pathway in elicitor treated, cultured tomato cells. *The Proceedings of the National Academy of Sciences of the United States of America*, **92**, 3166–3170.

Gou, J.Y., Yu, X.H. & Liu, C.J. (2009) A hydroxycinnamoyltransferase responsible for synthesizing suberin aromatics in Arabidopsis. *The Proceedings of the National Academy of Sciences of the United States of America*, **106**, 18855–18860.

Graziana, A., Boudet, A. & Boudet, A.-M. (1980) Association of the quinate: NAD+-oxidoreductase with one dehydroquinate-hydrolyase isoenzyme in corn seedlings. *Plant Cell and Physiology*, **21**, 1163–1174.

Guo, W., Kong, E. & Meydani, M. (2009) Dietary polyphenols, inflammation, and cancer. *Nutrition and Cancer*, **61**, 807–810.

Hartmann, T. (2007) From waste products to ecochemicals: 50 years research on plant secondary metabolism. *Phytochemistry*, **68**, 2831–2846.

Haslam, E. (2007) Vegetable tannins—lessons of a phytochemical lifetime. *Phytochemistry*, **68**, 2713–2721.

Hatfield, R.D., Marita, J.M., Frost, K., *et al.* (2009) Grass lignin acylation: *p*-coumaroyl transferase activity and cell wall characteristics of C3 and C4 grasses. *Planta*, **229**, 1253–1267.

Hawkins, A.R., Moore, J.D. & Adeokun, A.M. (1993) Characterization of the 3-dehydroquinase domain of the pentafunctional AROM protein, and the quinate dehydrogenase from *Aspergillus nidulans*, and the overproduction of the type II 3-dehydroquinase from *Neurospora crassa*. *Biochemical Journal*, **296**, 451–457.

Herrmann, K.M. & Weaver, L.M. (1999) The shikimate pathway. *Annual Review of Plant Physiology*, **50**, 472–503.

Hoegger, P.J., Kilaru, S., James, T.Y., Thacker, J.R. & Kues, U. (2006) Phylogenetic comparison and classification of laccase and related multicopper oxidase protein sequences. *FEBS Journal,* **273**, 2308–2326.

Hoffmann, L., Besseau, S., Geoffroy, P., *et al.* (2004) Silencing of hydroxycinnamoyl-coenzyme A shikimate/quinate hydroxycinnamoyltransferase affects phenylpropanoid biosynthesis. *Plant Cell,* **16**, 1446–1465.

Hooper, L., Kroon, P.A., Rimm, E.B., *et al.* (2008) Flavonoids, flavonoid-rich foods, and cardiovascular risk: A meta-analysis of randomized controlled trials. *American Journal of Clinical Nutrition*, **88**, 38–50.

Howles, P.A., Sewalt, V.J.H., Paiva, N.L., *et al.* (1996) Overexpression of L-phenylalanine ammonia-lyase in transgenic tobacco plants reveals control points for flux into phenylpropanoid biosynthesis. *Plant Physiology*, **112**, 1617–1624.

Jensen, R.A. (1986) The shikimate arogenate pathway—link between carbohydrate metabolism and secondary metabolism. *Physiology Plantarum*, **66**, 164–168.

Kang, X. & Scheibe, R. (1993) Purification and characterisation of the quinate: oxidoreductase from *Phaseolus mungo* sprouts. *Phytochemistry,* **33**, 769–773.

Kawaoka, A., Matsunaga, E., Endo, S., *et al.* (2003) Ectopic expression of a horseradish peroxidase enhances growth rate and increases oxidative stress resistance in hybrid aspen. *Plant Physiology*, **132**, 1177–1185.

Kenrick, P. & Crane, P.R. (1997) The origin and early evolution of plants on land. *Nature*, **369**, 33–39.

Knekt, P., Kumpulainen, J., Jarvinen, R., *et al.* (2002) Flavonoid intake and risk of chronic diseases. *American Journal of Clinical Nutrition*, **76**, 560–568.

Krämer, M., Bongaerts, J., Bovenberg, R., *et al.* (2003) Metabolic engineering for microbial production of shikimic acid. *Metabolic Engineering*, **5**, 277–283.

Kroken, S.B., Graham, L.E. & Cook, M.E. (1996) Occurrence and evolutionary significance of resistant cell walls in charophytes and bryophytes. *American Journal of Botany*, **83**, 1241–1254.

Lacombe, E., Hawkins, S., Van Doorsselaere, J., *et al.* (1997) Cinnamoyl CoA reductase, the first committed enzyme of the lignin branch biosynthetic pathway: cloning, expression and phylogenetic relationships. *The Plant Journal*, **11**, 429–441.

Lakshman, D.K., Liu, C., Mishra, P.K. & Tavantzis, S. (2006) Characterization of the arom gene in *Rhizoctonia solani*, and transcription patterns under stable and induced hypovirulence conditions. *Current Genetics*, **49**, 166–177.

Lamb, H.K., Newton, G.H., Levett, L.J., Cairns, E., Roberts, C.F. & Hawkins, A.R. (1996) The QUTA activator and QUTR repressor proteins of Aspergillus nidulans interact to regulate transcription of the quinate utilization pathway genes. *Microbiology*, **142**, 1477–1490.

Lamb, H.K., Van Den Hombergh, J.P., Newton, G.H., Moore, J.D., Roberts, C.F. & Hawkins, A.R. (1992) Differential flux through the quinate and shikimate pathways. Implications for the channelling hypothesis. *Biochemical Journal*, **284**, 181–187.

Larsen, J., Østergaard Petersen, M., Thirup, L., Wen Li, H. & Krogh Iversen, F. (2008) The IBUS process—lignocellulosic bioethanol close to a commercial reality. *Chemical Engineering and Technology*, **31**, 765–772.

Lee, W.K. & Lee, H.J. (2006) The roles of polyphenols in cancer chemoprevention. *Biofactors*, **26**, 105–121.

Lehfeldt, C., Shirley, A.M., Meyer, K., *et al.* (2000) Cloning of the SNG1 gene of Arabidopsis reveals a role for a serine carboxypeptidase-like protein as an acyltransferase in secondary metabolism. *The Plant Cell*, **12**, 1295–1306.

Lenschner, C., Herrmann, K.M. & Schultz, G. (1995) The Metabolism of Quinate in Pea Roots (Purification and Partial Characterization of a Quinate Hydrolyase). *Plant Physiol.* **108**, 319–325.

Levi, F. (1999) Cancer prevention: epidemiology and perspectives. *European Journal of Cancer*, **35**, 1046–1058.

Li, J., Ou-Lee, T.M., Raba, R., Amundson, R.G. & Last, R.L. (1993) *Arabidopsis* flavonoid mutants are hypersensitive to UV-B irradiation. *The Plant Cell*, **5**, 171–179.

Ligrone, R., Carafa, A., Duckett, J.G., Renzaglia, K.S. & Ruel, K. (2008) Immunocytochemical detection of lignin-related epitopes in cell walls in bryophytes and the charalean alga Nitella. *Plant Systematics and Evolution*, **270**, 257–272.

Lillo, C., Lea, U.S. & Ruoff, P. (2008) Nutrient depletion as a key factor for manipulating gene expression and product formation in different branches of the flavonoid pathway. *Plant, Cell & Environment*, **31**, 582–601.

Liu, R.H. (2004) Potential synergy of phytochemicals in cancer prevention: mechanism of action. *The Journal of Nutrition*, **134**, 3479–3485.

Logan, K.J. & Thomas, B.A. (1985) Distribution of lignin derivatives in plants. *New Phytologist*, **99**, 571–585.

Lotfy, S., Negrel, J. & Javelle, F. (1994) Formation of ω-feruloyloxypalmitic acid by an enzyme from wound-healing potato tuber discs. *Phytochemistry*, **35**, 1419–1424.

Lowry, B., Hebant, C. & Lee, D. (1980) The origin of land plants—a new look at an old problem. *Taxon*, **29**, 183–197.

Lynn, D.G., Chen, R.H., Manning, K.S. & Wood, H.N. (1987) The structural characterization of endogenous factors from Vinca rosea crown gall tumors that promote cell division of tobacco cells. *The Proceedings of the National Academy of Sciences of the United States of America*, **84**, 615–619.

MacDonald, M.J. & D'Cunha, G.B. (2007) A Modern view of phenylalanine ammonia lyase. *Biochemistry and Cell Biology*, **85**, 273–282.

Maeda, H., Shasany, A.K., Schnepp, J., *et al.* (2010) RNAi suppression of arogenate dehydratase1 reveals that phenylalanine is synthesized predominantly via the arogenate pathway in petunia petals. *The Plant Cell*, **22**, 832–849.

Mäki-Arvela, P., Holmbom, B., Salmi, T. & Murzin, D Yu. (2007) Recent progress in synthesis of fine and specialty chemicals from wood and other biomass by heterogeneous catalytic processes. *Catalysis Reviews*, **49**, 197–340.

Malien-Aubert, C. & Amiot-Carlin, M.J. (2006) Pigments phénoliques—structures, stabilité, marché des colorants naturels et effets sur la santé. In: *Les polyphénols en agroalimentaire* (eds. P. Sarni Manchado & V. Cheynier), pp. 295–360. Tech & Doc, Londres.

Marinova, K., Pourcel, L., Weder, B., *et al.* (2007) The *Arabidopsis* MATE transporter TT12 acts as a flavonoid/H -antiporter active in proanthocyanidinaccumulating cells off the seed coat. *The Plant Cell*, **19**, 2023–2038.

Marita, J.M., Ralph, J., Hatfield, R.D. & Chapple, C. (1999) NMR characterization of lignins in Arabidopsis altered in the activity of ferulate 5-hydroxylase. *The Proceedings of the National Academy of Sciences of the United States of America*, **96**, 12328–12332.

Marsh, K.B., Boldingh, H.L., Shilton, R.S. & Laing, W.A. (2009) Changes in quinic acid metabolism during fruit development in three kiwifruit species. *Plant Biology*, **36**, 463–470

Martínez, A.T., Ruiz-Dueñas, F.J., Martínez, M.J., Del Río, J.C. & Gutiérrez, A. (2009) Enzymatic delignification of plant cell wall: from nature to mill. *Current Opinion in Biotechnology*, **20**, 348–357.

Martone, P.T., Estevez, J.M., Lu, F., *et al.* (2009) Discovery of lignin in seaweed reveals convergent evolution of cell-wall architecture. *Current Biology*, **19**, 169–175.

Matsuno, M., Compagnon, V., Schoch, G.A., *et al.* (2009) Evolution of a novel phenolic pathway for pollen development. *Science*, **325**, 1688–1692.

Miksche, G.E. & Yasuda, S. (1978) Lignin of 'giant' mosses and some related species. *Phytochemistry*, **17**, 503–504.

Minamikawa, T. (1977) Quinate:NAD+-oxidoreductase of germinating *Phasaeolus mungo* seeds: partial purification and some properties. *Plant Cell Physiology*, **18**, 743–752.

Moffit, M.C., Louie, G.V., Bowman, M.E., Pence, J., Noel, J.P. & Moore, B.S. (2007) Discovery of two cyanobacterial phenylalanine ammonia lyases: kinetic and structural characterization. *Biochemistry*, **46**, 1004–1012.

Moglia, A., Lanteri, S., Comino, C., Acquadro, A., de Vos, R. & Beekwilder, J. (2008) Stress-induced biosynthesis of dicaffeoylquinic acids in globe artichoke. *Journal of Agricultural and Food Chemistry*, **56**, 8641–8649.

Molina, I., Li-Beisson, Y., Beisson, F., B. Ohlrogge, J. & Pollard, M. (2009) Identification of an Arabidopsis Feruloyl-Coenzyme A Transferase Required for Suberin Synthesis. *Plant Physiology*, **151**, 1317–1328.

Moore, B.S., Hertweck, C., Hopke, J.N., *et al.* (2002) Plant-like biosynthetic pathways in bacteria: from benzoic acid to chalcone 1. *Journal of Natural Products*, **65**, 1956–1962.

Moubarik, A., Charrier, B., Allal, A., Charrier, F. & Pizzi, A. (2010) Development and optimization of a new formaldehyde-free cornstarch and tannin wood adhesive. *European Journal of Wood Products*, **68**, 167–177.

Orcaray, L., Igal, M., Marino, D., Zabalza, A. & Royuela, M. (2010) The possible role of quinate in the mode of action of glyphosate and acetolactate synthase inhibitors. *Pest Management Science*, **66**, 262–269.

Orlova, I., Marshall-Colón, A., Schnepp, J., *et al.* (2006) Reduction of benzenoid synthesis in petunia flowers reveals multiple pathways to benzoic acid and enhancement in auxin transport. *The Plant Cell*, **18**, 3458–3475.

Ossipov, V., Bonner, C., Ossipova, S. & Jensen, R. (2000) Broad-specificity quinate (shikimate) dehydrogenase from *Pinus taeda* needles. *Plant Physiology and Biochemistry*, **38**, 923–928.

Ossipov, V., Chernov, A., Zrazhevskaya, G. & Shein, I. (1995) Quinate:NAD(P)$^+$-oxidoreductase from *Larix sibirica*: purification, characterisation and function. *Trees*, **10**, 46–51.

Pilate, G., Guiney, E., Holt, K., *et al.* (2002) Field and pulping performances of transgenic trees with altered lignification. *Nature Biotechnology*, **20**, 607–612.

Raven, J.A. (2000) Land plant biochemistry. *Philosophical Transaction of the Royal Society B: Biological Sciences*, **355**, 833–846.

Rice-Evans, C.A., Miller, N.J., Bolwell, P.G., Bramley, P.M. & Pridham, J.B. (1995) The relative antioxidant activities of plant-derived polyphenolic flavonoids. *Free Radical Research*, **22**, 375–383.

Rippert, P. & Matringe, M. (2002) Molecular and biochemical characterization of an Arabidopsis thaliana arogenate dehydrogenase with two highly similar and active protein domains. *Plant Molecular Biology*, **48**, 361–368.

Rippert, P., Puyaubert, J., Grisollet, D., Derrier, L. & Matringe, M. (2009) Tyrosine and phenylalanine are synthesized within the plastids in *Arabidopsis*. *Plant Physiology*, **149**, 1251–1260

Rohde, M., Waldmann, R. & Lunderstadt, J. (1996) Induced defence reaction *in* the phloem of spruce (*Picea abies*) and larch (*Larix decidua*) after attack by *Ips typographus* and *Ips cembrae*. *Forest Ecology and Management*, **86**, 51–59.

Ruiz-Dueñas, F.J., Morales, M., Pérez-Boada, M., *et al.* (2007) Manganese oxidation site in *Pleurotus eryngii* versatile peroxidase: a site-directed mutagenesis, kinetic, and crystallographic study. *Biochemistry*, **46**, 66–77.

Scalbert, A., Manach, C., Morand, C., Rémésy, C. & Jiménez, L. (2005) Dietary Polyphenols and the Prevention of Diseases. *Critical Reviews in Food Science and Nutrition*, **45**, 287–306.

Scher, J.M., Zapp, J. & Becker, H. (2003) Lignan derivatives from the liverwort *Bazzania trilobata*. *Phytochemistry*, **62**, 769–777.

Schmid, J. & Amrhein, N. (1999) The shikimate pathway. In: *Plant Amino Acids: Biochemistry and Biotechnology* (ed. B.K. Singh), pp. 147–169. M. Dekker, New York.

Schulze-Siebert, D., Heineke, D., Scharf, H. & Schultz, G. (1984) Pyruvate-derived amino acids in spinach chloroplasts. *Plant Physiology*, **76**, 465–471

Serrano, J., Puupponen-Pimiä, R., Dauer, A., Aura, A.M. & Saura-Calixto, F. (2009) Tannins: current knowledge of food sources, intake, bioavailability and biological effects. *Molecular Nutrition and Food Research*, **53**(Suppl. 2), 310–329.

Seshime, Y., Juvvadi, P.R., Fujii, I. & Kitamoto, K. (2005) Genomic evidences for the existence of a phenylpropanoid metabolic pathway in *Aspergillus oryzae*. *Biochemical and Biophysical Research Communications*, **337**, 747–751.

Shoseyov, O., Shani, Z. & Levy, I. (2006) Carbohydrate binding modules: biochemical properties and novel applications. *Microbiology and Molecular Biology Reviews*, **70**, 283–295.

Simmons, B.A., Loque, D. & Blanch, W.H. (2008) Next-generation biomass feedstocks for biofuel production. *Genome Biology*, **9**, 242.1–242.6.

Singh, S.A. & Christendat, D. (2006) Structure of Arabidopsis dehydroquinate dehydratase-shikimate dehydrogenase and implications for metabolic channeling in the shikimate pathway. *Biochemistry*, **45**, 7787–7796.

Smith, A.T., Doyle, W.A., Dorlet, P. & Ivancich, A. (2009) Spectroscopic evidence for an engineered, catalytically active Trp radical that creates the unique reactivity of lignin peroxidase. *The Proceedings of the National Academy of Sciences of the United States of America*, **106**, 16084–16089.

Suzuki, S. & Umezawa, T. (2007) Biosynthesis of lignans and norlignans. *Journal of Wood Science*, **53**, 273–284.

Tazaki, K., Minamikawa, T. & Yoshida, S. (1974) Alicyclic acid metabolism in plants. 6. Metabolism of quinate in the epicotyl of pea seedlings. *The Botanical Magazine Tokyo*, **87**, 61–68.

Tzin, V., Malitsky, S., Aharoni, A. & Galili, G. (2009) Expression of a bacterial bi-functional chorismate mutase/prephenate dehydratase modulates primary and secondary metabolism associated with aromatic amino acids in *Arabidopsis*. *The Plant Journal*, **60**, 156–167.

Ulanov, A., Lygin, A., Duncan, D., Widholm, J. & Lozovaya, V. (2009) Metabolic effects of glycophosate change the capacity of maize culture to regenerate plants. *Journal of Plant Physiology*, **166**, 978–987.

Umezawa, T. (2003) Diversity in lignan biosynthesis. *Phytochemistry Reviews*, **2**, 371–390.

Vogt, T. (2010) Phenylpropanoid biosynthesis. *Molecular Plant*, **3**, 2–20.

Weber, C., Farwick, A., Benisch, F., *et al.* (2010) Trends and challenges in the microbial production of lignocellulosic bioalcohol fuels. *Applied Microbiology and Biotechnology*, **87**, 1303–1315.

Weinstein, L.H., Porter, C.A. & Laurencot, H.J. Jr. (1959) Quinic acid as a precursor in aromatic biosynthesis in the rose. *Contributions from Boyce Thompson Institute*, **20**, 121–134.

Weinstein, L.H., Porter, C.A. & Laurencot, Jr. H.J. (1961) Role of quinic acid in aromatic biosynthesis in higher plants. *Contributions from Boyce Thompson Institute*, **21**, 201–214.

Weng, J.K., Akiyama, T., Bonawitz, N.D., Li, X., Ralph, J. & Chapple, C. (2010a) Convergent evolution of syringyl lignin biosynthesis via distinct pathways in the lycophyte Selaginella and flowering plants. *The Plant Cell*, **22**, 1033–1045.

Weng, J.K. & Chapple, C. (2010) The origin and evolution of lignin biosynthesis. *New Phytologist*, **187**, 273–285.

Weng, J.K., Li, X., Stout, J. & Chapple, C. (2008) Independent origins of syringyl lignin in vascular plants. *The Proceedings of the National Academy of Sciences of the United States of America*, **105**, 7887–7892.

Weng, J.K., Mo, H. & Chapple, C. (2010b) Over-expression of F5 H in COMT-deficient Arabidopsis leads to enrichment of an unusual lignin and disruption of pollen wall formation. *The Plant Journal*, **64**, 898–911.

Xiang, L. & Moore, B.B. (2005) Biochemical characterization of a prokaryotic phenylalanine ammonia lyase. *The Journal of Bacteriology*, **187**, 4286–4289.

Yamada, T., Matsuda, F., Kasai, K., *et al.* (2008) Mutation rice gene encoding a phenylalanine biosynthetic enzyme results in accumulation of phenylalanine and tryptophan. *The Plant Cell*, **20**, 1316–1329.

Yoshida, S., Tazaki, K. & Minamikawa, T. (1975) Occurrence of shikimate and quinic acids in angiosperms. *Phytochemistry*, **14**, 195–197.

Zakzeski, J., Bruijnincx, P.C.A., Jongerius, A.L. & Weckhuysen, B.M. (2010) The catalytic valorization of lignin for the production of renewable chemicals. *Chemical Reviews*, **110**, 3552–3599.

Zhao, J. & Dixon, R. (2009) Mate transporter facilitate vacuolar uptake of epicatechin 3′-O-glucoside for proanthocyanidin biosynthesis in Medicago truncatula and Arabidopsis. *The Plant Cell*, **21**, 2323–2340.

Chapter 3
Fifty Years of Polyphenol–Protein Complexes

Ann E. Hagerman

Abstract: Reaction with protein is the hallmark characteristic of high molecular weight polyphenolics known as "tannins." Many chemical and physical approaches have been used to characterize this reaction, with a strong emphasis on the noncovalent, reversible complexes that form rapidly in aqueous solution. Broadly speaking, polyphenolics bind tightly to large, proline-rich proteins that have relatively little secondary/tertiary structure. Less is known about how polyphenolic characteristics influence the reaction with protein, although size and polarity seem to be important characteristics. In addition to summarizing five decades of work on tannin–protein interactions, this chapter suggests fruitful new avenues for research, and highlights unanswered questions. Preliminary work on characterizing the covalent complexes that form under conditions that promote spontaneous oxidation is also described.

Keywords: Tannin–protein interaction; proline-rich protein; astringency; haze formation; tannin oxidation; covalent phenol–protein adduct

3.1 Introduction

This review places current studies of polyphenol–protein complexes in the context of pioneering studies initiated 50 years ago, and provides a perspective on areas that demand additional attention in the future. The introduction links early studies to current studies of protein complexation. In subsequent sections, insoluble and soluble complexes are described with an emphasis on methodology for study of the complexes. Interactions with proline-rich proteins are described in some detail, followed by a description of mechanism and stoichiometry of binding. Studies emphasizing tannin and protein structures and conformations are summarized. Section 3.8 focuses on recent studies of covalent polyphenol–protein interactions.

Recent Advances in Polyphenol Research, Volume 3, First Edition. Edited by Véronique Cheynier, Pascale Sarni-Manchado and Stéphane Quideau.

Protein precipitation by plant polyphenols has been exploited by humans since at least the fourth century BC, when Greeks first recorded methods for tanning hides with oak galls (Van-Driel Murray, 2000). Modern understanding of this reaction was initiated by the work of Bate-Smith, who formulated a chemical definition of tannins in 1962. Bate-Smith & Swain (1962) proposed that tannins were "...water-soluble phenolic compounds having molecular weights between 500 and 3000... [with] special properties such as the ability to precipitate alkaloids, gelatin and other proteins...". The lower molecular weight limit of 0.5 kD (kilodaltons) corresponds to the smallest compounds that fulfil the protein precipitation constraint. While flavan-3-ols such as epigallocatechin (EGC, 306 D) may precipitate protein at sufficiently high concentration, the galloylated flavan-3-ols such as epigallocatechin gallate (EGCg, 458 D) readily form insoluble complexes with even highly soluble proteins such as serum albumins. The higher molecular weight limit proposed by Bate-Smith has been extended to include very high molecular weight proanthocyanidins such as the recently characterized persimmon tannin, which comprises polymers in the 10–15 kD range (Li *et al.*, 2010). Protein precipitation is the critical characteristic that distinguishes the tannins from other natural polyphenols. Precipitation of hide proteins to yield leather (Bickley, 1992) is no longer commercially important, but haze formation in beverages including teas, wine, beer, and fruit juices continues to plague food scientists (Siebert, 2009). In contrast, precipitation of alkaloids (Mejbaum *et al.*, 1959) is rarely considered in modern studies of polyphenols.

Two influential review articles summarize our early understanding of the chemistry of tannin–protein interactions. Goldstein and Swain (1965) examined precipitation of β-glucosidase by tannic acid, by measuring residual enzyme activity in the supernatant after removing the precipitated protein. They examined the effects of pH, buffer, ratio of polyphenol to protein, and recovery of active enzyme from the precipitate. They found that detergents were effective for regenerating the enzyme, and proposed that the tannin–protein complexes were stabilized by noncovalent bonds that could be disrupted. They did note less active enzymes could be recovered from wattle tannin precipitates than from tannic acid, leading to speculation about other, more stable types of bonds with protein for some tannins. Loomis and Battaile (1966) investigated interactions between polyphenols and proteins from the perspective of plant enzymology, with an interest in preventing interactions during isolation of enzymes from phenolic-rich tissues. They discovered that insoluble polyvinyl pyrrolidone (PVP) could be exploited to preferentially bind tannins even in protein-rich extracts. Like Goldstein and Swain, they explored the effects of pH, ionic strength and detergents on polyphenol–protein interactions, using a phenolic-rich extract of peppermint leaves and measuring enzyme activities in the extract to monitor protein complexation.

These two classic papers provide the basis for much of our current understanding of polyphenol–protein complexation. Tight binding of tannins to PVP and polyamides such as nylon provided these authors with good evidence for the importance of hydrogen bonding between amide and phenolic groups. The ability of detergents and organic solvents to disrupt the complexes established them as noncovalent interactions. The effects of pH and ionic strength led to speculation about specificity of interactions based on differences between tannin preparations or protein characteristics.

The desire to understand the molecular details of tannin–protein interactions has been driven by interests of the ecological and botanical communities, and by needs of the human health and food science communities. P. Feeny's influential ecological paradigm (Feeny, 1976) was in part inspired by the work of Goldstein & Swain and Loomis & Battaile. Feeny proposed a special role for tannins as "quantitative" defenses against herbivores. He noted the relationships between tannin content of oaks and caterpillar growth and did experiments with crude tannin-containing plant extracts and proteins that suggested that tannin is a nonspecific protein precipitating agent that inhibits protein utilization (Feeny & Bostock, 1968; Feeny, 1968, 1969). His data led him to speculate that tannin might be different than the "qualitative" specific toxins such as alkaloids—tannins might deter herbivores by reducing general protein digestibility when present in sufficient quantity, while toxins are effective against specific targets at minimal doses. At least two decades of research in chemical ecology, especially that focused on insect herbivory, were dominated by the idea that tannins had nonspecific effects determined only by their quantity in the diet (Bernays *et al.*, 1989). Eventually, investigation of interactions between tannins and proteins in the specialized conditions of the insect gut suggested that for insect herbivores, tannins do not affect protein digestibility (Martin *et al.*, 1985). The role of tannins in insect herbivory remains an active area of research (Forkner *et al.*, 2004), but recently attention has been focused on the oxidative potential of tannin-rich insect diets (Barbehenn *et al.*, 2006).

Although Feeny was primarily interested in insect herbivores, his ideas proved to be more useful for understanding diet choices and performance of domestic and wild mammalian ruminants (Robbins *et al.*, 1987; Mueller-Harvey, 2006). Studies of the role of tannins for mammalian herbivores have developed in close parallel to human physiological work on tannin–protein interactions because of the importance of salivary proline rich proteins (PRP) to mammals including humans (Austin *et al.*, 1989). The role of PRPs in astringency and taste has provided much of the incentive for mechanistic studies of tannin–protein interactions (Noble, 2002), while the role of PRPs in formation of undesirable hazes has been the focus of applied research (Siebert, 2009). An emerging area that relies on understanding tannin–protein interactions is the role of polyphenols in human health (Santos-Buelga & Scalbert, 2000; Evans *et al.*, 2006; Khan & Mukhtar, 2007; Dangles & Dufour, 2008). Exploiting polyphenols or their derivatives as drugs to target specific diseases (Li *et al.*, 2007) will be possible only when tannin–protein interactions are understood in detail at the molecular level.

A major constraint on early work was the limited understanding of the molecular details of the polyphenolic components. That shortcoming was addressed vigorously by Haslam, who pioneered the use of structurally well-defined polyphenols in his studies. For example, he extended Goldstein and Swain's work on β-glucosidase, but used a series of galloyl glucoses and defined procyanidin oligomers as the precipitating agents instead of tannic acid and wattle tannin (Haslam, 1974). He found that the most important feature of the polyphenol was presence of multiple phenolic sites leading to crosslinking of the protein, and that larger polyphenols were better precipitating agents than smaller molecules. Additional structure–activity series established the importance of *ortho*-diphenolic substitution, and noted that addition of excess protein or of acetone reversed the precipitation reaction

(McManus *et al.*, 1981). A model in which polyphenols formed hydrogen bond-stabilized complexes at the surface of the protein, reducing surface polarity and ultimately inducing hydrophobic aggregation, was the first attempt to describe the steps and molecular forces responsible for protein precipitation by tannins. Haslam's work highlighted the importance of polyphenol structure in the interaction with protein (Haslam, 2007). More recent authors have found that precipitation can often be related directly to polyphenolic polarity, described as the octanol–water partition coefficient (K_{ow}), (Tanaka *et al.*, 1997; Hagerman *et al.*, 1998; Kajiya *et al.*, 2001; Poncet-Legrand *et al.*, 2006; Mueller-Harvey *et al.*, 2007). Examination of other physical parameters for complex molecules in solution, such as the critical micelle concentration (CMC), might provide further insight into how polyphenol structure influences precipitation (Pianet *et al.*, 2009).

3.2 Precipitable complexes

Nephelometry (optical assessment of turbidity) is one of the oldest tools for examining haze formation in the beverage industry, and provides a direct window into precipitation reactions. The method continues to be useful to beverage manufacturers, with recent introduction of flow systems to facilitate analysis (Carvalho *et al.*, 2004). Nephelometry and light scattering generally do not provide stoichiometric or molecular detail about complex formation. However, nephelometry provides insights into conditions required to form hazes, colloids, and insoluble complexes, and is particularly suitable for kinetic studies. Kinetic analysis is an excellent tool for probing reaction mechanism, but has not been fully exploited for tannin–protein interactions. The few studies that have been published suggest that the nature of the protein (de Freitas & Mateus, 2002) and of the polyphenol (Poncet-Legrand *et al.*, 2006) influence rates of precipitation. Using nephelometry under conditions suitable for detailed kinetic analysis of the data could yield important new information about the mechanisms of interaction between tannin and protein.

Improved technologies for purifying tannins prompted numerous groups to develop new methods for examining tannin–protein interactions. Analysis of precipitates that could be isolated by centrifugation served as a proxy for directly measuring binding. Questions that were addressed included further definitions of the conditions required for precipitation by various polyphenols, and determinations of stoichiometry.

An early simple method for determining precipitated phenols takes advantage of the susceptibility of the precipitate to detergent and high pH, and the ease of selectively determining the phenol as its violet iron–phenolate complex (Hagerman & Butler, 1978). The method is robust and can be used with any polyphenol that has *ortho*–diphenolic substitution. It can be used for crude extracts or pure compounds and gives a good estimate of the amount of precipitable phenol in a sample, although unfortunately the results are sometimes misinterpreted as a measure of the amount of protein precipitated. Modifications of the method have been employed in the wine industry (Harbertson *et al.*, 2008). The protein precipitable phenolics method was used to establish the pH dependence of precipitation by tannins for a variety of globular proteins (Hagerman & Butler, 1978). There is a strong dependence for precipitation on the isoelectric point (pI) of the protein, with maximum precipitation at pH values near the pI, presumably because the isoelectric form of the protein aggregates more readily than charged forms found at other pH values. The dependence on pH is less important for proteins with relatively few ionizable amino acids such as PRP.

Tannin in the precipitated complex can also be determined following chromatographic separation of the tannin from the protein and complex. HPLC was exploited by Kawamoto and coworkers in a series of papers that established details of the interaction between galloyl glucoses and serum albumin (Kawamoto *et al.*, 1995; Kawamoto *et al.*, 1996; Kawamoto *et al.*, 1997; Kawamoto & Nakatsubo, 1997a; Kawamoto & Nakatsubo, 1997b). Like Haslam, this group concluded that the reaction involved two stages, binding followed by aggregation to form the precipitate. The serum albumin appeared to be irreversibly denatured by interaction with galloyl glucoses, with limited solubility even after the removal of all the tannin. This early indication of substantial conformational changes induced by interaction with polyphenols has recently been confirmed for casein–EGCg interactions using more sophisticated physical techniques (Jobstl *et al.*, 2006).

For purified gallotannins ranging from tri- to pentagalloyl glucose (PGG, 1,2,3,4,6-penta-O-galloyl-β-D-glucopyranose), a higher degree of galloylation led to more efficient precipitation (Kawamoto *et al.*, 1995), and a similar trend was noted for the larger galloyl esters found in tannic acid (Verzele *et al.*, 1986). The unique composition of each preparation of tannic acid tested in the precipitation assays was reflected in the unique precipitation profiles obtained, although the general trend was towards more effective precipitation as degree of galloylation increased from tetra- to dodecagalloyl glucose. HPLC methods similar to those used by Kawamoto have been of limited utility for studies of proanthocyanidin (PA)–protein complexes, because of the difficulties of resolving high oligomers and polymers of PAs. The approach was improved by employing thiolytic degradation of the precipitated PA, followed by HPLC determination (Sarni-Manchado *et al.*, 1999). For the oligomers that could be tested, PA precipitability increased with increased degree of polymerization (da Silva *et al.*, 1991; Sarni-Manchado *et al.*, 1999).

Quantitation of polyphenol-precipitated protein has proved to be difficult in part because many standard methods for determining protein are susceptible to interference by polyphenols (A_{280}, Lowry assay, Bradford assay, bicinchoninic acid assay). Some investigators have assumed that enzymes are inactivated by complexation by polyphenols, and that protein binding can be indirectly determined by assessing enzyme activity. However, at least some enzymes' interaction with polyphenol does not inhibit activity (Juntheikki & Julkunen-Tiitto, 2000), making it difficult to use activity to monitor interaction. Furthermore, constraints on the conditions for measuring enzyme activity such as optimal pH for activity limit the application of enzyme-based approaches. Ninhydrin has been used to determine precipitated protein (Makkar *et al.*, 1987) but the method is inconvenient.

The most effective methods for determining protein in the presence of tannin rely on labeled protein. Early methods used proteins such as haemoglobin, with an intrinsic label that can be determined without interference from the polyphenol (Bate-Smith, 1973; Porter & Woodruffe, 1984), but better analytical results were obtained with extrinsically labeled protein. Radiolabeling with ^{125}I provides protein that is not significantly perturbed and can be detected with high sensitivity (Hagerman & Butler, 1980a). Radiolabeling with ^{14}C or with blue dye (Asquith & Butler, 1985; Asquith & Butler, 1986) has been explored, but neither method is as useful as the iodine labeling. Although, in principle, any protein can be labeled, most authors have focused on bovine serum albumin. Labeling other globular proteins, especially those with well-understood structural features, would lead to new insights into the role of the protein in the precipitation reaction. Using ^{125}I-labeled bovine serum albumin, Hagerman examined precipitation by two purified tannins, pentagalloyl glucose (940 D) and a high molecular weight procyanidin (5 kD) from *Sorghum* grain (*Sorghum* PC) (Hagerman *et al.*, 1998). Both polyphenols were efficient protein precipitating agents, but the procyanidin was about ten times more active on a molar basis. Stoichiometry of the saturated insoluble complexes was established by using HPLC or iron reagent to determine the phenolic component and the radiolabel to determine protein. At saturation, the protein (66 kD) bound about 40 moles of pentagalloyl glucose or about 20 moles of procyanidin per mole of protein. Precipitation by the pentagalloyl glucose was temperature sensitive and disrupted by traces of organic solvent, suggesting a hydrophobic component to the precipitation reaction that was not detected for the procyanidin. This study, along with other direct comparisons of structurally defined tannins (Frazier *et al.*, 2010), clearly established that most features of the interaction between tannin and protein are tannin-specific, with precipitation efficiency, binding stoichiometry, and basis for complexation all unique for a given polyphenol. The widely employed structural categories for tannins (proanthocyanidin, ellagitannin, gallotannin) are not useful for predicting precipitability, which is a manifestation of the unique molecular characteristic of a given polyphenol including its polarity, molecular weight, degree of galloylation and flexibility (Soares *et al.*, 2007).

3.3 Soluble complexes

Polyphenol–protein interactions do not always result in precipitation, and more recent studies have often focused on soluble complexes. Manipulation of protein–to–tannin ratios in stoichiometric studies provided some early insights into formation of soluble complexes, with precipitation dominating when phenol is in excess and soluble complexes when protein is in excess (Hagerman & Robbins, 1987; Luck *et al.*, 1994). A model was developed to describe the conditions that drove formation of soluble, rather than insoluble, complexes (Silber *et al.*, 1998). It has been a significant experimental challenge to devise ways to distinguish complexes from uncomplexed protein or tannin, so initially our understanding of soluble complexes lagged behind that of the precipitable complexes. However, soluble complexes can be probed with many of the physical tools used in structural biology, yielding detailed molecular descriptions of the complexes.

Equilibrium dialysis is a classic method for probing small molecule–protein interactions, and some early attempts were made to use the method for polyphenols. Gallotannin and ellagitannin interactions with serum albumin were evaluated at low pH (McManus *et al.*, 1985), and it was noted that the relatively inflexible ellagitannins had lower affinity for serum albumin than the more flexible galloyl esters. Because the dialysis membrane can bind substantial amounts of the polyphenol and thus bias the results (Feldman *et al.*, 1999), the method has been largely abandoned.

An early quantitative study of soluble complexes employed ^{125}I-labeled protein as a tracer in competitive binding assays (Hagerman & Butler, 1981). Although the analytical method did rely on the precipitability of the tannin–tracer complex, interaction with the competing protein was detected even if it formed a soluble complex. The method is analogous to competitive immunoassays such as radioimmunoassay, with the tracer (^{125}I-bovine serum albumin) and the precipitating agent (tannin) present in fixed amount in a pH 4.9 buffer, and the competing protein or peptide added in variable amounts. A critical control for the system is the demonstration that unlabeled tracer protein and the tracer have the same affinity for the binding agent. The amount of competitor required to inhibit tracer precipitation by 50% is the relative affinity. Because the reactions involve both solid and liquid phases, true affinity constants cannot be obtained, and results are expressed as relative affinity. Some variations on the original method have been proposed, such as using an enzyme as the tracer instead of the labeled serum albumin (Fickel *et al.*, 1999). Attempts to develop a method to compare different tannins by direct competition have been less successful (Bacon & Rhodes, 1998), because effective methods to label the polyphenol have not been developed.

The relative affinities of a range of globular proteins and polyamino acids for *Sorghum* PC were established with the competitive binding assay (Hagerman & Butler, 1981). In general, the procyanidin had higher affinity for proteins that were larger and that had relatively open structures. For example, bovine serum albumin (66 kD) had higher affinity than either lysozyme (14.4 kD) or ribonuclease A (13.7 kD). The latter are not only small, but are also tightly coiled, compact proteins compared to serum albumin, which has a relatively open structure (Sugio *et al.*, 1999). Highest affinities were noted for synthetic polyproline and gelatin, a PRP manufactured by partial digestion of collagen. It was proposed that proline enhanced the affinity of proteins for tannin by promoting random coil structures and by forming very favorable hydrogen bonds through the tertiary amide of the proline–peptide bond. A search for natural proline-rich proteins showed that two major classes were the seed proteins known as prolamines and mammalian salivary PRPs.

High affinity for PRPs appears to be a common characteristic of many tannins. Using competitive binding assays, a wide variety of polyphenols, including several different PAs, gallotannins including PGG, and ellagitannins all had very high relative affinity for PRPs (Asquith & Butler, 1986; Hagerman & Klucher, 1986; Hofmann *et al.*, 2006).

If we consider the stoichiometric studies and the affinity studies that have been conducted with a range of polyphenols and proteins, an important general principle is clearly revealed. The ability to precipitate protein and affinity for protein are two distinct manifestations of the polyphenol–protein interaction. While PGG precipitates protein very inefficiently compared to *Sorghum* PC (Hagerman *et al.*, 1998), both PGG and the procyanidin have high affinity for PRP and a lower affinity for serum albumin (Hofmann *et al.*, 2006).

Protein size, proline content, and secondary structure are important traits that govern affinity for tannin, but it is not clear what other structural features of proteins control affinity. Relatively little work has been done to pursue the seed prolamines (Hagerman & Butler, 1980b), in part because of their poor solubility, so they offer a new class of proteins for future investigations. Several reports of high affinity by a group of salivary proteins known as histatins (Yan & Bennick, 1995), which are not proline rich, suggest a second class of proteins that could be more fully explored. Our understanding of protein structure has far outpaced our examination of affinity of proteins for tannin. An expanded study of globular proteins that takes advantage of protein structure classifications such as that provided by CATH (http://www.cathdb.info/) or other data bases could yield many new insights into factors that influence the interaction between polyphenols and proteins.

3.4 Proline-rich proteins

The discovery that tannins have very high affinity for PRP like those found in mammalian saliva (Hagerman & Butler, 1981) has influenced subsequent studies of tannins from all perspectives. The potential role of salivary PRP as a defence mechanism that allows ingestion of dietary tannins (Hagerman & Robbins, 1993) inspired ecological studies of mammalian herbivores from ruminants (Mueller-Harvey, 2006) to primates (Mau *et al.*, 2009). In general, interaction of dietary tannins with salivary PRPs appears to improve nitrogen balance in herbivores. PRPs, which comprise primarily nonessential amino acids, are expended to protect valuable dietary nitrogen from polyphenols (Mehansho *et al.*, 1987). In some rodents, salivary PRP production is induced by dietary tannin (Mehansho *et al.*, 1983), suggesting a role for tannin in controlling gene expression. Additional studies are needed to elucidate the mechanism of induction and to evaluate whether dietary polyphenols regulate the expression of other proteins, perhaps providing clues to their role in health and disease prevention (Dangles & Dufour, 2008).

In humans, salivary PRPs are a complex family of proteins that are phenotypically variable among individuals (Bennick, 2002). The salivary PRPs include acidic, basic and glycosylated subgroups of proteins, with highest affinity for tannins associated with the basic proteins (Lu & Bennick, 1998). Although the acidic and glycosylated PRPs have important roles in dental health, the basic salivary PRPs have no role outside their ability to bind polyphenols. It is widely believed that the interaction between dietary polyphenols and basic salivary PRPs plays a role in the taste sensation known as astringency (Bajec & Pickering, 2008). Taste perception is complex, and explicit mechanistic links between taste and polyphenol–salivary PRP interactions have not been easy to formulate (Hofmann *et al.*, 2006; Obreque-Slier *et al.*, 2010).

In addition to improving protein nutrition and mediating taste, salivary PRPs can influence the bioavailability and bioactivity of dietary polyphenols. The polyphenol–salivary PRP complex is robust and stable to digestion, as indicated by the high levels of PRP and PGG excreted in the feces of rats that were consuming a PGG-supplemented diet (Skopec *et al.*, 2004). Polyphenol–PRP complexes have limited *in vitro* digestibility (Lu & Bennick, 1998) and PRPs inhibit uptake of PGG in the Caco cell model (Cai *et al.*, 2006). Complexing with the model PRP gelatin changed the kinetics of radical quenching and antioxidant lifetime of polyphenols (Riedl & Hagerman, 2001; Riedl *et al.*, 2002). As interest in the health-related activities of polyphenols continues to grow, it will be essential to undertake further investigation of whether salivary PRPs limit the utility of oral routes for administration of beneficial polyphenols or alter the *in vivo* activities of the compounds.

Proline-rich regions of proteins have broad biological importance, serving as molecular switches or as binding sites. As a consequence, proline-rich sequences are found in key proteins associated with biochemical functions ranging from well-understood molecular processes such as G-coupled receptors to poorly understood pathologies such as prion diseases (Reiersen & Rees, 2001). There is currently little evidence to link polyphenols to PRP-regulated intercellular processes. However, it has been proposed that amyloid fibril formation may be inhibited by polyphenols via interactions with PRP (Carver *et al.*, 2010).

3.5 Mechanisms of binding

PRPs have unique structural characteristics, forming either random coils or, if there are more than four adjacent proline residues, forming the polyproline helix (Reiersen & Rees, 2001). Early studies of polyphenol–PRP interactions focused on the strong hydrogen bonding between the proline-containing peptide amide backbone and phenolic hydroxyls. High proline content promotes hydrogen bonding by both imposing secondary structures that expose the peptide backbone, and by serving as a very strong hydrogen bond via the tertiary amide bonds (Hagerman, 1992). In numerous subsequent studies, a wide range of physical methods have been employed in an attempt to further elucidate the interaction on the molecular level.

NMR was first employed by Haslam's group in the early 1990s to examine the interaction between PGG and a 20 amino acid peptide derived from mouse salivary PRP (Murray *et al.*, 1994). In this and many subsequent studies, haze formation significantly interfered with data collection, so mixed solvents, for example 10% DMSO (dimethyl sulfoxide), were employed. The predicted hydrogen bonding interactions between the random coil peptide and the polyphenol were unexpectedly augmented by hydrophobic interactions that involved stacking of polyphenolic aromatic rings with the proline pyrrolidone ring. These observations were developed into a model for polyphenol–PRP interactions that emphasized hydrophobic forces and downplayed the importance of hydrogen bonding (Luck *et al.*, 1994). The model was further refined after examining the 2D NMR spectra of the same salivary peptide with a procyanidin dimer and several phenolic monomers (Baxter *et al.*, 1997). Although the data was all collected with a small peptide and low molecular weight phenols, the authors proposed that aromatic stacking on proline and other surface nonpolar residues was a general principle governing tannin–protein interactions, and this idea has been widely promulgated in many subsequent studies by many researchers.

There are a few studies that challenge the dogma that proline–phenol stacking interactions are a general feature of polyphenol–protein interactions. However, in a detailed examination of the interaction between the 14 amino acid salivary PRP, IB7$_{14}$, and the procyanidin dimer B3 (catechin-(4α→8)-catechin), NMR data were obtained that suggested a hydrogen bond-stabilized complex with no proline–aromatic stacking interactions (Simon *et al.*, 2003). Supporting evidence for the role of hydrogen bonds was provided by ESI–MS. The authors argued that the DMSO that was routinely added to minimize haze in earlier NMR studies could have altered the solvent to favor hydrophobic interactions that are less important in aqueous systems. Similar evidence for a predominant role for hydrogen bonds was recently provided by ESI–MS analysis of the complex between the larger salivary PRP,

IB5, and EGCg (Canon *et al.*, 2009). Additional studies are needed to clearly establish how conditions such as solvent, temperature, and reactant concentrations determine the relative importance of hydrogen bonding and hydrophobic and stacking interactions for a wide range of proteins and tannins. Different models may need to be developed to describe specific conditions and combinations of tannin and protein.

There have been relatively few investigations of phenols interacting with peptides that are not derived from salivary PRP. Bradykinin, a nine amino acid peptide containing three prolines, appears to interact with PGG via hydrophobic stacking interactions and hydrogen bonds (Verge *et al.*, 2002a). Neurotensin, a 30 amino acid peptide that is not proline-rich, interacts with PGG mainly at the more hydrophobic C-terminal region of the peptide (Richard *et al.*, 2005). Hydrophobic interactions between PGG and nonpolar amino acids isoleucine and leucine were postulated rather than the typical proline–aromatic ring stacking interactions. Larger proteins have not been explored because of the limitations of NMR, but recent developments in NMR may provide tools needed to further probe polyphenol–protein interactions. For example, improved pulse sequences (Balayssac *et al.*, 2009) may allow interactions to be examined at lower concentrations, eliminating some problems with haze and self-association of polyphenols. Solid state NMR may provide tools to analyze insoluble complexes in more detail, as well as providing tools for examining membrane-bound proteins that are inaccessible to solution chemistry methods (Yu *et al.*, 2011). High field NMR (Markley *et al.*, 2009) could be employed to examine full-sized proteins instead of peptides, and to examine interactions with globular proteins to complement existing data on the highly repetitive, limited amino acid PRPs.

Calorimetry is an attractive method for examining binding interactions, although data interpretation is subject to a number of artifacts (Poncet-Legrand *et al.*, 2007). In a careful study using isothermal titration calorimetry (ITC), interactions between polyproline and several different polyphenols were classified based on the dominance of hydrogen bonding (smaller phenols) or hydrophobic interactions (larger phenols) (Poncet-Legrand *et al.*, 2007). ITC can be applied to larger proteins and globular proteins, illustrated by several studies of polyphenol interactions with serum albumin or gelatin (Frazier *et al.*, 2003; Frazier *et al.*, 2010). Because several of the polyphenol preparations used in those studies were mixtures, rather than purified compounds, it was difficult to develop detailed molecular models, but the studies did establish that binding interactions for globular proteins are distinct from those for an unstructured PRP such as gelatin. The authors argued that there was a trend for less flexible polyphenols to bind more tightly to the more flexible protein, gelatin, while the more flexible polyphenols had similar affinities for gelatin and for more highly structured proteins like serum albumin.

3.6 Stoichiometry of binding

Precipitation assays suggested that the stoichiometry of binding varied with tannin and with protein (Hagerman & Carlson, 1998), but establishing the stoichiometry of binding for soluble complexes has been difficult. Cautious use of ITC established a stoichometry of 1 phenolic group per $\sim$10 proline residues for polyproline interactions with small tannins such as EGCg or PA tetramers (Poncet-Legrand *et al.*, 2007). Similar studies could be done with other proteins, although an advantage of polyproline for the analysis was its rigid structure, which minimized artifactual signals due to conformational changes.

Size exclusion chromatography was used to establish the sizes of polyphenol–bovine serum albumin complexes, but the complexes that were obtained were multimers of protein stabilized by polyphenols, not individual polyphenol–protein complexes (Hatano *et al.*, 2003). SDS–PAGE is not useful for noncovalent complexes since the SDS dissociates the polyphenol–protein complex.

Analytical ultracentrifugation is a well-established physical method for examining proteins and protein adducts that are stable to the centrifugal forces. In a successful application of analytical ultracentrifugation, 1:1 complexes of histatin with PGG were identified, but the very small size of the histatin facilitated the analysis by ensuring that there was a significant change in size associated with binding (Wroblewski *et al.*, 2001). When applied to a mixtures of dephosphorylated casein, a model PRP, and EGCg, analytical ultracentrifugation revealed changes in the hydrodynamic parameters of the protein but did not provide stoichiometries of binding (Jobstl *et al.*, 2004). EGCg complexed with the protein to form dimers, aggregates and very high molecular weight species with unknown amounts of polyphenol incorporated into the complexes.

Mass spectrometry has proved to be the most useful technique for obtaining stoichiometries. Surprisingly, adducts of small tannins and peptides are stable to electrospray ionization mass spectrometry (ESI–MS) (Sarni-Manchado & Cheynier, 2002). An 11 amino acid PRP formed 1:1 complexes with EGCg and PA dimers, as did bradykinin (Verge *et al.*, 2002b). These stoichiometries are similar to those obtained in aqueous solution with polyproline using ITC (Poncet-Legrand *et al.*, 2007). A similar ratio was obtained using ESI–MS to examine interactions between the 65 amino acid salivary PRP IB5 and EGCg, which yielded variable stoichiometry adducts containing up to five EGCg per peptide (Canon *et al.*, 2009). A disadvantage of ESI–MS is that the sample must enter the gas phase, so large proteins are inaccessible.

Matrix assisted laser desorption ionization-time of flight mass spectrometry (MALDI–TOF MS) uses a soft ionization technique and provides structural information for biomolecules including proteins (Canas *et al.*, 2006) and intact tannins, particularly PA (Monagas *et al.*, 2010). Complexes of serum albumin with up to four PGG per protein (66 kD) were detected with MALDI–TOF MS (Chen & Hagerman, 2004a). It was noted that although precipitation of serum albumin by PGG is inhibited by acetone, the soluble complexes form in the presence of acetone, suggesting that the two-step model involving phenol binding followed by aggregation does accurately describe the interaction of PGG with this globular protein. In a more comprehensive study of tannin–protein complexes by MALDI–TOF MS, several proteins were examined (Mane *et al.*, 2007). Myoglobin and β-lactoglobulin did not form complexes that could be detected, but both casein and bovine serum albumin complexes could be analyzed by MALDI–TOF MS. For serum albumin, the stoichiometry of the tannin–protein complex increased as the amount of tannin in the reaction mixture was increased, with values up to five moles of tannin per mole of protein when tannin was in a tenfold excess over protein. The method worked well with several PA fractions and with a mixed gallotannin. MALDI–TOF MS offers good potential as a method to further evaluate tannin:protein stoichiometries for both PRP and globular proteins, and a variety of tannins.

3.7 Protein conformation

It is well documented that protein secondary structure can be changed significantly when small molecules bind to a native protein (Ascenzi & Fasano, 2010). For polyphenol–protein interactions, the nature of the protein appears to dictate whether structural changes take place or not. The characteristic helical structure of synthetic polyproline peptides does not change when phenolics bind (Poncet-Legrand *et al.*, 2007). Although small peptides derived from salivary PRPs may be random coil rather than helical, these peptides also do not undergo structural change when polyphenols bind (Murray *et al.*, 1994; Simon *et al.*, 2003). However, bradykinin undergoes a large conformational change when procyanidin dimer B3 binds (Richard *et al.*, 2001). Larger PRPs such as the 65-amino acid IB5 may "wrap around" the polyphenol in a cooperative mode of binding (Charlton *et al.*, 1996) with an accompanying increase in peptide order (Pascal *et al.*, 2007). Dephosphorylated casein (23 kD), a protein with extended structure and comprising 15% proline, becomes more compact when EGCg binds the protein in the first step of a three-stage binding and aggregation process (Jobstl *et al.*, 2004). Atomic force microscopy and single angle X-ray scattering (SAXS) confirmed that casein "shrinks" as it binds EGCg (Jobstl *et al.*, 2006; Zanchi *et al.*, 2008b). A general model in which tannin site specifically interacts with PRPs, causing the protein to wrap around the polyphenol, has been proposed (Zanchi *et al.*, 2008a).

Globular proteins, which have a high degree of secondary and tertiary structure, may be irreversibly denatured by reaction with polyphenols under conditions that lead to haze formation or precipitation (Kawamoto *et al.*, 1995). However, if the reaction with polyphenol is controlled so that crosslinking and precipitation does not take place, the protein may not change conformation. For example, EGCg interactions with human serum albumin were probed with circular dichroism, fluorescence and FTIR spectroscopy (Maiti *et al.*, 2006). The fluorescence data suggested that the initial interactions were hydrophobic, with involvement of nonpolar residues such as tryptophan, but that hydrogen bonding stabilized the complex at more polar residues. The protein did not change structure as the EGCg "docked" to a single binding site between domains IIa and IIIa. A similar study with ribonuclease A and tea catechins including EGCg suggested that the phenolics bound at a specific docking site, and that the phenolic binding altered the solvent accessible surface areas of some residues (Ghosh *et al.*, 2007). The site-specific interactions between proteins and tannins are a nascent area that should receive more attention as evidence accumulates for specific pharmacological effects of polyphenols (Zhang *et al.*, 2009).

Understanding the parameters that control site-specific binding by polyphenols requires information about polyphenol conformation and flexibility. Structural studies on PAs have lagged behind structural studies on proteins, at least in part because lack of crystals has delayed acquisition of X-ray structures. Recently, NMR methods have been used to establish solution structures for procyanidin dimers and trimers (Pianet *et al.*, 2009). Initial studies indicate that the more extended procyanidin dimers bind more tightly to small salivary PRPs (Cala *et al.*, 2010). This conclusion complemented an earlier study with bradykinin, in which more flexible polyphenols had higher affinity for the peptide (Richard *et al.*, 2006).

In contrast to PAs, molecular modelling has provided tertiary structures for several ellagitannins (Quideau *et al.*, 2004). Recent application of surface plasmon resonance to evaluate binding of several ellagitannins to purified topoisomerase suggested that less flexible tannins bind to globular proteins (Douat-Casassus *et al.*, 2009). The authors propose a model in which flexible tannins have higher affinities for flexible proteins such as PRPs, while constrained tannins have higher affinities for tightly coiled globular proteins.

3.8 Covalent tannin–protein complexes

Although noncovalent, reversible tannin–protein complexes have been the subjects of extensive studies, there are several rationales for extending our work to include covalently stabilized complexes. At neutral to alkaline pH, phenols readily oxidize to form radicals and quinones, and these species may react with protein to form stable adducts. Even at low pH, phenols may spontaneously form radicals, or may react with biological free radicals or reactive oxygen species to form phenolic radicals that may ultimately be quenched by reaction with protein. Experimentally, generation of covalently stabilized polyphenol–protein complexes would provide a convenient reagent for examining the biological activity of tannin–protein complexes.

In early studies, Mason and Peterson established that quinones reacted with side-chains of lysine (amine) and cysteine (thiol), but not with other reactive amino acids such as histidine, arginine, tryptophan or the amide functional group (Mason & Peterson, 1965). Subsequently, W.S. Pierpoint examined reactions of *ortho*-quinones derived from plant phenols, and established that they reacted quickly with lysine and/or cysteine (Pierpoint, 1969a). Reactions of phenols with serum albumin under oxidizing conditions were complex, with cysteine reacting with the quinone when the protein was in excess and lysine reacting when the quinone was in excess. This clearly established lysine and cysteine as the residues most likely to react with phenols under conditions promoting quinone formation (Pierpoint, 1969b).

More recent characterizations of the products of oxidative reaction between proteins and small phenolics such as dopamine (Kerwin, 1997) or chlorogenic acid (Prigent *et al.*, 2007) have confirmed and extended Pierpoint's early conclusions. *N*-Acetyl dopamine can be oxidized to a mixture of quinones and quinone methides, which crosslink in the absence of protein to form polymers (Kerwin, 1997). In the presence of a ten amino acid peptide, products that could be characterized by ESI–MS were obtained. The model peptide did not contain lysine or cysteine, but other amino acids including the *N*-terminal aspartate, isoleucine and valine reacted with the quinone. This work suggests that salivary PRPs could form covalent complexes with polyphenols, despite the lack of cysteine and lysine in these proteins (Bennick, 1982). Chlorogenic acid reacted with the proteins β-lactalbumin, lysozyme, or bovine serum albumin under oxidizing conditions (Prigent *et al.*, 2007). MALDI–TOF MS analysis of the products showed that the phenolic dimers reacted with the proteins at lysine residues to yield products with altered solubility and surfactant properties. Similar oxidative crosslinking reactions that yield the marine mussel adhesive take place at dihydroxyphenylalanine residues in the protein itself (Wilker, 2010).

There have been relatively few studies of covalent reactions between tannins and proteins in part because the highly cross-linked adducts that are usually produced are insoluble and difficult to characterize. Tannin–protein complexation studies are generally carried out at slightly acidic to neutral pH to prevent spontaneous oxidation of the tannin and to inhibit formation of covalent complexes. Surprisingly, even under slightly acidic conditions, some phlorotannins spontaneously form covalent adducts with bovine serum albumin (Stern *et al.*, 1996). Phlorotannins are found only in marine brown algae, and are related to phloroglucinol (1,3,5-trihydroxybenzene). Although *ortho*-diphenolics are generally thought to be more susceptible to spontaneous oxidation than phenolics with *meta*-orientation, phlorotannins are far more likely to spontaneously form oxidation products than either PAs or hydrolysable tannins. Further examination of these very interesting phenolic natural products can be justified based both on their novel chemistry, and on their potential to serve as selective protein-binding agents with pharmacological activity (Jung *et al.*, 2010).

Gallotannins appear to form covalent complexes with protein only when an external oxidizing agent is used to oxidize the phenol. Potassium periodate is a convenient oxidizing agent, because it readily oxidizes phenolics but not protein (Fatiadi, 1974). The stoichiometry of complexes formed by periodate oxidation was evaluated using double labelling studies with ^{125}I-labeled bovine serum albumin and ^{14}C-labeled PGG (Chen & Hagerman, 2004b). By using an excess of protein, the reaction could be controlled to form stable soluble complexes, while excess polyphenol converted the products to recalcitrant precipitates (Chen & Hagerman, 2004b). Manipulating reaction conditions such as pH led to a proposed reaction mechanism in which quinone oxidation products of the polyphenol reacted with the protein (Chen & Hagerman, 2005). At low pH and in the absence of protein, PGG formed radical intermediates that polymerized to an insoluble phenolic material, but protein drove the reaction towards quinone intermediates and protein adducts even at low pH. Similar reaction products were obtained with periodate oxidation in the presence of protein if EGCg or *Sorghum* PC were substituted for the gallotannin.

In the absence of an external oxidizing agent, PGG reacts with protein to form unstable, noncovalent complexes (Chen & Hagerman, 2004a; Kusuda *et al.*, 2006). However, EGCg is highly susceptible to spontaneous oxidation (Hagerman *et al.*, 2003). At neutral to slightly basic pH, EGCg spontaneously reacts with serum albumin to form covalently stabilized complexes that are resistant to dissociation by SDS (Kusuda *et al.*, 2006). A similar reaction takes place if EGCg is incubated with the enzyme glyceraldehyde dehydrogenase (GAPDH) at pH 7.4 (Ishii *et al.*, 2008). That reaction could be inhibited by addition of dithiothreitol, suggesting reaction at cysteine to form an *S*-quinonyl moiety. Model reactions with peptides confirmed that EGCg reacted spontaneously with cysteine at slightly alkaline pH to give characteristic ESI–MS products. The radical-mediated reaction took place on the B-ring of the flavan-3-ol ester, consistent with the relative tendency of the B- and D-rings to oxidize (Hagerman *et al.*, 2003).

Under mild conditions at slightly alkaline pH, EGCg reacts spontaneously with bovine serum albumin to form a mixture of soluble and insoluble products (Kusuda *et al.*, 2006;

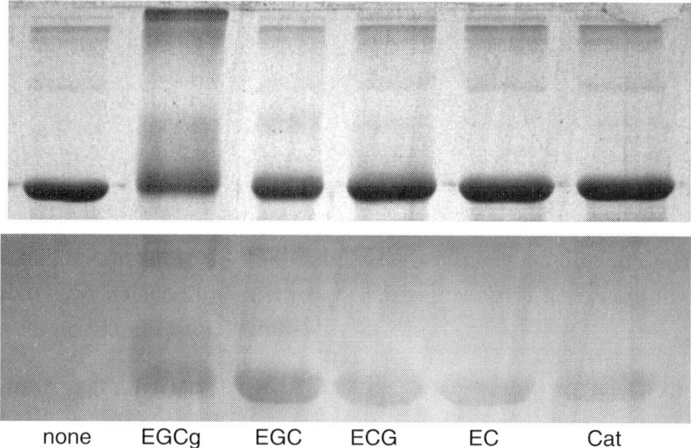

Fig. 3.1 Coomassie-stained SDS–PAGE (top panel) and NBT-stained blot (bottom panel) for bovine serum albumin with no additions (none) and the reaction products between the protein and EGCg, EGC, epicatechin gallate (ECG), epicatechin (EC), and catechin (Cat).

Trombley & Hagerman, unpublished). The soluble products are not dissociated by SDS, suggesting that they are stabilized by covalent bonds. On SDS–PAGE, the soluble EGCg reaction products include a major product with slightly lower mobility than native serum albumin, a poorly resolved set of lower mobility products, and some high molecular weight material that does not enter the gel (Plate 3.1). All the products react with nitroblue tetrazolium (NBT) after blotting to nitrocellulose, indicating that the protein has been modified by quinones (Paz *et al.*, 1991). EGC yields a similar NBT-reactive product, but very little of the high molecular weight product, suggesting covalent reaction but little crosslinking (Fig. 3.1). Epicatechin gallate, epicatechin and catechin barely react with serum albumin under these conditions (Fig. 3.1). These observations suggest that the reaction requires the trihydroxylated B-ring to form covalent adducts, and the gallate ester D-ring to form highly cross-linked reaction products.

When EGCg is added in excess over protein, the products are brown in color, and about 10 phenolics are bound per protein based on MALDI–TOF MS analysis (Fig. 3.2). The reaction products generate broad mass peaks, suggestive of heterogeneous products, with an average *m/z* about 5 kD larger than the native protein (71 kD vs. 66 kD). Although the reaction products are soluble, they could not be resolved on RP-HPLC, so we developed capillary electrophoresis methods for separating serum albumin, EGCg oxidation products, and the EGCg–serum albumin adduct (Fig. 3.3) (Trombley *et al.*, 2011). We were able to resolve several populations of products. We believed that the product with shortest elution time was partially modified protein, and the product with longer elution time was fully modified protein. Increasing the polyphenol:protein ratio in the reaction mixture yielded a larger amount of the highly modified protein (Fig. 3.4). Production of oxidized forms of EGCg (Sang *et al.*, 2007) was minimized by using only slight molar excesses of EGCg over protein. Oxidation products could be removed by DEAE (diethyl aminoethyl) ion exchange chromatography (Trombley *et al.*, 2011).

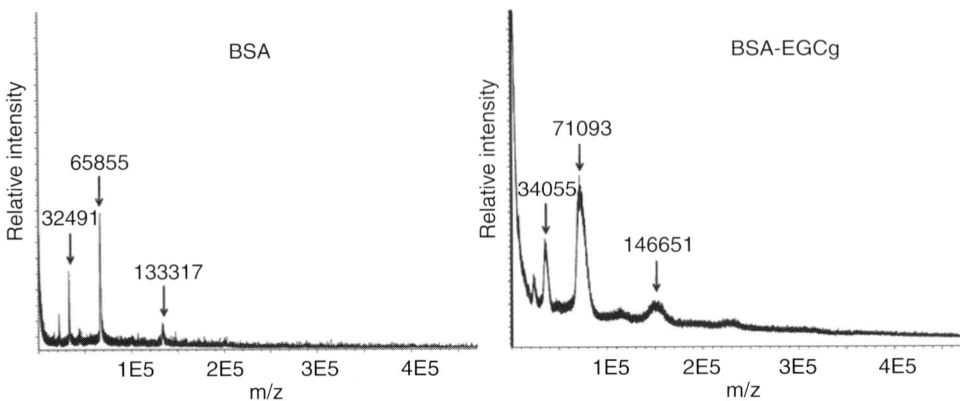

Fig. 3.2   MALDI–TOF mass spectroscopic analysis of bovine serum albumin (left panel) and the reaction product between the protein and EGCg (right panel).

The pH dependence of the reaction (Fig. 3.5) suggests a role for the EGCg anion (pKa 7.8) (Jovanovic *et al.*, 1995) rather than the EGCg semiquinone radical (pKa 4.8) (Hagerman *et al.*, 2003). Presumably, the deprotonated EGCg oxidizes to the quinone, which covalently reacts with the protein at cysteine or lysine residues (Pierpoint, 1969b). EGCg appeared to react specifically with the active site cysteine on GAPDH (Ishii *et al.*, 2008). Native bovine serum albumin (66 kD, 581 amino acids) has only one free thiol (Peters, 1975), but reacts with at least ten EGCg per protein (Fig. 3.2), so we attempted to identify additional sites of reaction. EGCg does not reduce disulfides such as glutathione (Trombley & Hagerman, unpublished), so is unlikely to have reduced any of the 17 disulfides in serum albumin to yield additional thiol groups for covalent reaction. About 80% of the 59 lysines in serum albumin are on the surface of the protein, making lysine a good candidate for reaction to form *N*-quinonyls or Schiff's bases with EGCg (Bittner, 2006). The isoelectric point of EGCg-modified serum albumin is more acidic than that of native serum albumin, supporting the hypothesis that lysines have been modified. Acetylating the surface lysines of serum albumin reduced reactivity with EGCg, but did not completely prevent the reaction, suggesting that some EGCg may penetrate the protein and react with internal residues, or some EGCg may react with residues other than lysine such as histidine (Bittner, 2006).

We treated the EGCg–protein adduct with cyanogen bromide in order to identify modified residues by MALDI–TOF MS analysis of the peptides. Unfortunately, the products of the reaction aggregated in the sample matrix and did not yield signals. Electrophoretic analysis of the peptides indicated that the major (41 kD) cyanogen bromide product contained EGCg, as did several of the other smaller peptides. Efforts to obtain useful MALDI–TOF MS spectra are ongoing.

We attempted to establish whether EGCg formed a Schiff's base or an *N*-quinonyl with primary amines using 1,6-hexane diamine as a model amine. Under our standard reaction conditions (pH 7.0 phosphate buffered saline, 37°C, 24 h), EGCg formed a recalcitrant precipitate with the diamine. We used Magic Angle Spinning solid state ^{13}C NMR to analyze the insoluble product, obtaining a poorly resolved spectrum with peak broadening in the

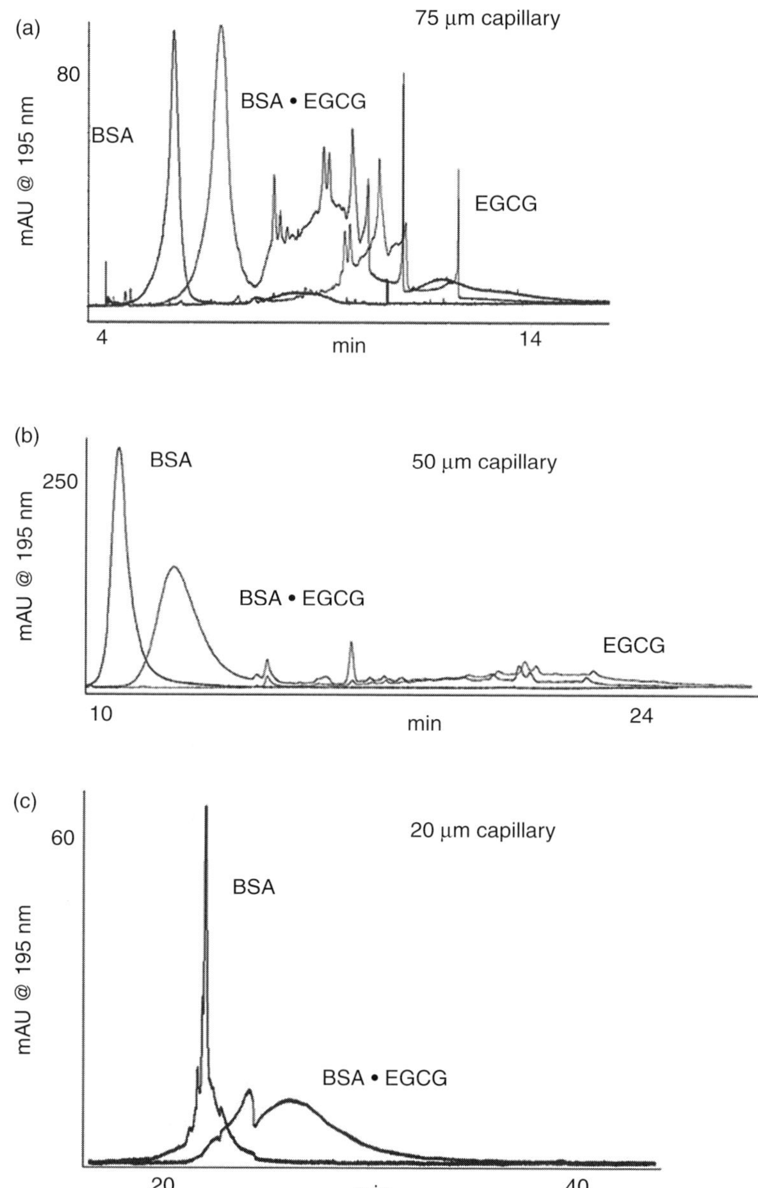

Fig. 3.3 Capillary zone electrophoresis separation methods for EGCg–bovine serum albumin reaction products. (a) Method optimized to detect EGCg oxidation products; (b) Method optimized to resolve protein and EGCg–protein adduct; (c) Method optimized to resolve heterogeneous reaction products.

methylene signals that was consistent with reaction between the amine and the polyphenolic, but not useful for structure determination (Fig. 3.6). Attempts to obtain a soluble product for solution phase NMR using different reaction conditions, or epigallocatechin, or other small amines, were unsuccessful.

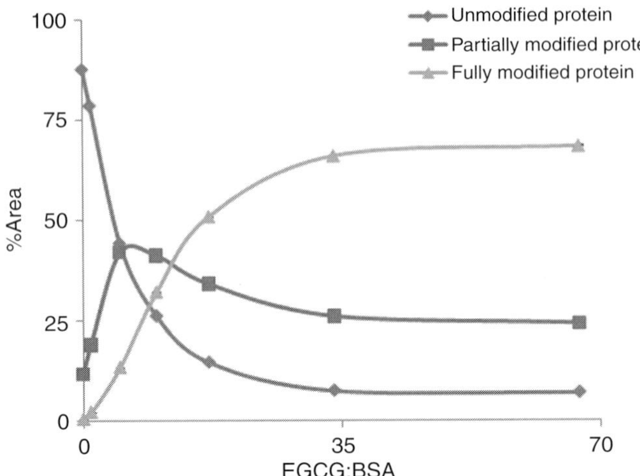

Fig. 3.4   EGCg was mixed with bovine serum albumin at molar ratios ranging from 1:1 to 67:1 and allowed to react. The products were separated with the 20 mm capillary zone electrophoresis method that allowed resolution of two populations of products. Fully modified protein that eluted more slowly from the capillary was the predominant product when EGCg was in large excess.

Our evidence to date suggests that EGCg reacts at surface lysines of serum albumin to make a heterogeneous mixture of products containing on average 10 EGCg per protein in the soluble products. We hypothesize that the reaction is at specific lysines that are adjacent to the usual binding sites for EGCg on serum albumin under nonoxidizing conditions, but we do not have data to support that hypothesis. Controlling reaction conditions more

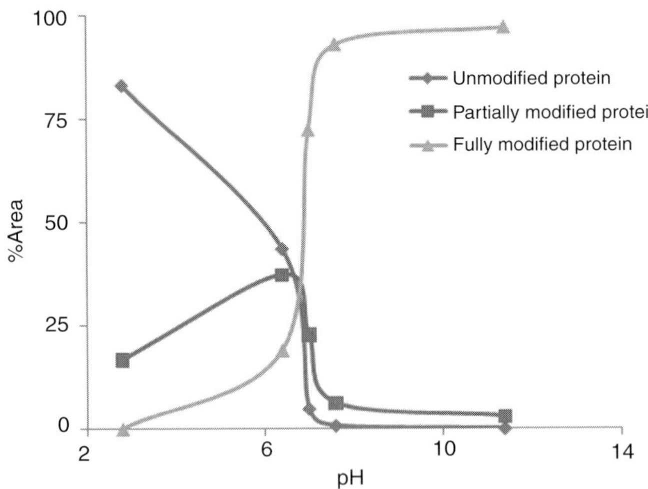

Fig. 3.5   EGCg was mixed with bovine serum albumin at at pH 2–12 and allowed to react. The products were separated with the 20 mm capillary zone electrophoresis method that allowed resolution of two populations of products. Fully modified protein that eluted more slowly from the capillary was the predominant product at pH values > 7.2.

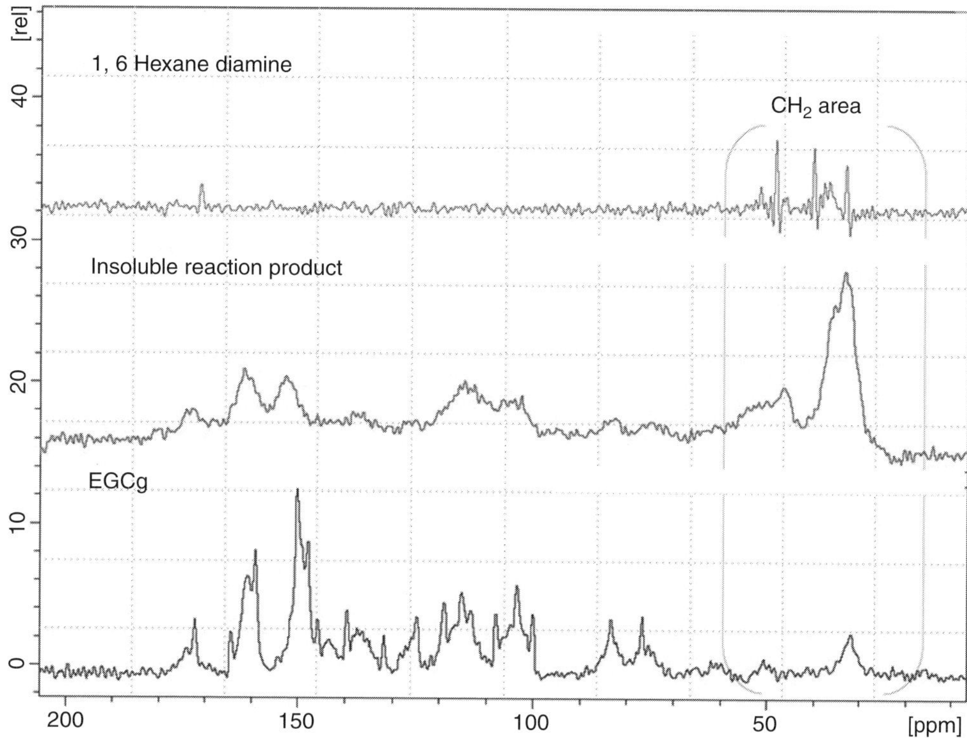

Fig. 3.6 Magic angle spinning solid state ^{13}C NMR of 1,6-hexane diamine, EGCg, and the insoluble reaction product. Peak broadening in the characteristic methylene signals of the hexane diamine suggest covalent reaction between the amine and the polyphenol.

carefully to obtain only a single reaction product with a lower ratio of phenol to protein might yield a product that is more easily characterized.

The potential for polyphenols to covalently react with proteins at pH values less than 7 has been largely ignored. While reactions involving quinones do not generally occur at lower pH values, polyphenols and proteins can participate in free radical reactions at acidic pH. Polyphenolics such as EGCg spontaneously form semiquinone radicals at pH 4-5, and in the presence of a radical stabilizing agent such as Zn^{2+}, those radicals covalently react with protein (Hagerman *et al.*, 2003). The products of those reactions have not been characterized.

An alternate route for formation of covalent tannin–protein complexes involves oxidized protein. Protein oxidation products may react with phenolics to yield new products not obtained with native proteins. Reactive oxygen species or high-energy radiation promote protein oxidation (Davies *et al.*, 1995). In preliminary studies, we photolyzed ^{125}I-labeled serum albumin to form protein hydroperoxides. We determined the amount of native protein and protein hydroperoxide precipitated at pH 4.9 by various polyphenol preparations. For the *Sorghum* PC or blueberry procyanidin, there was no difference between the native protein and protein hydroperoxide. But for EGCg, galloylated procyanidins from grape seed, gallotannins, and ellagitannins, the protein hydroperoxide was precipitated much

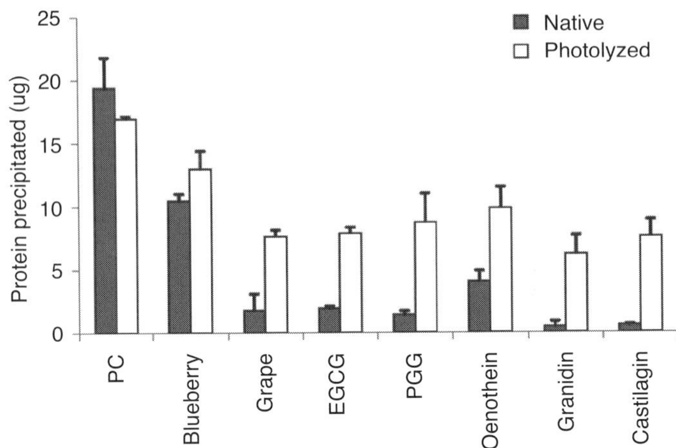

Fig. 3.7 Precipitation of ^{125}I-labeled bovine serum albumin by simple procyanidins (*Sorghum* PC and blueberry), galloylated procyanidin (grape) and EGCg, PGG, and three ellagitannins (Oenothein, granidin, castilagin). The protein was photolyzed by visible light in the presence of Bengal rose to produce protein hydroperoxide. The polyphenols containing galloyl esters precipitated more of the photolyzed protein, suggesting a different reaction between the protein hydroperoxide and the galloyl group compared to the native protein.

more efficiently (Fig. 3.7). We speculate that galloylated polyphenols are more susceptible to oxidation and thus more reactive with the protein hydroperoxide. The products of the reactions have not been characterized.

3.9 Conclusions

Interactions between high molecular mass polyphenols and proteins have been studied intensively by individuals interested in both practical and theoretical aspects of the problem. Fifty years ago, it was sufficient to postulate mechanisms of interaction for "tannins" with protein, with little knowledge of the detailed structure of either player. We now realize that reaction outcomes reflect the unique features of both polyphenol and protein, and that it is as important to study purified, structurally well-defined polyphenols as it is to use defined, purified proteins. While many features of polyphenol–protein interactions are well understood, important questions remain unanswered. We do not understand the interplay between hydrophobic interactions and hydrogen bonding, or how flexibility of the protein and the polyphenolic dictate the interaction. A broader understanding of the interactions could be achieved by exploring more types of polyphenols, such as the high molecular mass proanthocyanidins and phlorotannins. Only a limited number of globular proteins have been examined, and special classes of proteins such as prolamines and histatins have been neglected. Interactions with PRPs have been intensively studied, but we are not well informed about the possible roles of polyphenols in mediating PRP-related diseases or their role in controlling PRP gene expression. With few exceptions, covalently bonded species that form as a consequence of spontaneous oxidation of the protein-bound phenol have not been carefully characterized. Given the lively interest in polyphenols in human health and disease, and their importance in the beverage and food industry, it is likely that polyphenol–protein interactions will continue to be an active area of investigation for many more years.

Acknowledgments

This work was supported by NIH-NIDDKR15 DK069285 and by Agricultural Research Services Specific Cooperative Agreement Number 58-1932-6-634 with Miami University.

References

Ascenzi, P. & Fasano, M. (2010) Allostery in a monomeric protein: the case of human serum albumin. *Biophysical Chemistry*, **148**, 16–22.

Asquith, T.N. & Butler, L.G. (1985) Use of dye-labeled protein as spectrophotometric assay for protein precipitants such as tannin. *Journal of Chemical Ecology*, **11**, 1535–1544.

Asquith, T.N. & Butler, L.G. (1986) Interactions of condensed tannins with selected proteins. *Phytochemistry*, **25**, 1591–1593.

Austin, P.J., Suchar, L.A., Robbins, C.T. & Hagerman, A.E. (1989) Tannin binding proteins in the saliva of deer and their absence in the saliva of sheep and cattle. *Journal of Chemical Ecology*, **15**, 1335–1347.

Bacon, J.R. & Rhodes, M.J.C. (1998) Development of a competition assay for the evaluation of the binding of human parotid salivary proteins to dietary complex phenols and tannins using a peroxidase-labeled tannin. *Journal of Agricultural and Food Chemistry*, **46**, 5083–5088.

Bajec, M.R. & Pickering, G.J. (2008) Astringency: mechanisms and perception. *Critical Reviews in Food Science and Nutrition*, **48**, 858–875.

Balayssac, S., Delsuc, M., Gilard, V., Prigent, Y. & Malet-Martino, M. (2009) Two-dimensional DOSY experiment with excitation sculpting water suppression for the analysis of natural and biological media. *Journal of Magnetic Resonance*, **196**, 78–83.

Barbehenn, R.V., Jones, C.P., Hagerman, A.E., Karonen, M. & Salminen, J. (2006) Ellagitannins have greater oxidative activities than condensed tannins and galloyl glucoses at high pH: Potential impact on caterpillars. *Journal of Chemical Ecology*, **32**, 2253–2267.

Bate-Smith, E.C. (1973) Haemanalysis of tannins: the concept of relative astringency. *Phytochemistry*, **12**, 970–912.

Bate-Smith, E.C. & Swain, T. (1962) Flavonoid compounds. In: *Comparative Biochemistry* (eds. H.S. Mason & A.M. Florkin), vol. 3, pp. 755–809. Academic Press, New York.

Baxter, N.J., Lilley, T.H., Haslam, E. & Williamson, M.P. (1997) Multiple interactions between polyphenols and a salivary proline-rich protein repeat result in complexation and precipitation. *Biochemistry*, **36**, 5566–5577.

Bennick, A. (1982) Salivary proline-rich proteins. *Molecular and Cellular Biochemistry*, **45**, 83–99.

Bennick, A. (2002) Interaction of plant polyphenols with salivary proteins. *Critical Reviews in Oral Biology and Medicine*, **13**, 184–196.

Bernays, E.A., Cooper-Driver, G. & Bilgener, M. (1989) Herbivores and plant tannins. *Advances in Ecological Research*, **19**, 263–302.

Bickley, J.C. (1992) Vegetable tannins and tanning. *Journal of the Society of Leather Technologists and Chemists*, **76**, 1–5.

Bittner, S. (2006) When quinones meet amino acids: Chemical, physical and biological consequences. *Amino Acids*, **30**, 205–224.

Cai, K., Hagerman, A.E., Minto, R.E. & Bennick, A. (2006) Decreased polyphenol transport across cultured intestinal cells by a salivary proline-rich protein. *Biochemical Pharmacology*, **71**, 1570–1580.

Cala, O., Fabre, S., Fouquet, E., Dufourc, E.J. & Pianet, I. (2010) NMR of human saliva protein/wine tannin complexes. towards deciphering astringency with physico-chemical tools. *Comptes Rendus Chimie*, **13**, 449–452.

Canas, B., Lopez-Ferrer, D., Ramos-Fernandez, A., Camafeita, E. & Calvo, E. (2006) Mass spectrometry technologies for proteomics. *Briefings in Functional Genomics & Proteomics*, **4**, 295–320.

Canon, F., Pate, F., Meudec, E., *et al.* (2009) Characterization, stoichiometry, and stability of salivary protein-tannin complexes by ESI-MS and ESI-MS/MS. *Analytical and Bioanalytical Chemistry*, **395**, 2535–2545.

Carvalho, E., Mateus, N. & deFreitas, V. (2004) Flow nephelometric analysis of protein-tannin interactions. *Analytica Chimica Acta*, **513**, 97–101.

Carver, J.A., Duggan, P.J., Ecroyd, H., Liu, Y., Meyer, A.G. & Tranberg, C.E. (2010) Carboxymethylated-kappa-casein: a convenient tool for the identification of polyphenolic inhibitors of amyloid fibril formation. *Bioorganic & Medicinal Chemistry*, **18**, 222–228.

Charlton, A.J., Baxter, N.J., Lilley, T.H., Haslam, E., McDonald, C.J. & Williamson, M.P. (1996) Tannin interactions with a full-length human salivary proline-rich protein display a stronger affinity than with single proline-rich repeats. *FEBS letters*, **382**, 289–292.

Chen, Y. & Hagerman, A.E. (2004a) Characterization of soluble non-covalent complexes between bovine serum albumin and beta-1,2,3,4,6-penta-O-galloyl-D-glucopyranose by MALDI-TOF MS. *Journal of Agricultural and Food Chemistry*, **52**, 4008–4011.

Chen, Y. & Hagerman, A.E. (2004b) Quantitative examination of oxidized polyphenol-protein complexes. *Journal of Agricultural and Food Chemistry*, **52**, 6061–6067.

Chen, Y. & Hagerman, A.E. (2005) Reaction pH and protein affect the oxidation products of beta-pentagalloyl glucose. *Free Radical Research*, **39**, 117–124.

Dangles, O. & Dufour, C. (2008) Flavonoid-protein binding processes and their potential impact on human health. *Recent Advances in Polyphenol Research*, **1**, 67–87.

Da Silva, J.M.R., Cheynier, V., Souquet, J.M., Moutounet, M., Cabanis, J.C. & Bourzeix, M. (1991) Interaction of grape seed procyandins with various proteins in relation to wine fining. *Journal of the Science of Food and Agriculture*, **57**, 111–125.

Davies, M.J., Fu, S. & Dean, R.R. (1995) Protein hydroperoxides can give rise to reactive free radicals. *Biochemical Journal*, **305**, 643–649.

De Freitas, V. & Mateus, N. (2002) Nephelometric study of salivary protein-tannin aggregates. *Journal of the Science of Food and Agriculture*, **82**, 113–119.

Douat-Casassus, C., Chassaing, S., Di Primo, C. & Quideau, S. (2009) Specific or nonspecific protein-polyphenol interactions? Discrimination in real time by surface plasmon resonance. *ChemBioChem*, **10**, 2321–2324.

Evans, D.A., Hirsch, J.B. & Dushenkov, S. (2006) Phenolics, inflammation, and nutrigenomics. *Journal of the Science of Food and Agriculture*, **86**, 2503–2509.

Fatiadi, A.J. (1974) New applications of periodic acid and periodates in organic and bioorganic chemistry. *Synthesis*, **1974**, 229–272.

Feeny, P. (1976) Plant apparency and chemical defense. *Recent Advances in Phytochemistry*, **10**, 1–40.

Feeny, P.P. (1968) Effect of oak leaf tannins on larval growth of the winter moth *Operophtera brumata*. *Journal of Insect Physiology*, **14**, 805–817.

Feeny, P.P. & Bostock, H. (1968) Seasonal changes in the tannin content of oak leaves. *Phytochemistry*, **7**, 871.

Feeny, P.P.M. (1969) Inhibitory effect of oak leaf tannins on the hydrolysis of proteins by trypsin. *Phytochemistry*, **8**, 2119–2126.

Feldman, K.S., Sambandam, A., Lemon, S.T., *et al.* (1999) Binding affinities of gallotannin analogs with bovine serum albumin: Ramifications for polyphenol-protein molecular recognition. *Phytochemistry*, **51**, 867–872.

Fickel, J., Pitra, C., Joest, B.A. & Hofmann, R.R. (1999) A novel method to evaluate the relative tannin-binding capacities of salivary proteins. *Comparative Biochemistry and Physiology C-Pharmacology Toxicology & Endocrinology*, **122**, 225–229.

Forkner, R.E., Marquis, R.T. & Lill, J.T. (2004) Feeny revisited: condensed tannins as antiherbivore defences in leaf-chewing herbivore communities of *Quercus*. *Ecological Entomology*, **29**, 174–187.

Frazier, R.A., Deaville, E.R., Green, R.J., *et al.* (2010) Interactions of tea tannins and condensed tannins with proteins. *Journal of Pharmaceutical and Biomedical Analysis*, **51**, 490–495.

Frazier, R.A., Papadopoulou, A., Mueller-Harvey, I., Kissoon, D. & Green, R.J. (2003) Probing protein-tannin interactions by isothermal titration microcalorimetry. *Journal of Agricultural and Food Chemistry*, **51**, 5189–5195.

Ghosh, K.S., Maiti, T.K., Debnath, J. & Dasgupta, S. (2007) Inhibition of ribonuclease A by polyphenols present in green tea. *Proteins: Structure, Function, and Bioinformatics*, **69**, 566–580.

Goldstein, J.L. & Swain, T. (1965) The inhibition of enzymes by tannins. *Phytochemistry*, **4**, 185–192.

Hagerman, A.E. (1992) Tannin-protein interactions. *ACS Symposium Series*, **506**, 236–247.

Hagerman, A.E. & Butler, L.G. (1978) Protein precipitation method for the determination of tannins. *Journal of Agricultural and Food Chemistry*, **26**, 809–812.

Hagerman, A.E. & Butler, L.G. (1980a) Determination of protein in tannin-protein precipitates. *Journal of Agricultural and Food Chemistry*, **28**, 944–947.

Hagerman, A.E. & Butler, L.G. (1980b) Condensed tannin purification and characterization of tannin-associated proteins. *Journal of Agricultural and Food Chemistry*, **28**, 947–952.

Hagerman, A.E. & Butler, L.G. (1981) The specificity of proanthocyanidin-protein interactions. *Journal of Biological Chemistry*, **256**, 4494–4497.

Hagerman, A.E. & Carlson, D.M. (1998) Biological responses to dietary tannins and other polyphenols. *Recent Research Developments in Agricultural & Food Chemistry*, **2**, 689–704.

Hagerman, A.E., Dean, R.T. & Davies, M.J. (2003) Radical chemistry of epigallocatechin gallate and its relevance to protein damage. *Archives of Biochemistry and Biophysics*, **414**, 115–120.

Hagerman, A.E., Klucher, K.M. (1986) Tannin-protein interactions. In: *Plant Flavonoids in Biology and Medicine: Biochemical, Pharmacological and Structure Activity Relationships.* (eds. V. Cody, E. Middleton & J. Harborne), pp. 67–76. Alan R. Liss, Inc, New York.

Hagerman, A.E., Rice, M.E. & Ritchard, N.T. (1998) Mechanisms of protein precipitation for two tannins, pentagalloyl glucose and epicatechin$_{16}$(4→8)catechin (procyanidin) *Journal of Agricultural and Food Chemistry*, **46**, 2590–2595.

Hagerman, A.E. & Robbins, C.T. (1987) Implications of soluble tannin-protein complexes for tannin analysis and plant defense mechanisms. *Journal of Chemical Ecology*, **13**, 1243–1254.

Hagerman, A.E. & Robbins, C.T. (1993) Specificity of tannin binding salivary proteins relative to diet selection by mammals. *Canadian Journal of Zoology*, **71**, 628–633.

Harbertson, J.F., Hodgins, R.E., Thurston, L.N., *et al.* (2008) Variability of tannin concentration in red wines. *American Journal of Enology and Viticulture*, **59**, 210–214.

Haslam, E. (1974) Polyphenol-protein interactions. *Biochemical Journal*, **139**, 285–288.

Haslam, E. (2007) Vegetable tannins – Lessons of a phytochemical lifetime. *Phytochemistry*, **68**, 2713–2721.

Hatano, T., Hori, M., Hemingway, R.W. & Yoshida, T. (2003) Size exclusion chromatographic analysis of polyphenol-serum albumin complexes. *Phytochemistry*, **63**, 817–823.

Hofmann, T., Glabasnia, A., Schwarz, B., Wisman, K.N., Gangwer, K.A. & Hagerman, A.E. (2006) Protein binding and astringent taste of a polymeric procyanidin, 1,2,3,4,6-penta-O-galloyl-beta-D-glucopyranose, castalagin, and grandinin. *Journal of Agricultural and Food Chemistry*, **54**, 9503–9509.

Ishii, T., Mori, T., Tanaka, T., *et al.* (2008) Covalent modification of proteins by green tea polyphenol (-)-epigallocatechin-3-gallate through autoxidation. *Free Radical Biology & Medicine*, **45**, 1384–1394.

Jobstl, E., Howse, J.R., Fairclough, J.P.A. & Williamson, M.P. (2006) Noncovalent cross-linking of casein by epigallocatechin gallate characterized by single molecule force microscopy. *Journal of Agricultural and Food Chemistry*, **54**, 4077–4081.

Jobstl, E., O'Connell, J., Fairclough, J.P.A. & Williamson, M.P. (2004) Molecular model for astringency produced by polyphenol/protein interactions. *Biomacromolecules*, **5**, 942–949.

Jovanovic, S.V., Hara, Y., Steenken, S. & Simic, M.G. (1995) Antioxidant potential of gallocatechins. A pulse radiolysis and laser photolysis study. *Journal of the American Chemical Society*, **117**, 9881–9888.

Jung, H.A., Oh, S.H. & Choi, J.S. (2010) Molecular docking studies of phlorotannins from *eisenia bicyclis* with BACE1 inhibitory activity. *Bioorganic & Medicinal Chemistry Letters*, **20**, 3211–3215.

Juntheikki, M. & Julkunen-Tiitto, R. (2000) Inhibition of beta-glucosidase and esterase by tannins from *betula*, *salix*, and *pinus* species. *Journal of Chemical Ecology*, **26**, 1151–1165.

Kajiya, K., Kumazawa, S. & Nakayama, T. (2001) Steric effects on interaction of tea catechins with lipid bilayers. *Bioscience, Biotechnology, and Biochemistry*, **65**, 2638–2643.

Kawamoto, H., Mizutani, K. & Nakatsubo, F. (1997) Binding nature and denaturation of protein during interactions with galloylglucose. *Phytochemistry*, **46**, 473–478.

Kawamoto, H. & Nakatsubo, F. (1997a) Effects of environmental factors on two-stage tannin-protein co-precipitation. *Phytochemistry*, **46**, 479–483.

Kawamoto, H. & Nakatsubo, F. (1997b) Solubility of protein complexed with galloylglucoses. *Phytochemistry*, **46**, 485–488.

Kawamoto, H., Nakatsubo, F. & Murakami, K. (1995) Quantitative determination of tannin and protein in the precipitates by high-performance liquid chromatography. *Phytochemistry*, **40**, 1503–1505.

Kawamoto, H., Nakayama, M. & Murakami, K. (1996) Stoichiometric studies of tannin-protein co-precipitation. *Phytochemistry*, **41**, 1427–1431.

Kerwin, J.L. (1997) Profiling peptide adducts of oxidized N-acetyldopamine by electrospray mass spectrometry. *Rapid Communications in Mass Spectrometry*, **11**, 557–566.

Khan, N. & Mukhtar, H. (2007) Tea polyphenols for health promotion. *Life Sciences*, **81**, 519–533.

Kusuda, M., Hatano, T. & Yoshida, T. (2006) Water-soluble complexes formed by natural polyphenols and bovine serum albumin: Evidence from gel electrophoresis. *Bioscience, Biotechnology, and Biochemistry*, **70**, 152–160.

Li, C., Leverence, R., Trombley, J.D., *et al.* (2010) High molecular weight persimmon (*Diospyros kaki L.*) proanthocyanidin: a highly galloylated, A-linked tannin with an unusual flavonol terminal unit, myricetin. *Journal of Agricultural and Food Chemistry*, **58**, 9033–9042.

Li, X., Zhang, W., Qiao, X. & Xu, X. (2007) Prediction of binding for a kind of non-peptic HCV NS3 serine protease inhibitors from plants by molecular docking and MM-PBSA method. *Bioorganic and Medicinal Chemistry*, **15**, 220–226.

Loomis, W.D. & Battaile, J. (1966) Plant phenolic compounds and the isolation of plant enzymes. *Phytochemistry*, **5**, 423–438.

Lu, Y. & Bennick, A. (1998) Interaction of tannin with human salivary proline-rich proteins. *Archives of Oral Biology*, **43**, 717–728.

Luck, G., Liao, H., Murray, N.J., *et al.* (1994) Polyphenols, astringency and proline-rich proteins. *Phytochemistry*, **37**, 357–371.

Maiti, T.K., Ghosh, K.S. & Dasgupta, S. (2006) Interaction of (-)-epigallocatechin-3-gallate with human serum albumin: Fluorescence, fourier transform infrared, circular dichroism, and docking studies. *Proteins: Structure, Function, and Bioinformatics*, **64**, 355–362.

Makkar, H.P.S., Dawra, R.K. & Singh, B. (1987) Protein precipitation assay for quantitation of tannins: determination of protein in tannin-protein complex. *Analytical Biochemistry*, **166**, 435–439.

Mane, C., Sommerer, N., Yalcin, T., Cheynier, V., Cole, R.B. & Fulcrand, H. (2007) Assessment of the molecular weight distribution of tannin fractions through MALDI-TOF MS analysis of protein-tannin complexes. *Analytical Chemistry*, **79**, 2239–2248.

Markley, J.L., Bahrami, A., Eghbalnia, H.R., *et al.* (2009) Macromolecular structure determination by NMR spectroscopy. In: *Structural Bioinformatics*. (eds. J. Gu & P.E. Bourne), pp. 93–142. John Wiley & Sons, Hoboken, NJ

Martin, M.M., Rockholm, D.C. & Martin, J.S. (1985) Effects of surfactants, pH and certain cations on precipitation of proteins by tannins. *Journal of Chemical Ecology*, **11**, 485–494.

Mason, H.S. & Peterson, E.W. (1965) Melanoproteins. I. Reactions between enzyme-generated quinones and amino acids. *Biochimica et Biophysica Acta*, **111**, 134–146.

Mau, M., Sudekum, K., Johann, A., Silwa, A. & Kaiser, T.M. (2009) Saliva of the graminivorous theropithecus gelada lacks proline-rich proteins and tannin-binding capacity. *American Journal of Primatology*, **71**, 663–669.

McManus, J.P., Davis, K.G., Beart, J.E., Gaffney, S.H., Lilley, T.H. & Haslam, E. (1985) Polyphenol interactions. Part 1. Introduction: some observations on the reversible complexation of polyphenols with proteins and polysaccharides. *Journal of the Chemical Society, Perkin Transactions 2*, **1985**, 1429–1438.

McManus, J.P., Davis, K.G., Lilley, T.H. & Haslam, E. (1981) The association of proteins with polyphenols. *Journal of the Chemical Society Chemical Communications*, **1981**, 309–311.

Mehansho, H., Butler, L.G. & Carlson, D.M. (1987) Dietary tannins and salivary proline-rich proteins: interactions, induction, and defense mechanisms. *Annual Review of Nutrition*, **7**, 423–440.

Mehansho, H., Hagerman, A., Clements, S., Butler, L., Rogler, J. & Carlson, D.M. (1983) Modulation of proline-rich protein biosynthesis in rat parotid glands by sorghums with high tannin levels. *Proceedings of the National Academy of Sciences, U.S.A.*, **80**, 3948–3952.

Mejbaum, K.,W., Dobryszycka, W., B.-Jaworska, J. & Morawiecka, B. (1959) Regeneration of protein from insoluble protein-tannin complexes. *Nature*, **184**, 1799–1800.

Monagas, M., Quintanilla-Lopez, J.E., Gomez-Cordoves, C., Bartolome, B. & Lebron-Aguilar, R. (2010) MALDI-TOF MS analysis of plant proanthocyanidins. *Journal of Pharmaceutical and Biomedicial Analysis*, **51**, 358–372.

Mueller-Harvey, I. (2006) Unravelling the conundrum of tannins in animal nutrition and health. *Journal of the Science of Food and Agriculture*, **86**, 2010–2037.

Mueller-Harvey, I., Mlambo, V., Sikosana, J.L.N., Smith, T., Owen, E. & Brown, R.H. (2007) Octanol-water partition coefficients for predicting the effects of tannins in ruminant nutrition. *Journal of Agricultural and Food Chemistry*, **55**, 5436–5444.

Murray, N.J., Williamson, M.P., Lilley, T.H. & Haslam, E. (1994) Study of the interaction between salivary proline-rich proteins and a polyphenol by ^{1}H-NMR spectroscopy. *European Journal of Biochemistry*, **219**, 923–935.

Noble, A.C. (2002) Astringency and bitterness of flavonoid phenols. In: *Chemistry of Taste: Mechanisms, Behaviors, and Mimics*. (eds. P. Given & D. Paredes), pp. 192–201. ACS Symposium Series; American Chemical Society: Washington, DC.

Obreque-Slier, E., Pena-Neira, A. & Lopez-Solis, R. (2010) Enhancement of both salivary protein-enological tannin interactions and astringency perception by ethanol. *Journal of Agricultural and Food Chemistry*, **58**, 3729–3735.

Pascal, C., Poncet-Legrand, C., Imberty, A., *et al.* (2007) Interactions between a non glycosylated human proline-rich protein and flavan-3-ols are affected by protein concentration and polyphenol/protein ratio. *Journal of Agricultural and Food Chemistry*, **55**, 4895–4901.

Paz, M.A., Fluckiger, R., Boak, A., Kagan, H.M. & Gallop, P.M. (1991) Specific detection of quinoproteins by redox-cycling staining. *Journal of Biological Chemistry*, **266**, 689–692.

Peters, T. (1975) Serum albumin. In: *The Plasma Proteins*. (ed. F.W. Putnam), pp. 133–181. Academic Press, New York.

Pianet, I., Barathieu, K., Tarascou, I., *et al.* (2009) 3D-structure, colloidal behavior and human saliva protein recognition of wine tannins determined by NMR. *Acta Horticulturae*, **841**, 389–396.

Pierpoint, W.S. (1969a) *O*-Quinones formed in plant extracts. Their reactions with amino acids and peptides. *Biochemical Journal*, **112**, 609–616.

Pierpoint, W.S. (1969b) *O*-Quinones formed in plant extracts. Their reaction with bovine serum albumin. *Biochemical Journal*, **112**, 619–629.

Poncet-Legrand, C., Edelmann, A., Putaux, J.-., Cartalade, D., Sarni-Manchado, P. & Vernhet, A. (2006) Poly(-proline) interactions with flavan-3-ols units: Influence of the molecular structure and the polyphenol/protein ratio. *Food Hydrocolloids*, **20**, 687–697.

Poncet-Legrand, C., Gautier, C., Cheynier, V. & Imberty, A. (2007) Interactions between flavan-3-ols and poly(L-proline) studied by isothermal titration calorimetry: Effect of the tannin structure. *Journal of Agricultural and Food Chemistry*, **55**, 9235–9240.

Porter, L.J. & Woodruffe, J. (1984) Haemanalysis: The relative astringency of proanthocyanidin polymers. *Phytochemistry*, **23**, 1255–1256.

Prigent, S.V.E., Voragen, A.G.J., Visser, A.J.W.G., van Koningsveld, G.A. & Gruppen, H. (2007) Covalent interactions between proteins and oxidation products of caffeoylquinic acid (chlorogenic acid) *Journal of the Science of Food and Agriculture*, **87**, 2502–2510.

Quideau, S., Varadinova, T., Karagiozova, D., *et al.* (2004) Main structural and stereochemical aspects of the antiherpetic activity of nonahydroxyterphenoyl-containing C-glycosidic ellagitannins. *Chemistry & Biodiversity*, **1**, 247–258.

Reiersen, H. & Rees, A.R. (2001) The hunchback and its neighbours: Proline as an environmental modulator. *Trends in Biochemical Sciences*, **26**, 679–684.

Richard, T., Lefeuvre, D., Descendit, A., Quideau, S. & Monti, J.P. (2006) Recognition characters in peptide-polyphenol complex formation. *Biochimica et Biophysica Acta*, **1760**, 951–958.

Richard, T., Verge, S., Berke, B., Vercauteren, J. & Monti, J.P. (2001) NMR and simulated annealing investigations of bradykinin in presence of polyphenols. *Journal of Biomolecular Structure and Dynamics*, **18**, 627–637.

Richard, T., Vitrac, X., Merillon, J.M. & Monti, J.P. (2005) Role of peptide primary sequence in polyphenol-protein recognition: an example with neurotensin. *Biochimica et Biophysica Acta*, **1726**, 238–243.

Riedl, K.M., Carando, S., Alessio, H.M., McCarthy, M. & Hagerman, A.E. (2002) Antioxidant activity of tannins and tannin-protein complexes: Assessment in vitro and in vivo. *ACS Symposium Series*, **807**, 188–200.

Riedl, K.M. & Hagerman, A.E. (2001) Tannin-protein complexes as radical scavengers and radical sinks. *Journal of Agricultural and Food Chemistry*, **49**, 4917–4923.

Robbins, C.T., Hanley, T.A., Hagerman, A.E., *et al.* (1987) Role of tannins in defending plants against ruminants: reduction in protein availability. *Ecology*, **68**, 98–107.

Sang, S., Yang, I., Buckley, B., Ho, C. & Yang, C.S. (2007) Oxidative quinone formation in vitro and metabolite formation in vivo from tea polyphenol (-)-epigallocatechin-3-gallate: Studied by real-time mass spectrometry combined with tandem mass ion mapping. *Free Radical Biology and Medicine*, **43**, 362–371.

Santos-Buelga, C. & Scalbert, A. (2000) Proanthocyanidins and tannin-like compounds – nature, occurrence, dietary intake and effects on nutrition and health. *Journal of the Science of Food and Agriculture*, **80**, 1094–1117.

Sarni-Manchado, P. & Cheynier, V. (2002) Study of non-covalent complexation between catechin derivatives and peptides by electrospray ionization mass spectrometry. *Journal of Mass Spectrometry*, **37**, 609–616.

Sarni-Manchado, P., Cheynier, V. & Moutounet, M. (1999) Interactions of grape seed tannins with salivary proteins. *Journal of Agricultural and Food Chemistry*, **47**, 42–47.

Siebert, K.J. (2009) Haze in beverages. *Advances in Food and Nutrition Research*, **57**, 53–86.

Silber, M.L., Davitt, B.B., Khairutdinov, R.F. & Hurst, J.K. (1998) A mathematical model describing tannin-protein association. *Analytical Biochemistry*, **263**, 46–50.

Simon, C., Barathieu, K., Laguerre, M., *et al.* (2003) Three-dimensional structure and dynamics of wine tannin-saliva protein complexes. A multitechnique approach. *Biochemistry*, **42**, 10385–10395.

Skopec, M.M., Hagerman, A.E. & Karasov, W.H. (2004) Do salivary proline-rich proteins counteract dietary hydrolyzable tannin in laboratory rats? *Journal of Chemical Ecology*, **30**, 1679–1692.

Soares, S., Mateus, N. & De Freitas, V. (2007) Interaction of different polyphenols with bovine serum albumin (BSA) and human salivary alpha-amylase (HSA) by fluorescence quenching. *Journal of Agricultural and Food Chemistry*, **55**, 6726–6735.

Stern, J.L., Hagerman, A.E., Steinberg, P.D. & Mason, P.K. (1996) Phlorotannin-protein interactions. *Journal of Chemical Ecology*, **22**, 1877–1899.

Sugio, S., Kashima, A., Mochizuki, S., Noda, M. & Kobayashi, K. (1999) Crystal structure of human serum albumin at 2.5 A resolution. *Protein Engineering Design and Selection*, **12**, 439–446.

Tanaka, T., Zhang, H., Jiang, Z.H. & Kouno, I. (1997) Relationship between hydrophobicity and structure of hydrolyzable tannins, and association of tannins with crude drug constituents in aqueous solution. *Chemical and Pharmaceutical Bulletin*, **45**, 1891–1897.

Trombley, J.D., Loegel, T.N., Danielson, N.D. & Hagerman, A.E. (2011) Capillary electrophoresis methods for the determination of covalent polyphenol-protein complexes. *Analytical and Bioanalytical Chemistry*, **401**, 1523–1529.

Van-Driel Murray, C. (2000) Leatherwork and skin products. In: *Ancient Egyptian Materials and Technology.* (eds. P.T. Nicholson & I. Shaw), pp. 299–319. Cambridge University Press, Cambridge.

Verge, S., Richard, T., Moreau, S., *et al.* (2002a) First observation of solution structures of bradykinin-penta-O-galloyl-D-glucopyranose complexes as determined by NMR and simulated annealing. *Biochimica et Biophysica Acta*, **1571**, 89–101.

Verge, S., Richard, T., Moreau, S., *et al.* (2002b) First observation of non-covalent complexes for a tannin-protein interaction model investigated by electrospray ionisation mass spectroscopy. *Tetrahedron Letters*, **43**, 2363–2366.

Verzele, M., Delahaye, P. & Van Damme, F. (1986) Determination of the tanning capacity of tannic acids by high-performance liquid chromatography. *Journal of Chromatography, B: Biomedical Sciences and Applications*, **362**, 363–374.

Wilker, J.J. (2010) Marine bioinorganic materials: Mussels pumping iron. *Current Opinion in Chemical Biology*, **14**, 276–283.

Wroblewski, K., Muhandiram, R., Chakrabartty, A. & Bennick, A. (2001) The molecular interaction of human salivary histatins with polyphenolic compounds. *European Journal of Biochemistry*, **268**, 4384–4397.

Yan, Q. & Bennick, A. (1995) Identification of histatins as tannin-binding proteins in human saliva. *The Biochemical Journal*, **311**, 341–347.

Yu, X., Chu, S., Hagerman, A.E. & Lorigan, G.A. (2011) Probing the interaction of polyphenols with lipid bilayers by solid-state NMR spectroscopy. *Journal of Agricultural and Food Chemistry*, **59**, 6783–6789.

Zanchi, D., Narayanan, T., Hagenmuller, D., *et al.* (2008a) Tannin-assisted aggregation of natively unfolded proteins. *Europhysics Letters*, **82**, 58001/1–58001/5.

Zanchi, D., Poulain, C., Konarev, P., Tribet, C. & Svergun, D.I. (2008b) Colloidal stability of tannins: astringency, wine tasting and beyond. *Journal of Physics: Condensed Matter*, **20**, 494224/1–494224/6.

Zhang, J., Li, L., Kim, S., Hagerman, A.E. & Lu, J. (2009) Anti-cancer, anti-diabetic and other pharmacologic and biological activities of penta-galloyl-glucose. *Pharmaceutical Research*, **26**, 2066–2080.

Chapter 4
Chemistry of Flavonoids in Color Development

Kumi Yoshida, Kin-ichi Oyama and Tadao Kondo

Abstract: In plant coloration, anthocyanins, colored flavonoids, play a considerably important role. Anthocyanins distribute in various plant tissues, including flowers, fruits, leaves, stems, tubers and seed coats, and exhibit a wide variety of colors ranging from red to purple and even blue. However, the number of anthocyanidins that are chromophores is few, with only six major ones in nature. Furthermore, the structural differences between each are subtle. Therefore, the question of how similar chromophores can develop such a wide variety of colors is still incompletely known. Recent progress in synthetic studies on anthocyanins and related polyphenols opened the door for understanding the chemical mechanisms of color development by anthocyanins. In this chapter, synthetic studies on anthocyanins and polyphenolic copigments, such as glycosylated flavones and acylquinic acid derivatives, are described first, followed by a discussion of chemical mechanisms of color development in these synthetic compounds.

Keywords: anthocyanin synthesis; biosynthesis; glycosylated flavones; acylquinic acid; blue flower-coloration; metalloanthocyanin; hydrangea; commelinin; protodelphin; supramolecule.

4.1 Introduction

In plant coloration, anthocyanins (colored flavonoids) play a considerably important role (Brouillard, 1988; Harborne & Grayer, 1988; Goto & Kondo, 1991; Brouillard & Dangles, 1994; Harborne & Williams, 2000; Andersen & Jordheim, 2006; Nigel & Grayer, 2008; Yoshida *et al.*, 2009a; Brouillard *et al.*, 2010). Anthocyanins are present in various plant tissues, including flowers, fruits, leaves, stems, tubers, and seed coats and exhibit a wide variety of colors from red to purple and blue. However, the number of chromophore

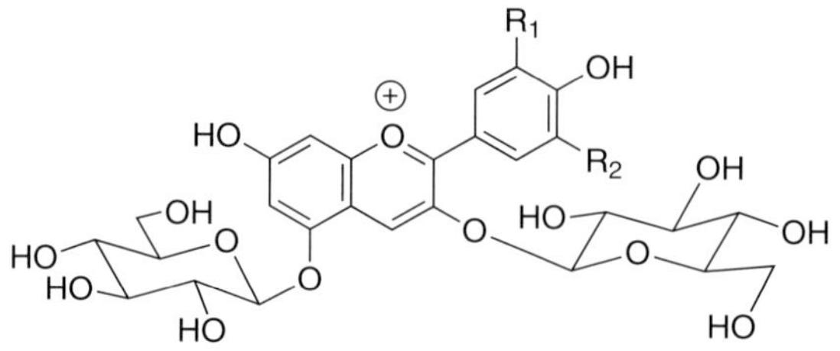

R_1	R_2	anthocyanin	aglycon
H	H	pelargonin (**1**)	pelargonidin (**7**)
OH	H	cyanin (**2**)	cyanidin (**8**)
OCH_3	H	peonin (**3**)	peonidin (**9**)
OH	OH	delphin (**4**)	delphinidin (**10**)
OCH_3	OH	petunin (**5**)	petunidin (**11**)
OCH_3	OCH_3	malvin (**6**)	malvidin (**12**)

Fig. 4.1   Structure of common anthocyanins (**1–6**) and their aglycons (**7–12**).

structures in anthocyanins (and anthocyanidins) is small, with only six major ones found in nature (Fig. 4.1) (Goto, 1987; Brouillard, 1988; Goto & Kondo, 1991; Yoshida *et al.*, 2009a). Furthermore, the structural differences between them are subtle and essentially based on the substitution pattern and the number of hydroxyl and methoxyl groups on the B-ring. In plant cells, anthocyanins are usually dissolved in a central vacuole that occupies over 90% of the total cell volume (Yoshida *et al.*, 2009a). Anthocyanin in aqueous solution changes color, depending on pH, like litmus paper: red in strong acid, purple in neutral, and blue in alkaline solution (Fig. 4.2). The vacuolar pH is generally controlled around slightly acidic values by the activity of two vacuolar-type proton pumps, V-ATPase and V-PPase (Martinoia *et al.*, 2007). Therefore, the mechanism of various flower colorations and their stability under weakly acidic aqueous solutions has remained a mystery. Especially puzzling has been the stable development of blue color in living petals. From a chemical perspective, any anhydrobase (or quinonoidal base) anion form of anthocyanins can simply develop blue color in aqueous solutions. Therefore, the question is how to stabilize the blue anhydrobase anion form in plant vacuoles, where the pH value is usually weakly acidic to neutral.

Research on color development of anthocyanins started more than 100 years ago, and several theories explaining flower color variation and stability have been proposed as a result. These theories include the pH theory (Willstätter & Everest, 1913; Willstätter & Mallison, 1915), the metal-complex theory (Shibata *et al.*, 1919), the copigmentation theory (Robinson & Robinson, 1931), and the self-association theory (Asen *et al.*, 1972), among others. From the end of the 1970s, the equilibrium constant between each form of anthocyanidin chromophore in aqueous solution has been studied in detail, furthering the understanding of mechanisms of color change and decoloration (Brouillard & Delaporte, 1977; Brouillard, 1988; Brouillard *et al.*, 2010). At the same time, intramolecular

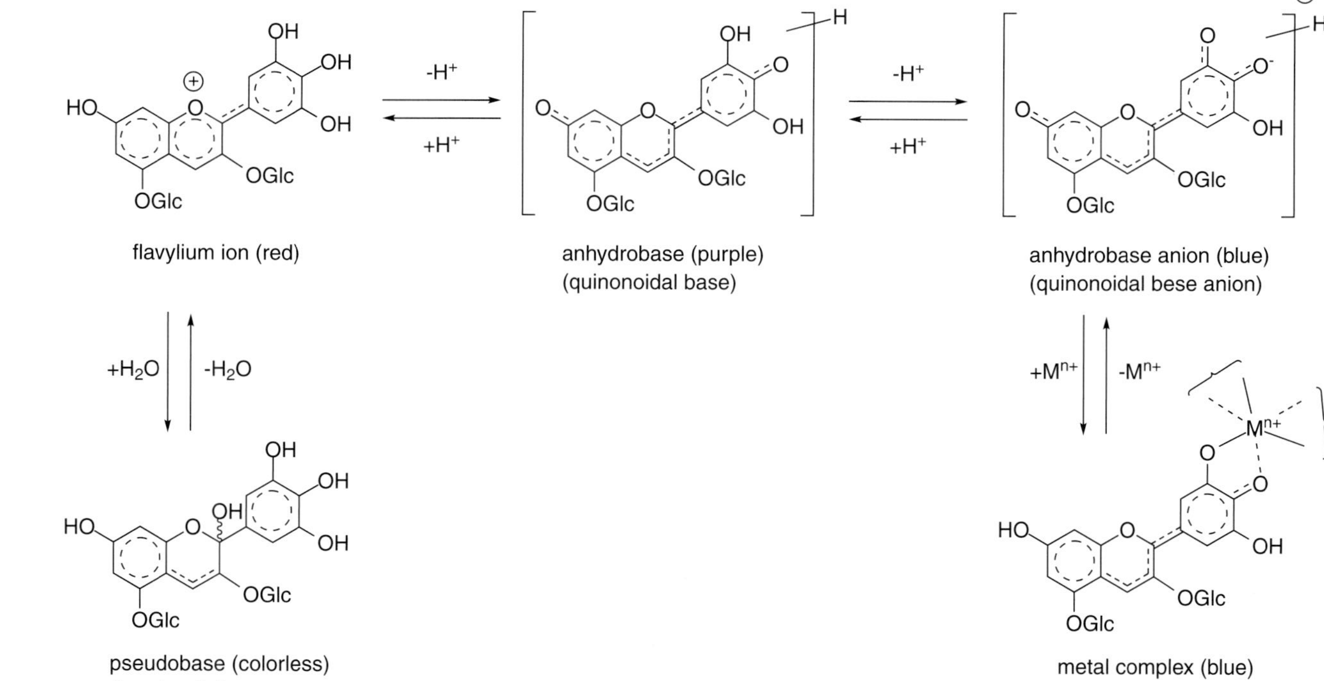

Fig. 4.2 Structural changes in the anthocyanidin, a chromophore of anthocyanin, and their pH-dependent color changes in aqueous solution.

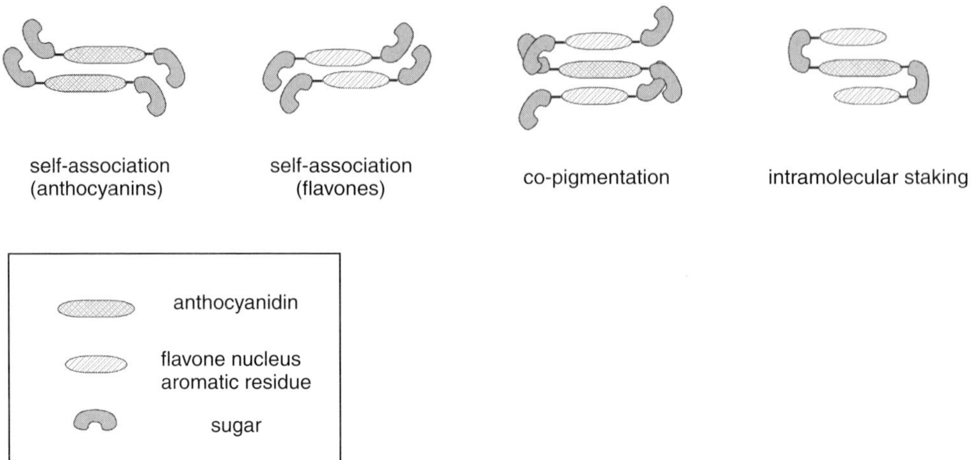

Fig. 4.3   Schematic structure of molecular stackings of anthocyanins.

copigmentation for development and stabilization of the blue-purple color was suggested by Saito and his group (Saito *et al.*, 1971). Over the next decade, the molecular stacking theory, involving hydrophobic π–π interactions, was then proposed (Fig. 4.3) (Goto *et al.*, 1982, 1986; Goto *et al.*, 1987; Goto & Kondo, 1991). This concept encompassed three previously presented theories (copigmentation, self-association, and intramolecular copigmentation) and has gradually become accepted with continued experimental support (Hoshino *et al.*, 1980; Hoshino & Goto, 1990; Yoshida, 1992). In 1992, a crucial result was provided for this theory by the elucidation of the X-ray crystallographic structure of commelinin (**13**) (Kondo *et al.*, 1992), a blue pigment found in the petals of the dayflower, *Commelina communis* (Hayashi *et al.*, 1958). This achievement also provided support for the metal-complex theory, as commelinin is a self-assembled supramolecular metal-complex pigment composed of anthocyanin, flavone and Mg^{2+} ions.

It is now understood that plants primarily use four strategies to create blue flowers: (1) generating a highly hydroxylated flavylium chromophore, particularly delphinidin (3,5,7,3′,4′,5′-hexahydroxy-2-phenyl-benzopyrilium) (**10**); (2) increasing the vacuolar pH; (3) forming a metal-complex with anthocyanins; and (4) generating vertical π–π molecular stacking of aromatic residues with the anthocyanidin chromophore. In plant tissues, two or more of these strategies typically occur concurrently, allowing anthocyanins to form a supramolecular structure resulting in color modification and stability (Yoshida *et al.*, 2009a). Therefore, supramolecular chemistry concerning molecular interactions with non-covalent bonds and/or metal-complexes is very important to understanding coloration. To elucidate the fine chemical structure of these supramolecules, "*in vivo* natural product chemical methodology," including synthetic approaches, should be very powerful tools. However, the synthesis of anthocyanins is still a challenging task in flavonoid research. This chapter describes recent chemical studies on color development in flavonoids and correlate them with synthetic studies on flavonoids and related polyphenols.

4.2 Synthetic studies on anthocyanins toward polyacylated pigments

In 1913, Willstätter isolated a flower pigment, cyanin (**2**), from *Centaurea cyanus* and reported the structure (Willstätter & Everest, 1913). The structure of cyanin (**2**) was then confirmed by Robinson's synthesis of cyanidin 3,5-di-β-D-glucoside (**2**) (Robinson & Todd, 1932). Although synthetic studies on anthocyanins declined after Robinson's era, the structures of more than 500 naturally occurring anthocyanins and related compounds were elucidated (Andersen & Jordheim, 2006) using reliable analytical methodologies developed by Goto and Kondo (1991). Nowadays, a great number of new properties have been discovered for these compounds, including health protective effects and possible application as materials for dyes. However, to continue advancing chemical and functional research on anthocyanins, synthetic studies are needed.

All of the naturally occurring anthocyanins are glycosylated, and these sugars are further substituted with other sugars and/or various acyl moieties (Harborne, 1988; Strack & Wray, 1994; Andersen & Jordheim, 2006). Furthermore, these modifications give anthocyanins attractive activities, such as stable colors and biological functions. Although there are many reports on the synthesis of anthocyanidins, aglycones of anthocyanins, synthetic studies on anthocyanins remains scarce. Moreover, there has been no report of the synthesis of polyacylated anthocyanins. This should be one of the reasons why chemical studies on color development and supramolecular pigments did not proceed so much. Fortunately, new synthetic strategies and improvements on previous procedures for anthocyanin synthesis are now actively being pursued (Oyama *et al.*, 2011). In this section, these recently reported synthetic studies on anthocyanins are reviewed. In addition, the synthesis of copigments and the chemical mechanism of coloration are discussed in terms of these newly synthesized compounds.

4.2.1 Previously reported syntheses of anthocyanins

Synthetic studies on anthocyanins and anthocyanidins before 1982 have been comprehensively summarized by Iacobucci and Sweeny (1983). On the synthesis of anthocyanins, the earliest sound study was that reported by Robinson and coworkers (Robertson & Robinson, 1927; Murakami *et al.*, 1931; Robinson & Todd, 1932; Robinson, 1934). They synthesized the 3-monoglucoside and the 3,5-diglucoside of peralgonidin (**7**), cyanidin (**8**), delphinidin (**10**), petunidin (**11**), and malvidin (**12**) using acidic aldol condensation between phloroglucinaldehyde derivatives and 2-glycosyloxyacetophenone derivatives (Fig. 4.4). This methodology is still useful and may be applied for the preparation of the aforementioned simple anthocyanins. However, because both the reaction conditions of the acidic aldol condensation and the final deprotection step under basic reaction conditions are harsh, polyacylated anthocyanins are not attainable by this method.

Based on the classical Robinson aldol method, a synthesis of pelargonidin 3-glucoside (**14**) was reported (Dangles & El Hajji, 1994). A glycosylation of 4′-acetoxy-2-hydroxyacetophenone (**15**) using peracetylglucosyl bromide (**16**) in the presence of Hg(CN)$_2$ gave 2-β-glucosyloxyacetophenone (**17**) in 40% yield in a β-selective manner.

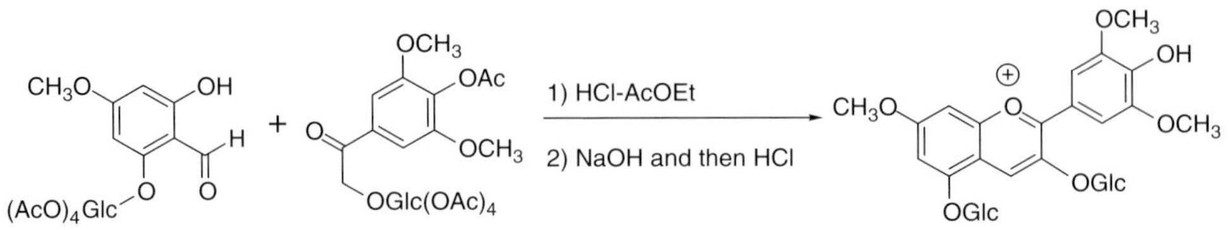

Fig. 4.4 Synthetic route to anthocyanin formulated by Robinson.

The glycoside **17** and 2,4-acetoxy-6-hydroxybenzaldehyde (**18**) were condensed for two days in HCl–EtOAc at –20°C to afford the pelargonidin 3-glucoside (**14**) in approximately 40% yield (Fig. 4.5). Using the same strategy, 4′,7-dihydoxy-3′-glucosyloxyflavylium ion (**22**) was synthesized from 3,4-dihydroxyacetophenone (**19**) in four steps (Fig. 4.6) (El Hajji *et al.*, 1997).

It was already evident in the early twentieth century that the reduction of flavonols and flavonol glucosides gave the corresponding anthocyanidins and anthocyanins, respectively. Because the metal-mediated reductive transformation of naturally occurring flavonol and flavone glycosides to the corresponding anthocyanins and 3-deoxyanthocyanins using Clemmensen's reaction was practically convenient, various related experiments were conducted (Everest, 1914a, 1914b; Shibata *et al.*, 1919; Asahina *et al.*, 1929). Everest obtained cyanidin **8**, which was produced by hydrolyzing a sugar moiety from quercitrin (quercetin 3-*O*-rhamnoside) (**23**) by reduction with zinc in 2 N HCl (Everest, 1914a, 1914b). Additionally, Willstätter and Marison (1914) obtained a red-colored solution by

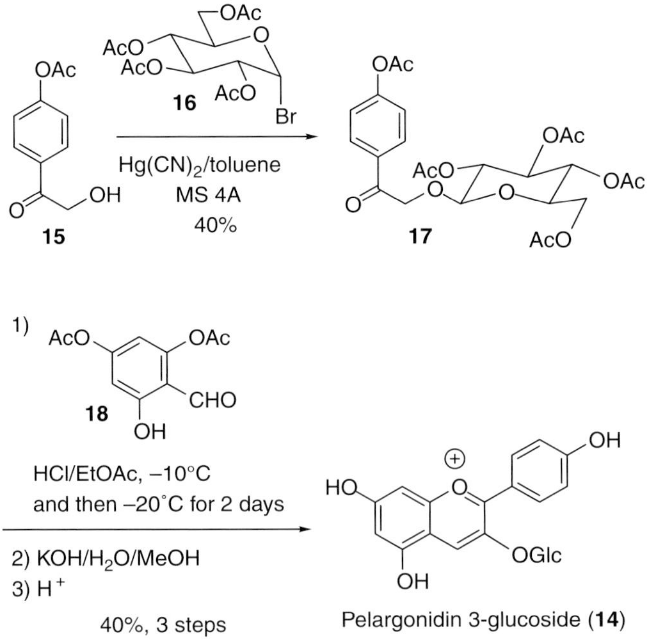

Fig. 4.5 Synthesis of pelargonidin 3-glucoside (**14**) (Dangles & El Hajji, 1994).

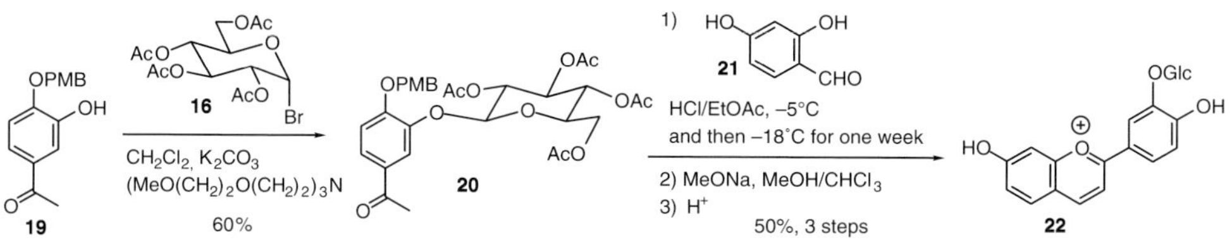

Fig. 4.6   Synthesis of 4′,7-dihydroxy-3′-glucosylflavylium ion (**22**) (El Hajji *et al.*, 1997).

reduction of quercetin with Zn or Mg in EtOH–HCl. Meanwhile, Shibata and coworkers reduced myricitrin (myricetin 3-*O*-rhamnoside) (**24**) with Mg in EtOH–AcOH and obtained a blue-colored solution (Shibata *et al.*, 1919; Asahina *et al.*, 1929). Shibata *et al.* concluded that the reduction of **24** with Mg likely afforded delphinidin 3-*O*-rhamnoside (**25**) and that a blue color might have been a result of the complexation of anthocyanin and Mg^{2+} with weak organic acids. These findings brought the metal-complex theory (Shibata *et al.*, 1919) for blue flower colors in sharp contrast to the pH theory (Willstätter & Everest, 1913).

Transformations from flavonols to anthocyanidin, and from glycosylflavonol to anthocyanins, have been reported multiple times. However, in all of the reports, the yield of the reductive transformation was generally less than 20%. Only one report, using an improved Zn–Hg method instead of Zn or Mg, afforded cyanidin 3-rutinoside (**26**) from rutin (**27**) with 60% yield (Fig. 4.7) (Elhabiri *et al.*, 1995).

4.2.2 Synthesis of anthocyanin using biomimetic oxidation

The former half-part of the biosynthetic pathway of anthocyanins (from chalcone to anthocyanidin) has been clarified as shown in Fig. 4.8 (Heller & Forkmann, 1994; Davies & Schwinn, 2006; Grotewold, 2006). The structural genes of each transformation have been cloned (Holton & Cornish, 1995; Springob *et al.*, 2003), and nowadays, genetic breeding technique has been applied to the production of "blue carnations" (Fukui *et al.*, 2003) and "blue roses" (Katsumoto *et al.*, 2007). In the biosynthetic pathway, the last and key step from a colorless flavan-3,4-diol (or leucoanthocyanidin) to a colored anthocyanidin is catalyzed by anthocyanidin synthase (ANS), a family of 2-oxoglutarate-dependent oxygenases requiring molecular O_2 and ferrous ion for oxidation (Fig. 4.8) (Heller &

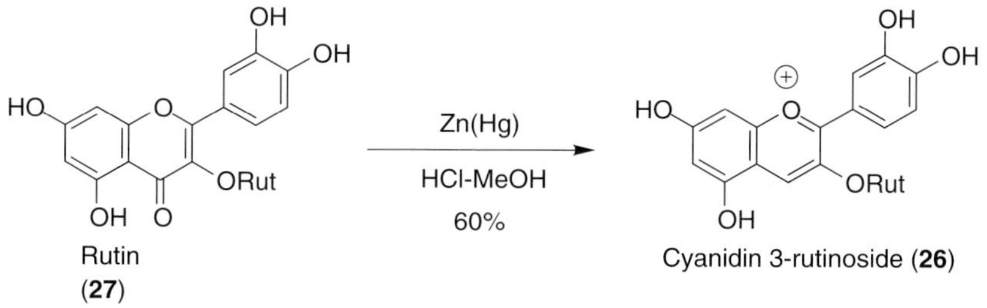

Fig. 4.7   Reduction of rutin (**27**) to cyanidin 3-rutinoside (**26**) by Zn–Hg (Elhabiri *et al.*, 1995).

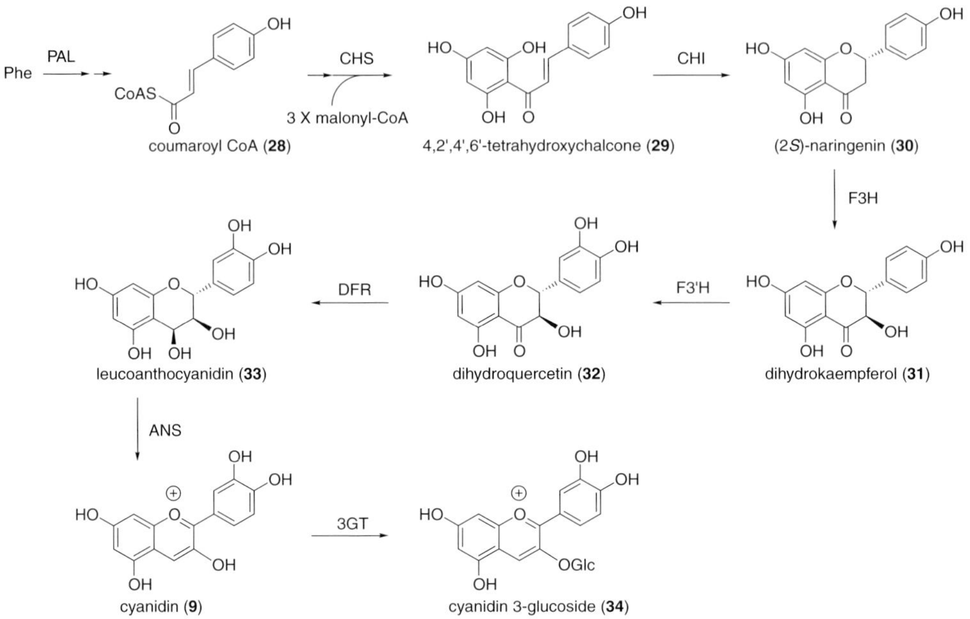

Fig. 4.8 Biosynthetic pathway of anthocyanin.

Forkman, 1994; Nakajima *et al.*, 2001; Turnbull *et al.*, 2004; Davies & Schwinn, 2006; Grotewold, 2006). After this, anthocyanidin 3-*O*-glucosyltransferase (3GT) acts to give anthocyanins. However, there has been no attempt to synthesize anthocyanins using this oxidative route from a leucoanthocyanidin or chemical equivalents thereof. Only the observation of a red coloration of the reaction mixture and the detection of the anthocyanidin nucleus from leucoanthocyanidin have been previously described (Sweeny & Iacobucci, 1977a, 1977b).

The first total synthesis of an anthocyanin using a biomimetic oxidative reaction was reported in 2006 (Fig. 4.9) (Kondo *et al.*, 2006). The starting material, (+)-catechin (**35**), was tetrabenzylated and then the remaining 3-OH group was β-glucosylated. After the benzyl groups were replaced with TBS groups, the C-4 position was oxidized by DDQ to give the 3,4-*cis*-4-hydroxycatechin **39**, which has the same substitution pattern as that of a leucoanthocyanidin, the biosynthetic intermediate. However, various trials to oxidize **39** to the corresponding anthocyanin failed. The reaction gave a reddish mixture, but chemical analysis indicated only traces of anthocyanin. The authors concluded that, under treatment with HCl, a cation might be generated at the C-4 position of **39**, followed by polymerization. Therefore, they converted **39** to the flav-3-en-3-ol (**40**) by dehydration using MsCl and *i*-Pr$_2$NEt at 80°C. The obtained compound **40** had the same oxidation state as the 3,4-*cis*-leuco compound, but could not generate any cation at the C-4 position. After deacetylation with MeONa, the enol was exposed to dried air in 1% HCl (gas)–MeOH solution for 8 hours to yield cyanidin 3-β-D-glucoside (**34**) (Fig. 4.9). This result strongly indicated that mild air oxidative conversion of an enol to a flavylium could be applicable to the synthesis of acylated anthocyanins.

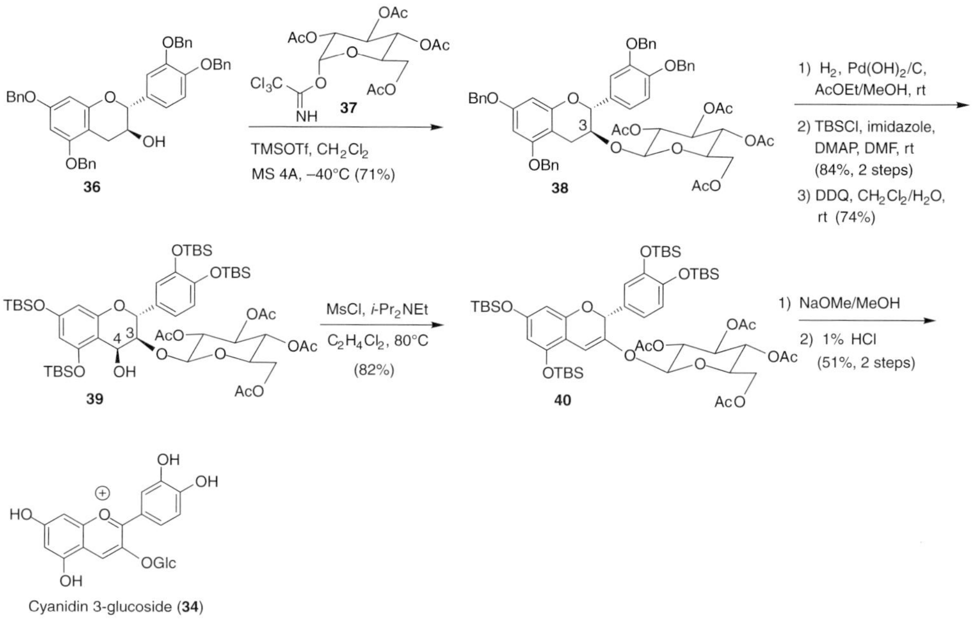

Cyanidin 3-glucoside (**34**)

Fig. 4.9   Total synthesis of cyanidin 3-glucoside (**34**) (Kondo *et al.*, 2006).

4.2.3 Transformation of flavonol derivatives to anthocyanins via a flavenol glycoside

As described in the previous section, the metal-mediated reduction of flavonol glycosides can give the corresponding anthocyanins in one step (Elhabiri *et al.*, 1995). Therefore, this was a practically convenient and promising method to obtain anthocyanins of complex structure. However, there were several problems: the yield of this transformation is generally low and preparation of glycosylated flavonols as a starting material is not trivial. The first total synthesis of pelargonidin 3-*O*-6″-*O*-acetylglucopyranoside (an acylated anthocyanin) (**41**) via metal reduction as the final step was recently reported (Fig. 4.10) (Oyama *et al.*, 2007). The authors synthesized a flavonol glycoside and applied the metal-reductive conversion to the corresponding anthocyanin. From phloroglucinol (**42**), the α-ketoalcohol **43** was prepared, then glucosylated. To subsequently apply the Baker–Venkataraman cyclization, **45** was esterified with the benzoic acid derivative **46**, representative of the B-ring of flavonol. Treatment of this ester under basic conditions gave the kaempferol 3-*O*-glucoside derivative **48**. Deprotection followed by metal reduction using Zn–Hg under HCl/MeOH gave the corresponding anthocyanin **41**. The 6″-*O*-acetyl residue of the glucoside in this anthocyanin is acid labile, and half of the acetyl groups were hydrolyzed under these conditions. Despite the hydrolysis, the desired acylated anthocyanin was obtained for the first time, albeit in low yield (Oyama *et al.*, 2007). Similarly, Bjorøy *et al.* reported the conversion of nine *C*-glycosylflavones (**50–58**) to the corresponding *C*-glycosylanthocyaninidins **59–67** in 14–32% yield (Fig. 4.11) (Bjorøy *et al.*, 2009).

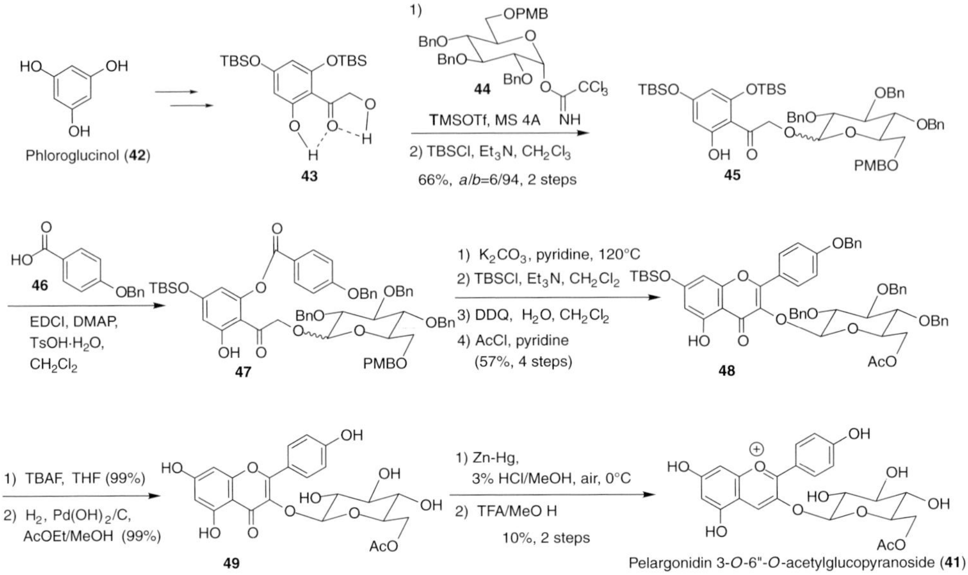

Fig. 4.10   Total synthesis of pelargonidin 3-*O*-6″-*O*-acetylglucopyranoside (**41**) (Oyama *et al.*, 2008).

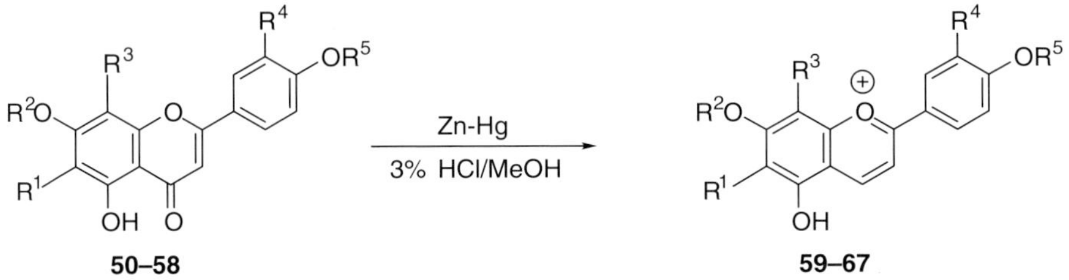

Flavone	R^1	R^2	R^3	R^4	R^5	Anthocyanin	Yield (%)
50	Glc	Me	H	H	H	**59**	20
51	Glc	Me	H	OH	H	**60**	21
52	Soph	Me	H	H	H	**61**	25
53	Soph	Me	H	H	Me	**62**	14
54	H	H	Glc	H	H	**63**	15
55	H	H	Rut	H	H	**64**	15
56	H	H	4-AcRut	H	H	**65**	32
57	Glc	H	Glc	H	H	**66**	18
58	Glc	H	Glc	OH	Me	**67**	22

Glc: glucoside, Soph: sohproside, Rut: rutinoside,
AcRut: 2-(4-acetylrhamnosyl)glucoside

Fig. 4.11   Conversion of *C*-glucosylflavones (**50–58**) to anthocyanins (**59–67**) (Bjorøy *et al.*, 2009).

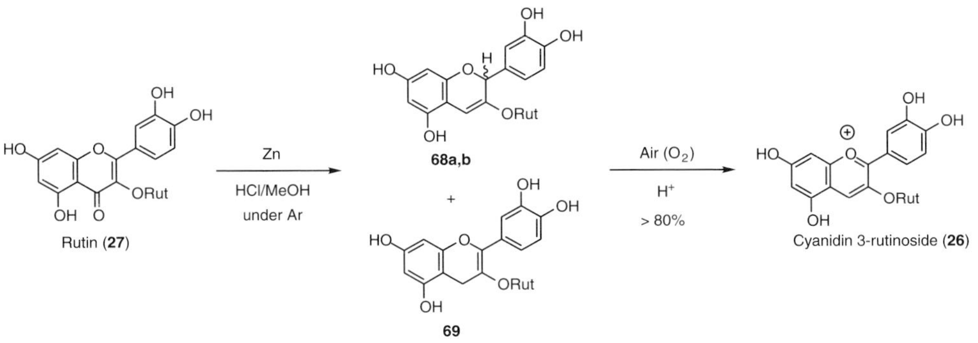

Fig. 4.12   Mechanism of transformation of flavonol to anthocyanin using metal reduction (Kondo *et al.*, 2009).

The reaction mechanism of the metal reduction was reinvestigated by Kondo *et al.* in detail and it was reported that metal reduction of glycosylflavonols did not give the corresponding anthocyanin in one step, but rather in two steps (Kondo *et al.*, 2008; Yoshida *et al.*, 2009b). When glycosylflavonols were treated with Zn or Zn–Hg in HCl/MeOH under anaerobic conditions, a mixture of flavenol derivatives was obtained. These flavenol derivatives were very much liable to air (O_2) and quickly oxidized to the corresponding anthocyanins under acidic conditions (Kondo *et al.*, 2008; Yoshida *et al.*, 2009b). Using this procedure, rutin (**27**) was converted to cyanidin 3-rutinoside (**26**) with >80% yield (Kondo *et al.*, 2009) (Fig. 4.12). Remarkably, this reductive transformation process to anthocyanins also passed through flavenols (**68a/b** and **69**) as an intermediate, which had the same oxidation state as that of the leucoanthocyanidin biosynthetic intermediate.

4.3 Synthesis of copigments for studying blue color development

For stable color development by anthocyanins, the coexistence of colorless compounds plays a very important role (Brouillard, 1988; Goto & Kondo, 1991; Brouillard & Dangles, 1994; Andersen & Jordheim, 2006; Yoshida *et al.*, 2009a; Brouillard *et al.*, 2010). This concept was first studied and proposed by R. Robinson and his group (1931) as the copigment effect or copigmentation, and compounds that expressed a bathochromic shift in the absorption spectrum and stabilized the chromophore of anthocyanins were named copigments. At the end of Robinson's era, not only were colorless flavonoids and aromatic compounds thought to be copigments, but also amino acid and sugars, mainly interacting through hydrogen bonds. Nowadays, copigments are recognized to be compounds with aromatic parts in their structures, so they can be stacked with the anthocyanidin chromophore through hydrophobic π–π interactions (Goto *et al.*, 1982, 1983a; Brouillard, 1988; Goto & Kondo, 1991; Brouillard & Dangles, 1994; Andersen & Jordheim, 2006; Yoshida *et al.*, 2009a; Brouillard *et al.*, 2010). This molecular stacking then gives a blue shift due to a charge transfer effect (Pina, 1998).

In addition to this stacking, the anthocyanidin chromophore is hence surrounded by hydrophobic motifs that prevent hydration, further stabilizing the color development. Therefore, clarification of the mechanisms of copigmentation is one of the very important tasks in the understanding of color development. In these studies, the synthesis of natural and unnatural copigments also made great contributions to this understanding. In this section, we will describe two synthetic studies on copigments for investigating blue flower color development using commelinin (**13**) from the blue dayflower (*C. communis*) and for sepal color development of the hydrangea (*Hydrangea macrophylla*).

4.3.1 Copigmentation in metalloanthocyanins

Glycosyl flavonoids have a strong copigment effect and have been studied in many flower colors (Asen *et al.*, 1970, 1972; Markham *et al.*, 1997; Yabuya *et al.*, 1997). The direct structural evidence for copigmentation was obtained from X-ray crystallographic analysis of commelinin (**13**) in 1992 (Kondo *et al.*, 1992), revealing a metalloanthocyanin in the blue dayflower (Fig. 4.13). The X-ray structure of **13** proved that this supramolecular pigment was composed of six molecules of anthocyanins, six molecules of flavones, and two Mg^{2+} ions. In this supramolecular association, three kinds of important molecular stacking interactions, self-association of anthocyanins (A), self-association of flavones (B), and anthocyanin–flavone copigmentation (C), take place within the metal complex of the anthocyanin and Mg^{2+} (Fig. 4.14) (Kondo *et al.*, 1992; Nakagawa, 1993; Yoshida *et al.*, 2009a). Furthermore, all of these molecular stackings are chiral; in the two self-associations, each component is stacked in an anticlockwise manner, and in the copigmentation, the components are stacked together in a clockwise fashion. This phenomenon was attributed to the fact that all the flavonoid components in commelinin (**13**) are glycosylated (Fig. 4.15) (Tamura *et al.*, 1986), so the chirality of the sugar units probably play a role in the induction the chirality in molecular stackings (Goto *et al.*, 1986; Kondo *et al.*, 2001; Yoshida *et al.*, 2009a).

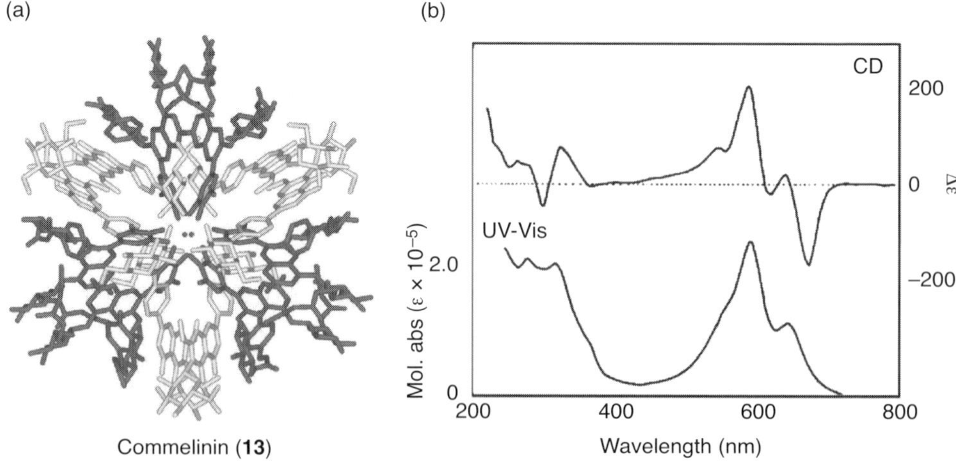

Fig. 4.13 X-ray crystallographic structure of commelinin (**13**) (a) and UV/VIS absorption spectrum and CD of commelinin (**13**) in aqueous solution (b) (Kondo *et al.*, 1992, with permission).

(a) (b) (c)

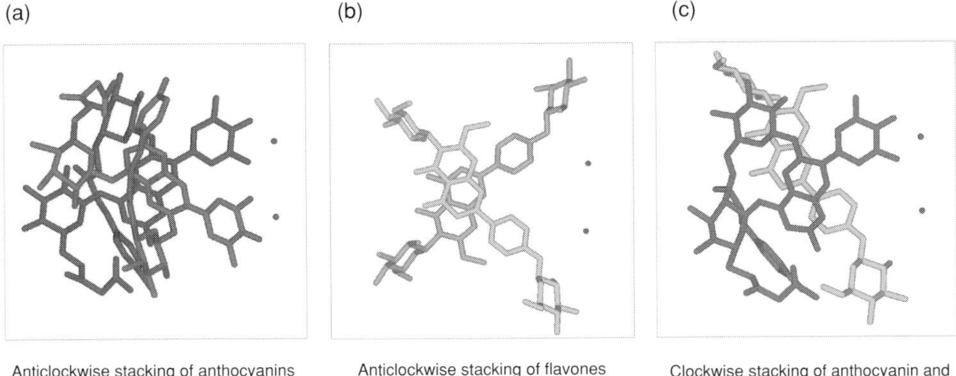

Fig. 4.14 A side view of a left-handed stacking of two molecules of malonylawobanin (**72**) that coordinate to different Mg^{2+}. (a) Side view of a copigmentation of Ma (**72**) and Fc (**73a**) in a right-handed stacking arrangement. (b) Side view of a left-handed stacking of two Fcs (**73a**). (c) Grayed thick line: Ma (**72**); white thick line: Fc (**73a**); black dots: Mg^{2+}. (Kondo *et al.*, 1992, with permission.)

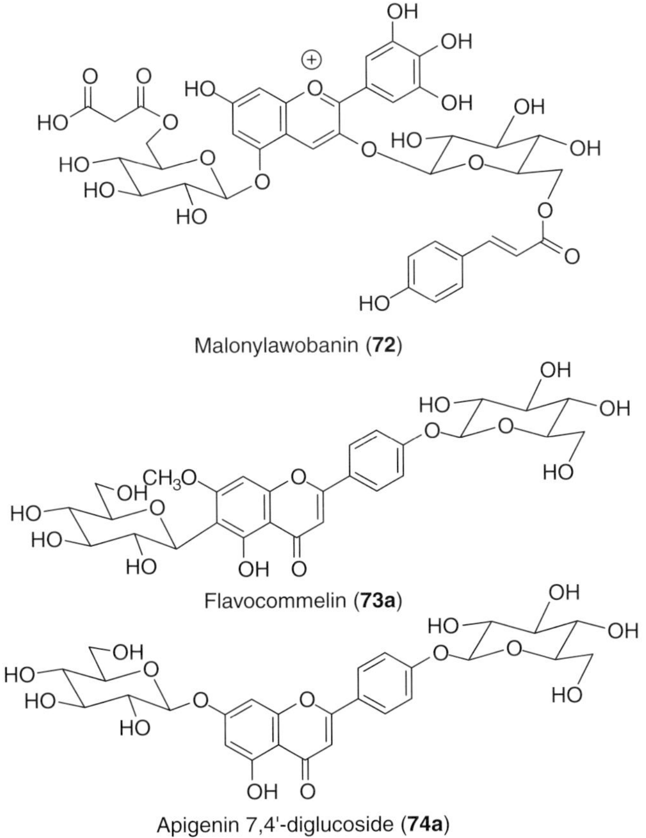

Fig. 4.15 Structure of the components of commelinin (**13**) and protodelphin (**71**).

In other metalloanthocyanins, such as protocyanin (**70**) (Kondo *et al.*, 1994, 1998; Shiono *et al.*, 2005) from blue cornflower petals (*C. cyanus*), the same chiral stacking was observed in circular dichroism (CD) measurements, NMR studies (Kondo *et al.*, 1994, 1998), and X-ray structure (Shiono *et al.*, 2005). Protodelphin (**71**) (Takeda *et al.*, 1994; Kondo *et al.*, 2001) from blue salvia petals (*Salvia patens, Salvia uliginosa*) was also shown to have similar molecular associations (Kondo *et al.*, 2001; Mori *et al.*, 2008). To clarify the copigmentation phenomenon and the mechanism of chiral molecular stacking, synthetic studies on glycosylated flavones have proven to be invaluable (Ellestad, 2006; Yoshida *et al.*, 2009a).

4.3.2 Synthesis of glycosylated flavones

Commelinin (**13**) and protodelphin (**71**) contains the same anthocyanin pigment, malonylawobanin (**72**) (Goto *et al.*, 1983b), but the flavone components are different. The former contains flavocommelin (**73a**), a *C*-glucosylflavone (Takeda *et al.*, 1966), and the latter contains apigenin-7,4′-*O*-di-glucoside (**74a**) (Fig. 4.15) (Takeda *et al.*, 1994). Both flavones have a similar structure in that two β-D-glucosyl residues are located at the right and left sides of the molecule. However, flavocommelin (**73a**) is highly water soluble and has a strong copigmentation effect (Hoshino *et al.*, 1980; Goto *et al.*, 1990). In aqueous solution, **73a** stacks with itself in a counterclockwise manner, which was proven by a strong negative Cotton effect in the CD spectrum (Goto *et al.*, 1990; Goto & Kondo, 1991). Therefore, the chiral stacking should be derived from the chirality of the D-glucosyl residues.

For chiral recognition studies, general and effective synthetic procedures for flavones containing antipodal glycosides are necessary. However, phenolic hydroxyl residues are relatively unreactive under glycosylation conditions, so only a few syntheses of poly-*O*-glycosyl-flavones with satisfactory yields have been reported. Among the phenolic hydroxyl residues in the flavonoid skeleton, the nucleophilicity of the 4′-hydroxyl group on the B-ring is the poorest (Oyama *et al.*, 2008). Therefore, there are very few examples of direct 4′-*O*-glycosylation (Demetzos *et al.*, 1990). Compared to *O*-glycosylation, the reports of *C*-glycosylation are even more scarce and require multistep routes (Frick *et al.*, 1989; Mahling *et al.*, 1995; Kumazawa *et al.*, 2000, 2001; Lee *et al.*, 2003). To overcome these problems, a reliable glycosylation method had to be developed.

The low reactivity to glycosylation of the 4′-OH residues in flavones was overcome by a new Lewis acid/base-promoted glycosylation (Oyama & Kondo, 1999) using peracetyl-glucosyl fluoride (**75**) (Hayasi et al., 1984). Using the combination of this method and the Koenigs–Knorr glycosylation, the flavone component in blue salvia petals, apigenin 7,4′-di-glucoside (**74a**), was synthesized (Fig. 4.16) (Kondo *et al.*, 2001, Oyama & Kondo, 2004a). In the synthetic studies, natural and unnatural apigenin 7,4′-di-*O*-β-glucosides, substituted with D- and/or L-glucose, were prepared. Briefly, condensation of naringenin (**76**) with peracetyl-D-glucosyl bromide (**16**) and consequent oxidization with DDQ gave the apigenin 7-*O*-peracetyl-β-D-glucoside (**77**). Treatment of **77** with peracetyl-D-glucosyl fluoride (**75**) in the presence of $BF_3 \bullet OEt_2$, 2,6-di-*tert*-butyl-4-methylpyridine (DTBMP) and 1,1,3,3-tetramethylguanidine (TMG) afforded apigenin 7,4′-*O*-di-β-D-glucoside as an acetate. Subsequent deprotection provided apigenin 7,4′-di-β-D-glucoside (**74a**). Using the

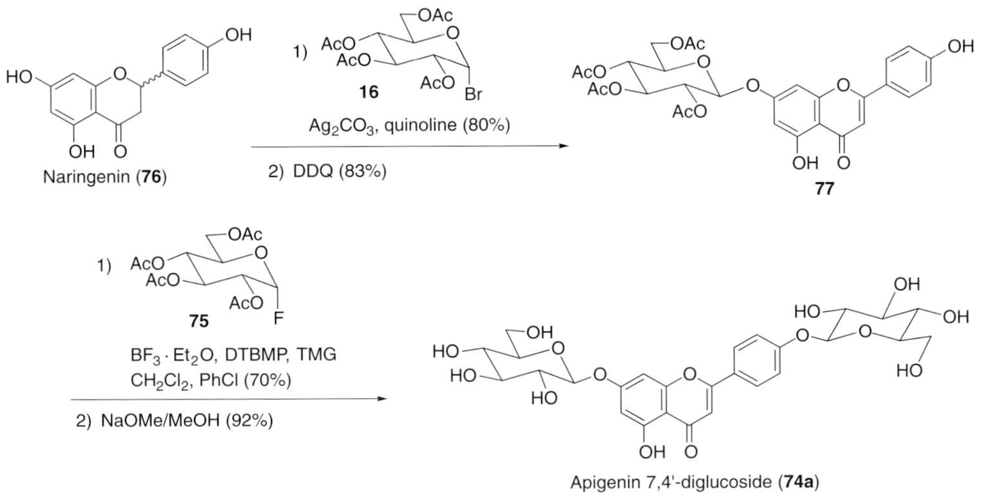

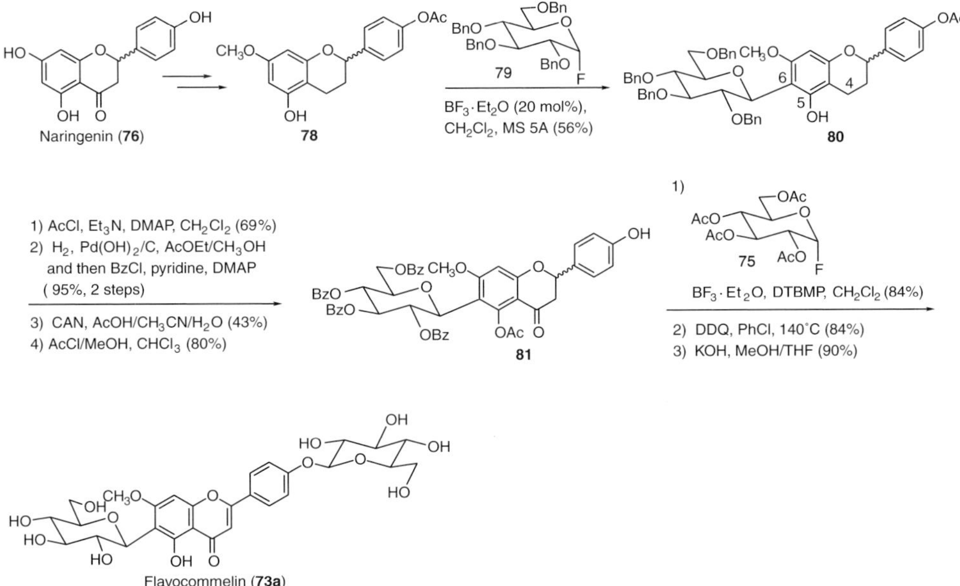

Apigenin 7,4'-diglucoside (**74a**)

Fig. 4.16   Total synthesis of apigenin 7,4′-*O*-di-glucoside (**74a**) (Kondo *et al.*, 2001; Oyama & Kondo, 2004a).

same procedure, a series of unnatural apigenin 7,4′-di-*O*-β-glucosides (**74b–d**) containing L-glucose and either 7-OH or 4′-OH monoglucosides were prepared.

Next, flavocommelin (7-*O*-methylapigenin 6-*C*-,4′-*O*-di-glucoside) (**73a**), the flavone component in commelinin from blue dayflower petals, was synthesized (Fig. 4.17) (Oyama, 2004; Oyama & Kondo, 2004b). In this synthesis, potent *C*-glucosylation methodology had to be developed. In the authors' strategy, 6-*C*-glucosylation was done first using a

Flavocommelin (**73a**)

Fig. 4.17   Total synthesis of flavocommelin (**73a**) (Oyama, 2004; Oyama & Kondo, 2004b).

7-*O*-methylflavan derivative. However, when a 4-carbonyl group existed in the flavonoid skeleton (flavone and flavanone), no direct glycosylation to the 5-*O* or the 6-*C* position occurred. Therefore, the C-ring of naringenin was reduced to flavan **78** and the reaction conditions were optimized. *C*-glycosylation of the flavan **78** was then carried out with glucosyl fluoride (**79**) using $BF_3 \bullet Et_2O$ as a promoter. The glycosylation reaction gave a mixture of 6-*C*-β-, 8-*C*-β-, 5-*O*-α-, and 5-*O*-β-glucosides. Despite the mixture, the desired 6-*C*-β-glucoside was produced in 56% yield as the major product. After oxidation at C-4 by cerium ammonium nitrate (CAN), the obtained flavanone derivative was then glucosylated at the 4′-OH using a combination of $BF_3 \bullet Et_2O$ and DTBMP to give the desired 4′-*O*-β-glucosyl compound in a high yield. Oxidation to the flavone skeleton by DDQ and deprotection gave flavocommelin (**73a**) in high yield (12 steps, 6.2% overall yield). Using the same strategy, unnatural L-glucosylated flavocommelins (**73b–d**) and monoglucosylated derivatives were prepared (Oyama, 2004). This first direct *C*-glycosylation of a flavonoid nucleus provided a promising method for the synthesis of other potentially naturally occurring *O*- and *C*-glycosyl flavonoids and their chiral analogs.

4.3.3 Chiral recognition in metalloanthocyanin formation

As mentioned previously, natural flavone components and their unnatural enantiomers, chiral analogs and mono-deglucosyl derivatives were synthesized and the stereochemistry of the self-association of these flavones was analyzed. To conduct this investigation, synthesized derivatives were dissolved in aqueous solution at a high concentration and the CD was then measured. As shown in Fig. 4.18, each pair of enantiomers exhibited opposite Cotton effects. These observations indicated that the chirality of stacking interactions during self-association of glucosylflavones could be explained by the chirality of the sugar residues.

To clarify the chiral recognition effects of the structural components of flavones during the formation of metalloanthocyanins, reconstruction of protodelphin (**71**) and commelinin (**13**) was investigated. For the synthesis of protodelphin (**71**), natural malonylawobanin (D-glucosylated) (**72**) (Goto *et al.*, 1983b) and the previously synthesized flavone components (**74b–d, 82**, and **83**), were mixed in a buffered solution with Mg^{2+} and the UV/VIS and CD spectra were recorded. As shown in Fig. 4.19, natural apigenin 7,4′-di-β-D-glucoside (**74a**) gave a stable blue solution, but its antipode, apigenin 7,4′-di-β-L-glucoside (**74b**), gave an unstable purple solution. The UV/VIS spectrum and CD of the solution obtained by mixing malonylawobanin (D-glucosylated) (**74a**) and apigenin 7,4′-di-β-L-glucoside (**74b**) did not show typical spectra of metalloanthocyanins, indicating that protodelphin was not formed; the purple color might be developed by malonylawobanin (**72**) being weakly copigmentated with **74b**.

Interestingly, D-glucosylated malonylawobanin (**72**) recognized the chirality of the flavone component. As such, apigenin 7-β-L-glucoside-4′-β-D-glucoside (**74d**) and apigenin 4′-β-D-glucoside (**74a**) also gave a blue solution of protodelphin-like metalloanthocyanin, although the stability of the color was low compared with that of the natural protodelphin (**71**). Furthermore, the enantiomer (**74b**), apigenin 4′-*O*-β-L-glucoside (**83b**), was investigated for complex formation, but no metalloanthocyanin was generated.

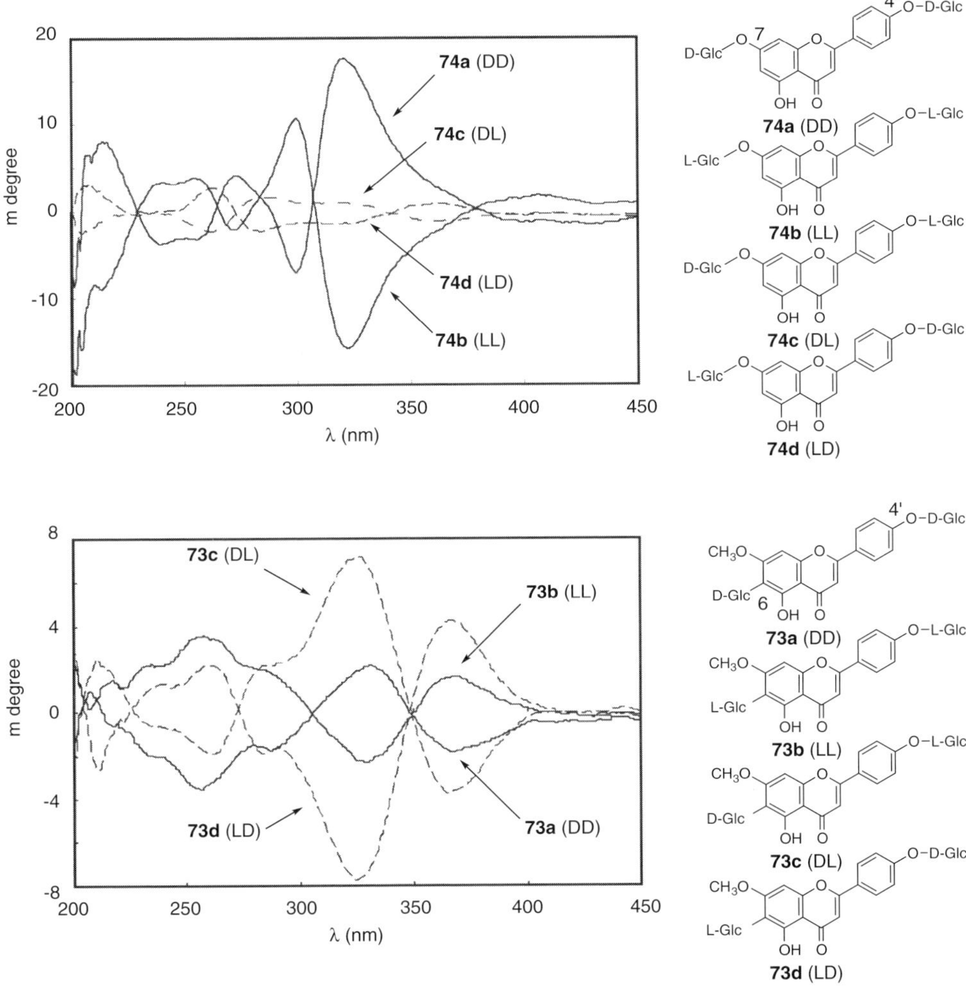

Fig. 4.18 Chiral stacking of natural (**74a, 73a**) and synthesized unnatural glycosylflavones (**73b, c, 74b, c**). CD spectra of apigenin di-glucosides (**74a, b**) in aqueous solution (5 × 10^{-4} M, 1.0-mm cell) (a). CD spectra of flavocommelins (**73a, b**) (in acetate buffer at pH 5.0, 5 × 10^{-3} M, 0.1-mm cell (b) (Oyama, 2004).

These results indicated that the D-glucosyl residue at the 4′-OH is indispensable for the formation of a metal-complex pigment. In addition, the D-glucose at 7-OH could stabilize the molecular association, while an L-glucose destabilized the complex due to steric hindrance (Kondo *et al.*, 2001; Oyama, 2004).

Furthermore, reconstruction experiments using a 1:1 mixture of apigenin 7,4′-di-β-D-glucoside (**74a**) and 7,4′-di-β-L-glucoside (**74b**) with Mg^{2+} was carried out. The obtained metalloanthocyanin was purified by GPC-LC, then the composition of the blue-black amorphous mass was analyzed by chiral-HPLC. The result was that contamination from the indicated L,L-diglucoside (**74b**) was completely excluded because the natural anthocyanin

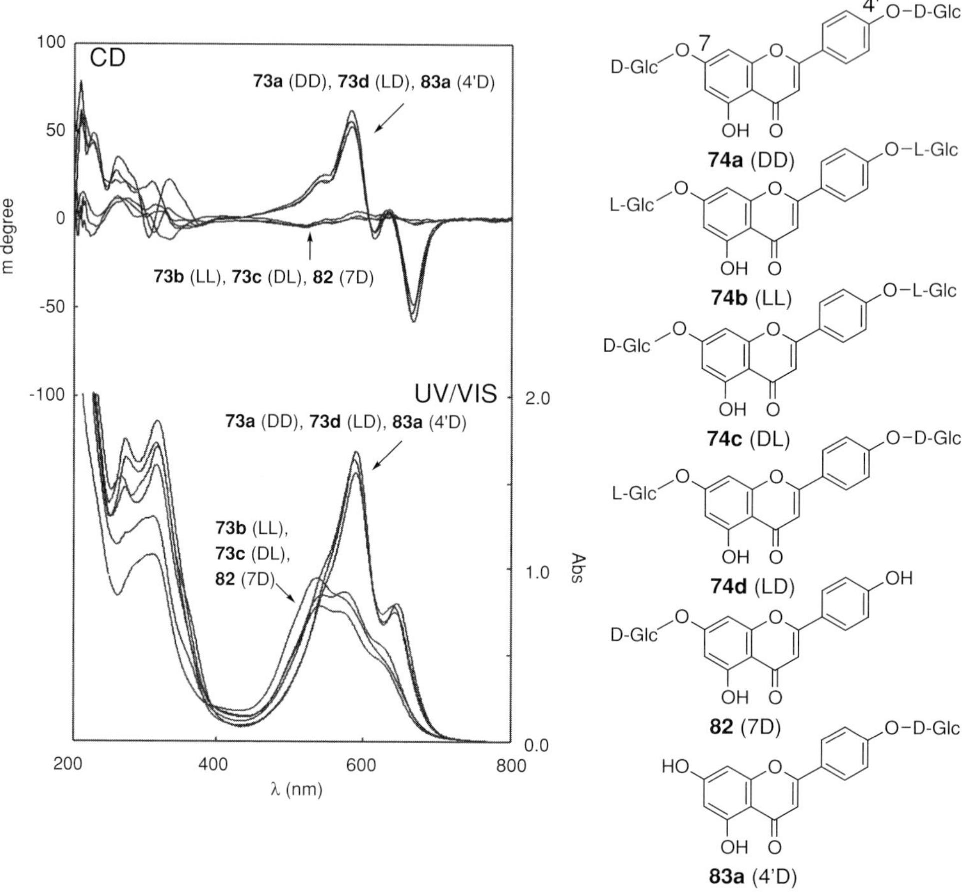

Fig. 4.19 Chiral recognition in the formation of protodelphin-like supramolecular pigments by mixing various apigenin glycosides (**74a–d, 82, 83a**) with malonylawobanin (**72**) and Mg^{2+}. CD and UV/VIS spectra of the solutions in acetate buffer at pH 6.0 (5×10^{-4} M, 1.0-mm cell). (Kondo *et al.*, 2001 with permission, Oyama & Kondo, 2004a.)

molecule, with a D-glucosyl functionality, chose only the D-series of glucosylflavones as partners to form metalloanthocyanins.

The same experiment was carried out in the reconstruction of commelinin. Natural 6-β-D-glucosyl-4′-β-D-glucoside (**73a**), the enantiomer 6-β-L-glucosyl-4′-β-L-glucoside (**73b**), and the diastereomers 6-β-D-glucosyl-4′-β-L-glucoside (**73c**) and 6-β-L-glucosyl-4′-β-D-glucoside (**73d**), were used for the commelinin formation. As shown in Fig. 4.20, natural flavocommelin (**73a**) and 6-β-L-glucosyl-4′-β-D-glucoside (**73d**) afforded a metalloanthocyanin (Oyama, 2004). The results were the same as those of experiments on the formation of protodelphin (**71**).

The high enantioselectivity with which the supramolecular association occurred could thus be caused by the chiral stacking arrangement of the sugar moieties in metalloanthocyanins. This phenomenon was further examined with the coordination data of commelinin

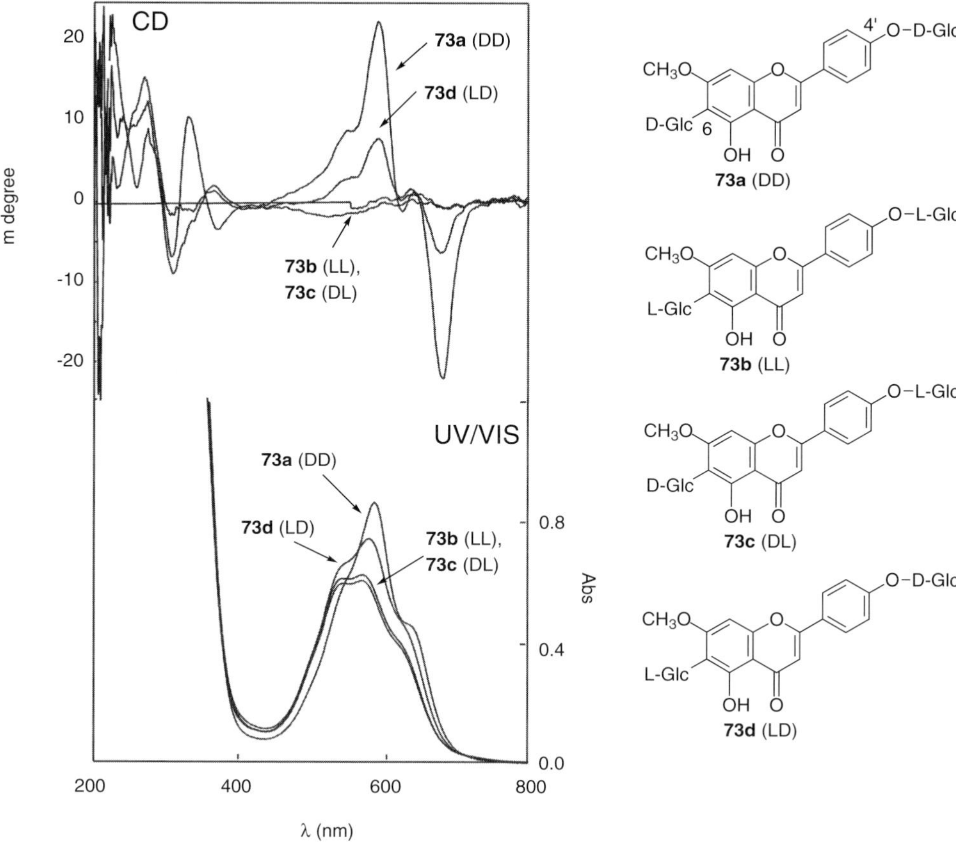

Fig. 4.20   Chiral recognition in the formation of commelinin-like supramolecular pigments by mixing various flavocommelin derivatives (**73a–d**) with malonylawobanin (**72**) and Mg^{2+}. (a) CD and (b) UV/VIS spectra of the solutions in acetate buffer at pH 6.0 (2.5×10^{-5} M, 1.0-mm cell) (Oyama, 2004).

(**13**) obtained by X-ray crystallographic analysis (Fig. 4.13). The three flavone molecules in commelinin (**13**) associated to form a *M* (minus) helical structure, just like a propeller with three blades. They were found bound at the pivot point by a strong hydrogen-bond network among hydroxyl groups at C-2 and C-3 of the 4′-*O*-β-D-glucopyranosides. Additionally, in the case of protodelphin (**71**), the two sets of *M*-helical flavones fit closely into the vacant space formed from the metal complex of six molecules of **72** and two ions of Mg^{2+}. Replacement of D-glucose with L-glucose at the 4′-OH of apigenin changed the helicity to the *P* (plus) form. As a result, **74b**, **74c**, and **83b** could not fit within the vacant space (Fig. 4.21). Thus, the *M*-helicity formed by the three molecules containing the D-glucosyl residue at the 4′-OH played a key role in the formation of metalloanthocyanins (Oyama, 2004). In conclusion, malonylawobanin (**72**) recognizes only the D-chirality of the 4′-*O*-glucosyl residue in apigenin 7,4′-diglucoside (**74a**) and flavocommelin (**73a**) to form the stoichiometric supramolecular metal-complex pigments, protodelphin (**71**) and commelinin (**13**), respectively (Kondo *et al.*, 2001). This strict chiral and structural

Three flavone molecules of **73a** (DD)
M-helical structure

Three flavone molecules of **73b** (LL)
P-helical structure

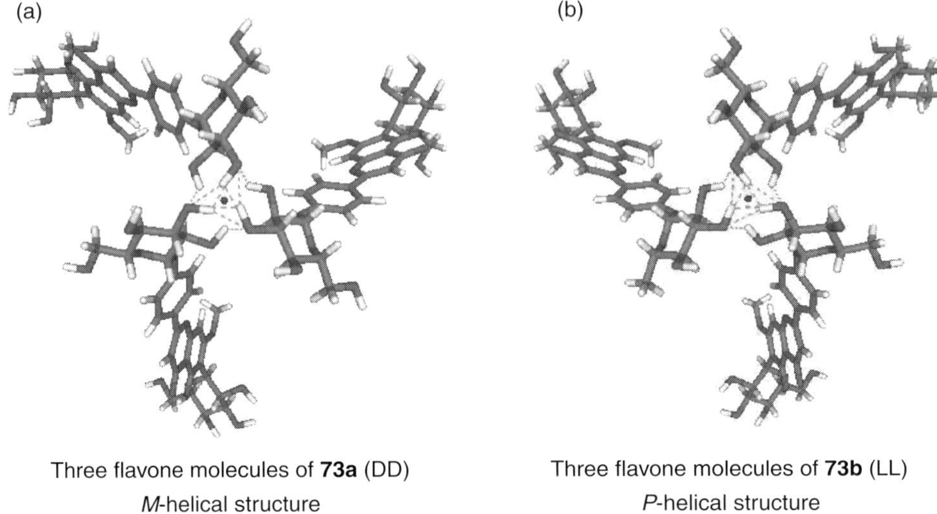

Fig. 4.21 Mechanism of chiral recognition in the formation of protodelphin (**71**). The *M*-helical chiral stacking arrangement of the three apigenin diglucosides (**74a**) in protodelphin (**71**) stabilized by a hydrogen-bond network (a). The *P*-helical chiral stacking arrangement of di-L-glucosyl apigenins (**74b**) (b).

recognition governed the entire self-assembly of metalloanthocyanins and is responsible for the beautiful blue flower color.

4.3.4 Synthesis of acylquinic acid derivatives for studies on hydrangea coloration

Coloration studies on sepals of *Hydrangea macrophylla* have a very long history extending into the nineteenth century. The most famous character of hydrangea is that their color changes readily (Allen, 1943). In addition to this, not only the blue color but also all of the other colors observed in hydrangeas are formed by one simple anthocyanin, delphinidin 3-glucoside (**84**) (Fig. 4.22) (Lawrence *et al.*, 1938; Robinson, 1939; Hayashi & Abe, 1953; Asen *et al.*, 1957). Furthermore, the three major copigments, 5-*O*-caffeoylquinic acid (neochlorogenic acid) (**85**), 5-*O*-*p*-coumaroylquinic acid (**86**), and 3-*O*-caffeoylquinic

Delphinidin 3-glucoside (**84**)

Neochlorogenic acid (**85**): R_1 = H, R_2 = caffeoyl
p-Coumaroylquinic acid (**86**): R_1 = H, R_2 = *p*-coumaroyl
Chlorogenic acid (**87**): R_1 = caffeoyl, R_2 = H

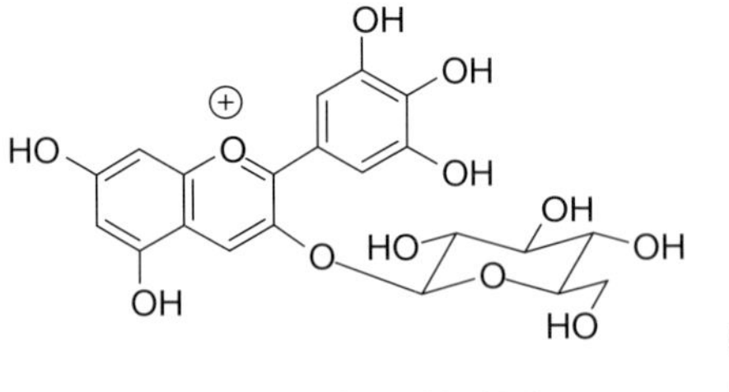

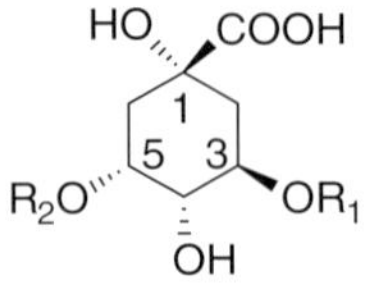

Fig. 4.22 Structure of the pigment components in blue hydrangea sepals.

acid (chlorogenic acid) (**87**) are also contained in all of the colored sepals (Hayashi & Abe, 1955; Asen *et al.*, 1957). Moreover, a correlation between the blue coloration and Al^{3+} has been suggested since the 1930s (Allen, 1932; Allen, 1943; Chenery 1937). In regard to the blue coloration, Takeda *et al.* (1985a, 1985b, 1990) reported that a stable blue solution could be obtained by mixing **84**, **85** or **86** with Al^{3+}. As suggested earlier, reconstruction experiments to realize the same color by mixing proposed components was a very powerful tool in the investigation of hydrangea color development.

In 2003, we reported that vacuolar pH also exerted some effect on color variation in hydrangea sepals (Yoshida *et al.*, 2003). Nevertheless, the chemical structure of the blue pigment in hydrangea sepals is still unclear, because the blue pigment is unstable and exists only in aqueous solutions (Yoshida *et al.*, 2003; Kondo *et al.*, 2005; Toyama-Kato *et al.*, 2007; Ito *et al.*, 2009). To obtain structural information on this elusive blue pigment, we designed and synthesized various copigment analogs (Toyama-Kato, 2003; Kondo *et al.*, 2005; Toyama-Kato *et al.*, 2007) and then carried out blue color reproduction experiments by mixing them with delphinidin 3-glucoside (**84**) and Al^{3+} *in vitro*.

In hydrangea sepal coloration, the copigments are acylquinic acid derivatives. Therefore, to clarify the essential structural part of the copigments, we designed several analogs that had structural modifications in the following six parts: (A) the position of the acyl moiety on the 3-OH or 5-OH, (B) the necessity of the hydroxyl group(s) at the aromatic acyl moiety, (C) the effect of the length of the conjugate plane of the aromatic acyl moiety, (D) the necessity of the 1-COOH of the quinic acid nucleus, (E) the necessity of the 1-OH of the quinic acid nucleus and (F) the necessity of the 5-ester group. The designed copigments are shown in Table 4.1.

Among the acylquinic acid derivatives, chlorogenic acid (**87**), neochlorogenic acid (**85**), and isochlorogenic acid (**88**) are commonly found in various plants, including edible vegetables, and attract attention for their anticancer properties (Cilla *et al.*, 2009; Carvalho *et al.*, 2010). Therefore, the development of a synthetic route to them is of importance, not only in coloration studies but also for a wide range of polyphenol research. In 2001, Sefkow *et al.* reported the synthesis of neochlorogenic acid (**85**) in high yield, but not its derivatives (Sefkow *et al.* 2001).

To realize a convergent and versatile synthesis of the target acylquinic acids, we planned the synthetic strategy shown in Fig. 4.23 as the first-generation route. The 1-COOH of (-)-quinic acid (**89**) was converted to the methyl ester (**90**) and then the 3,4-dihydroxyl group was protected using a previously reported procedure (Montchamp *et al.*, 1996). The protected compound **91** was dissolved in pyridine and treated with various acid chlorides to give the 5-*O*-acyl derivatives. All of the protecting groups were then removed under acidic conditions (2 N HCl–CH$_3$CN) to give the target acylquinic acids (**85**, **86**, **92–94**). Other compounds (**95–98**) were prepared after some modification of the synthetic route (Kondo *et al.*, 2005; Toyama-Kato *et al.*, 2007).

However, this synthetic route had several problems. For 5-acylation, acid chlorides must be prepared, and the acylation yield was not always good. Furthermore, the yield of the last deprotection was sometimes poor, because hydrolysis of the methyl ester of the 1-COOH was competitive with the hydrolysis reaction of the 5-ester, even under acidic conditions. Therefore, we developed a second-generation route for the synthesis of 5-acylquinic acids

Table 4.1 Formation of blue metal-complex solutions by mixing delphinidin 3-glucoside (**84**) with various copigments (**85–87, 92–98**) in the presence of Al^{3+} ion.

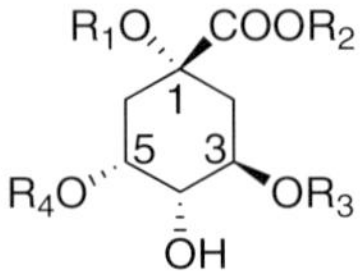

Compound no.	R_1	R_2	R_3	R_4	Copigmentation effect*
85	H	H	H	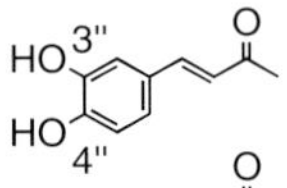 (caffeoyl, 3″,4″-diOH)	○
86	H	H	H	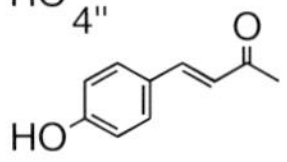 (p-coumaroyl, 4-OH)	○
87	H	H	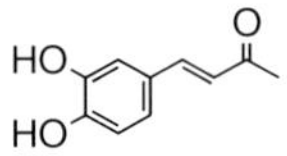 (caffeoyl, 3,4-diOH)	H	✕
92	H	H	H	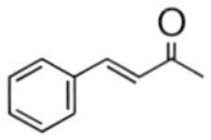 (cinnamoyl)	○
93	H	H	H	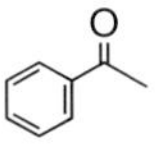 (acetophenone/benzoyl)	△
94	H	H	H	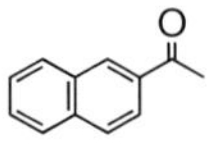 (naphthoyl)	◎
95	H	H	H	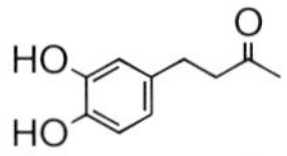 (dihydrocaffeoyl, 3,4-diOH)	△
96	H	CH_3	H	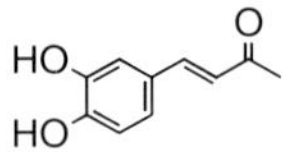 (caffeoyl, 3,4-diOH)	✕
97	CH_3	H	H	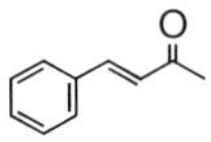 (cinnamoyl)	✕
98	H	H	H	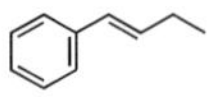 (cinnamyl)	✕

Copigmentation effect was examined by the color and the stability of the reproduction solution by mixing delphinidin 3-glucoside (**84**) : copigment : $Al^{3+} = 1$ mM : 3 mM : 1 mM ratio in buffer at pH 4.0. ◎, very stable blue solution; ○, stable blue solution; △, blue solution was obtained first, but after 1 day at room temperature a small amount of precipitate appeared; ✕, blue-black precipitate was produced quickly.

(Fig. 4.24). In this route, the protecting group for the 1-COOH was changed to a *p*-methoxybenzyl (PMB) ester, because this protecting group could be easily removed under acidic condition (Yan & Kahne, 1995; Shoji *et al.*, 2004; Berry *et al.*, 2009). For 5-acylation, we developed a modified Tanabe's esterification reaction. As such, Tanabe *et al.* reported a direct esterification between free carboxylic acids and alcohols using tosyl chloride and *N*-methylimidazole (Wakasugi *et al.* 2003). Tanabe's original reaction conditions gave only low

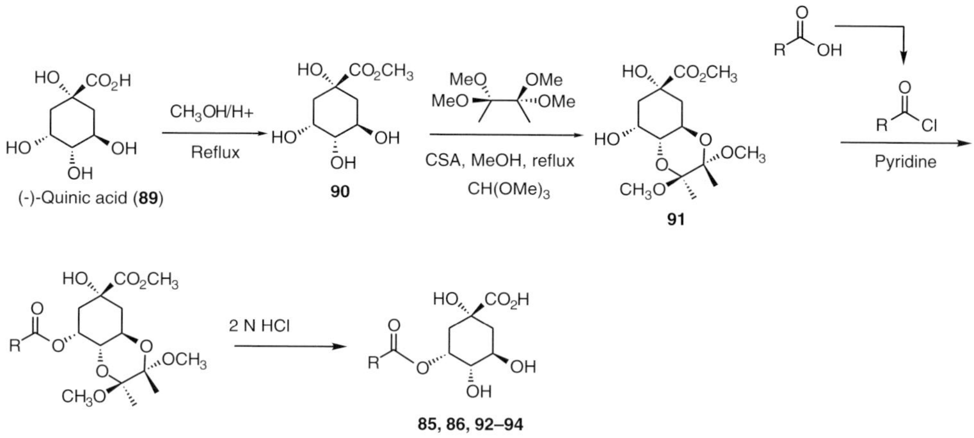

Fig. 4.23   First-generation synthetic route for acylquinic acids (**85, 86, 92–94**).

yield in our hands due to high steric hindrance, but the addition of di-*iso*-propylethylamine drastically enhanced the reaction rate and afforded the desired acylated compounds in high yields. Global deprotection was then carried out using trifluoroacetic acid–dichloromethane at room temperature (Yan & Kahne 1995; Shoji *et al.*, 2004; Berry *et al.*, 2009). Using this improved synthetic route, 5-*O*-acylquinic acids (**85, 86, 92–94**) were obtained from quinic acid (**89**) in 35–59% overall yield.

Using the natural copigments and synthesized analogs, experiments on the reproduction of the blue sepal color of the hydrangea were carried out. To determine the experimental conditions, the vacuolar pH value, the concentration of delphinidin 3-glucoside (**84**), the three copigment components, and the Al^{3+} in colored cells of hydrangea sepals were analyzed (Yoshida *et al.*, 2003; Ito *et al.*, 2009). The vacuolar pH of colored sepal cells was around 3–4 and the concentration of **84** in the colored vacuole was about 5–10 mM. Furthermore, the ratio of copigments **85, 86**, and **87** to **84** was between 1 and 10 eq. and the content of Al^{3+} varied depending on the sepal color. Specifically, the molar ratio of Al^{3+} to **84** in blue-colored cells was approximately one equivalent or higher (Yoshida *et al.*,

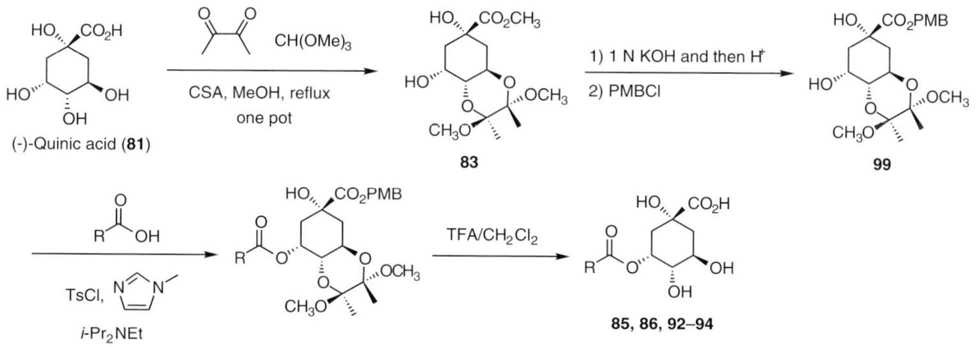

Fig. 4.24   Second-generation synthetic route for acylquinic acids (**85, 86, 92–94**).

2003; Ito *et al.*, 2009). Therefore, the ratio of mixing for the reproduction of the blue color was determined to be **84** : copigment : Al^{3+} =1 mM : 1–5 eq. : 1/3–3 eq. at pH 4. All the components were mixed in a buffer solution and the UV/VIS spectrum and CD were measured in a quartz cuvette (1.0 mm path length). The solution was held at room temperature and the stability of the color was recorded.

A mixture of **84** and Al^{3+} (1/3–3 eq.) at pH 4.0 without copigment gave a blue-colored solution at first, but the color was unstable and a blue-black precipitate quickly appeared, suggesting that the aluminum complex of delphinidin 3-glucoside (**84**) was insoluble in water. Furthermore, chlorogenic acid (**87**), a natural 3-*O*-acyl quinic acid, did not have any solubilizing or stabilizing effect on this complex; the precipitate formed immediately upon mixing and was composed of only anthocyanin and Al^{3+}. In contrast to this result, addition of >3 eq. of neochlorogenic acid (**85**), a natural 5-*O*-acyl quinic acid, to the anthocyanin solution gave a stable blue solution. Combining the results of all the experiments on reproducing the hydrangea blue color (Table 4.1), it was concluded that the 5-*O*-ester, the 1-COOH, and the 1-OH in the copigment were essential for obtaining a blue solution. In addition, the dihydroxyl group in the aromatic acyl moiety did not have any effect on the formation of the blue solution, but the aromatic plane of the 5-*O*-acyl moiety of quinic acid significantly contributed to stabilization of the blue color. These results strongly indicated that Al^{3+} may have chelated with the *ortho*-dihydroxyl group of the B-ring of **84**, but that this blue **84**–Al^{3+} complex was barely soluble in weakly acidic aqueous solutions. The 5-*O*-acylquinic acid derivatives had a copigmentation effect from a hydrophobic interaction between the anthocyanidin nucleus of **84** and the aromatic plane, solubilizing the complex and stabilizing the blue color in the solution, a mechanism that was indicated from CD measurements.

The CD of the suspension of blue protoplasts and the solution obtained by mixing **84**, **85**, and Al^{3+} were similar, all showing a single peak at 590 nm. Additionally, all stable blue solutions showed the same CD spectra as the one obtained for the **84** + **85** + Al^{3+} mixture. However, the CD of the mixture with **87** gave a negative exciton-type Cotton effect around the λ_{max}, indicating self-association of the chromophores of **84**. The mixture of **84** and Al^{3+} also showed the same CD spectrum. Furthermore, the copigment analogs (**96, 97, 98**) that did not give a blue solution, but did precipitate, also expressed negative exciton-type Cotton effects in their CD spectra. For the elaboration of the stable blue solution, it was confirmed that the 5-*O*-ester, the free 1-OH, and 1-carboxyl groups were essential. These findings suggested that the hydrophobic aromatic acyl moiety in effective copigments (**85, 86, 92, 94**) might insert in the self-associated anthocyanidin nuclei of the **84**–Al^{3+} complex, and stack with the chromophore. At that point, some coordination between the copigment and Al^{3+} might assist the stacking (Fig. 4.25) (Kondo *et al.*, 2005; Toyama-Kato *et al.*, 2007).

4.4 Conclusion

Stable color development of anthocyanins is never achieved without molecular stacking, except for polyacylated anthocyanins, which can stack intramolecularly. Including these intramolecular associations, all the anthocyanin coloration is endowed from formation

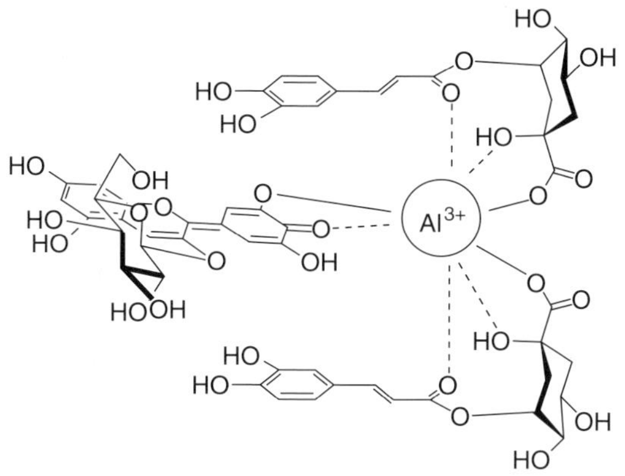

Fig. 4.25　Proposed gross structure of the blue hydrangea pigment.

of supermolecules spontaneously constructed when sufficient components and conditions are present. To clarify these chemical mechanisms, the development of organic synthetic methods for flavonoids and polyphenols is essential. Recently, synthetic studies on these compounds were refocused. As a result, versatile and practical synthetic routes for the production of flavonoid skeletons and the introduction of functionalization, such as glycosylation, acylation, and polymerization, will be established in the very near future. This progress not only should assist flower color research but also can open doors to polyphenol chemistry and investigation of biological functions.

Abbreviations

Ac	acetyl
AcCl	acetyl chloride
Bn	benzyl
BzCl	benzoyl chloride
CAN	cerium (IV) ammonium nitrate
CSA	(*1R*)-(−)-camphor-10-sulfonic acid
DDQ	2,3-dichloro-5,6-dicyano-*p*-benzoquinone
DMAP	dimethylaminopyridine
DMF	*N,N*-dimethylformamide
DTBMP	2,6-di-*tert*-butyl-4-methylpyridine
EDCI	1-ethyl-3(3-dimethylaminopropyl)carbodiimide hydrochloride
Et$_3$N	triethylamine
i-Pr$_2$Net	*N,N*-diisopropylethylamine (Hünig's base)
Me	methyl
TBAF	tetra-*n*-butylammonium fluoride
TBS	*tert*-butyldimethylsilyl
TBSCl	*tert*-butyldimethylsilyl chloride
TMG	tetramethylguanidine

TMSOTf	trimethylsilyl trifluoromethanesulfonate
TsCl	*p*-toluenesulfonyl chloride
TsOH	*p*-toluenesulfonic acid
MS 4A	molecular sieves 4 Å
MS 5A	molecular sieves 5 Å
Ms Cl	methanesulfonyl chloride
PhCl	phenylchloride
PMB	*p*-methoxybenzyl
PBMCl	*p*-methoxybenzyl chloride
TFA	trifluoroacetic acid

Acknowledgments

We are grateful to our coworkers and for financial support by a Grant-in-Aid for Scientific Research in the Global COE in Chemistry, Nagoya University, Creative Scientific Research (B) and (C) from The Ministry of Education, Culture, Sports, Science and Technology, Japan.

References

Allen, R.C. (1932) Factors influencing the color of hydrangea. *Proceedings of the American Society for Horticultural Science*, **28**, 410–412.

Allen, R.C. (1943) Influence of aluminum on the flower color of hydrangea macrophylla. *Boyce Thompson Institute*, **13**, 221–242.

Andersen, Ø.M. & Jordheim, M. (2006) The anthocyanins. In: *Flavonoids—Chemistry, Biochemistry and Applications* (eds. Ø.M. Andersen & K.R. Markham), pp. 471–551. CRC Press, Boca Raton, FL.

Asahina, Y. Nakagome, G. & Mototaro, I. (1929) Über die Reduktion der flavon- und flavanon-derivate (V. Mitteilung über die flavanon-glucoside). *Berichte der Deutschen Chemischen Gesellschaft*, **62B**, 3016–3021.

Asen, S., Siegelman, H.W. & Stuart, N.W. (1957) Anthocyanins and other phenolic compounds in red and blue sepals of hydrangea macrophylla var. merveille. *Proceedings of the American Society for Horticultural Science*, **69**, 561–569.

Asen, S., Stewart, R.N. & Norris, K.H. (1972) Co-pigmentation of anthocyanins in plant tissues and its effect on color. *Phytochemistry*, **11**, 1139–1144.

Asen, S., Stewart, R.N., Norris, K.H. & Massie, D.R. (1970) A stable blue non-metallic co-pigment complex of delphinidin and C-glycosylflavones in Prof. Blaauw Iris. *Phytochemistry*, **9**, 619–627.

Berry, N.G., Iddon, L., Iqbal, M., *et al.* (2009) Synthesis, transacylation kinetics and computational chemistry of a set of arylacetic acid 1*β*-*O*-acyl glucuronides. *Organic and Biomolecular Chemistry*, **7**, 2525–2533.

Bjorøy, Ø., Rayyan, S., Fossen, T., Kalberg, K. & Andersen, Ø.M. (2009) *C*-glycosylanthocyanidins synthesized from *C*-glycosylflavones. *Phytochemistry*, **70**, 278–287.

Brouillard, R. & Dangles, O. (1994) Flavonoids and flower colour. In: *The Flavonoids—Advances in Research Since 1986* (ed. J.B. Harborne), pp. 565–588. Chapman & Hall, London.

Brouillard, R., Chassaing, S., Isorez, G., Kueny-Stotz, M. & Figueiredo, P. (2010) The visible flavonoids or anthocyanins: from research to applications. In: *Recent Advances in Polyphenol*

Research (eds. C. Santos-Buelga M.T. Escibano-Bailon & V. Lattanzio), pp. 1–22. Wiley-Blackwell, Oxford.

Brouillard, R. (1988) Flavonoids and flower colour. In: *The Flavonoids—Advances in Research Since 1980* (ed. J.B. Harborne), pp. 525–538. Chapman & Hall, London.

Brouillard, R. & Delaporte, B. (1977) Chemistry of anthocyanin pigments. 2. Kinetic and thermo-dynamic study of proton transfer, hydration, and tautomeric reactions of malvidin 3-glucoside. *Journal of the American Chemical Society*, **99**, 8461–8468.

Carvalho, M., Jerónimo, C., Valentão, P., Andrade, P.B. & Silva, B.M. (2010) Green tea: a promising anticancer agent for renal cell carcinoma. *Food Chemistry*, **122**, 49–54.

Chenery, E.M. (1937) The problem of the blue hydrangea. *Journal of the Royal Horticultural Society*, **62**, 604–620.

Cilla, A., González-Sarrías, A., Tomás-Barberán. F.A., Espín, J.C. & Barberá, R. (2009) Availability of polyphenols in fruit beverages subjected to *in vitro* gastrointestinal digestion and their effects on proliferation, cell-cycle and apoptosis in human colon cancer Caco-2 cells. *Food Chemistry*, **114**, 813–820.

Dangles, O. & El Hajji, H. (1994) Synthesis of 3-methoxy- and 3-(β-D-glucopyranosyloxy)flavylium ions. Influence of the flavylium substitution pattern on the reactivity of anthocyanins in aqueous solution. *Helvetica Chimica Acta*, **77**, 1595–1610.

Davies, K.M. & Schwinn, K.E. (2006) Molecular biology and biotechnology of flavonoid biosyn-thesis. In: *Flavonoids—Chemistry, Biochemistry and Applications* (eds. Ø.M. Andersen & K.R. Markham), pp. 143–218. CRC Press, Boca Raton, FL.

Demetzos, C., Skaltsounis, A.-L., Tillequin, F. & Koch, M. (1990) Phase-transfer-catalyzed synthesis of flavonoid glycosides. *Carbohydrate Research*, **207**, 131–137.

El Hajji, H., Dangles, O., Figueiredo, P. & Brouillard, R. (1997) 3′-(β-D-glycopyranosyloxy)flavylium ions: synthesis and investigation of their properties in aqueous solution. Hydrogen bonding as a mean of color variation. *Helvetica Chimica Acta*, **80**, 398–413.

Elhabiri, M., Figueiredo, P., Fougerousse, A. & Brouillard, R. (1995) A convenient method for conversion of flavonols into anthocyanins. *Tetrahedron Letters*, **36**, 4611–4614.

Ellestad, G.A. (2006) Structure and chiroptical properties of supramolecular flower pigments. *Chirality*, **18**, 134–144.

Everest, A.E. (1914a) The Production of anthocyanins and anthocyanidins. *Proceedings of the Royal Society, Series B*, **87**, 444–452.

Everest, A.E. (1914b) The Production of anthocyanins and anthocyanidins. Part II. *Proceedings of the Royal Society, Series B*, **88**, 326–332.

Frick, W., Jung, K.-H. & Schmidt, R.R. (1989) Einfache Synthese von C-β-D-Glucopyranosylaromaten—Synthese des 5,7,4′-Tri-*O*-methylvitexins. *Justus Liebigs Annalen der Chemie*, **10**, 565–570.

Fukui, Y., Tanaka, Y., Kusumi, T., Iwashita, T. & Nomoto, K. (2003) A rationale for the shift in colour towards blue in transgenic carnation flowers expressing the flavonoid 3′,5′-hydroxylase gene. *Phytochemistry*, **63**, 15–23.

Goto, T. (1987) Structure, stability and color variation of natural anthocyanins. *Progress in the Chemistry of Organic Natural Products*, **52**, 113–158.

Goto, T. & Kondo, T. (1991) Structure and molecular stacking of anthocyanins Flower color variation. *Angewandte Chemie International Edition*, **30**, 17–33.

Goto, T., Kondo, T., Tamura, H., Imagawa, H., Iino, A. & Takeda, K. (1982) Structure of gentiodelphin, an acylated anthocyanin isolated from *Gentiana makinoi* that is stable in dilute aqueous solution. *Tetrahedron Letters*, **23**, 3695–3698.

Goto, T., Kondo, T., Tamura, H. & Kawahori, K. (1983a) Structure of platyconin, a diacylated anthocyanin isolated from the Chinese bell-flower *Platycodon grandiflorum. Tetrahedron Letters*, **24**, 2181–2184.

Goto, T., Kondo, T., Tamura, H. & Takase, S. (1983b) Structure of malonylawobanin, the real anthocyanin present in blue-colored flower petals of *Commelina communis. Tetrahedron Letters*, **24**, 4863–4866.

Goto, T., Tamura, H., Kawai, T., Hoshino, T., Harada, N. & Kondo, T. (1986) Chemistry of metalloanthocyanins. *Annals of the New York Academy of Sciences*, **471**, 155–173.

Goto, T., Tamura, H. & Kondo, T. (1987) Chiral stacking of cyanin and pelargonin. *Tetrahedron Letters*, **28**, 5907–5908.

Goto, T., Yoshida, K., Yoshikane, M. & Kondo, T. (1990) Chiral stacking of a natural flavone, flavocommelin, in aqueous solutions. *Tetrahedron Letters*, **31**, 713–716.

Grotewold, E. (2006) The genetics and biochemistry of floral pigments. *Annual Review of Plant Biology*, **57**, 761–780.

Harborne, J.B. & Grayer, R.J. (1988) The anthocyanins. In: *The Flavonoids—Advances in Research Since 1980* (ed. J. B. Harborne), pp. 1–20. Chapman & Hall, London.

Harborne, J.B. & Williams, C.A. (2000) Advances in flavonoid research since 1992. *Phytochemistry*, **55**, 481–504.

Hayashi, K. & Abe, Y. (1953) Studien über Anthocyane, XXIII. Papier-chromatographische übersicht der anthocyane im pflanzeneich I. *Miscellaneous Reports of the Research Institute for Natural Resources*, **29**, 1–8.

Hayashi, K. & Abe, Y. (1955) Anthocyanins. XXVII. Paper chromatographic observation on anthocyanins in plant pigments in autumn colored leaves. *Botanical Magazine (Tokyo)*, **68**, 299–308.

Hayashi, K., Abe, Y. & Mitsui, S. (1958) Blue anthocyanin from the flowers of commelina, the crystallisation and some properties there of studies on anthocyanins. XXX. *Proceedings of the Japan Academy*, **34**, 373–378.

Hayashi, M.H., Shunichi; Noyori, Ryoji. (1984) Simple synthesis of glycosyl fluorides. *Chemistry Letters*, 1747–1750.

Heller, W. & Forkmann, G. (1994) Biosynthesis of flavonoids. In: *The Flavonoids—Advances in Research Since 1986* (ed. J.B. Harborne), pp. 499–535. Chapman & Hall, London.

Holton, T.A. & Cornish, E.C. (1995) Genetics and biochemistry of anthocyanin biosynthesis. *Plant Cell*, **7**, 1071–1083.

Hoshino, T. & Goto, T. (1990) Effects of pH and concentration on the self-association of malvin quinonoidal base – electronic and circular dichroic studies. *Tetrahedron Letters*, **31**, 1593–1596.

Hoshino, T., Matsumoto, U. & Goto, T. (1980) The stabilizing effect of the acyl group on the co-pigmentation of acylated anthocyanins with *C*-glucosylflavones. *Phytochemistry*, **19**, 663–667.

Iacobucci, G.A. & Sweeny, J.G. (1983) The chemistry of anthocyanins, anthocyanidins and related flavylium salts. *Tetrahedron*, **39**, 3005–3038.

Ito, D., Shinkai, Y., Kato, Y., Kondo, T. & Yoshida, K. (2009) Chemical studies on different color development in blue- and red-colored sepal cells of *Hydrangea macrophylla*. *Bioscience, Biotechnology, and Biochemistry*, **49**, 1054–1059.

Katsumoto, K., Fukuchi-Mizutani, M., Fukui, Y., *et al.* (2007) Engineering of the rose flavonoid biosynthetic pathway successfully generated blue-hued flowers accumulating delphinidin. *Plant and Cell Physiology*, **48**, 1589–1600.

Kondo, T., Oyama, K.-I., Bjorøy, Ø. *et al.* (2008) New synthetic methodologies of anthocyanin, flower pigment. In: 50th Symposium on the Chemistry of Natural Products, Fukuoka, Japan (ed. T. Katsuki), vol. 50, pp. 725–730. Organizing Committee of 50th Symposium on the Chemistry of Natural Products, Fukuoka, Japan.

Kondo, T., Oyama, K.-I, Nakamura, S. Yamakawa, D., Tokuno, K. & Yoshida, K. (2006) Novel and efficient synthesis of cyanidin 3-*O*-*β*-D-Glucoside from (+)-catechin via a flav-3-en-3-ol as a key intermediate. *Organic Letters*, **8**, 3609–3612.

Kondo, T., Oyama, K.-I. & Yoshida, K. (2001) Chiral molecular recognition on formation of a metalloanthocyanin: a supramolecular metal complex pigment from blue flower of *Salvia patens*. *Angewandte Chemie International Edition*, **40**, 894–897.

Kondo, T., Toyama-Kato, Y. & Yoshida, K. (2005) Essential structure of co-pigment for blue sepal-color development of hydrangea. *Tetrahedron Letters*, **46**, 6645–6649.

Kondo, T., Ueda, M., Isobe, M. & Goto, T. (1998) A new molecular mechanism of blue color development with protocyanin, a supramolecular pigment from cornflower, *Centaurea cyanus*. *Tetrahedron Letters*, **39**, 8307–8310.

Kondo, T., Ueda, M., Tamura, H., Yoshida, K., Isobe, M. & Goto, T. (1994) Composition of protocyanin, a self-assembled supramolecular pigment from the blue cornflower, *Centaurea cyanus*. *Angewandte Chemie International Edition*, **33**, 978–979.

Kondo, T., Yoshida, K., Nakagawa, A., Kawai, T., Tamura, H. & Goto, T. (1992) Structural basis of blue-color development in flower petals from *Commelina communis*. *Nature*, **358**, 515–518.

Kumazawa, T., Kimura, T., Matsuba, S., Sato, S. & Onodera, J.-I. (2001) Synthesis of 8-*C*-glucosylflavones. *Carbohydrate Research*, **334**, 183–193.

Kumazawa, T., Minatogawa, T., Matsuba, S., Sato, S. & Onodera, J.-I. (2000) An effective synthesis of isoorientin: the regioselective synthesis of a 6-*C*-glucosylflavone. *Carbohydrate Research*, **329**, 507–513.

Lawrence, W.J.C., Price, J.R., Robinson, G.M. & Robinson, R. (1938) CCXV. A survey of anthocyanins. V. *Biochemical Journal*, **32**, 1661–1667.

Lee, D.Y.W., Zhang, W.-Y. & Karnati, V.V.R. (2003) Total synthesis of puerarin, an isoflavone *C*-glycoside. *Tetrahedron Letters*, **44**, 6857–6859.

Mahling, J.-A., Jung, K.-H. & Schmidt, R.R. (1995) Synthesis of flavone *C*-glycosides vitexin, isovitexin, and isoembigenin. *Liebigs Annalen der Chemie*, 461–466.

Markham, K.R., Mitchell, K.A. & Boase, M.R. (1997) Malvidin-3-*O*-glucoside-5-*O*-(6-acetylglucoside) and its color manifestation in 'Johnson's Blue' and other 'Blue' geraniums. *Phytochemistry*, **45**, 417–423.

Martinoia, E., Maeshima, M. & Neuhaus, H.E. (2007) Vacuolar transporters and their essential role in plant metabolism. *Journal of Experimental Botany*, **58**, 83–102.

Montchamp, J.-L. Tian, F., Hart, M.E. & Frost, J.W. (1996) Butane 2,3-bisacetal protection of vicinal diequatorial diols. *Journal of Organic Chemistry*, **61**, 3897–3899.

Mori, M., Kondo, T. & Yoshida, K. (2008) Cyanosalvianin, a supramolecular blue metalloanthocyanin, from petals of *Salvia uliginosa*. *Phytochemistry*, **69**, 3151–3158.

Murakami, S., Robertson, A. & Robinson, R. (1931) Synthesis of anthocyanins. VI. Synthesis of chrysanthemin chloride. *Journal of the Chemical Society*, 2665–2671.

Nakagawa, A. (1993) X-ray structure determination of commelinin from *Commelina communis* and its blue-color development. *Journal of the Crystallographic Society of Japan*, **35**, 327–333.

Nakajima, J.-i., Tanaka, Y., Yamazaki, M. & Saito, K. (2001) Reaction mechanism from leucoanthocyanidin to anthocyanidin 3-glucoside, a key reaction for coloring in anthocyanin biosynthesis. *Journal of Biological Chemistry*, **276**, 25797–25803.

Nigel, C.V. & Grayer, R.J. (2008) Flavonoids and their glycosides, including anthocyanins. *Natural Product Reports*, **25**, 555–611.

Oyama, K.-I. & Kondo, T. (1999) Highly efficient β-glucosylation of the acidic hydroxyl groups, phenol and carboxylic acid, with an peracetylated glucosyl fluoride using a combination of $BF_3 \cdot Et_2O$ and DTBMP as a promoter. *Synlett*, 1627–1629.

Oyama, K.-I. (2004) Synthesis of glycosylated flavones and their chiral recognition in formation of metalloantocyanins. Doctoral Thesis, Nagoya University, Nagoya, Japan.

Oyama, K.-I., Kawaguchi, S., Yoshida, K. & Kondo, T. (2007) Synthesis of pelargonidin 3-*O*-6″-*O*-acetyl-β-D-glucopyranoside, an acylated anthocyanin, via the corresponding kaempferol glucoside. *Tetrahedron Letters*, **48**, 6005–6009.

Oyama, K.-I. & Kondo, T. (2004a) Total synthesis of apigenin 7,4′-di-*O*-β-glucopyranoside, a component of blue flower pigment of *Salvia patens*, and seven chiral analogs. *Tetrahedron*, **60**, 2025–2034.

Oyama, K.-I. & Kondo, T. (2004b) Total Synthesis of flavocommelin, a component of the blue supramolecular pigment from *Commelina communis*, on the basis of direct 6-*C*-glycosylation of flavan. *Journal of Organic Chemistry*, **69**, 5240–5246.

Oyama, K.-I., Kondo, T. & Yoshida, K. (2008) Synthesis of oriented anti-virus 7-*O*-substituted apigenins. *Heterocycles,* **76**, 1607–1615.

Oyama, K.-I., Kondo, T. & Yoshida, K. (2011) Recent progress in the synthesis of flavonoids: from monomers to supra-complex molecules. *Current Organic Chemistry*, **15**, 2567–2607.

Pina, F. (1998) Caffeine interaction with synthetic flavylium salts. A flash photolysis study for the adduct involving 4′,7-dihydroxyflavylium. *Journal of Photochemistry and Photobiology A: Chemistry*, **117**, 51–59.

Robertson, A. & Robinson, R. (1927) Synthesis of anthocyanins. II. Synthesis of 3- and 7-glucosidoxyflavylium salts. *Journal of the Chemical Society*, 242–247.

Robinson, G.M. (1939) The colloid chemistry of leaf and flower pigments and the precursors of the anthocyanins. *Journal of the American Chemical Society*, **61**, 1606–1607.

Robinson, G.M. & Robinson, R. (1931) CLXXXII. A survey of anthocyanins. I. *Biochemical Journal*, **25**, 1687–1705.

Robinson, R. (1934) Synthesis of anthocyanins. *Chemische Berichte*, **67A**, 85–105.

Robinson, R. & Todd, A.R. (1932) Experiments on the synthesis of anthocyanins. Part XV. A synthesis of hirsutin chloride. *Journal of the Chemical Society*, 2293–2299.

Saito, N., Osawa, Y. & Hayashi, K. (1971) Platyconin, a new acylated anthocyanin in chinese bell-flower, *Platycodon grandiflorum*. *Phytochemistry*, **10**, 445–447.

Sefkow, M., Kelling, A. & Schilde, U. (2001) First efficient syntheses of 1-,4-, and 5-caffeoylquinic acid. *European Journal of Organic Chemistry*, 2735–2742.

Shibata, K., Shibata, Y. & Kasiwagi, I. (1919) Anthocyanins: color-variation in anthocyanins. *Journal of the American Chemical Society*, **41**, 208–220.

Shiono, M., Matsugaki, N. & Takeda, K. (2005) Structure of the blue cornflower pigment. *Nature*, **436**, 791.

Shoji, M., Uno, T. & Hayashi, Y. (2004) Stereoselective total synthesis of *ent*-EI-1941-2 and epi-*ent*-EI-1941-2. *Organic Letters*, **6**, 4535–4538.

Springob, K., Nakajima, J., Yamazaki, M. & Saito, K. (2003) Recent advances in the biosynthesis and accumulation of anthocyanins. *Natural Product Reports*, **20**, 288–303.

Strack, D. & Wray, V. (1994) The anthocyanins. In: *The Flavonoids – Advances in Research since 1986* (ed. J.B. Harborne), pp. 1–22. Chapman & Hall, London.

Sweeny, J.G. & Iacobucci, G.A. (1977a) Synthesis of anthocyanidins. I. The oxidative generation of flavylium cations using benzoquinones. *Tetrahedron*, **33**, 2923–2926.

Sweeny, J.G. & Iacobucci, G.A. (1977b) Synthesis of anthocyanidins. II. The synthesis of 3-deoxyanthocyanidins from 5-hydroxy-flavanones. *Tetrahedron*, **33**, 2927–2932.

Takeda, K., Kariuda, M. & Itoi, H. (1985a) Blueing of sepal colour of *Hydrangea macrophylla*. *Phytochemistry*, **24**, 2251–2254.

Takeda, K., Kubota, R. & Yagioka, C. (1985b) Copigments in the blueing of sepal colour of Hydrangea macrophylla. *Phytochemistry*, **24**, 1207–1209.

Takeda, K., Mitsui, S. & Hayashi, K. (1966) Structure of a new flavonoid in the blue complex molecule of commelinin. Studies on anthocyanins, LIV. *Botanical Magazine (Tokyo)*, **79**, 578–587.

Takeda, K., Yamashita, T., Takahashi, A. & Timberlake, C.F. (1990) Stable blue complexes of anthocyanin-aluminium-3-p-coumaroyl- or 3-caffeoyl-quinic acid involved in the blueing of *Hydrangea* flower. *Phytochemistry*, **29**, 1089–1091.

Takeda, K., Yanagisawa, M., Kifune, T., Kinoshita, T. & Timberlake, C.F. (1994) A blue pigment complex in flowers of *Salvia patens*. *Phytochemistry*, **35**, 1167–1169.

Tamura, H., Kondo, T. & Goto, T. (1986) The composition of commelinin, a highly associated metalloanthocaynin present in the blue flower petals of *Commelina communis*. *Tetrahedron Letters*, **27**, 1801–1804.

Toyama-Kato, Y. (2003) Chemical studies on blue flower color development of hydrangea using a molecular designed synthetic co-pigments. Doctoral Thesis, Nagoya University, Nagoya, Japan.

Toyama-Kato, Y., Kondo, T. & Yoshida, K. (2007) Synthesis of designed acylquinic acid derivatives involved in blue color development of hydrangea and their co-pigmentation effect. *Heterocycles*, **72**, 239–254.

Turnbull, J.J., Nakajima, J.-I., Richard, W.D., Yamazaki, M., Saito, K. & Schofield, C.J. (2004) Mechanistic studies on three 2-oxoglutarate-dependent oxygenases of flavonoid biosynthesis: anthocyanidin synthase, flavonol synthase, and flavanone 3β-hydroxylase. *Journal of Biological Chemistry*, **279**, 1206–1216.

Wakasugi, K., Iida, A., Misaki, T., Nishii, Y. & Tanabe, Y. (2003) Simple, mild, and practical esterification, thioesterification, and amide formation utilizing *p*-toluenesulfonyl chloride and *N*-methylimidazole. *Advanced Synthesis & Catalysis*, **345**, 1209–1214.

Willstätter, R. & Everest, A.E. (1913) Untersuchungen uber die Anthocyane. I. Uber den Farbstoff der Kornblume. *Justus Liebigs Annalen der Chemie*, **401**, 189–232.

Willstätter, R. & Mallison, H. (1915) Untersuchungen uber die Anthocyane; X. Uber Variationen der Blutenfarben. *Justus Liebigs Annalen der Chemie*, **408**, 147–162.

Willstätter, R. & Marison, H. (1914) The relationship between anthocyans and flavones. *Sitzungsberichte der Königlich Preussischen Akademie der Wissenschaften zu Berlin*, 769–777.

Yabuya, T., Nakamura, M., Iwashina, T., Yamaguchi, M. & Takehara, T. (1997) Anthocyanin-flavone copigmentation in bluish purple flowers of Japanese garden iris (Iris ensata Thunb.) *Euphytica*, **98**, 163–167.

Yan, L. & Kahne, D. (1995) *p*-Methoxybenzyl ethers as acid-labile protecting groups in oligosaccharide synthesis. *Synlett*, 523–524.

Yoshida, K. (1992) Molecular stacking of anthocyanins with fine structure recognition. Doctoral Thesis, Nagoya University, Nagoya, Japan.

Yoshida, K., Kondo, T. & Oyama, K.-I. (2009b) Process for the preparation of anthocyanidin derivative. Patent No. JP 2009137904.

Yoshida, K., Mori, M. & Kondo, T. (2009a) Blue flower color development by anthocyanins: from chemical structure to cell physiology. *Natural Product Reports*, **26**, 884–915.

Yoshida, K., Toyama-Kato, Y., Kameda, K. & Kondo, T. (2003) Sepal color variation of *Hydrangea macrophylla* and vacuolar pH measured with a proton-selective microelectrode. *Plant and Cell Physiology*, **44**, 262–268.

Chapter 5
Colouring up Plant Biotechnology

*Cathie Martin, Yang Zhang, Laurence Tomlinson, Kalyani Kallam,
Jie Luo, Jonathan D.G. Jones, Antonio Granell, Diego Orzaez
and Eugenio Butelli*

Abstract: We review the anthocyanin biosynthetic pathway describing the enzymatic
steps involved in anthocyanidin formation and the proteins involved in transport of
anthocyanins to the vacuole. Regulatory proteins interacting in a MYB-bHLH-WDR
(MBW) complex control the transcription of the genes encoding the enzymes and
associated proteins involved in anthocyanin biosynthesis. Using this system, we
have developed vectors that allow for *in vivo* screening of transformed plant cells
(marked by their production of anthocyanin). These vectors can be used as *cis*-genic
screenable markers for different plant species. This system can also be used as an
in vivo reporter of gene expression that is sensitive and does not involve destruction of
tissues, staining or use of fluorescence microscopes. We have produced tomato plants
in which the fruit make high levels of purple anthocyanins. This provides a research
tool for marking silenced areas following agro-injection of VIGS vectors in fruit.
This facilitates high-throughput analysis of genes of unknown function expressed
during fruit ripening.

Keywords: anthocyanins; pigments; copigmentation; synthesis; metabolic engineering; gene reporter; VIGS; health-promoting.

5.1 Introduction

The accumulation of anthocyanin pigments in plants is one of the most familiar processes
of secondary metabolism and also one of the best understood, in terms of the biochemistry
of anthocyanin synthesis, the cellular mechanisms of sequestration of anthocyanins and the
mechanisms of regulation of the biosynthetic pathway. Anthocyanin pigments are produced
in flowers to attract pollinators and in fruits to attract seed dispersers. Anthocyanins are
made in vegetative tissues under a wide variety of stress conditions and are believed to
offer protection against photo-oxidative damage. Anthocyanins also accumulate in the

Recent Advances in Polyphenol Research, Volume 3, First Edition. Edited by Véronique Cheynier,
Pascale Sarni-Manchado and Stéphane Quideau.

senescing leaves of many deciduous trees, giving them their autumn colours. Anthocyanins are water-soluble pigments with colours which range from orange through purple to blue, dependent on their chemistry, their interactions with other compounds and their chemical environments. Recently, they have also been suggested to have health-promoting properties when consumed in the diet in significant amounts, although it remains unclear whether these properties result from their activities as antioxidants, or from more specific effects on signalling pathways in animals.

5.2 Plant production of anthocyanins

5.2.1 Synthesis of anthocyanins

Most higher plants have the ability to synthesise anthocyanins, but the precise nature of the anthocyanins that form may differ widely in different plant species due to the activity of specific decorating enzymes which add sugars, methyl and acyl residues to the basic anthocyanidin skeleton. Because mutants affecting anthocyanin biosynthesis are easy to identify, it is known that the synthesis of anthocyanidins is catalysed by a core pathway of about nine conserved enzymes; phenylalanine ammonia lyase (PAL), cinnamate 4-hydroxylase (C4H) and p-coumaroyl 4-CoA ligase (4CL) of the general phenyl propanoid pathway, and chalcone synthase (CHS), chalcone isomerase (CHI), flavanone 3-hydroxylase (F3H), dihydroflavonol 4-reductase (DFR) and anthocyanidin synthase (ANS; also known as leucoanthocyanidin dioxygenase or LDOX) of the flavonoid pathway (Fig. 5.1).

5.2.2 Transport of anthocyanins

Most anthocyanidins are highly unstable and accumulate in cells only if stabilised by glycosylation. Usually, this glycosylation is catalysed by a glucosyl transferase that adds glucose to the 3 position of the C-ring of the flavylium nucleus (R4 in Fig. 5.2).

All anthocyanins found in plants are glycosylated on the 3 position of the C ring, and usually the first sugar is glucose, but occasionally a different sugar is involved such as galactose in lisianthus (*Eustoma grandiflora*) (Schwinn *et al.*, 1997).

Glycosylation of anthocyanidins is normally accompanied by transfer to the vacuole, a process which is believed to involve a glutathione S-transferase (GST) (Marrs *et al.*, 1995) and transporters, which may be of a range of different types including ABC and MATE transporters (Yu *et al.*, 1997; Goodman *et al.*, 2004; Gomez *et al.*, 2008). No mutations of genes encoding transporters that eliminate anthocyanin accumulation in vacuoles have been identified, suggesting that there might be multiple routes to the vacuole. The GST probably acts as a molecular chaperone for transporting anthocyanins to the vacuole (Edwards *et al.*, 2000) rather than glutathionylating anthocyanins as part of the transport process.

5.2.3 Decoration of anthocyanins

Following the core pathway for anthocyanin biosynthesis, many anthocyanins are further decorated by addition of methyl, acyl or glycosyl groups. These additional modifications are often species-specific and give the huge range of anthocyanins found in different plant

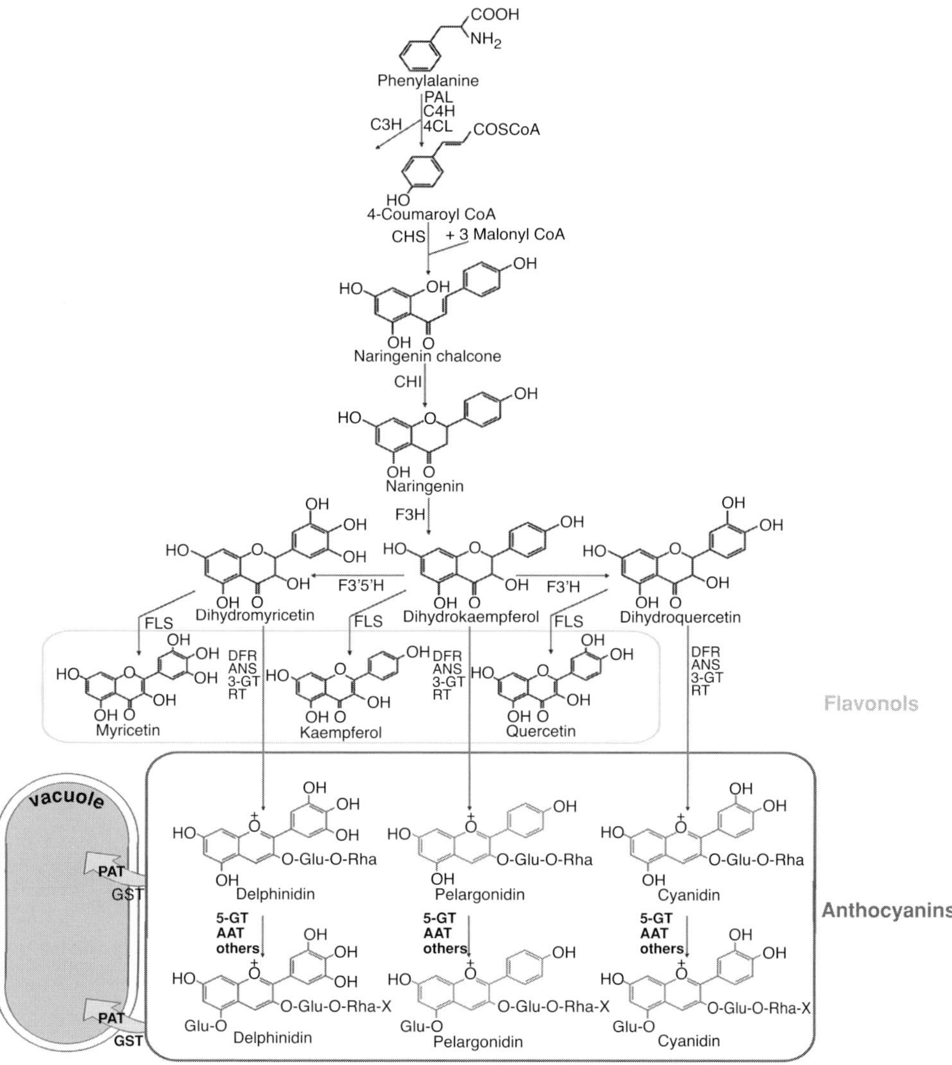

Fig. 5.1 Anthocyanin biosynthetic pathway; the full names for enzymes abbreviated in this figure are given in the text. DFR and ANS activities are essential for anthocyanin formation. 3-GT refers to UDPGlucose 3-*O*-glucosyl transferase that normally glycosylates the 3-OH group of the C-ring. RT refers to UDP Rhamnose 6''-*O*-rhamnosyl transferase that adds rhamnose to the 3-*O*-glucoside to form the rutinoside in many species. Further modifications of anthocyanins are possible such as those catalysed by anthocyanin 5-*O*-glucosyl transferase (5-GT) and anthocyanin acyl transferases (AAT), which impact the colour and stability of anthocyanins in some species.

species, estimated as at least 600 different types in 2006 (Andersen & Jordheim, 2006). Anthocyanins can be classified into various groups relating to the position of attachment and number of sugar residues; 3-monosides, 3-biosides, 3,5-diglycosides and 3,7-diglycosides are the most common. The most common glycosides are (in descending order of frequency): glucose, rhamnose, galactose, xylose, arabinose, glucuronic acid and apiose. Glycosylation of the B-ring (at positions R1, R2 and R3 in Fig. 5.2) is also found. The nature of the sugar also affects anthocyanin stability.

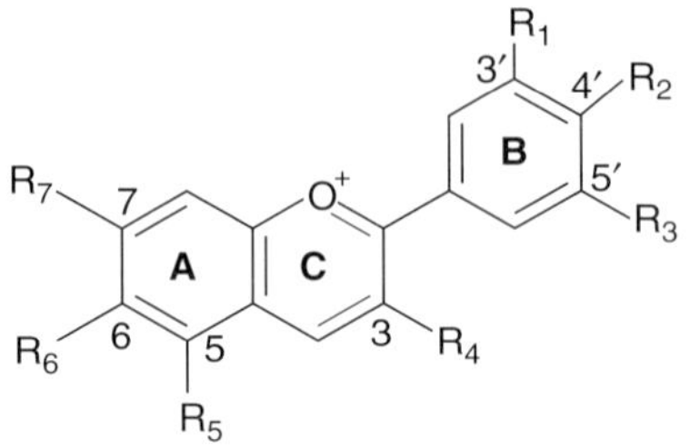

Fig. 5.2   Flavylium nucleus.

Many anthocyanins are further decorated by methylation of the hydroxyl groups of the anthocyanin (usually, the B-ring), acylation (involving both aliphatic acyl groups such as malonyl and aromatic groups such as caffeoyl and coumaroyl) and further glycosylation of either sugars or other hydroxyl groups on the anthocyanidin itself. A very common glycosylation is the addition of rhamnose by a rhamnosyl transferase onto the 3-*O*-glucoside to form a 3-*O*-rutinoside. Most side-chain decorations are thought to occur in the cytoplasm prior to transport to the vacuole. However, some decorations occur in the vacuole, such as addition of a sinapoyl group to 3-*O*-(6-*O*-*p*-coumaroyl-2-*O*-β-D-xylopyranosyl-β-D-glucopyranosyl)-5-*O*-(6-*O*-malonyl-β-D-glucopyranosyl)cyanidin by sinapoyl–glucose:anthocyanin acyltransferase in *Arabidopsis*, a member of the serine carboxypeptidase family of proteins (Fraser *et al.*, 2007). This activity contrasts to the more usual acyl transferases that decorate anthocyanins, which are cytoplasmic and belong to the BAHD family of CoA-dependent acyl transferases (D'Auria *et al.*, 2006; Luo *et al.*, 2007).

5.2.4 *Factors affecting the colour of anthocyanins*

Major differences in the colours of anthocyanins are determined in many cases by the degree of hydroxylation of the B-ring. Two cytochrome P450 mono-oxygenases known as flavonoid 3′ hydroxylase (F3′H) and flavonoid 3′5′ hydroxylase (F3′5′H) catalyse the addition of one (3′ position) or two (3′ and 5′ positions) hydroxyl groups to the B-ring, respectively (Fig. 5.1). Many species, such as Antirrhinum, rose and carnation lack the activity of F3′5′H and are unable to synthesise blue or purple colours. Consequently, addition of the F3′5′H activity to species that lack it has formed the centre-piece of strategies to modify the colour of flowers, such as in the Moonseries™ of purple carnations and the 'Blue rose' produced by Florigene (Chandler & Tanaka, 2007).

Decoration of anthocyanins can also modify their colour. Aromatic acylation also causes a bathochromic shift such that the absorption maximum is lowered and the pigment appears bluer (Luo *et al.*, 2007). It has been suggested that glycosylation can make anthocyanins bluer because the RT mutation of Petunia, which knocks out the rhamnosyl transferase activity, accumulates anthocyanins with a redder hue (Kroon *et al.*, 1994). However, the anthocyanins of Petunia flowers are acylated with a coumaroyl group attached to the rhamnose. In the absence of RT activity, the anthocyanins are no longer acylated, which probably accounts for the colour shift.

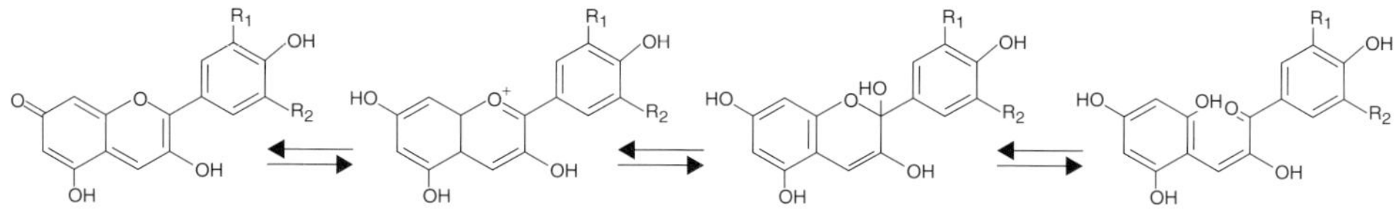

Fig. 5.3 Structural changes of anthocyanins with pH.

It is the sugar residues of anthocyanins that are acylated with aromatic acids (*p*-coumaric, caffeic, ferulic, sinapic, gallic or *p*-hydroxybenzoic acids) or aliphatic acids (malonic, acetic, malic, succinic, tartaric and oxalic acids). Acylation of anthocyanins increases their stability, probably through intra- and intermolecular copigmentation. In some plants, the acyl groups are, themselves, glycosylated. Some of the most complex anthocyanins have alternating glycosyl and acyl groups, which generally lead to increased stability in solution. Since aromatic acylation shifts the colour of the pigments towards blue, some of the most intense blues in flowers (such as those in morning glory and lobelia) are conferred by highly acylated anthocyanins.

Anthocyanins change colour according to pH, because they can exist in four pH-dependent forms (Fig. 5.3): the flavylium cation, which exists at low pH, is the most stable form in solution. It is the lack of stability in the colour of anthocyanins, both the reversible pH-dependent changes and the irreversible changes that occur at higher pHs, that causes instability in the colours they confer. At higher pH values, the quinoidal bases (which are blue) form. These can be stabilised through copigmentation. In intramolecular copigmentation, the conversion of the carbinol pseudo-base/chalcone, which involves hydration, is inhibited so that the formation of quinoidal bases from the flavylium ions is favoured, and colour is maintained.

5.2.5 Copigmentation

Intramolecular copigmentation involves side chain substitutions of anthocyanins, particularly acyl groups, which stack, sandwich-like, with the ring structures of the anthocyanins (Fig. 5.4). These molecular stacks are promoted by intramolecular bonding. Glycosylation improves stability in its own right and also because acylation occurs on the glycosyl residues. Acylation promotes anthocyanin stability primarily through intramolecular copigmentation. In intermolecular copigmentation, quinoidal bases are stabilised by stacking, but this time with other molecules, often other flavonoids such as flavones and flavonols (Plate 5.1).

5.2.6 Transcriptional regulation of anthocyanin biosynthesis

Anthocyanin production is regulated at the transcriptional level by a conserved regulatory complex, the MYB-bHLH-WDR (MBW) complex comprise three types of transcriptional regulator (Koes *et al.*, 2005; Ramsay & Glover, 2005). Proteins involved in regulating

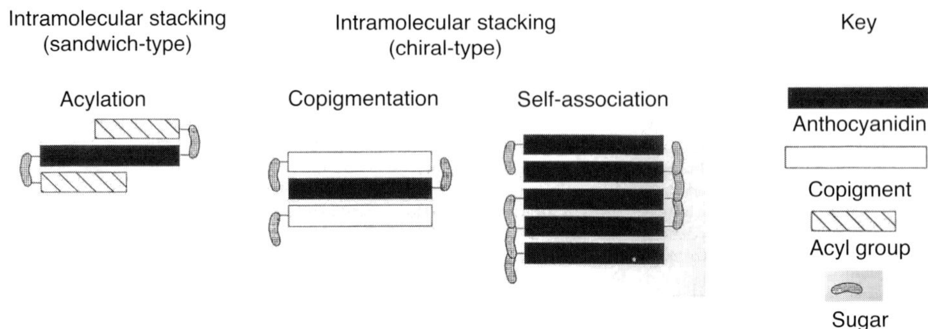

Fig. 5.4   Diagram illustrating how intra- and- intermolecular stacking stabilises anthocyanins (Henry, 1996).

anthocyanin biosynthesis are conserved between all higher plant species. The WDR protein is believed to interact with the bHLH protein in the complex. In Arabidopsis, the WDR protein is encoded by the *TTG1* locus (Walker *et al.*, 1999), in *Petunia hybrida* by the *An11* locus (de Vetten *et al.*, 1999) and in maize by the *pac1* locus (Carey *et al.*, 2004). The bHLH partners that regulate anthocyanin biosynthesis belong to a specific subgroup of plant bHLH proteins, which have a highly conserved *N*-terminal domain in addition to their bHLH domains. Examples of bHLH proteins involved in the transcriptional regulation of anthocyanin biosynthesis are R/B/Lc in maize (Ludwig *et al.*, 1989), Delila in *Antirrhinum majus* (Goodrich *et al.*, 1992), and An1 in *Petunia hybrida* (Spelt *et al.*, 2000). These proteins interact with the MYB proteins in the complex through a conserved *N*-terminal domain (Goff *et al.*, 1992), and with the WDR protein. They may also interact with each other and participate as dimers in the MBW complex. Although the MYB and bHLH partners can interact directly with DNA, the role of the WDR protein is probably to stabilise this interaction to provide a functional trans-activation complex (Baudry *et al.*, 2004).

The MYB proteins, which are involved in the MBW complex, belong to the R2R3MYB family of transcription factors (Paz-Ares *et al.*, 1987). These are usually the proteins that limit the activity of the complex (Schwinn *et al.*, 2006) and functionally similar proteins are often encoded by complex loci resulting from recent gene duplications (Schwinn *et al.*, 2006; Jung *et al.*, 2009). In maize, MYB proteins regulating anthocyanin biosynthesis are encoded by the *C1* and *Pl* genes (Paz-Ares *et al.*, 1987; Cone *et al.*, 1993), in *Antirrhinum majus* by the *Rosea1, Rosea2* and *Venosa* genes (Schwinn *et al.*, 2006) and in *Petunia hybrida* by the *An2* and *An4* genes (Quattrocchio *et al.*, 1998, 1999; Kroon, 2004). The MYB proteins contain a signature motif of amino acids in the R3 repeat of their DNA binding domains that is indicative of interaction with bHLH proteins (Zimmermann *et al.*, 2004).

There remains debate concerning which genes are targets for the MBW complex in the regulation of anthocyanin biosynthesis. Analysis of mutations in regulatory genes suggests that these regulate only the late biosynthetic genes (LBGs) in the anthocyanin biosynthetic pathway (Martin *et al.*, 1991; Quattrocchio *et al.*, 1993; Pelletier *et al.*, 1997), although the precise targets of the regulatory genes differ between different plant species. For example, C1 and R in maize are thought to regulate biosynthetic genes from CHS to the GST (Dooner *et al.*, 1991; although CHI may not be a target gene), whereas Rosea1 and Delila regulate

F3H, DFR, ANS and 3GT in *Antirrhinum majus* (Martin *et al.*, 1991; Schwinn *et al.*, 2006) and *DFR, ANS, 3GT* and GST are targets for MBW regulation in *Petunia hybrida* (Quattrocchio *et al.*, 1993). The problem with identification of target genes through mutant analysis, is that expression of true targets may not be down regulated in regulatory gene mutants, if there are other regulators that can compensate for the loss of function of the MBW complex. This may well be the case for the early biosynthetic genes (EBGs: CHS and CHI) in flowers, which are also regulated by transcriptional regulators of flavonol metabolism, such as AtMYB11, 12 and 111 and their functional homologues in other species that are involved in inducing flavonol biosynthesis (Mehrtens *et al.*, 2005; Stracke *et al.*, 2007). Indeed, over-expression of MYB and bHLH proteins together often results in the activation of genes of general phenylpropanoid metabolism (PAL, C4H, 4CL) and EBG as well as LBG (Bovy *et al.*, 2002; Butelli *et al.*, 2008). An alternative explanation is that EBGs and the genes of general phenylpropanoid metabolism are not targets of the MBW but that at high levels of MBW they are activated as an 'artefact' (Gonzalez *et al.*, 2007). MYB transcription factors have been reported to show a dose dependency and to activate different targets at different concentrations (Andersson *et al.*, 1999; Jin *et al.*, 2000), so target genes identified through over-expression may not be artefacts, but rather represent lower affinity targets of the regulatory complex.

The ability of the MBW complex to regulate LBGs including genes encoding decorating enzymes has allowed a number of new genes involved in glycosylation and acylation of anthocyanins to be identified through coexpression analysis (Tohge *et al.*, 2005; Luo *et al.*, 2007; Yonekura-Sakakibara *et al.*, 2008; Butelli *et al.*, 2008). In tomato, a gene encoding a putative anthocyanin transporter of the MATE family is induced in response to the MYB and bHLH proteins inducing anthocyanin biosynthesis (Butelli *et al.*, 2008).

5.3 Engineering anthocyanin production in plants

Using genes encoding the regulatory proteins one can switch on anthocyanin production in plant tissues by design and this can provide powerful tools for biotechnology. Using a combination of both the gene encoding the bHLH transcription factor and the gene encoding the MYB transcription factor (expressed constitutively under the control of the CaMV 35S promoter) one can induce anthocyanin accumulation in any cell of the plant, even in tissues where they are never normally made such as vegetative tissues of tobacco.

5.3.1 *An* in vivo *reporter of promoter activity*

One can use the gene encoding the MYB transcription factor as a reporter for the activity of any promoter used to drive its expression. In this way, we have created an *in vivo* reporter system for monitoring gene expression.

The promoter of a gene of unknown function, which is expressed in tomato fruit, has been used to drive the expression of an anthocyanin-regulating MYB protein in tomato. This promoter drives the production of anthocyanin in the stems and leaves of tomato. In fruit, it promotes the formation of anthocyanin in the hairs (trichomes) of the fruit and internally at low levels in pericarp and in seed coats (Plate 5.2). This anthocyanin reporter requires no staining and could be used to provide interesting new pigmentation patterns for plants and foods. In addition, if promoters responsive to abiotic stress were used to drive expression of the MYB protein it could be used to sense changing environments in field-grown plants, for example.

5.3.2 Biofortified crops

Motivated by the suggested health benefits of anthocyanins, we have used the MYB and bHLH regulatory genes to engineer very high levels of anthocyanins in tomato fruits (Butelli *et al.*, 2008). These experiments have provided us with 'model foods' that have allowed us to test the health benefits of consumption of high levels of anthocyanins in a standardised food matrix in preclinical animal studies. Consumption of a diet supplemented with high anthocyanin tomatoes led to highly tumorigenic mice (*Trp53$^{-/-}$*) experiencing an extension of 30% to their life spans compared with those fed diets supplemented with control, red tomatoes. We have started to extend the analysis of the health-promoting properties of the tomatoes to other types of disease and shown that anthocyanins in tomato juice can suppress significantly the inflammatory response in an animal model of inflammatory bowel disease. We are now using the regulatory genes to engineer high levels of a range of polyphenols in tomato fruit so that we can make comparative assays of the efficacy of different polyphenol phytonutrients in a common food matrix in preventing disease or ameliorating the effects of disease in a range of preclinical animal studies.

5.3.3 Visually traceable system for VIGS analysis of gene function

The purple, high anthocyanin tomato lines also provide a very suitable system for fruit-specific silencing of genes by viral-induced gene-silencing (VIGS) for determining the roles of unknown genes in fruit development and maturation. In most plant species, including tomato, VIGS works but the penetration of the silencing phenotype is only partial. The inhibition of production of purple anthocyanins by VIGS using sequences from the *Del* and *Ros1* genes in the Tobacco Rattle Virus 2 vector has been used in *Del/Ros1* tomato fruit to develop a visually traceable system for VIGS silencing in fruit. Incorporation of additional gene sequences in addition to *Del* and *Ros1* in the TRV2 vector shows complete coincidence of silencing of anthocyanin production and the additional gene of interest. As the silencing of anthocyanin production involves inhibition of induction of a pathway not normally active in tomato fruit, the red silenced sectors can be viewed as 'wild type' with respect to their metabolism and development. This system has proved to be particularly useful for the analysis of genes of unknown function involved in the later stages of fruit ripening, particularly genes associated with different branches of metabolism in fruit (Orzaez *et al.*, 2009; Ballester *et al.*, 2009).

Plants expressing *Del* and *Ros1* in other tissues can be used to monitor VIGS elsewhere. For example, the line shown in Plate 5.3 can be used to monitor VIGS induced by agro-infection of TRV2 in leaves to assay the function of genes involved in leaf and flower development. These methods can be adapted for high-throughput assays of gene function in target plant species (Orzaez *et al.*, 2009).

5.4 Conclusions

The high level of understanding of how anthocyanin pigments are synthesised in plants, and the availability of genes encoding biosynthetic enzymes, transporters and transcription factors mean that almost all the tools and technical knowledge required to engineer pigment production in plants are available. Anthocyanin production can be harnessed to a number of important biotechnological outputs through genetic modification or through breeding. The tools available will allow for the production of anthocyanins in cell cultures, which would provide a source of these pigments for medical research and also as natural food colourants. Production in cell cultures would allow their use for producing C^{13}-labelled anthocyanins for bioavailability studies. In addition, the ability to engineer anthocyanin production in crop plants provides an important resource for nutritional research into the role of dietary anthocyanins in promoting health and protecting against chronic disease, resources that can be used in cellular studies, preclinical studies and taken forward to human studies. The future for anthocyanins in biotechnological research is not only bright, but likely to be very colourful.

Acknowledgements

AG and DO were supported by the Spanish MCIN $^1/_2$AQ2 (grant nos. BIO2005-01015 and BIO2008-03434), the Trilateral GenMetFrutQual ERA-PG and by the MEC Ramón y Cajal Program. CM and EB were supported by the European Union FP6 FLORA project (grant no. FOOD-CT-01730) and CM, EB, JL and KK were supported by the core strategic grant by BBSRC to JIC. LT and JDGJ were supported by a Follow-on Fund from BBSRC.

References

Andersen, Ø.M. & Jordheim, M. (2006) The anthocyanins. In: *Flavonoids – Chemistry, Biochemistry and Applications* (eds. Ø.M. Andersen & K.R. Markham), pp. 471–551. CRC Press, Taylor & Francis Group, Boca Raton, FL.

Andersson, K.B., Berge, T., Matre, V. & Gabrielsen, O.S. (1999) Sequence selectivity of c-Myb in vivo. Resolution of a DNA target specificity paradox. *Journal of Biological Chemistry*, **274**, 21986–21994.

Ballester, A.-R., Molthoff, J., de Vos, R., *et al.* (2009) Biochemical and molecular analysis of pink tomatoes: deregulated expression of the gene encoding transcription factor SlMYB12 leads to pink tomato fruit color. *Plant Physiology*, **152**, 71–84.

Baudry, A., Heim, M.A., Dubreucq, B., Caboche, M., Weisshaar, B. & Lepiniec, L. (2004) TT2, TT8, and TTG1 synergistically specify the expression of *BANYULS* and proanthocyanidin biosynthesis in *Arabidopsis thaliana*. *Plant Journal*, **39**, 366–380.

Bovy, A., de Vos, R., Kemper, M., *et al.* (2002) High-flavonol tomatoes resulting from the heterologous expression of the maize transcription factor genes *LC* and *C1*. *Plant Cell*, **14**, 2509–2526.

Burbulis, I.E. & Winkel-Shirley, B. (1999) Interactions among enzymes of the *Arabidopsis* flavonoid biosynthetic pathway. *Proceedings of the National Academy of Sciences of the United States of America*, **96**, 12929–12934.

Butelli, E., Titta, L., Giorgio, M., *et al.* (2008) Induced anthocyanin biosynthesis in tomato results in purple fruit with increased antioxidant and dietary, health-protecting properties. *Nature Biotechnology*, **26**, 1301–1308.

Carey, C.C., Strahle, J.T., Selinger, D.A. & Chandler, V.L. (2004) Mutations in the *pale aleurone color1* regulatory gene of the *Zea mays* anthocyanin pathway have distinct phenotypes relative to the functionally similar *TRANSPARENT TESTA GLABRA1* gene in *Arabidopsis thaliana*. *Plant Cell*, **16**, 450–464.

Chandler, S. & Tanaka, Y. (2007) Genetic modification in floriculture. *Critical Reviews in Plant Sciences*, **26**, 169–197.

Cone, K.C., Cocciolone, S.M., Burr, F.A. & Burr, B. (1993) Maize anthocyanin regulatory gene *pl* is a duplicate of *c1* that functions in the plant. *Plant Cell*, **5**, 1795–1805.

D'Auria, J.C. (2006) Acyltransferases in plants: a good time to be BAHD. *Current Opinion in Plant Biology*, **9**, 331–340.

de Vetten, N., ter Horst, J., van Schaik, H.P., de Boer, A., Mol, J. & Koes, R. (1999) A cytochrome b5 is required for full activity of flavonoid 3′, 5′-hydroxylase, a cytochrome P450 involved in the formation of blue flower colors. *Proceedings of the National Academy of Sciences of the United States of America*, **96**, 778–783.

Dooner, H.K., Robbins, T.P. & Jorgensen, R.A. (1991) Genetic and developmental control of anthocyanin biosynthesis. *Annual Review of Genetics*, **25**, 173–199.

Edwards, R., Dixon, D.P. & Walbot, V. (2000) Plant glutathione *S*-transferases: enzymes with multiple functions in sickness and in health. *Trends in Plant Science*, **5**, 193–198.

Estornell, L.H., Orzáez, D., López-Peña, L., *et al.* (2009) A multisite gateway-based toolkit for targeted gene expression and hairpin RNA silencing in tomato fruits. *Plant Biotechnology Journal*, **7**, 298–309.

Fraser, C.M., Thompson, M.G., Shirley, A.M., *et al.* (2007) Related *Arabidopsis* serine carboxypeptidase-like sinapoylglucose acyltransferases display distinct but overlapping substrate specificities. *Plant Physiology*, **144**, 1986–1999.

Goff, S.A., Cone, K.C. & Chandler, V.L. (1992) Functional analysis of the transcriptional activator encoded by the maize B gene: Evidence for a direct functional interaction between two classes of regulatory proteins. *Genes Development*, **6**, 864–875.

Gomez, C., Terrier, N., Torregrosa, L., *et al.* (2008) Grapevine MATE-type proteins act as vacuolar H^+-dependent acylated anthocyanin transporters. *Plant Physiology*, **150**, 402–415.

Gonzalez, A., Zhao, M., Leavitt, J.M. & Lloyd, A.M. (2007) Regulation of the anthocyanin biosynthetic pathway by the TTG1/bHLH/Myb transcriptional complex in Arabidopsis seedlings. *Plant Journal*, **53**, 814–827.

Goodman, C.D., Casati, P. & Walbot, V. (2004) A multidrug resistance-associated protein involved in anthocyanin transport in *Zea mays*. *Plant Cell*, **16**, 1812–1826.

Goodrich, J., Carpenter, R. & Coen, E.S. (1992) A common gene regulates pigmentation pattern in diverse plant species. *Cell*, **68**, 955–964.

Henry, B.S. (1996) Natural food colors. In: *Natural Food Colorants* (eds. G.A.F. Hendry & J.D. Houghton), pp. 40–79. Chapman & Hall, London.

Jin, H., Cominelli, E., Bailey, P., *et al.* (2000) Transcriptional repression by AtMYB4 controls production of UV-protecting sunscreens in *Arabidopsis*. *EMBO Journal*, **19**, 6150–6161.

Jung, C.S., Griffiths, H.M., De Jong, D.M., *et al.* (2009) The potato *developer* (*D*) locus encodes an R2R3 MYB transcription factor that regulates expression of multiple anthocyanin structural genes in tuber skin. *Theoretical and Applied Genetics*, **120**, 45–57.

Koes, R., Verweij, W. & Quattrocchio, F. (2005) Flavonoids: a colorful model for the regulation and evolution of biochemical pathways. *Trends in Plant Science*, **10**, 236–242.

Kroon, A. (2004) Transcription regulation of the anthocyanin pathway in *Petunia hybrida*. Doctoral thesis, Vrije Universiteit, Amsterdam.

Kroon, J., Souer, E., de Graaff, A., Xue, Y., Mol, J. & Koes, R. (1994) Cloning and structural analysis of the anthocyanin pigmentation locus *Rt* of *Petunia hybrida*: characterization of insertion sequences in two mutant alleles. *Plant Journal*, **5**, 69–80.

Ludwig, S.R., Habera, L.F., Dellaporta, S.L. & Wessler, S.R. (1989) Lc, a member of the maize R gene family responsible for tissue-specific anthocyanin production, encodes a protein similar to transcriptional activators and contains the myc-homology region. *Proceedings of the National Academy of Sciences of the United States of America*, **86**, 7092–7096.

Luo, J., Nishiyama, Y., Fuell, C., *et al.* (2007) Convergent evolution in the BAHD family of acyl transferases; identification and characterisation of anthocyanin acyl transferases from *Arabidopsis thaliana*. *Plant Journal*, **49**, 810–828.

Marrs, K.A., Alfenito, M.R., Lloyd, A.M. & Walbot, V. (1995) A glutathione S-transferase involved in vacuolar transfer encoded by the maize gene *Bronze-2*. *Nature*, **375**, 397–400.

Martin, C., Prescott, A., Mackay, S., Bartlett, J. & Vrijlandt, E. (1991) Control of anthocyanin biosynthesis in flowers of *Antirrhinum majus*. *Plant Journal*, **1**, 37–49.

Mehrtens, F., Kranz, H., Padnarek, P. & Weisshaar, B. (2005) The *Arabidopsis* transcription factor MYB12 is a flavonol-specific regulator of phenylpropanoid biosynthesis. *Plant Physiology*, **138**, 1083–1096.

Orzaez, D., Medina, A., Torrez, S., *et al.* (2009) A visual reporter system for virus-induced gene silencing in tomato fruit based on anthocyanin accumulation. *Plant Physiology*, **150**, 1122–1134.

Paz-Ares, J., Ghosal, D., Wienand, U., Peterson, P.A. & Saedler, H. (1987) The regulatory c1 locus of Zea mays encodes a protein with homology to myb proto-oncogene products and with structural similarities to transcriptional activators. *EMBO Journal*, **6**, 3553–3558.

Pelletier, M.K., Murrell, J.R. & Shirley, B.W. (1997) Characterization of flavonol synthase and leucoanthocyanidin dioxygenase genes in Arabidopsis - Further evidence for differential regulation of "early" and "late" genes. *Plant Physiology*, **113**, 1437–1445.

Quattrocchio, F., Wing, J., van der Woude, K., *et al.* (1999) Molecular analysis of the *anthocyanin2* gene of petunia and its role in the evolution of flower color. *Plant Cell*, **11**, 1433–1444.

Quattrocchio, F., Wing, J.F., Leppen, H.T.C., Mol, J.N.M. & Koes, R.E. (1993) Regulatory genes controlling anthocyanin pigmentation are functionally conserved among plant species and have distinct sets of target genes. *Plant Cell*, **5**, 1497–1512.

Quattrocchio, F., Wing, J.F, van der Woude, K., Mol, J.N.M. & Koes, R.E. (1998) Analysis of bHLH and MYB domain proteins: species specific regulatory differences are caused by divergent evolution of target anthocyanin genes. *Plant Journal*, **13**, 475–488.

Ramsay, N.A. & Glover, B.J. (2005) MYB-bHLH-WD40 protein complex and the evolution of cellular diversity. *Trends in Plant Science*, **10**, 63–70.

Schwinn, K.E., Davies, K.M., Deroles, S.C., *et al.* (1997) Expression of an *Antirrhinum majus* UDP-glucose:flavonoid-3-O-glucosyltransferase transgene alters flavonoid glycosylation and acylation in lisianthus (*Eustoma grandiflorum* Grise.) *Plant Science*, **125**, 53–61.

Schwinn, K., Venail, J., Shang, Y., *et al.* (2006) A small family of MYB-regulatory genes controls floral pigmentation intensity and patterning in the genus Antirrhinum. *Plant Cell*, **18**, 831–851.

Spelt, C., Quattrocchio, F., Mol, J.N. & Koes, R.E. (2000) Anthocyanin1 of petunia encodes a basic helix-loop-helix protein that directly activates transcription of structural anthocyanin genes. *Plant Cell*, **12**, 1619–1632.

Stracke, R., Ishihara, H., Huep, G., *et al.* (2007) Differential regulation of closely related R2R3-MYB transcription factors controls flavonol accumulation in different parts of the *Arabidopsis thaliana* seedling. *Plant Journal*, **50**, 660–677.

Tohge, T., Nishiyama, Y., Hirai, M.Y., *et al.* (2005) Functional genomics by integrated analysis of metabolome and transcriptome of Arabidopsis plants over-expressing an MYB transcription factor. *Plant Journal*, **42**, 218–235.

Walker, A.R., Davison, P.A., Bolognesi-Winfield, A.C., *et al.* (1999) The TRANSPARENT TESTA GLABRA1 locus, which regulates trichome differentiation and anthocyanin biosynthesis in Arabidopsis, encodes a WD40 repeat protein. *Plant Cell*, **11**, 1337–1350.

Yonekura-Sakakibara, K., Tohge, T., Matsuda, F., *et al.* (2008) Comprehensive flavonol profiling and transcriptome coexpression analysis leading to decoding gene–metabolite correlations in *Arabidopsis*. *Plant Cell*, **20**, 2160–2176.

Yu, Y.P., Li, Z.S. & Rea, P.A. (1997) *AtMRP1* gene of *Arabidopsis* encodes a glutathione *S*-conjugate pump: Isolation and functional definition of a plant ATP-binding cassette transporter gene. *Proceedings of the National Academy of Sciences of the United States of America*, **94**, 8243–8248.

Zimmermann, I.M., Heim, M.A., Weisshaar, B. & Uhrig, J.F. (2004) Comprehensive identification of *Arabidopsis thaliana* MYB transcription factors interacting with R/B-like BHLH proteins. *Plant Journal*, **40**, 22–34.

Chapter 6

Anthocyanin Biosynthesis, Regulation, and Transport: New Insights from Model Species

Lucille Pourcel, Andrés Bohórquez-Restrepo, Niloufer G. Irani and Erich Grotewold

Abstract: Anthocyanins belong to the flavonoid family of phytochemicals. They are produced in plants, often developmentally controlled and induced in response to biotic and abiotic stresses, playing a major role as defensive molecules. The pigmentation and protection properties of anthocyanins have made them ideal phytochemicals for studying many biological principles in plants such as maize, *Arabidopsis thaliana*, petunia, and snapdragon. In this chapter, we describe the very significant knowledge that has been gained over the past few years with regards to the chemistry, biosynthesis, regulation, and transport of anthocyanins that resulted in this being one of the best described biosynthetic pathways and regulatory networks in plants. Also, we present the recent advances on intracellular transport and biosynthetic regulation of anthocyanins.

Keywords: metabolism; *Arabidopsis*; maize; regulation; transcription factors; transport; anthocyanic vacuolar inclusion; vacuole; flavonoids

6.1 Anthocyanins and related pigments in model plant species

6.1.1 General characteristics of anthocyanins

Anthocyanins are aromatic heterocyclic compounds derived from the phenylpropanoid and subsequent flavonoid biosynthetic pathways (Forkmann, 1991; Winkel, 2008). A C15 (C6-C3-C6 ring structure) flavonoid carbon skeleton (Fig. 6.1) provides the anthocyanidin (nonglycosylated anthocyanin) flavylium cation, a conjugated ring structure, that results in a diffuse positive charge that brings in continuum the conjugated nature of the rings that it connects, resulting in color. Depending on the organization and modifications of the three

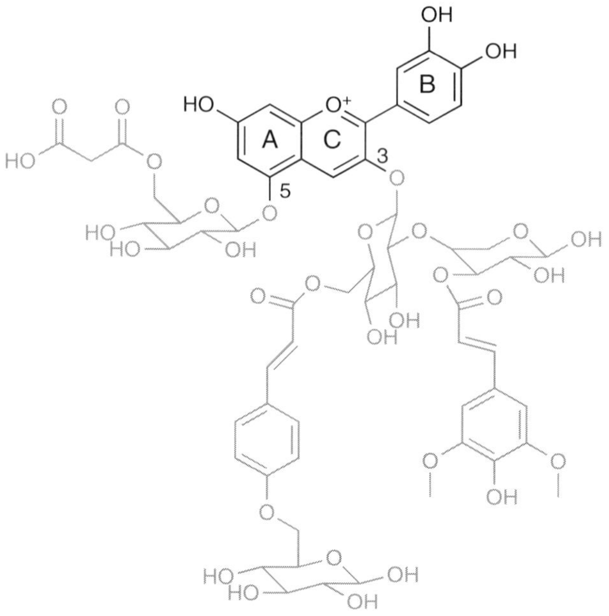

Fig. 6.1 Structure of 3–*O*–[2″–*O*–(6‴–*O*–{(sinapoyl) xylosyl} 6″–*O*–(*p*–*O*–(glucosyl)–*p*–coumaroyl) glucoside] 5–*O*–(6⁗–*O*–malonyl) glucoside (A11). (Adapted from Tohge *et al.*, 2005.) The names of the three rings (A, B, and C) common to all flavonoids are indicated.

rings, in addition to anthocyanins, flavonoids can be classified into a number of different subclasses, which include the flavones, the flavonols, the isoflavones, and the proanthocyanidins (PAs, condensed tannins). The most conspicuous function of anthocyanins is to provide the bright red and blue colors that serve as attractants for pollinators and seed dispersants (Grotewold, 2006). Indeed, anthocyanin pigments have appeared quite recently in the evolution of plants (Stafford, 1991). Other flavonoid compounds, already present in early photosynthetic plants, also play central roles in plant biology. These functions, extensively reviewed elsewhere (Koes *et al.*, 1994; Shirley, 1996; Taylor & Grotewold, 2005; Pourcel & Grotewold, 2009), include photoprotection, communication in plant–microbe interactions, hormone signaling, and male fertility.

The pigmentation provided by anthocyanins and other flavonoid-derived compounds has made this pathway a favorite for genetic studies since the early days of modern genetics. For example, flavonoid pigments contributed to understanding the basis of dominance by Mendel, the mutation theory (de Vries, 1901), the chemical basis of genetics (Wheldale, 1907), correlated traits (Emerson, 1911), variegation (Emerson, 1917), allelic diversity (Anderson, 1924), the discovery of transposable elements (McClintock, 1950a, 1950b, 1951, 1955, 1958), and to the discovery and elucidation of the molecular mechanisms underlying RNA silencing (Napoli *et al.*, 1990) and paramutation (Brink, 1958; Stam & Mittelsten Scheid, 2005; Alleman *et al.*, 2006). Today, mutants and the corresponding genes for a large number of flavonoid biosynthetic enzymes or regulatory proteins are available, making it among the best described plant biosynthetic pathways. Major contributions to the understanding of anthocyanin accumulation were made in model systems such as maize, petunia, snapdragon, and *Arabidopsis* (Dooner *et al.*, 1991; Mol *et al.*, 1998; Winkel-Shirley, 2001).

Anthocyanin biosynthesis does not only provide a beautiful system to investigate the molecular basis of several fundamental plant cellular processes, but they are also phytochemicals with tremendous agricultural significance. The floricultural and ornamental plant industries are largely dependent on the formation of new colors, or colors in novel patterns, for increased marketing (Tanaka *et al.*, 2005). Anthocyanins and related compounds also confer a number of health benefits, presumed to be largely a consequence of their antioxidant properties. A good example is provided by the "French Paradox," which attempts to explain the beneficial effects of red wine. The recent generation of tomatoes accumulating high levels of health-promoting anthocyanins resulting from the fruit-specific expression of pathway regulators further highlights the potential of these phytochemicals as value-added complements of our diet (Butelli *et al.*, 2008).

6.1.2 *Anthocyanin biosynthetic enzymes*

The first committed step in the formation of flavonoids is the conjugation of malonyl-CoA and coumaroyl-CoA molecules (derived from the general phenylpropanoid pathway) to form chalcones, catalyzed by conserved chalcone synthase (CHS) enzymes (Table 6.1). CHS enzymes belong to the polyketide synthase family (Austin & Noel, 2003) and crystallization of CHS demonstrated that it functions as a dimer (Ferrer *et al.*, 1999). Chalcones are converted to flavanones by the action of chalcone flavanone isomerase (CHI). Although originally believed to correspond to a fold unique to the plant kingdom (Jez *et al.*, 2000), subsequent studies revealed a much more ancient origin for this enzyme (Gensheimer and Mushegian, 2004). In fact, structural homologues were found in bacteria and fungi, and perhaps these enzymes are related to a function different from flavonoid biosynthesis (Ferrer *et al.*, 2008), or flavonoid degradation by these microorganisms (Herles *et al.*, 2004). The phenotype of chalcone isomerase mutants in *Arabidopsis* (*TT5*) suggests that the isomerization of chalcone to flavanone is not necessarily spontaneous under *in vivo* conditions in this

Table 6.1   Enzyme and genes involved in anthocyanin biosynthesis in model plant species.

Enzyme	Maize gene(s)	*Arabidopsis* gene(s)	Petunia gene(s)
Chalcone synthase (CHS)	*C2, Whp*	*TT4*	*chsA, chsB, chsG,* and *chsJ*
Chalcone isomerase (CHI)	*CHI1*	*TT5*	*Po*
Flavanone 3-hydroxylase (F3H)	*F3H*	*TT6*	*An3*
Flavonoid 3′-hydroxylase (F3′H)	*Pr1*	*TT7*	*Ht1*
Dihydroflavonol 4-reductase (DFR)	*A1*	*TT3*	*An6*
Anthocyanidin synthase (ANS, LDOX)	*A2*	*TT18*	*Ant17*
O-Methyltransferase (OMT)			*Mt1, Mt2*
3-*O*-Glucosyl transferase (3GT)	*Bz1*	*At5g17050*	*PGT8*
5-*O*-Glucosyl transferase (5GT)		*At4g14090*	*PH1*
Rhamnosyl transferase (RT)			*Rt*
Flavonoid 3′,5′-hydroxylase (F3′5′H)			*Hf, Hf2*

plant, despite this reaction having been shown to occur spontaneously *in vitro* (Mol *et al.*, 1985). In fact, anthocyanins and 3-deoxy flavonoids form in maize Black Mexican Sweet (BMS) cells in the absence of detectable CHI activity (Grotewold *et al.*, 1998), suggesting that the reaction does occur spontaneously under *in vivo* conditions in maize. Flavanones are the precursors of all flavonoids (except aurones, whose precursors are the chalcones), including flavones, flavonols, isoflavones, anthocyanins, and the phlobaphene pigments of maize and other grasses (Grotewold, 2006). For anthocyanin accumulation, flavanones are converted to dihydroflavonols by the action of flavanone 3-hydroxylase (F3H). F3H is one of four enzymes in the flavonoid pathway that belongs to the 2-oxoglutarate-dependent dioxygenase (2-ODD) family (Winkel, 2008). In *Arabidopsis*, *TT6* encodes F3H, and *TT6* mutants have been called "leaky" as a consequence of the remaining seed coat pigmentation and the lack of accumulation of the naringenin precursor (Wisman *et al.*, 1998; Peer *et al.*, 2001). However, recent studies attributed this leaky phenotype to the action of other 2-ODD enzymes including FLS (flavonol synthase) and ANS (anthocyanidin synthase, also called leucoanthocyanidin dioxygenase, LDOX), resulting in the formation of 3-deoxy flavonoids in the seed coat, which are not characteristic of *Arabidopsis* (Owens *et al.*, 2008b); however it has been shown that *F3H* mutants in *Antirrhinum majus* do not present this phenotype (Martin *et al.*, 1991). In maize, the *F3H* function was characterized based on homology (Deboo *et al.*, 1995), yet remains one of few maize anthocyanin biosynthetic genes for which no mutants have yet been described. Dihydroflavonols serve as substrates for dihydroflavonol 4-reductase (DFR) for the last common step leading to anthocyanins and PAs, generating leucoanthocyanidins (flavan-3,4-diols). The subsequent step is the formation of anthocyanidins by the action of the enzyme ANS that has been assayed *in vitro* and crystallized, showing a stereoselective C-3 hydroxylation (Wilmouth *et al.*, 2002). *Arabidopsis* ANS, encoded by tannin-deficient seed (*TDS4*) has been characterized, and studies showed that this step precedes *BAN* in PAs formation (Abrahams *et al.*, 2003). However, the formation of the phlobaphene pigments in maize involves the direct conversion of flavanones to flavan-4-ols, providing one example of the numerous branch points present in the flavonoid pathway.

The contribution of branched pathways to the formation of different flavonoid compounds from the various possible precursors is poorly established. Evidence that *Arabidopsis* flavonoid biosynthetic enzymes physically interact with each other (Burbulis & Winkel-Shirley, 1999; Owens *et al.*, 2008a) suggests the formation of different complexes that may contribute to the biosynthesis of each group of flavonoids (Winkel-Shirley, 1999). Similar complexes have been identified in other plants (Liu & Dixon, 2001), suggesting the broader existence of one or several "flavonoid metabolons." However, that such metabolic complexes are solely formed by protein–protein interactions is partially inconsistent with the finding that divergent flavonoid metabolic enzymes can complement mutants in evolutionary distant plants (Meyer *et al.*, 1987; Dong *et al.*, 2001).

6.1.3 *Anthocyanins in* Arabidopsis

Recent studies identified a surprising complexity in the *Arabidopsis* anthocyanin contents. At least 11 distinct anthocyanins, derived from the anthocyanidin cyanidin, are present in plants ectopically expressing the MYB transcriptional regulator PAP1, or in seedlings induced for anthocyanin accumulation (Tohge *et al.*, 2005; Pourcel *et al.*, 2010). The most decorated of these anthocyanins, cyanidin 3-*O*-[2″-*O*-(6‴-*O*-{(sinapoyl) xylosyl} 6″-*O*-(*p*-*O*-(glucosyl)-*p*-coumaroyl) glucoside] 5-*O*-(6″″-*O*-malonyl) glucoside (Fig. 6.1), is referred to here as A11 (Tohge *et al.*, 2005). The formation of A11 from cyanidin starts by glycosylation at the 3-*O* position by the UDP-glucose:flavonoid 3-*O*-glucosyltransferase (3GT, *At5g17050*) (Tohge *et al.*, 2005; Pourcel *et al.*, 2010). This modification is necessary for the 5-*O* glycosylation, catalyzed by UDP-glucose:cyanidin 5-*O*-glucosyltransferase (5GT, *At4g14090*) (Tohge *et al.*, 2005), since *3GT* mutants have almost undetectable levels of anthocyanins (Pourcel *et al.*, 2010). Some of the factors that incorporate the additional modifications present in the most decorated anthocyanins have been described (Shirley & Chapple, 2003; Luo *et al.*, 2007), and current efforts on identifying the enzymes responsible for the most complex anthocyanins are in progress through omics-guided integrative approaches (Nakabayashi *et al.*, 2010).

6.2 Transcriptional regulation of anthocyanin biosynthetic genes

6.2.1 Maize

The control of anthocyanin accumulation provides one of the best described regulatory networks in plants. In maize, anthocyanin accumulation is controlled by two classes of regulatory proteins: an R2R3-MYB-domain containing class (C1 or PL; Paz-Ares *et al.*, 1987; Cone *et al.*, 1993) and a basic helix-loop-helix (bHLH)-domain containing class (members of the *R/B* gene families (Ludwig *et al.*, 1989)). Anthocyanin production requires the interaction of a member of the MYB-domain C1/PL family and a member of the bHLH-domain R/B family (Goff *et al.*, 1992). The pattern of anthocyanin pigmentation in any particular plant part is controlled by the combinatorial, tissue-specific expression of these regulatory genes. In addition to 3-hydroxy flavonoids and anthocyanins, maize and its close relatives (e.g., sorghum, wheat and rice), accumulate 3-deoxy flavonoids and derived pigments, which include the phlobaphenes (Styles & Ceska, 1975, 1989). Together with the anthocyanins, the phlobaphenes are responsible for the striking kernel pigmentation patterns found in "Indian Corn." A single known genetic factor, *P1* controls 3-deoxy flavonoid and phlobaphene accumulation (Styles & Ceska, 1975, 1977, 1989). Similar to *C1*, *P1* encodes an R2R3-MYB transcription factor (Grotewold *et al.*, 1991). However, unlike C1, the P1 regulatory function is independent of R/B (Grotewold *et al.*, 1994).

C1 controls the expression of the *A1* gene (encoding DFR) by specifically binding to two *cis*-regulatory elements (CREs) located within 110 base pairs upstream of the transcription start site (TSS). These CREs are also recognized by P1, and termed the high- and low-affinity P1 binding sites ([ha]PBS and [la]PBS, respectively) (Grotewold *et al.*, 1994; Sainz *et al.*, 1997) separated by the anthocyanin regulatory element [ARE (Lesnick & Chandler, 1998)]. In transient expression experiments, these three elements contribute to the regulation of *A1* by P1 or by C1+R (Grotewold *et al.*, 1994; Tuerck & Fromm, 1994;

Sainz *et al.*, 1997), and transposon insertions and mutations in the ARE differentially affect the *in vivo* regulation of *A1* by P1 or C1+R (Pooma *et al.*, 2002; Hernandez *et al.*, 2004). The [ha]PBS and [la]PBS sites fit well the C1 DNA-binding consensus, identified by systematic evolution of ligands by exponential enrichment (SELEX) (Sainz *et al.*, 1997), and which is very similar to the consensus binding site for P1 (Grotewold *et al.*, 1994; Williams & Grotewold, 1997) corresponding to $ACC^{T}/_{A}ACC$. While P1 recognizes *in vitro* the [ha]PBS and [la]PBS with dissociation constants (K_d) around 30–50 and 250–400 nm, respectively, C1 has an intrinsic inability to bind DNA with high affinity and it binds both sites with K_d of 300–500 nM (Sainz *et al.*, 1997). However, this low affinity for DNA is not what makes the C1 regulatory activity R-dependent, since a *C1* mutant that binds DNA with an affinity comparable to P1 is still dependent on R for function (Hernandez *et al.*, 2004).

In spite of the fact that the MYB domains of P1 and C1 are ~70% identical (Grotewold *et al.*, 1994) and that they recognize very similar DNA sequences *in vivo* and *in vitro* (Sainz *et al.*, 1997), C1 has an absolute requirement for the bHLH factor R to activate transcription of the anthocyanin biosynthetic genes (R-dependent transcription). In contrast, P1 controls gene expression independently of R (R-independent transcription). Although C1 binds to the [ha]PBS, it requires R to activate transcription of a promoter containing these sites (Grotewold *et al.*, 1994). Thus, one function of R is to modulate the activity of C1, independently of DNA contacts other than the DNA-binding of C1. The substitution of six residues in the MYB domain of P1 with the corresponding residues from C1 results in the P1* protein, which physically interacts with R (Grotewold *et al.*, 2000). Similar to P1, P1* activates *A1* but not *Bz1* in the absence of R, indicating that the DNA-binding properties of P1* have not been altered. However, interestingly, in the presence of R, the activation of *A1* by P1* is enhanced (R-enhanced transcription), and P1* provides a robust activation of the *Bz1* promoter (Grotewold *et al.*, 2000). From these findings, it was concluded that R has at least two functions on C1: (1) it alters the regulatory potential of C1, and (2) it provides C1 (or P1*) the ability to activate anthocyanin specific promoters, such as *Bz1*. Thus, R and R-like proteins are central to explain how R2R3-MYB transcription factors with very similar DNA-binding domain regulate specific sets of target genes.

In addition to anthocyanin biosynthesis, recent studies suggested a role for C1 and R in the regulation of flavonol accumulation, through the direct activation of *ZmFLS1* (Falcone Ferreyra *et al.*, 2010). Besides C1 and R, *ZmFLS1* is also regulated by P1, resembling the mechanism by which *FLS* genes are controlled in *Arabidopsis* by the production of flavonol glycosides (*PFG1-3*) genes, encoding three R2R3-MYB transcription factors most related to P1 from maize (Mehrtens *et al.*, 2005; Stracke *et al.*, 2007, 2009).

The analysis of several other R2R3-MYB and bHLH interactions permitted the identification of a consensus sequence in the R2 MYB domain responsible for the interaction with R-like factors (Zimmermann *et al.*, 2004), providing a powerful tool for predicting protein–protein interactions in other members of the large R2R3-MYB family (Dubos *et al.*, 2010).

Recent studies also identified R-interacting factor 1 (RIF1) specifically interacting with the bHLH region of R (Hernandez *et al.*, 2007), making it the first identified partner for this conserved region of R-like bHLH transcription factors. RIF1 harbors an EMSY *N*-terminal (ENT) domain and an AGENET domain, which belongs to the Royal Family domains

(Maurer-Stroh *et al.*, 2003), implicated in the control of chromatin remodeling. The ENT domain was first found in the human protein EMSY, which interacts with the *N*-terminal activation domain of the breast cancer 2 (BRCA2) protein (Hughes-Davies *et al.*, 2003). Deletion of this region in BRCA2 or amplification of the *EMSY* gene associates with breast and ovarian cancers (Yao & Polyak, 2004). *RIF1* is expressed in most maize tissues, and the protein accumulates in large nuclear foci or speckles (Hernandez *et al.*, 2007). Knocking-down *RIF1* in BMS cells using a double-stranded RNA (dsRNA) significantly impairs the ability of C1+R to induce anthocyanin formation, but had no effect on the activation of a reporter gene (luciferase) driven by the *A1* gene promoter, introduced transiently by bombardment (Hernandez *et al.*, 2007).

In addition to the MYB and bHLH factors, regulatory proteins containing conserved WD-repeats (WDR) participate in anthocyanin regulation in maize (PAC1; Carey *et al.*, 2004), Petunia (AN11; de Vetten *et al.*, 1997), *Arabidopsis* (TTG1; Walker *et al.*, 1999) and Perilla (Sompornpailin *et al.*, 2002). While the mechanisms by which these WDR proteins function is unclear, in yeast they appear to stabilize the interaction between the TT2 (R2R3-MYB) and TT8 (bHLH) regulators and enhance the activation of the *BAN* gene required for condensed tannin (PA) accumulation (Baudry *et al.*, 2004). Indeed, it has been recently demonstrated that TTG1 is part of the GL3/GL1 regulatory complex on various trichome developmental genes (Zhao *et al.*, 2008), in the *TTG1* mutants, the pathway is blocked at the DFR step (Shirley *et al.*, 1995). Additionally, repressors for the anthocyanins biosynthetic pathway have been reported, and include *Arabidopsis* AtMYB4 (Jin *et al.*, 2000) and MYBL2 (Dubos *et al.*, 2008). *Petunia,* PhMYBx (Koes *et al.*, 2005; Quattrocchio *et al.*, 2006a) and maize ZmIN1 (Burr *et al.*, 1996).

6.2.2 Arabidopsis

The control of anthocyanins in *Arabidopsis* involves similar players as in maize. The R2R3-MYB function is largely provided by MYB75/PAP1 (*production of anthocyanin pigment*), identified through screening of activation tagging mutants (Borevitz *et al.*, 2000). PAP1 corresponds to a major locus involved in sugar-induced anthocyanin accumulation in *Arabidopsis* (Teng *et al.*, 2005), and *PAP1-Dominant* plants (*PAP1-D*), overexpressing PAP1 constitutively, accumulate anthocyanin purple pigment in all vegetative tissues. In *PAP1-D* plants, some structural genes for anthocyanin biosynthesis are activated, including those corresponding to the phenylalanine ammonia lyase (*PAL*) and *CHS* early pathway genes (Borevitz *et al.*, 2000). Transcriptome analyses of *PAP1-D* plants, compared to the wild-type, revealed new genes involved in the anthocyanin pathway, and also ones regulated by the PAP1 transcription factor, such as the 3GT and 5GT glycosyltransferases, an acyltransferase, a glutathione *S*-transferase (GST), sugar transporters, and transcription factors (Tohge *et al.*, 2005).

PAP2, another R2R3-MYB transcription factor, shares high sequence similarity with PAP1, and also control anthocyanin accumulation (Borevitz *et al.*, 2000). Overexpression of PAP2 in tobacco activates production of purple pigment in leaves (Borevitz *et al.*, 2000). However, PAP1 RNAi lines result in an obvious loss of anthocyanin, whereas PAP2 RNAi have normal anthocyanin levels, under normal growth conditions (Gonzalez *et al.*, 2008).

Recently, two additional MYB transcription factors, AtMYB113 and AtMYB114, were identified and shown to be able to control the anthocyanin pathway (Gonzalez *et al.*, 2008). Similarly to PAP1 and PAP2, overexpression of AtMYB113 or AtMYB114 induced the same level of anthocyanin in plant tissues. However, AtMYB113 and AtMYB114 regulate only the late genes of the pathway, i.e., *DFR* and *LDOX*, but not the *CHS* and *PAL* early genes. AtMYB113 and AtMYB114 induction of anthocyanins is dependent on the presence of the TTG1 WD40-regulatory protein, as well as the TT8, GL3, and EGL3 bHLH-transcription factors (Gonzalez *et al.*, 2008). The authors proposed that TTG1-dependent MYBs and bHLHs of *Arabidopsis* regulate late anthocyanin pathway genes, starting with *F3H*. They also suggested that early gene expression changes observed in *PAP1-D* could be a consequence of secondary effects, such as a metabolite feedback phenomenon resulting from strong upregulation of the late pathway genes and increased flux through the flavonoid pathway (Gonzalez *et al.*, 2008). A question that remains is what is the participation and relative contribution of each one of these apparently partially redundant R2R3-MYBs in anthocyanin production in *Arabidopsis*.

6.2.3 Petunia

As is the case in maize and *Arabidopsis*, the regulation of anthocyanin biosynthesis in Petunia also involves R2R3-MYB, bHLH, and WD40 factors (Ramsay & Glover, 2005; Koes *et al.*, 2005). The R2R3-MYB function is provided by AN2, which physically interacts with the bHLH factors AN1 or JAF13 (Spelt *et al.*, 2000). Indeed, variations at the *AN2* locus have been proposed to underlie major color differences in Petunia species (Quattrocchio *et al.*, 1999). AN1 and JAF13 can form homo- and heterodimers, and while *AN1* mutants lack petal anthocyanin pigmentation (Spelt *et al.*, 2002), JAF13 contributes to pigmentation as well (Quattrocchio *et al.*, 2006a). AN11 corresponds to the WD40 function (de Vetten *et al.*, 1997). Similar to *Arabidopsis TT2*, AN11 was suggested to interact with the *AN4*, presumably encoding a MYB factor similar to AN2. However, in similar yeast two-hybrid experiments, no interaction between AN11 and AN2 was observed (Quattrocchio *et al.*, 2006a). AN1 expression is controlled by AN2 and AN4, presumably through the interaction with other bHLH factors (Spelt *et al.*, 2000).

An interesting difference between the function of the regulators in Petunia and other plants is that the AN1 factor is also a key determinant of vacuolar pH, therefore impacting anthocyanin pigmentation by controlling both the levels of these compounds and their ability to function as pigments (Spelt *et al.*, 2002). Indeed, the *PH6* allele (Chuck *et al.*, 1993) corresponds to an allele of the *AN1* locus (Spelt *et al.*, 2002). Vacuolar pH is controlled by the interaction of AN1 with the R2R3-MYB factor PH4 (Quattrocchio *et al.*, 2006b). Similar R2R3-MYB factors regulating vacuolar pH are yet to be identified in some of the other plants currently used as model systems to understand anthocyanin pigmentation.

6.2.4 Snapdragon

Many of the flavonoid biosynthetic genes have also been identified in snapdragon (*A. majus*), thanks to a wealth of transposon insertion mutants that alter pigment accumulation (Almeida *et al.*, 1989). In snapdragon, three R2R3-MYB factors, ROSEA1 (ROS1), ROSEA2 (ROS2), and VENOSA (VEN), control anthocyanin floral patterning and pigmentation (Schwinn *et al.*, 2006). The bHLH function is provided by the DELILA (DEL) protein (Goodrich *et al.*, 1992). DEL harbors significant amino acid differences with other similar bHLH factors, differences that were exploited in directed evolution experiments to change the ability of related factors to activate transcription and interact with R2R3-MYB factors (Pattanaik *et al.*, 2006, 2008). The expression of ROS1 and DEL in tomato was recently used to successfully generate tomato fruits with very high levels of health-promoting anthocyanins (Butelli *et al.*, 2008).

6.3 Anthocyanin transport and subvacuolar localization

The very significant knowledge that has been gained over the past few years with regards to the chemistry (Harborne, 1988; Stafford, 1990; Andersen & Jordheim, 2006), biosynthesis (Grotewold, 2006; Winkel, 2008), and regulation (Quattrocchio *et al.*, 2006a), contrasts with how little is still understood with regards to the intracellular transport of anthocyanins. Anthocyanins have been proposed to be synthesized on the cytoplasmic surface of the endoplasmic reticulum (ER) in a complex formed by various biosynthetic enzymes (see Section 6.1.2) (Winkel-Shirley, 1999; Winkel, 2004). However, anthocyanins are not pigmented until they reach the acidic pH (pH $\sim$ 5) of the vacuole (Forkmann, 1991), creating significant challenges in tracking how anthocyanins move from the ER to the vacuole.

Models that try to explain how flavonoids move from the surface of the ER to the tonoplast are usually divided into two types (Grotewold & Davies, 2008; Zhao and Dixon, 2010; Zhao *et al.*, 2010). We have referred to them before as the Ligandin Transporter and the Vesicular Transport models (Grotewold & Davies, 2008). The Ligandin Transporter model derives from the finding that mutations in the maize *BZ2* gene, which encodes a GST, prevent the vacuolar localization of anthocyanins, resulting in the accumulation of brown oxidation products (hence the *Bronze2* name) (Marrs *et al.*, 1995). Similar to *BZ2*, *Petunia AN9* encodes a GST and, despite low identity, BZ2 functionally complements *AN9* mutants (Alfenito *et al.*, 1998). Yet, the GST enzymatic activity of AN9 is not required for the AN9-dependent vacuolar sequestration of anthocyanins, resulting in the suggestion that AN9/BZ2 serve as "ligandins," necessary for escorting anthocyanins (e.g., cyanidin 3-*O*-glucoside, C3G) through the cytoplasm, from the ER surface to the tonoplast (Mueller *et al.*, 2000). The identification of the maize tonoplast-localized multidrug resistance-associated protein ZmMRP3, provided an additional player in this model involving carrier/ligandin and transporter proteins in the trafficking of anthocyanins from the ER surface to the vacuole (Goodman *et al.*, 2004). In *Arabidopsis*, *TRANSPARENT TESTA19* (*TT19*) mutations affect both vacuolar anthocyanin accumulation in vegetative tissues and vacuolar PA accumulation in seed coats (Kitamura *et al.*, 2010). Similar to BZ2 and AN9, TT19 encodes a GST, and AN9 complements the anthocyanin but not the PA defect of *TT19* mutants (Kitamura *et al.*, 2004). While TT19 and AN9/BZ2 may function similarly by stabilizing/escorting anthocyanins, *TT19* mutants have a distinctive phenotype in the seed coat, where PA precursors accumulate in membrane-wrapped cytoplasmic structures (Kitamura *et al.*, 2004,

2010). This contrasts with the phenotype of mutations in the *Arabidopsis* TT12 locus, encoding a MATE transporter involved in PA vacuolar sequestration. In TT12 seeds, there is a reduction in the soluble PAs and a complete absence of insoluble PAs, and in the seed coat PA precursors are evenly distributed in the cytoplasm (Debeaujon *et al.*, 2001; Marinova *et al.*, 2007a). Similar to ZmMRP3, TT12 localizes to the tonoplast and has the ability to transport C3G, but not the corresponding aglycones (Marinova *et al.*, 2007b). The Vesicular Transport model is largely based on the presence in many different plants [e.g., lisianthus (Zhang *et al.*, 2006); maize cultured cells (Grotewold *et al.*, 1998); *Arabidopsis* (Poustka *et al.*, 2007); sweet potato (Nozue & Yasuda, 1985); red cabbage (Small & Pecket, 1982)] of anthocyanin-filled structures in the cytoplasm. Most enlightening has been the recent description of the tapetosomes in *Brassica* tapetum cells. These ER-derived organelles accumulate several flavonoids, which are delivered to the pollen surface upon tapetum's cell death. In tapetal cells, flavonoids are found inside vesicles, which then come to form part of the tapetosome (Hsieh & Huang, 2005, 2007).

The expression of the C1 & R regulators of anthocyanin biosynthesis in maize BMS cells (BMS^{R+C1}) resulted in the induction of the entire anthocyanin pathway, and cells accumulated large quantities of pigments (Grotewold *et al.*, 1998). Transmission electron microscopy (TEM) analysis of BMS^{R+C1} showed the presence of dilated ER filled with large quantities of electron-dense material, absent in control cells. Interestingly, electron-dense bodies were found inside the ER, solely in BMS^{R+C1} cells, resulting in the speculation the ER might be an initial site of anthocyanin accumulation (Grotewold *et al.*, 1998), in the same way as flavonols accumulate inside the ER in tapetal cells (Hsieh & Huang, 2005, 2007). However, the inability to demonstrate directly that the electron-dense pigments corresponded to anthocyanins made it difficult to link these findings to anthocyanin transport.

A significant breakthrough was the identification of red fluorescence properties of *Arabidopsis* anthocyanins, which is compatible with subcellular protein markers tagged with green and yellow fluorescent proteins (GFP and YFP, respectively) (Poustka *et al.*, 2007). Using these fluorescent properties and the possibility to induce anthocyanin accumulation in *Arabidopsis* seedlings by the complementation of *TT5* mutants with the flavanone naringenin, it became clear that similar to proteins targeted to the protein storage vacuole (PSV), anthocyanins can be transported in ER-derived structures that share characteristics and cargo with vesicles that transport proteins from the ER to the PSV (Poustka *et al.*, 2007). Once accumulated into the vacuole, anthocyanin can be found uniformly distributed or as part of subvacuolar pigment bodies, called by a number of names that reflects their diversity. Anthocyanoplasts have been described in number of plants, and are usually membrane-bound structures that can fuse together to give larger bodies containing anthocyanins (Pecket & Small, 1980; Nozzolillo & Ishikura, 1988). Cyanoplasts found in sweet potato cells, in contrast, are described as membraneless structure (Nozue *et al.*, 1993). The term anthocyanic vacuolar inclusions (AVIs), originally used to describe membraneless structures accumulating anthocyanins inside the vacuoles of many plants (Nozue *et al.*, 1993; Markham *et al.*, 2000), have since been utilized to describe a variety of pigment structures in the vacuoles of many plants, including maize, grape, lisianthus, and *Arabidopsis* (Irani & Grotewold, 2005; Zhang *et al.*, 2006; Poustka *et al.*, 2007; Conn *et al.*, 2010). Although *Arabidopsis*, accumulates few AVIs under normal growth conditions, has been developed into a good model for studying the formation of these structures (Poustka *et al.*, 2007). When seedlings are grown in inductive conditions that include light and high-sucrose, anthocyanins accumulate in the epidermal cells of the abaxial side of the cotyledons with AVIs present in about 1% of the cells (Poustka *et al.*, 2007). The use of GST or ATP-dependent transporter inhibitors dramatically increases the percentage of cells containing AVIs (Poustka *et al.*, 2007). Thus, when the tonoplast uptake route involving GST and ABC transporters is inhibited, the anthocyanins are mainly sequestered into the vacuole via the vesicular transport pathway. AVIs preferentially accumulate specific anthocyanin types (Morita *et al.*, 2005; Pourcel *et al.*, 2010). Furthermore, inhibition of position-specific glycosylation of the core anthocyanidin, as found in the *Arabidopsis 5GT* mutant, forms AVIs in almost every epidermal cell of the cotyledons. Moreover, this mutant accumulates only C3G derivatives, and may be the type of anthocyanin selectively being transported into the *Arabidopsis* AVIs (Pourcel *et al.*, 2010).

AVIs purified from *Vitis vinifera* accumulate higher proportions of acylated anthocyanins compared to vacuolar anthocynains (Conn *et al.*, 2010). Further, the lipid profile of the grape AVIs show similarity with the tonoplast, suggesting that the AVIs may acquire their membrane encasement from a combination of vesicular trafficking (ER- or Golgi-derived) and tonoplast membranes *via* vesicular invagination/autophagy during AVI shuttling.

Packaging of anthocyanins into AVIs and vacuoles also influences how cells are pigmented. In flowers of Lisianthus, the dark pigmentation at the base of the petals is attributed to the increased number of AVIs as compared to the lighter purple outer zones where their number are reduced or absent (Markham *et al.*, 2000). The rose cultivar "Rhapsody in Blue" progressively accumulates AVIs as their petals age which cause a color shift from red-purple to bluish-purple (Gonnet, 2003). Light also influences the vacuolar and sub-vacuolar distribution of anthocyanins by changing the morphology of AVIs as seen red cabbage (Pecket & Small, 1980), in the floral whorls of maize flowers of the B-I/B-Peru line and cultured maize BMS^{R+Cl} cells hyperaccumulating anthocyanins (Irani & Grotewold, 2005). Exposure of maize BMS^{R+Cl} cells to light cause them to darken which cannot be attributed to a transcriptional response of their biosynthetic genes, an increase in the amount of anthocyanins or vacuolar pH but to the subcellular morphology of the AVIs and the way anthocyanins are distributed in the vacuole. Light triggers the vacuoles and AVIs to fuse and "spread" out the anthocyanins and pigment the vacuolar sap. The changing morphology of AVIs from round to tubular thread-like, further supports that these structures are membrane bound and indeed maybe formed by vacuolar invagination/ autophagy.

Thus, the type of anthocyanin produced and the type of transporters present would determine how plants accumulate and package anthocyanins into vacuoles and subvacuolar structures. The identification of the "anthocyanin-type specific" transporters, their localization, and the mechanisms of the vesicular transport pathway would greatly enhance our understanding of the transport and final sequestration compartment of not only the flavonoids but also biologically important small molecules that include plant hormones and provide the tools to genetically manipulate pharmaceutically important plant secondary metabolites.

6.4 Concluding remarks

The pigmentation provided by anthocyanins has served as a traditional system to understand basic concepts of genetics. During the past 20 years, anthocyanin pigmentation has significantly contributed to our understanding of the basic rules governing the control of plant gene expression and provided powerful examples to understand combinatorial plant gene regulation. The utilization of anthocyanins to understand pathways for the trafficking of specialized compounds and mechanisms of subcellular sequestration cannot be underestimated, and we anticipate that the next few years will witness a very significant advancement in this area, perhaps bringing to light the crosstalk between biosynthesis, transport, and regulation for this important group of pigments.

References

Abrahams, S., Lee, E., Walker, A., Tanner, G., Larkin, P. & Ashton, A. (2003) The *Arabidopsis* TDS4 gene encodes leucoanthocdyanidin dioxygenase (LDOX) and is essential for proanthocyanidin synthesis and vacuole development. *The Plant Journal*, **35**, 624–636.

Alfenito, M.R., Souer, E., Goodman, C.D., *et al.* (1998) Functional complementation of anthocyanin sequestration in the vacuole by widely divergent glutathione *S*-transferases. *Plant Cell*, **10**, 1135–1149.

Alleman, M., Sidorenko, L., McGinnis, K., *et al.* (2006) An RNA-dependent RNA polymerase is required for paramutation in maize. *Nature*, **442**, 295–298.

Almeida, J., Carpenter, R., Robbins, T.P., Martin, C. & Coen, E.S. (1989) Genetic interactions underlying flower color patterns in *Antirrhinum majus*. *Genes Development*, **3**, 1758–1767.

Andersen, Ø.M. & Jordheim, M. (2006) The Anthocyanins. In: *The Flavonoids—Chemistry, Biochemistry, and Applications* (eds. Ø.M. Andersen & K.R. Markham), pp. 471–551. CRC Press, Taylor & Francis Group, Boca Raton, FL.

Anderson, F.G. (1924) Pericarp studies in maize. II. The allelomorphism of a series of factors for pericarp color. *Genetics*, **9**, 442–453.

Austin, M.B. & Noel, J.P. (2003) The chalcone synthase superfamily of type III polyketide synthases. *Natural Product Reports*, **20**, 79–110.

Baudry, A., Heim, M., Dubreucq, B., Caboche, M., Weisshaar, B. & Lepiniec, L. (2004) TT2, TT8 and TTG1 synergistically specify the expression of *Banyuls* and proanthocyanidin biosynthesis in *Arabidopsis thaliana*. *The Plant Journal*, **39**, 366–380.

Borevitz, J.O., Xia, Y., Blount, J., Dixon, R.A. & Lamb, C. (2000) Activation tagging identifies a conserved Myb regulator of phenylpropanoid biosynthesis. *Plant Cell*, **12**, 2383–2394.

Brink, R.A. (1958) Paramutation at the *R* locus in maize. *Cold Spring Harbor Symposia on Quantitative Biology*, **23**, 379–391.

Burbulis, I.E. & Winkel-Shirley, B. (1999) Interactions among enzymes of the *Arabidopsis* flavonoid biosynthetic pathway. *Proceedings of the National Academy of Science of the United States of America*, **96**, 12929–12934.

Burr, F.A., Burr, B., Scheffler, B.E., Blewitt, M., Wienand, U. & Matz, E.C. (1996) The maize repressor-like gene *intensifier1* shares homology with the *r1/b1* multigene family of transcription factors and exhibits missplicing. Plant Cell, **8**, 1249–1259.

Butelli, E., Titta, L., Giorgio, M., *et al.* (2008). Enrichment of tomato fruit with health-promoting anthocyanins by expression of select transcription factors. *Nature Biotechnology*, **26**, 1301–1308.

Carey, C.C., Strahle, J.T., Selinger, D.A. & Chandler, V.L. (2004). Mutations in the *Pale Aleurone Color1* regulatory gene of the *Zea mays* anthocyanin pathway have distinct phenotypes relative to the functionally similar *Transparent Testa Glabra1* gene in *Arabidopsis thaliana*. *Plant Cell*, **16**, 450–464.

Chuck, G., Robbins, T., Nijjar, C., Ralston, E., Courtney-Gutterson, N. & Dooner, H.K. (1993). Tagging and cloning of a petunia flower color gene with the maize transposable element *activator*. *Plant Cell*, **5**, 371–378.

Cone, K.C., Cocciolone, S.M., Burr, F.A. & Burr, B. (1993). Maize anthocyanin regulatory gene *Pl* is a duplicate of *C1* that functions in the plant. *Plant Cell*, **5**, 1795–1805.

Conn, S., Franco, C. & Zhang, W. (2010). Characterization of anthocyanic vacuolar inclusions in *Vitis vinifera* l. cell suspension cultures. *Planta*, **231**, 1343–1360.

de Vries, H. (1901). *Die Mutationstheorie, Versuche und Beobachtungen über die Entstehung von Arten im Pflanzenreich*. Verlag von Veit & Comp., Leipzig.

de Vetten, N., Quattrocchio, F., Mol, J. & Koes, R. (1997). The *An11* locus controlling flower pigmentation in petunia encodes a novel WD-repeat protein conserved in yeast, plants, and animals. *Genes & Development*, **11**, 1422–1434.

Debeaujon, I., Peeters, A.J.M., Leon-Kloosterziel, K.M. & Koornneef, M. (2001). The *Transparent Testa12* gene of *Arabidopsis* encodes a multidrug secondary transporter-like protein required for flavonoid sequestration in vacuoles of the seed coat endothelium. *Plant Cell*, **13**, 853–871.

Deboo, G.B., Albertsen, M.C. & Taylor, L.P. (1995). Flavanone 3-hydroxylase transcripts and flavonol accumulation are temporally coordinated in maize anthers. *The Plant Journal*, **7**, 703–713.

Dong, X., Braun, E.L. & Grotewold, E. (2001). Functional conservation of plant secondary metabolic enzymes revealed by complementation of *Arabidopsis* flavonoid mutants with maize genes. *Plant Physiology*, **127**, 46–57.

Dooner, H.K., Robbins, T.P. & Jorgensen, R.A. (1991). Genetic and developmental control of anthocyanin biosynthesis. *Annual Review on Genetics*, **25**, 173–199.

Dubos, C., Le Gourrierec, J., Baudry, A., *et al.* (2008) MYBL2 is a new regulator of flavonoid biosynthesis in *Arabidopsis thaliana*. *The Plant Journal*, **55**, 940–953.

Dubos, C., Stracke, R., Grotewold, E., Weisshaar, B., Martin, C. & Lepiniec, L. (2010) Myb transcription factors in *Arabidopsis*. *Trends in Plant Science*, **15**, 573–581.

Emerson, R.A. (1911) Genetic correlation and spurious allelomorphism in maize. *Annual Report of the Nebraska Agricultural Experiment Station*, **24**, 59–90.

Emerson, R.A. (1917) Genetical studies of variegated pericarp in maize. *Genetics*, **2**, 1–35.

Falcone Ferreyra, M. , Rius, S., Emiliani, J., *et al.* (2010) Cloning and characterization of a uv-b inducible maize flavonol synthase. *The Plant Journal*, **62**, 67–91.

Ferrer, J.L., Austin, M.B., Stewart, C., Jr. & Noel, J.P. (2008) Structure and function of enzymes involved in the biosynthesis of phenylpropanoids. *Plant Physiology and Biochemistry*, **46**, 356–370.

Ferrer, J.L., Jez, J.M., Bowman, M.E., Dixon, R.A. & Noel, J.P. (1999) Structure of chalcone synthase and the molecular basis of plant polyketide biosynthesis. *Nature Structural Biology*, **6**, 775–784.

Forkmann, G. (1991) Flavonoids as flower pigments: the formation of the natural spectrum and its extension by genetic engineering. *Plant Breeding*, **106**, 1–26.

Gensheimer, M. & Mushegian, A. (2004) Chalcone isomerase family and fold: no longer unique to plants. *Protein Science*, **13**, 540–544.

Goff, S.A., Cone, K.C. & Chandler, V.L. (1992) Functional analysis of the transcriptional activator encoded by the maize *B* gene: evidence for a direct functional interaction between two classes of regulatory proteins. *Genes Development*, **6**, 864–875.

Gonnet, J.F. (2003) Origin of the color of cv. rhapsody in blue rose and some other so-called "blue" roses. *Journal of Agricultural and Food Chemistry*, **51**, 4990–4994.

Gonzalez, A., Zhao, M., Leavitt, J.M. & Lloyd, A.M. (2008) Regulation of the anthocyanin biosynthetic pathway by the TTG1/BHLH/MYB transcriptional complex in *Arabidopsis* seedlings. *The Plant Journal*, **53**, 814–827.

Goodman, C.D., Casati, P. & Walbot, V. (2004) A multidrug resistance-associated protein involved in anthocyanin transport in *Zea mays*. *Plant Cell*, **16**, 1812–1826.

Goodrich, J., Carpenter, R. & Coen, E.S. (1992) A common gene regulates pigmentation pattern in diverse plant species. *Cell*, **68**, 955–964.

Grotewold, E. (2006) The genetics and biochemistry of floral pigments. Annual Review on Plant Biology, **57**, 761–780.

Grotewold, E., Athma, P. & Peterson, T. (1991) Alternatively spliced products of the maize *p* gene encode proteins with homology to the dna-binding domain of Myb-like transcription factors. *Proceedings of the National Academy of Science of the United States of America*, **88**, 4587–4591.

Grotewold, E., Chamberlin, M., Snook, M., *et al.* (1998) Engineering secondary metabolism in maize cells by ectopic expression of transcription factors. *Plant Cell*, **10**, 721–740.

Grotewold, E. & Davies, K. (2008) Trafficking and sequestration of anthocyanins. *Natural Product Communications*, **3**, 1251–1258.

Grotewold, E., Drummond, B.J., Bowen, B. & Peterson, T. (1994) The Myb-homologous *P* gene controls phlobaphene pigmentation in maize floral organs by directly activating a flavonoid biosynthetic gene subset. *Cell*, **76**, 543–553.

Grotewold, E., Sainz, M.B., Tagliani, L., Hernandez, J.M., Bowen, B. & Chandler, V.L. (2000). Identification of the residues in the Myb domain of maize C1 that specify the interaction with the bHLH cofactor R. *Proceedings of the National Academy of Science of the United States of America*, **97**, 13579–13584.

Harborne, J.B. (1988) *The Flavonoids—Advances in Research since 1980*. Chapman & Hall, New York.

Herles, C., Braune, A. & Blaut, M. (2004) First bacterial chalcone isomerase isolated from *Eubacterium ramulus*. *Archives in Microbiology*, **181**, 428–434.

Hernandez, J., Heine, G., Irani, N.G., *et al.* (2004) Different mechanisms participate in the R-dependent activity of the R2R3 Myb transcription factor C1. *Journal of Biological Chemistry*, **279**, 48205–48213.

Hernandez, J.M., Feller, A., Morohashi, K., Frame, K. & Grotewold, E. (2007) The basic helix loop helix domain of maize r links transcriptional regulation and histone modifications by recruitment of an EMSY-related factor. *Proceedings of the National Academy of Science of the United States of America*, **104**, 17222–17227.

Hsieh, K. & Huang, A.H. (2005) Lipid-rich tapetosomes in *Brassica* tapetum are composed of oleosin-coated oil droplets and vesicles, both assembled in and then detached from the endoplasmic reticulum. *The Plant Journal*, **43**, 889–899.

Hsieh, K. & Huang, A.H. (2007) Tapetosomes in *Brassica tapetum* accumulate endoplasmic reticulum-derived flavonoids and alkanes for delivery to the pollen surface. *Plant Cell*, **19**, 582–596.

Hughes-Davies, L., Huntsman, D., Ruas, M., *et al.* (2003) EMSY links the BRCA2 pathway to sporadic breast and ovarian cancer. *Cell*, **115**, 523–535.

Irani, N.G. & Grotewold, E. (2005) Light-induced morphological alteration in anthocyanin-accumulating vacuoles of maize cells. *BMC Plant Biology*, **5**, 7.

Jez, J.M., Bowman, M.E., Dixon, R.A. & Noel, J.P. (2000) Structure and mechanism of the evolutionarily unique plant enzyme chalcone isomerase. *Nature Structural Biology*, **7**, 786–791.

Jin, H., Cominelli, E., Bailey, P., *et al.* (2000) Transcriptional repression by AtMYB4 controls production of uv-protecting sunscreens in *Arabidopsis*. *The EMBO Journal*, **19**, 6150–6161.

Kitamura, S., Matsuda, F., Tohge, T., *et al.* (2010) Metabolic profiling and cytological analysis of proanthocyanidins in immature seeds of *Arabidopsis thaliana* flavonoid accumulation mutants. *The Plant Journal*, **62**, 549–559.

Kitamura, S., Shikazono, N. & Tanaka, A. (2004) *TRANSPARENT TESTA 19* is involved in the accumulation of both anthocyanins and proanthocyanidins in *Arabidopsis*. *The Plant Journal*, **37**, 104–114.

Koes, R., Verweij, W. & Quattrocchio, F. (2005) Flavonoids: a colorful model for the regulation and evolution of biochemical pathways. *Trends in Plant Science*, **10**, 236–242.

Koes, R.E., Quattrocchio, F. & Mol, J.N.M. (1994) The flavonoid biosynthetic pathway in plants: function and evolution. *Bioessays*, **16**, 123–132.

Lesnick, M.L. & Chandler, V.L. (1998) Activation of the maize anthocyanin gene *A2* Is mediated by an element conserved in many anthocyanin promoters. *Plant Physiology*, **117**, 437–445.

Liu, C.J. & Dixon, R.A. (2001) Elicitor-induced association of isoflavone *o*-methyltransferase with endomembranes prevents the formation and 7-*O*-methylation of daidzein during isoflavonoid phytoalexin biosynthesis. *Plant Cell*, **13**, 2643–2658.

Ludwig, R., Habera, L.F., Dellaporta, S.L. & Wessler, S.R. (1989) *Lc*, a member of the maize *R* gene family responsible for tissue-specific anthocyanin production, encodes a protein similar to transcriptional activators and contains the *Myc*-homology region. *Proceedings of the National Academy of Science of the United States of America*, **86**, 7092–7096.

Luo, J., Nishiyama, Y., Fuell, C., *et al.* (2007) Convergent evolution in the BAHD family of acyl transferases: identification and characterization of anthocyanin acyl transferases from *Arabidopsis thaliana*. *The Plant Journal*, **50**, 678–695.

Marinova, K., Kleinschmidt, K., Weissenbock, G. & Klein, M. (2007a) Flavonoid biosynthesis in barley primary leaves requires the presence of the vacuole and controls the activity of vacuolar flavonoid transport. *Plant Physiology*, **144**, 432–444.

Marinova, K., Pourcel, L., Weder, B., *et al.* (2007b) The Arabidopsis mate transporter TT12 acts as a vacuolar flavonoid/H+—antiporter active in proanthocyanidin-accumulating cells of the seed coat. *Plant Cell*, **19**, 2023–2038.

Markham, K.R., Gould, K.S., Winefield, C.S., Mitchell, K.A., Bloor, S. J. & Boase, M.R. (2000) Anthocyanic vacuolar inclusions—their nature and significance in flower colouration. *Phytochemistry*, **55**, 327–336.

Marrs, K.A., Alfenito, M.R., Lloyd, A.M. & Walbot, V. (1995) A glutathione S-transferase involved in vacuolar transfer encoded by the maize gene *bronze-2*. *Nature*, **375**, 397–400.

Martin, C., Prescott, A., Mackay, S., Bartlett, J. & Vrijlandt, E. (1991) Control of anthocyanin biosynthesis in flowers of *Antirrhinum majus*. *The Plant Journal*, **1**, 37–49.

Maurer-Stroh, S., Dickens, N.J., Hughes-Davies, L., Kouzarides, T., Eisenhaber, F. & Ponting, C.P. (2003) The tudor domain 'royal family': TUDOR, plant AGENET, CHROMO, PWWP and MBT domains. *Trends in Biochemical Science*, **28**, 69–74.

McClintock, B. (1950a) Mutable loci in maize. *Carnegie Institution of Washington Year Book*, **49**, 157–167.

McClintock, B. (1950b) The origin and behavior of mutable loci in maize. *Proceedings of the National Academy of Science of the United States of America*, **36**, 344–355.

McClintock, B. (1951) Mutable loci in maize. *Carnegie Institution of Washington Year Book*, **50**, 174–181.

McClintock, B. (1955) Controlled mutation in maize. *Carnegie Institution of Washington Year Book*, **54**, 254–255.

McClintock, B. (1958) The *suppressor–mutator* system of control of gene action in maize. *Carnegie Institution of Washington Year Book*, **57**, 415–429.

Mehrtens, F., Kranz, H., Bednarek, P. & Weisshaar, B. (2005) The *Arabidopsis* transcription factor Myb12 is a flavonol-specific regulator of phenylpropanoid biosynthesis. *Plant Physiology*, **138**, 1083–1096.

Meyer, P., Heidmann, I., Forkmann, G. & Saedler, H. (1987) A new petunia flower colour generated by transformation of a mutant with a maize gene. *Nature*, **330**, 677–678.

Mol, J., Grotewold, E. & Koes, R. (1998) How genes paint flowers and seeds. *Trends in Plant Science*, **3**, 212–217.

Mol, J.N.M., Robbins, M.P., Dixon, R.A. & Veltkamp, E. (1985) Spontaneous and enzymic rearrangement of naringenin chalcone to flavanone. *Phytochemistry*, **24**, 2267–2269.

Morita, Y., Hoshino, A., Kikuchi, Y., *et al.* (2005) Japanese morning glory *dusky* mutants displaying reddish-brown or purplish-gray flowers are deficient in a novel glycosylation enzyme for anthocyanin biosynthesis, UDP-glucose:anthocyanidin 3-O-glucoside–2″-O-glucosyltransferase, due to 4-bp insertions in the gene. *The Plant Journal*, **42**, 353–363.

Mueller, L.A., Goodman, C.D., Silady, R.A. & Walbot, V. (2000) An9, a petunia glutathione S-transferase required for anthocyanin sequestration, is a flavonoid-binding protein. *Plant Physiology*, **123**, 1561–1570.

Nakabayashi, R., Yamazaki, M. & Saito, K. (2010) A polyhedral approach for understanding flavonoid biosynthesis in *Arabidopsis*. *Nature Biotechnology*, **27**, 829–836.

Napoli, C., Lemieux, C. & Jorgensen, R. (1990) Introduction of a chimeric chalcone synthase gene into petunia results in reversible co-suppression of homologous genes in *trans*. *Plant Cell*, **2**, 279–289.

Nozue, M., Nishimura, M., Katou, A., Hattori, C. & Usuda, N. (1993) Characterization of intravacuolar pigmented structures in anthocyanin-containing cells of sweet potato suspension cultures. *Plant and Cell Physiology*, **34**, 803–808.

Nozue, M. & Yasuda, H. (1985) Occurrence of anthocyanoplasts in cell suspension cultures of sweet potato. *Plant Cell Reports*, **4**, 252–255.

Nozzolillo, C. & Ishikura, N. (1988) An investigation of the intracellular site of anthocyanoplasts using isolated protoplasts and vacuoles. *Plant Cell Reports*, **7**, 389–392.

Owens, D.K., Alerding, A.B., Crosby, K.C., Bandara, A.B., Westwood, J.H. & Winkel, B.S. (2008a) Functional analysis of a predicted flavonol synthase gene family in *Arabidopsis*. *Plant Physiology*, **147**, 1046–1061.

Owens, D.K., Crosby, K.C., Runac, J., Howard, B.A. & Winkel, B.S. (2008b) Biochemical and genetic characterization of *Arabidopsis* flavanone 3β-hydroxylase. *Plant Physiology and Biochemistry*, **46**, 833–843.

Pattanaik, S., Xie, C.H., Kong, Q., Shen, K.A. & Yuan, L. (2006) Directed evolution of plant basic helix-loop-helix transcription factors for the improvement of transactivational properties. *Biochimica et Biophysica Acta*, **1759**, 308–318.

Pattanaik, S., Xie, C.H. & Yuan, L. (2008) The interaction domains of the plant Myc-like bHLH transcription factors can regulate the transactivation strength. *Planta*, **227**, 707–715.

Paz-Ares, J., Ghosal, D., Weinland, U., Peterson, P. A. & Saedler, H. (1987) The regulatory *C1* locus of *Zea mays* encodes a protein with homology to *Myb* Proto-oncogene products and with structural similarities to transcriptional activators. *EMBO Journal*, **6**, 3553–3558.

Pecket, C.R. & Small, C.J. (1980) Occurrence, location and development of anthocyanoplasts. *Phytochemistry*, **19**, 2571–2576.

Peer, W.A., Brown, D.E., Tague, B.W., Muday, G.K., Taiz, L. & Murphy, A.S. (2001) Flavonoid accumulation patterns of *TRANSPARENT Testa* mutants of *Arabidopsis*. *Plant Physiology*, **126**, 536–548.

Pooma, W., Gersos, C. & Grotewold, E. (2002) Transposon insertions in the promoter of the *Zea mays A1* gene differentially affect transcription by the Myb factors P and C1. *Genetics*, **161**, 793–801.

Pourcel, L. & Grotewold, E. (2009) Phytochemicals, plant development and growth—who is in control? In: *Plant-Derived Natural Products—Synthesis, Function and Application* (eds. A.E. Osbourn & V. Lanzotti), pp. 269–279. Springer, New York.

Pourcel, L., Irani, N.G., Lu, Y., Riedl, K., Schwartz, S. & Grotewold, E. (2010) The formation of anthocyanic vacuolar inclusions in *Arabidopsis thaliana* and implications for the sequestration of anthocyanin pigments. *Molecular Plant*, **3**, 78–90.

Poustka, F., Irani, N.G., Feller, A., *et al.* (2007) A trafficking pathway for anthocyanins overlaps with the endoplasmic reticulum-to-vacuole protein sorting route in Arabidopsis and contributes to the formation of vacuolar inclusions. *Plant Physiology*, **145**, 1323–1335.

Quattrocchio, F., Baudry, A., Lepiniec, L. & Grotewold, E. (2006a) The regulation of flavonoid biosynthesis. In: *The Science of Flavonoids* (ed. E. Grotewold), pp. 97–122. Springer, New York.

Quattrocchio, F., Verweij, W., Kroon, A., Spelt, C., Mol, J. & Koes, R. (2006b) Ph4 of petunia is an R2r3 Myb protein that activates vacuolar acidification through interactions with basic-helix-loop-helix transcription factors of the anthocyanin pathway. *Plant Cell*, **18**, 1274–1291.

Quattrocchio, F., Wing, J., van der Woude, K., *et al.* (1999) Molecular analysis of the *ANTHOCYANIN2* gene of petunia and its role in the evolution of flower color. *Plant Cell*, **11**, 1433–1444.

Ramsay, N.A. & Glover, B.J. (2005) Myb-Bhlh-Wd40 protein complex and the evolution of cellular diversity. *Trends in Plant Science*, **10**, 63–70.

Sainz, M.B., Grotewold, E. & Chandler, V.L. (1997) Evidence for direct activation of an anthocyanin promoter by the maize C1 protein and comparison of DNA binding by related Myb domain proteins. *Plant Cell*, **9**, 611–625.

Schwinn, K., Venail, J., Shang, Y., *et al.* (2006) A small family of Myb-regulatory genes controls floral pigmentation intensity and patterning in the genus *Antirrhinum*. *Plant Cell*, **18**, 831–851.

Shirley, A. & Chapple, C. (2003) Biochemical characterization of sinapoylglucose: c choline sinapoyl-transferase, a serine carboxypeptidase-like protein that functions as an acyltransferase in plant secondary metabolism. *The Journal of Biological Chemistry*, **278**, 19870–19877.

Shirley, B.W. (1996) Flavonoid biosynthesis: 'new' functions for an 'old' pathway. *Trends in Plant Science*, **1**, 377–382.

Shirley, B.W., Kubasek, W.L., Storz, G., *et al.* (1995) Analysis of *Arabidopsis* mutants deficient in flavonoid biosynthesis. *Plant Journal*, **8**, 659–671.

Small, C.J. & Pecket, R.G. (1982) The ultrastructure of anthocyanoplasts in red cabbage. *Planta*, **154**, 97–99.

Sompornpailin, K., Makita, Y., Yamazaki, M. & Saito, K. (2002) A Wd-repeat-containing putative regulatory protein in anthocyanin biosynthesis in *Perilla frutescens*. *Plant Molecular Biology*, **50**, 485–495.

Spelt, C., Quattrocchio, F., Mol, J. & Koes, R. (2002) Anthocyanin1 of petunia controls pigment synthesis, vacuolar ph, and seed coat development by genetically distinct mechanisms. *Plant Cell*, **14**, 2121–2135.

Spelt, C., Quattrocchio, F., Mol, J.N. & Koes, R. (2000) *ANTHOCYANIN1* of petunia encodes a basic helix-loop-helix protein that directly activates transcription of structural anthocyanin genes. *Plant Cell*, **12**, 1619–1632.

Stafford, H.A. (1990) *Flavonoid Metabolism*, CRC Press, Boca Raton, FL.

Stafford, H.A. (1991) Flavonoid evolution: an enzymic approach. *Plant Physiology*, **96**, 680–685.

Stam, M. & Mittelsten Scheid, O. (2005) Paramutation: an encounter leaving a lasting impression. *Trends in Plant Science*, **10**, 283–290.

Stracke, R., De Vos, R.C., Bartelniewoehner, L., *et al.* (2009) Metabolomic and genetic analyses of flavonol synthesis in Arabidopsis thaliana support the *in vivo* involvement of leucoanthocyanidin dioxygenase. *Planta*, **229**, 427–445.

Stracke, R., Ishihara, H., Huep, G., *et al.* (2007) Differential regulation of closely related R2R3-MYB transcription factors controls flavonol accumulation in different parts of the *Arabidopsis thaliana* seedling. *The Plant Journal*, **50**, 660–677.

Styles, E.D. & Ceska, O. (1975) Genetic control of 3-hydroxy- and 3-deoxy-flavonoids in *Zea mays*. *Phytochemistry*, **14**, 413–415.

Styles, E.D. & Ceska, O. (1977) The genetic control of flavonoid synthesis in maize. *Canadian Journal of Genetics and Cytology*, **19**, 289–302.

Styles, E.D. & Ceska, O. (1989) Pericarp flavonoids in genetic strains of *Zea Mays*. *Maydica*, **34**, 227–237.

Tanaka, Y., Katsumoto, Y., Bruglicra, F. & Mason, J. (2005) Genetic engineering in floriculture. *Plant Cell, Tissue and Organ Culture*, **80**, 1–24.

Taylor, L.P. & Grotewold, E. (2005) Flavonoids as developmental regulators. *Current Opinion Plant Biology*, **8**, 317–323.

Teng, S., Keurentjes, J., Bentsink, L., Koornneef, M. & Smeekens, S. (2005) Sucrose-specific induction of anthocyanin biosynthesis in Arabidopsis requires the Myb75/Pap1 gene. *Plant Physiology*, **139**, 1840–1852.

Tohge, T., Nishiyama, Y., Hirai, M.Y., *et al.* (2005) Functional genomics by integrated analysis of metabolome and transcriptome of *Arabidopsis* plants over-expressing an Myb transcription factor. *The Plant Journal*, **42**, 218–235.

Tuerck, J.A. & Fromm, M.E. (1994) Elements of the maize *A1* promoter required for transactivation by the anthocyanin *B/C1* or phlobaphene *P* regulatory genes. *Plant Cell*, **6**, 1655–1663.

Walker, A.R., Davison, P.A., Bolognesi-Winfield, A.C., *et al.* (1999) The transparent testa glabra1 locus, which regulates trichome differentiation and anthocyanin biosynthesis in *Arabidopsis*, encodes a WD40 repeat protein. *Plant Cell*, **11**, 1337–1349.

Wheldale, M. (1907) The inheritance of flower colour in *Antirrhinum majus*. *Proceedings of the Royal Society B*, **79**, 288–305.

Williams, C.E. & Grotewold, E. (1997) Differences between plant and animal Myb domains are fundamental for DNA-binding, and chimeric Myb domains have novel DNA-binding specificities. *Journal of Biological Chemistry*, **272**, 563–571.

Wilmouth, R., Turnbull, J., Welford, R., Clifton, I., Prescott, A. & Schofield, C. (2002) Structure and mechanism of anthocyanidin synthase from *Arabidopsis thaliana*. *Structure*, **10**, 93–103.

Winkel, B.S.J. (2004) Metabolic channeling in plants. *Annual Review in Plant Biology*, **55**, 85–107.

Winkel, B.S.J. (2008) The biosynthesis of flavonoids. In: *The Science of Flavonoids* (ed. E. Grotewold), pp. 71–95. Springer Science+Business Media, LLC, New York.

Winkel-Shirley, B. (1999) Evidence of enzyme complexes in the phenylpropanoid and flavonoid pathways. *Physiology Plantarum*, **107**, 142–149.

Winkel-Shirley, B. (2001) Flavonoid biosynthesis. A colorful model for genetics, biochemistry, cell biology and biotechnology. *Plant Physiology*, **126**, 485–493.

Wisman, E., Hartmann, U., Sagasser, M., *et al.* (1998) Knock-out mutants from an *En-1* mutagenized *Arabidopsis thaliana* population generate phenylpropanoid biosynthesis phenotypes. *Proceedings of the National Academy of Science of the United States of America*, **95**, 12432–12437.

Yao, J. & Polyak, K. (2004) Emsy links breast cancer gene 2 to the 'royal family'. *Breast Cancer Research*, **6**, 201–203.

Zhang, H., Wang, L., Deroles, S., Bennett, R. & Davies, K. (2006) New insight into the structures and formation of anthocyanic vacuolar inclusions in flower petals. *BMC Plant Biology*, **6**, 29.

Zhao, J. & Dixon, R.A. (2010) The 'ins' and 'outs' of flavonoid transport. *Trends in Plant Science*, **15**, 72–80.

Zhao, J., Pang, Y. & Dixon, R.A. (2010). The mysteries of proanthocyanidin transport and polymerization. *Plant Physiology*, **153**, 437–443.

Zhao, M., Morohashi, K., Hatlestad, G., Grotewold, E. & Lloyd, A. (2008) The ttg1-Bhlh-Myb complex controls trichome cell fate and patterning through direct targeting of regulatory loci. *Development*, **135**, 1991–1999.

Zimmermann, I.M., Heim, M.A., Weisshaar, B. & Uhrig, J.F. (2004) Comprehensive identification of *Arabidopsis thaliana* Myb transcription factors interacting with R/B-like Bhlh proteins. *The Plant Journal*, **40**, 22–34.

Chapter 7
Shedding Light on the Black Boxes of the Proanthocyanidin Pathway with Grapevine

Yung-Fen Huang, Véronique Cheynier and Nancy Terrier

Abstract: Grapevine (*Vitis vinifera*) proanthocyanidins (PAs) contribute to plant defence mechanisms against biotic stress and are also involved in organoleptic properties of wine. In this chapter, we first present the particularity of grape PAs and the molecular tools available on this species. Then, we relate the recent progresses on PA biosynthesis in plants while trying to demonstrate why grape appears as a good model to fill the remaining black boxes, in particular, mechanisms involved in polymerisation and nature of the substrates for this polymerisation, transport and storage, and regulation of the biosynthesis pathway.

Keywords: grapevine; proanthocyanidin; biosynthesis

7.1 Tools available on grape to study PA biosynthesis

Flavan-3-ols make up a large group of flavonoid compounds, encountered in several tissues of plants and involved in defence reactions against various biotic aggressions, such as attacks by microbial pathogens (bacteria and fungi), insects and larger herbivores (Dixon *et al.*, 2005). They comprise not only monomers but also oligomers and polymers, which are condensed tannins or proanthocyanidins (PAs). They are quantitatively the most abundant secondary metabolites of grape berries. Extracted during winemaking, PAs are a major qualitative factor in red wines because of their implication in colour stability, astringency and bitterness.

Despite numerous works aiming at deciphering the mechanisms by which PAs are synthesised in grape and in other plants, as recently reviewed by Braidot *et al.* (2008), Zhao and Dixon (2010) and Zhao *et al.* (2010), there are still gaps in understanding how they are

Recent Advances in Polyphenol Research, Volume 3, First Edition. Edited by Véronique Cheynier, Pascale Sarni-Manchado and Stéphane Quideau.

formed, polymerised, and stored, and how their biosynthesis is regulated. In this chapter, we present the particularity of grape PAs and the molecular tools available on this species and relate the recent progresses on PA biosynthesis in plants while trying to demonstrate why grape appears as a good model to fill the remaining black boxes.

7.1.1 Grape PAs

7.1.1.1 Grape PA structure

All flavonoids are based on a common C6-C3-C6 skeleton (Fig. 7.1a), which consists in two phenolic rings (A and B) linked *via* a heterocyclic pyran ring (C-ring). This large group is subdivided into several families based on the oxidation state of their C-ring. Flavan-3-ols exhibit a saturated C-ring hydroxylated in the 3-position. The A-ring of flavan-3-ols is generally hydroxylated in C5 and C7 and the B-ring in C4' (Fig. 7.1b, R, R' = H,

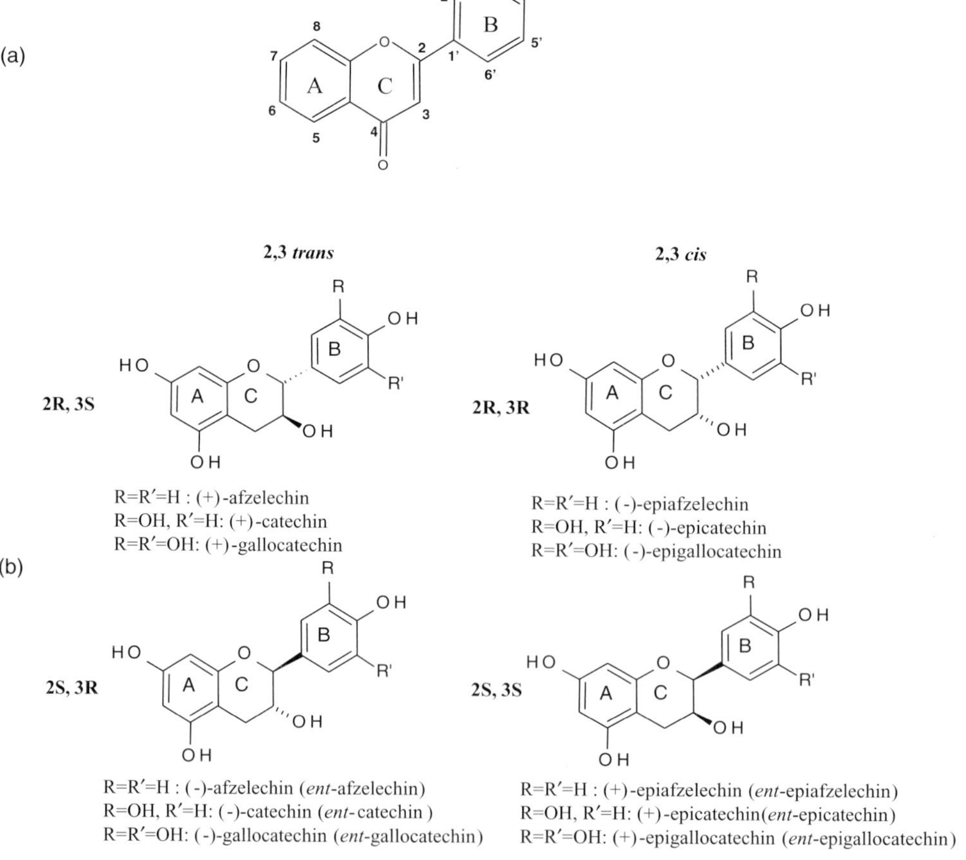

Fig. 7.1   Structure and nomenclature of the 2-phenylbenzopyrone (a) and structures of major flavan-3-ol monomers (b).

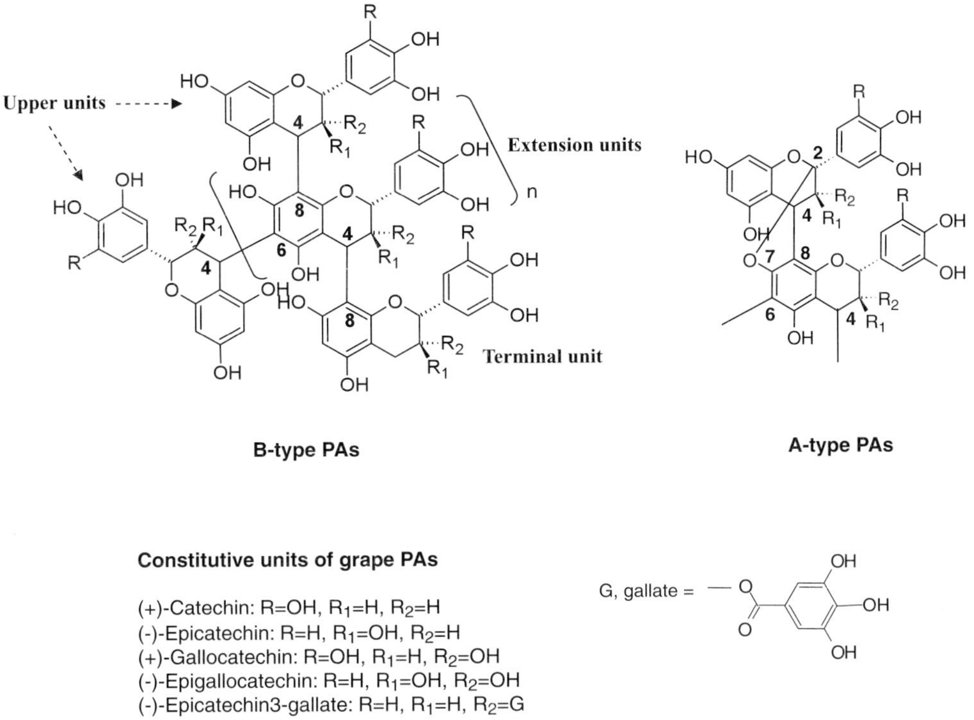

Fig. 7.2   Structures of proanthocyanidins.

afzelechin series). Diversity arises from the substitution pattern of the B-ring: the presence of an extra hydroxyl group in 3′ (R = OH, R′ = H) or two extra hydroxyl groups in 3′ and 5′ (R, R′ = OH) gives rise to the catechin and gallocatechin series, respectively. The presence of two asymmetric carbons (in C2 and C3) opens the possibility for different stereoisomers. The most abundant natural isomers show *2R* configuration, the *2R,3S* (2,3-*trans*) configuration being displayed in (+)-afzelechin, (+)-catechin or (+)-gallocatechin, while the *2R,3R* (2,3-*cis*) configuration is displayed in (−)-epiafzelechin, (−)-epicatechin or (−)-epigallocatechin. The corresponding *2S* isomers, respectively (−)-afzelechin, (−)-catechin, or (−)-gallocatechin in the *2S,3R* (2,3-*trans*) series, and (+)-epiafzelechin, (+)-epicatechin, or (+)-epigallocatechin in the *2S,3S* (2,3-*cis*) series, can be designated with the prefix *ent*. Diversity can also be increased by galloylation and glucosylation of the 3-hydroxyl group. In PAs, linkages between constitutive flavan-3-ol units are found between C4 and C6 or C4 and C8 in the case of B-type PAs (Fig. 7.2). A-type PAs are linked with additional C2-*O*-C7 or C2-*O*-C5 bonds. Substitution in the 4-position gives rise to another asymmetric centre on extension and upper units, but the usual configuration is 3,4-*trans* (i.e. *3S,4S* or *3R,4R*). The chain length of one polymer is described by the degree of polymerisation (DP), and the mean DP (mDP) is the mean degree of polymerisation of a heterogeneous population of polymers.

Major flavan-3-ol monomers in grape are (+)-catechin, (−)-epicatechin and (−)-epicatechin 3-gallate (Fig. 7.1b). All of these units are also found in grape PAs, along with

(–)-epigallocatechin and trace amounts of (+)-gallocatechin. Grape PAs are essentially of B-type, with more abundant C4–C8 linkages than C4–C6 ones. In addition to flavan-3-ol oligomers (Ribéreau-Gayon, 1964; Weinges & Piretti, 1971; Cheynier & Rigaud, 1986; Ricardo da Silva *et al.*, 1991), a wide range of polymeric forms are found in grape berries. Both their qualitative and quantitative differences depend on several factors described in the subsequent text.

7.1.1.2 Grape PA variations according to genotype, tissue and development

In grapevine, flavan-3-ols are present in various vegetative and reproductive organs and tissues, including wood (Boukharta *et al.*, 1988), leaves (Bogs *et al.*, 2005; Tesnière *et al.*, 2006), stems (Souquet *et al.*, 2000), inflorescence and fruit (Bogs *et al.*, 2005). The presence of PAs in vegetative organs confirms that PA accumulation is not toxic for the plant itself, suggesting storage in a dedicated compartment within the cell. Within the berry, they are particularly abundant in seeds and skins. PA composition in different grape vine organs is summarised in Table 7.1.

Seed PAs are based on (+)-catechin, (–)-epicatechin and (–)-epicatechin 3-gallate units (Prieur *et al.*, 1994). Mean degrees of polymerisation (mDP) from 3 (Mané *et al.*, 2007) to 16 (Cheynier *et al.*, 1998) have been reported for different grape varieties, while the percentage of galloylation can reach 23 (Verriès *et al.*, 2008). Major constituents of skin PAs are (–)-epicatechin and (–)-epigallocatechin as extension units with a lower percentage of (–)-epicatechin 3-gallate units (Souquet et al., 1996). Terminal subunits are mainly (+)-catechin and, to a lesser extent, (–)-epicatechin, although (epi)gallocatechin units are present as terminal units of PA dimers in wine (Fulcrand *et al.*, 1999). The mDP values are around 20–30, but higher values (up to 50) have been found in other cultivars (Monagas *et al.*, 2003; Souquet *et al.*, 2006). PAs from stems (Souquet *et al.*, 2000) and pulp (Mané *et al.*, 2007; Verriès *et al.*, 2008) are also a mix of B-ring di- and trihydroxylated units with lower proportions of (–)-epigallocatechin units and higher levels of galloylation than skin PAs. The mDP calculated for pulp PAs (about 20) are also intermediate between those in seeds and skins (Mané *et al.*, 2007; Verriès *et al.*, 2008). PA content and composition greatly vary according to genetic composition even for related genotypes. For example, the first large-scale study on PA composition in skin and seeds of a sibling population derived from

Table 7.1 Mean characteristics of grape PAs in several grapevine tissues and comparison to seed Arabidopsis PAs. Important variations can occur between genotypes.

Species	Tissue	mDP	Major terminal unit	Major elongation unit	% Of B-ring trihydroxylation	% Of galloylation
Grapevine	Skin	20–50	Catechin	Epi(gallo)catechin	+++ (5–40)	+ (1–5)
	Pulp	20–30	Catechin/Epicatechin	Epi(gallo)catechin	+ (2–15)	+ (1–5)
	Seed	3–16	Epicatechin/Catechin	Epicatechin	0	+++ (10–20)
	Leaves	10–50	Catechin/Epicatechin	Epi(gallo)catechin	++ (20)	++ (5)
Arabidopsis	Seed	≈5	Epicatechin	Epicatechin	0	0

Source: Data from Prieur *et al.* (1994), Bogs *et al.* (2005), Cheynier *et al.* (1998), Monagas *et al.* (2003), Tesnière *et al.* (2006), Souquet *et al.* (2006), Routaboul *et al.* (2006), Mané *et al.* (2007) and Verriès *et al.* (2008).

a cross between Syrah and Monestrell segregating population reported that concentration of skin PAs ranged from 1 to 11 mg g^{-1} of fresh weight and that of seed PAs from 15 to 85 mg g^{-1} (Hernandez-Jimenez *et al.*, 2009). They also reported great variability in composition among the different genotypes.

It should be emphasised that PAs are unstable molecules. Under the acidic conditions prevailing in vacuoles, especially in grape berry cells, they can undergo spontaneous cleavage of the interflavanic bonds, leading to formation of new PAs showing increased DP diversity (Haslam, 1980; Vidal *et al.*, 2002). They are also highly susceptible to oxidation, usually resulting in formation of quinones that proceed to new types of polymers (Guyot *et al.*, 1996; Pourcel *et al.*, 2005).

Bogs *et al.* (2005) showed that PAs are present in noticeable amounts in Syrah leaves from the earlier stages of development, with mDP around 10, presence of galloylated and B-ring trihydroxylated subunits and total content reaching 10 mg g^{-1} of fresh weight. Tesnière *et al.* (2006) reported higher values of mDP, reaching 50 in leaves from Portan cultivar.

Skin flavan-3-ols are synthesised mainly during a few weeks after flowering (Kennedy *et al.*, 2001; Downey *et al.*, 2003). PA contents remain constant after 'véraison' (onset of ripening) when expressed on a per berry basis. No qualitative changes were recorded during this period (Fournand *et al.*, 2006). In contrast, other authors have described a decrease in the content of flavan-3-ol monomers and PAs after véraison (Kennedy *et al.*, 2002; Downey *et al.*, 2003), and a concomitant increase (Kennedy *et al.*, 2001) or decrease (Downey *et al.*, 2003) of PA mDP. However, the apparent loss of PA may be due to lower extraction rate as a possible result of increased adsorption of PA on plant cell walls (Kennedy *et al.*, 2001; Downey *et al.*, 2003; Marles *et al.*, 2003; Bindon *et al.*, 2010), or *in vivo* conversion of PA for instance through laccase catalysed oxidation, as reported in *Arabidopsis thaliana* (Pourcel *et al.*, 2005), to new products that are difficult to analyse. In seeds, accumulation of flavan-3-ols is a bit delayed when compared to skin, and maximal concentration is reached a few weeks after véraison (Downey *et al.*, 2003; Bogs *et al.*, 2005).

Arabidopsis PAs contain only (–)-epicatechin units as building blocks, with DP reaching 9 at maximum and centred on 5 in samples analysed to date. They accumulate only in the endothelial cells of the seed coat at concentration not exceeding 8 mg g^{-1} of seed (Routaboul *et al.*, 2006). The model legume *Medicago truncatula* has also been extensively used to decipher molecular steps of PA biosynthesis. In this plant, PAs are also accumulated quite exclusively in the seed coat, composed of (–)-epicatechin as unique terminal and elongation units, with mean DP reaching values around 17 in the mature seed (Pang *et al.*, 2007). Barley was also an historical model to study PA biosynthesis. It was developed and used by the Calsberg's company, with the final industrial purpose to obtain barley varieties without PA to avoid haze formation in beer (reviewed by vonWettstein, 2007). PAs of the barley testa consist of dimers and trimers of (+)-catechin and (+)-gallocatechin (Jende-Strid & Møller, 1981). Therefore, grapevine offers considerable amount and diversity considering its PA composition when compared with other plants usually considered as model plants for research purpose. At the moment, the biological role of this PA complexity in grape is not fully understood, but both galloylation and mDP have been shown to impact antibacterial, antiviral and antioxidant activity of PA (Kajiya *et al.*, 2001, 2002; da Silva Porto *et al.*,

2003; Lizarraga *et al.*, 2007) and could contribute to defence mechanism against biotic stresses.

7.1.2 Grape genetic and genomic tools

The first genetic map of grapevine was published in the 1990s (Lodhi *et al.*, 1995). With the progress in molecular marker development, several grapevine linkage maps were developed since 2000, which enabled quantitative trait loci (QTL) mapping of traits of agronomic interests, such as flavour and colour variation (reviewed by Welter *et al.*, 2010). As grapevine is a perennial plant, it is a time-consuming work to identify the precise causal polymorphism of interesting trait variation from a QTL. Association mapping could be an alternative to rapidly and precisely identify the potential causal polymorphism by exploring the existent genetic diversity and involving genome sequence knowledge (Flint-Garcia *et al.*, 2003; Yu & Buckler, 2006; Rafalski, 2010). Combining QTL detection and association mapping appeared to be a relevant approach to identify causal polymorphism in grapevine, as shown by the original study of Fournier-Level *et al.* (2009), who identified a set of highly significant polymorphisms implicated in anthocyanin variation in grape berries.

Grapevine genomic tools began to be developed in the early 2000s, i.e. later than for model plants such as *Arabidopsis*, with large-scale EST sequencing (Terrier *et al.*, 2001; DaSilva *et al.*, 2005). This allowed several generations of microarray of growing size to be developed and used for higher and higher throughput transcriptomic analysis: the first custom 3 K oligoarray by MWG (Terrier *et al.*, 2005; Ageorges *et al.*, 2006) was followed by two parallel 14 K array available on the market, developed by Operon (Cutanda-Perez *et al.*, 2009; Terrier *et al.*, 2009) and Affymetrix (Deluc *et al.*, 2007). In 2007, *Vitis vinifera* became the fourth plant with complete genome sequence after *Arabidopsis*, rice and poplar (Jaillon *et al.*, 2007; Velasco *et al.*, 2007). This gave rise to complete genome array with the $\approx$30,000 genes represented on microarrays with Nimblegen or Combimatrix technology (M. Delledonne, Univ. Verona, Italy, personal communication).

One of the major problems while using grape as model plant to dissect a biosynthetic pathway is grape transformation for final gene functional validation. Grape can be transformed (Tesnière *et al.*, 2006); however, this woody perennial plant possesses a long life cycle and one has to wait for 4–5 years after transformation to obtain berries. To circumvent this problem, several authors used heterologous transformation with more easily transformable plants such as *Arabidopsis* or tobacco, which always raises the questions of protein recognition by a heterologous host, and, therefore, real significance of the results. Transgenic hairy roots, deriving from plantlet transformation with *Agrobacterium rhizogenes* (Torregrosa & Bouquet, 1997), were proved to be a useful tool for protein subcellular localisation (Gomez *et al.*, 2009), or function validation of transcription factor (Cutanda-Perez *et al.*, 2009; Terrier *et al.*, 2009). Recently, it was demonstrated that microvine, a dwarf mutant requiring only 1 year from vegetative to reproductive growth (Boss & Thomas, 2002) could be transformed and used for forward and reverse genetics (Chaïb *et al.*, 2010).

All those recently developed tools open the way to use grapevine as an alternative model to dissect the PA biosynthetic pathway.

7.2 Biosynthesis

7.2.1 Enzymes of the pathway

All three main classes of flavonoids, anthocyanins, flavonols and flavan-3-ols including PAs share the upstream of the biosynthetic pathway. Proteins involved in the anthocyanin branch of the pathway have been well and first characterised. Indeed, the colour biosynthesis pathways provide a natural visual screening system and gave rise to historical discoveries (Winkel-Shirley, 2001). Many structural and regulatory genes of the flavonoid pathway were first isolated from maize, petunia and snapdragon (Mol *et al.*, 1998; Winkel-Shirley, 2001; Koes *et al.*, 2005). The dissection of regulation and synthesis of PA branch came later due to the absence of an adequate reporter and the lack of high throughput methods for phenotyping. A pioneer work in PA pathway dissection was done on barley beginning in the 1970s, where seeds were screened with a vanillin coloration test and mutants, called *ant*, were revealed to be mutated for structural and regulatory genes (Jende-Strid, 1993). The recent advances in knowledge about PA pathway were principally obtained from *A. thaliana* thanks to *transparent testa (tt)* mutants and *tannin-deficient seeds (tds)* mutants (Koornneef, 1981, 1990; Abrahams *et al.*, 2002). As PAs accumulate in the seed coat of *Arabidopsis* and their oxidation leads to brown pigmentation, the seed coat colour was suggested to be a reporter for PA studies and used to detect *tt* mutants. *tds* mutants were screened according to their seed coat coloration deficiency with dimethylaminocinnamaldehyde (DMACA) (Koornneef, 1981, 1990; Abrahams *et al.*, 2002). In *tt* and *tds* mutants, there is complete or partial loss of seed coat pigmentation.

Starting from phenylalanine, flavonoid biosynthesis begins with three successive reactions catalysed by phenylalanine ammonia-lyase PAL (Koukol & Conn, 1961), cinnamate 4-hydroxylase (C4H) (Russell & Conn, 1967) and 4-coumaroyl CoA ligase (4CL) (Heller & Forkmann, 1988) to yield 4-coumaroyl CoA (Fig. 7.3). The typical flavonoid structure C6-C3-C6 is produced through the action of a chalcone synthase (CHS). The chalcone is formed by the condensation of *p*-coumaroyl-CoA with three molecules of malonyl-CoA (Kreuzaler & Hahlbrock, 1972). Chalcone isomerase (CHI) catalyses the conversion of chalcones to flavanones (Moustafa & Wong, 1967). The C-ring of this molecule is then hydroxylated by flavanone 3-hydroxylase (F3H) to form a dihydroflavonol (Forkmann *et al.*, 1980). This molecule, called dihydrokaempferol, can be hydroxylated on the B-ring either on 3′ position or on both 3′ and 5′ positions by a flavonoid 3′ or 3′-5′ hydroxylase (F 3′H or F 3′5′H) (Froemel *et al.*, 1985; Menting *et al.*, 1994) to produce respectively dihydroquercetin and dihydromyricetin.

Downstream, two other genes are still common to the anthocyanin and PA pathway. Dihydroflavonol reductase (DFR) catalyses the reaction leading from dihydroflavonols to leucoanthocyanidins, *syn* flavan-3,4-diols (Stafford & Lester, 1982), which are the substrates of anthocyanidin synthase (ANS), also called leucoanthocyanidin dioxygenase (LDOX), which forms anthocyanidins (Abrahams *et al.*, 2003). More recently, two enzymes involved in the synthesis of monomers and terminal units of flavan-3-ols were identified: leucoanthocyanidin reductase (LAR) and anthocyanidin reductase (ANR), catalysing, respectively, the synthesis of (+)-catechin and (−)-epicatechin (Devic *et al.*, 1999; Tanner *et al.*, 2003;

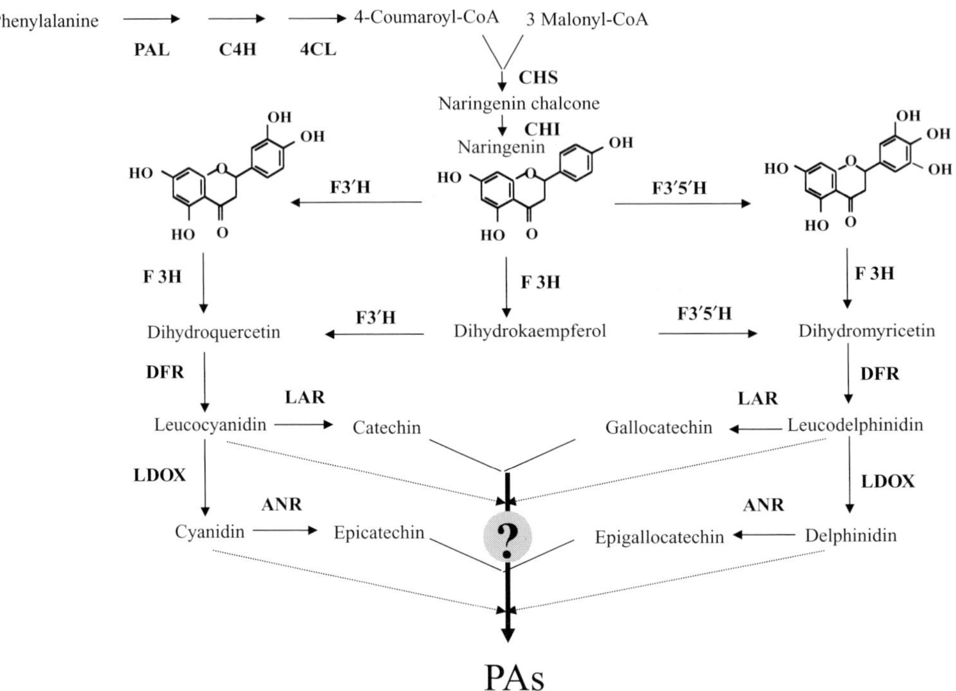

Fig. 7.3   The PA biosynthesis pathway. PAL, phenylalanine ammonia lyase; C4H, cinnamate 4-hydroxylase; 4CL, 4-coumarate CoA ligase; CHS, chalcone synthase; CHI, chalcone isomerase; F3′H, flavonoid 3′-hydroxylase; F3′5 ′H, flavonoid 3′,5′-hydroxylase; F3H, flavonone 3-hydroxylase; DFR, dihydroflavonol 4-reductase; LDOX, leucoanthocyanidin dioxygenase; LAR, leucoanthocyanidin reductase; ANR, anthocyanidin reductase.

Xie *et al.*, 2003). The function of ANR was first demonstrated through description of the *Arabidopsis ANR* mutant, called *ban* (Devic *et al.*, 1999). Its *in vitro* enzymatic activity of proteins of *Arabidopsis* and *Medicago* produced by heterologous expression and analysis of their *in vitro* reaction products (Xie *et al.*, 2003, 2004). LAR was identified after protein purification from *Desmodium uncinatum* leaves, followed by cDNA cloning and heterologous protein study (Tanner *et al.*, 2003). Its function is in accordance with the absence of *LAR* in the genome of *Arabidopsis*, which only contains (–)-epicatechin-based PAs (Abrahams *et al.*, 2003; Routaboul *et al.*, 2006). However, no catechin could be detected in *M. truncatula*, although there is one *LAR* gene in its genome (Pang *et al.*, 2007). To date, the LAR function still awaits *in vivo* validation. When tobacco plants are transformed with *ANR* from *Arabidopsis* or grape, anthocyanin content in the flowers is decreased and epicatechin is detected in the petals (Xie *et al.*, 2003; Bogs *et al.*, 2005), but when the same kind of experiment was performed with *MtLAR*, transgenic flowers did not contain any catechin, although the enzyme was active *in vitro* (Pang *et al.*, 2007).

New genes are also emerging from different screening experiments, but their precise function in PA synthesis is not yet clarified. For example, serine carboxypeptidase-like protein might be involved in PA accumulation, since correlation of its gene expression and accumulation of PAs has been reported at least twice, in persimmon fruits (*Diospyros kaki*) (Ikegami *et al.*, 2007) and in hairy roots of grape overexpressing MYB transcription factors (Terrier *et al.*, 2009).

Genes encoding 1-Cys Prx (*DkPrx*) and flavonoid 3-*O*-galT (*DkFGT*) have been identified through suppression subtractive hybridisation (SSH) between astringent and non-astringent persimmon fruits (Ikegami *et al.*, 2009). The abundance of transcripts encoding *DkPrx* and *DkFGT* correlates with the accumulation of soluble PAs in young fruits. As an antioxidant, *DkPrx* may prevent oxidation of PA by a laccase type flavonoid oxidase, and thus have a positive effect on accumulation of soluble PA. The recombinant DkFGt has been demonstrated to form anthocyanidin and flavonol galactosides *in vitro*. However, since genomic tools available for *Diospyros kaki* are scarce, the precise role of these genes in PAs *in vivo* accumulation remains to be elucidated.

In grapevine, the flavonoid pathway has been first described at the molecular level by Sparvoli *et al.* (1994). This work has allowed the cloning of several genes such as *PAL, CHS, CHI, F3H, DFR, LDOX* through a screening of a grapevine cDNA library with probes from *Antirrhinum majus* and maize. In 2006, Bogs *et al.* (2006), Jeong *et al.* (2006) and Castellarin *et al.* (2006) identified genes coding for F3′H and F3′5′H. Concerning the specific PA pathway, two isogenes of *LAR* (*LAR1* and *LAR2*) and one for *ANR* have been identified (Bogs *et al.*, 2005). Most of the genes of the flavonoid pathway are present in several copies in the grape genome in highly multigenic families (*PAL* (13 isogenes), *F3′5′H* (10)) or smaller ones (*CHS* (4), *F3H* (3), *FLS* (4), *CHI* (2), *LAR* (2)). Other genes are present in single copy (*C4H, 4CL, F3′H, DFR, LDOX, ANR*) (Velasco *et al.*, 2007; Jeong *et al.*, 2008). The significance of this redundancy is not fully understood, since only some of these multigenic families have been completely studied both at the expression and functional level. A multiple correlation between the expression of the isogenes of *CHS, CHI* and *F3H* and the flavonoid profiles in different parts of the plant at several stages revealed that the expression of some of the isogenes was specifically associated with one type of molecules such as that of *CHI2* with flavan-3-ols (Jeong *et al.*, 2008). *LAR2* expressed both in grape berry skin and seed during PA synthesis, while *LAR1* is specifically expressed in seeds (Bogs *et al.*, 2005; Fig. 7.4). The maximum of expression of *LAR2* in the skin occurs at the early stages of berry development, whereas it reaches its maximum at véraison in the seed (Fig. 7.4). On the other hand, *ANR* expression in skin and seed is parallel, exhibiting a continuous decrease (Fig. 7.4). Despite being incomplete, biochemical analysis of both recombinant proteins did not reveal any substrate specificity (Pfeiffer *et al.*, 2006).

7.2.2 Transport and storage of PAs

In 1974, Stafford postulated that most of the enzymes of the general phenylpropanoid pathway function as a multienzymatic complex (Stafford, 1974). In 1992, Hrazdina and Jensen proposed a model where this complex would be assembled at the level of the endoplasmic reticulum (ER) through weak interactions between proteins (Hrazdina & Jensen, 1992). Direct associations among flavonoid enzymes were first demonstrated in *Arabidopsis* where CHI and DFR, PAL and C4H, and CHI and F3H were shown to be physically interacting by yeast two-hybrid analysis (Burbulis & Winkel-Shirley, 1999;

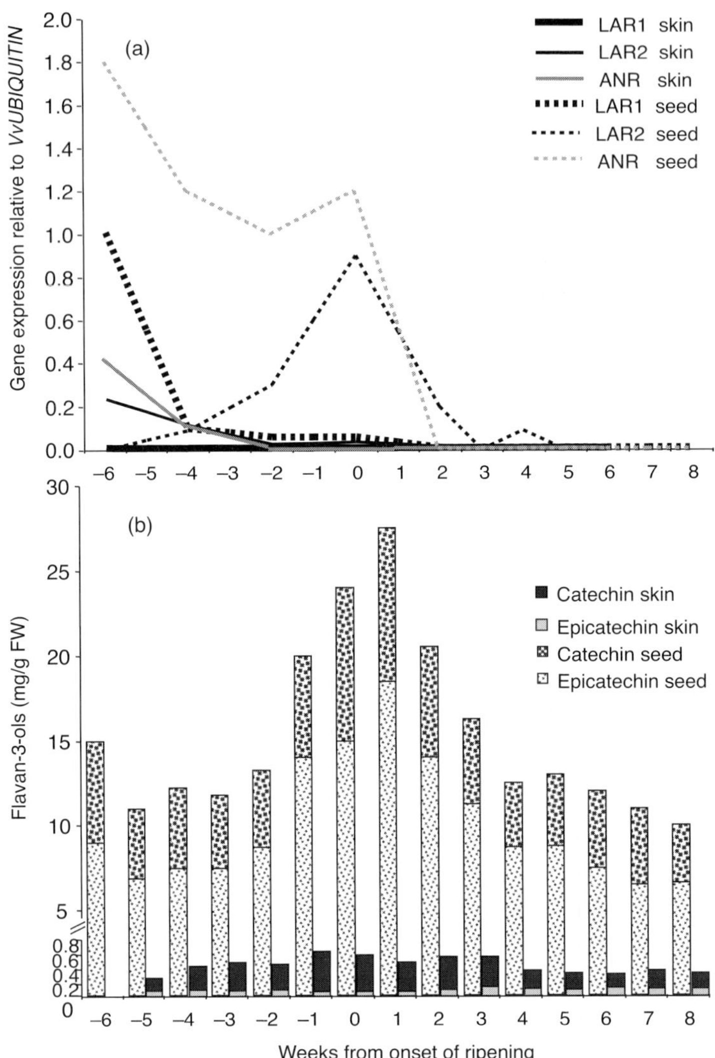

Fig. 7.4   Expression of LAR1, LAR2 and ANR (a) and concentration of monomers of catechin and epicatechin (b) in skin and seed along grape berry development. (Adapted from Bogs *et al.*, 2005.)

Achnine *et al.*, 2004) and CHI, CHS, DFR and F3H were localised at the cytoplasmic face of the ER (Saslowsky & Winkel-Shirley, 2001). Using the same strategy, OsCHS1 was demonstrated to interact physically with OsF3H, OsF3'H, OsDFR and OsANS1, suggesting the existence of a macromolecular complex (Shih *et al.*, 2008). Green fluorescent protein (GFP) fusion proteins of *M. truncatula* LAR and ANR were detected in the cytoplasm and, therefore, do not belong to this multienzymatic complex (Pang *et al.*, 2007).

On the other hand, PAs were found to be located mainly inside the vacuole in *Arabidopsis* and in grape (Abrahams *et al.*, 2002; Cadot *et al.*, 2006), which raises the question of how they cruise across the cytoplasm and are transported through the tonoplast. Two non-exclusive different mechanisms have been described to explain the transfer of the molecules from ER to the vacuole (Grotewold & Davies, 2008). The first hypothesis, called vesicular transport (VT), implies that PAs or their precursors reach the vacuole in vesicles derived from the ER. Several clues indicate an involvement of vesicular trafficking for PA accumulation in the vacuole. Ancient observations already identified PAs in small structures, distinct from the vacuole, probably originating from the rough ER (Chafe & Durzan, 1973; Baur & Walkinshaw, 1974; Parham & Kaustinen, 1977). In tea leaves, small vacuoles (0.5–3 μm) containing tannins were detected in the cytoplasm (Suzuki *et al.*, 2003). This kind of vesicles has also been observed for other flavonoids. Hsieh and Huang (2007) observed ER-derived structures, called tapetosomes, of around 2 μm allowing flavonol transport. Zhang *et al.* (2006) observed pre-vacuolar vesicles with various size ranging from 0.2 to a few μm, electron-dense and probably containing anthocyanins, in the cytoplasm of *Lisianthus* petal cells. These structures would further fuse with a larger vacuole in the cell. The second hypothesis suggests that flavonoids cross the cytoplasm escorted by a ligandin and is, therefore, called ligandin transport (LT). Several potential molecular actors of these mechanisms were identified during the last years.

In *M. truncatula* hairy roots, a GT (glucosyl transferase) was induced in parallel with PA accumulation (Pang *et al.*, 2008). After heterologous expression in *Escherichia coli*, the protein was shown to catalyse *in vitro* the formation of epicatechin 3′-glucoside, with (−)-epicatechin as substrate. Only a weak glucosylation activity was measured with (+)-catechin. Indeed, *M. truncatula* PAs are comprised primarily of (−)-epicatechin units. Epi-catechin 3′-glucoside was detected as a transient component of *Medicago* seed during PA accumulation, suggesting that this molecule was further deglucosylated. However, *Medicago* knock-out mutants for this gene did not exhibit significant modification in their PA content (Pang *et al.*, 2008), which indicates that there is probably another scenario of PA accumulation in *Medicago*.

tt12 mutants exhibited a transparent testa phenotype due to the lack of a multidrug and toxic extrusion (MATE) protein (Debeaujon *et al.*, 2001). This *tt12* mutant is affected for both PA and flavonol accumulation (Marinova *et al.*, 2007). A *M. truncatula* orthologue, called *MATE1*, was recently identified through a transcriptomic screening after PA over-accumulation (Pang *et al.*, 2008; Zhao & Dixon, 2009). In grape, CAO69962 transcript is a probable orthologue of MATE1 and TT12 and was identified in a transcriptomic screening between hairy roots over-accumulating PA and control (Terrier *et al.*, 2009). However, its functional validation is still lacking. *In vitro* transport experiments with vesicles purified from yeasts overexpressing *MATE1* or *TT12* demonstrated that this MATE protein acts as a flavonoid/H$^+$ secondary antiporter and mediates cyanidin 3-glucoside transport (Marinova *et al.*, 2007) but epicatechin 3′-*O*-glucoside appears to be the preferred substrate (Zhao & Dixon, 2009). Neither catechin nor epicatechin is the substrate of this protein (Marinova *et al.*, 2007; Zhao & Dixon, 2009). Recently, detection of significant amount of epicatechin glucoside in immature seeds of *tt12* mutant argues in the favour of the glucosylated form of epicatechin being an intermediate metabolite, substrate of the MATE transporter (Kitamura *et al.*, 2010). Decorations of metabolites were already mentioned as being of great importance as recognition tags by diverse studies: glycosylation would be specific for secondary transporters and glutathione tag for primary ABC transporters

(Yazaki, 2005). By the way, two anthoMATE genes, MATE-type transporters isolated from grape berries, were shown to transport only acylated anthocyanins (Gomez *et al.*, 2009).

Since MATE proteins seem to function as H^+/flavonoid antiporters, it implies that other actors of PA sequestration are involved in the creation of a proton gradient. In addition to traditional vacuolar proton pumps, V-ATPase and V-Ppase, usually found on plant tonoplast, another pump was found to be involved in the creation of a proton gradient allowing secondary transport. *AHA10* encodes one of the 11 plasma membrane P-type proton-pump ATPase found in the Arabidospis genome. The *aha10* mutant exhibits a transparent testa phenotype due to a dramatic reduction in seed PA content (Baxter *et al.*, 2005). Moreover, the cells in *aha10* seeds showed multiple smaller vacuoles instead of the large central vacuole observed in wild-type seeds. Accumulation of other flavonoids is not affected in this mutant, suggesting that specific genes could be involved in the transport of each of them. In addition, free epicatechin could be detected in significant amount in *aha10* mutant when compared to wild-type plant. This could indicate that AHA10 is more specifically dedicated to the transport of elongation units or that AHA10 creates pH condition necessary for polymerisation in specialised subcellular compartment or in the large central vacuole. Knowing whether AHA10 is localised on the membrane of pre-vacuolar compartments or on the tonoplast should provide helpful information on its role. A more recent study showed that *Petunia × hybrida PH5*, a putative orthologue of *AHA10*, is involved in flower colour determination and PA production in seeds (Verweij *et al.*, 2008); moreover, PH5-GFP fusion proteins were localised to the tonoplast (Verweij *et al.*, 2008). However, no vesicles or pre-vacuolar compartment containing anthocyanin could be seen on the presented pictures.

Does this P-type pump provide the proton gradient necessary for the functionality of the MATE transporter? The first requirement is that both kinds of proteins are located on the same membrane. The protein encoded by the *TT12* gene was found to be localised to the tonoplast (Marinova *et al.*, 2007). However, the localisation of the GFP fusion protein was performed in cells that did not accumulate PAs. In *Arabidopsis* tapetum cells, flavonoids were detected in ER-derived vesicles, but were present in the cytosol in *tt12* mutant, suggesting that this gene is also involved in transport of the flavonols inside these vesicles and therefore possibly located in their surrounding membranes (Hsieh & Huang 2007). However, the localisation of TT12 appears to be tissue- or flavonoid-specific, since the PA distribution pattern in *tt12* endothelial cells of the seed coat, at the outer side of the vacuoles, suggested that the uptake of PA precursors into endomembranes by TT12 would rather takes place at the level of the vacuole (Kitamura *et al.*, 2010). In single cell expressing *MATE1-GFP*, GFP signals were detected not only in the tonoplast but also in some of the pre-vacuole-like membrane vesicles as well (Zhao & Dixon, 2009). The localisation of the P-type pump involved in PA accumulation would rather be tonoplastic (Verweij *et al.*, 2008). Another way to test this hypothesis is to perform functional test on native membrane vesicles. Zhao and Dixon (2009) isolated vesicles from *Medicago* hairy roots with enhanced PA production and *MATE1* expression. They performed transport experiments with different substrates and inhibitors. Since vanadate, an inhibitor of ABC transporters (Pezza *et al.*, 2002), had no effect on substrate accumulation rates, they concluded that this transport was not linked to the presence of an ABC transporter. However, vanadate is also well known to be an inhibitor of P-type ATPase (Sze, 1985) and some particular V-ATPase (Müller *et al.*, 1999). It could be concluded that energisation of the MATE-mediated transport by a P-type ATPase is not compulsory for correct transport. However, unlike for petunia or *Arabidopsis*, no P-type ATPase was identified in transgenic hairy-roots over-accumulating PAs in *Medicago* (Pang *et al.*, 2008).

In the literature, several glutathione S-transferases (GST) have been shown to be compulsory for correct compartmentation of flavonoids inside the cells: *Bz2* and *AN9* for anthocyanins in maize and petunia, respectively (Marrs *et al.*, 1995; Alfenito *et al.*, 1998), and *tt19*, for both PAs and anthocyanins in *Arabidopsis* (Kitamura *et al.*, 2004). However, the GST enzymatic activity does not seem to be required for appropriate targeting of the compound to the vacuole, since no glutathionated flavonoid could be detected through *in vivo* or *in vitro* labelling experiments (Mueller *et al.*, 2000). This suggests that GST rather acts as a 'ligand,' allowing the flavonoid molecule to cruise through the cytosol from the ER to the tonoplast. Its subcellular localisation in the cytosol is in accordance with this function (Kitamura *et al.*, 2010). However, the observation that in double mutant *tt12 tt19*, PA derivates were localised, as in *tt12* mutant, on the cytoplasmic face of the tonoplast suggests that the activity of TT19 is not compulsory for correct targeting of the flavonoid from the site of synthesis towards the site of transport (Kitamura *et al.*, 2010). *tt19* immature seeds contained considerable amount of insoluble PAs with unusual structures when compared to wild-type seeds, which suggests that TT19 acts as a protectant against oxidation or any modification that could affect flavonoid structure during the transport (Kitamura *et al.*, 2010). In grape, a *GST* was systematically associated with anthocyanin accumulation (Terrier *et al.*, 2005; Ageorges *et al.*, 2006) and was shown to complement *bz2* mutants (Conn *et al.*, 2008). However, this *GST* is very unlikely to be involved in PA sequestration, since its expression was restricted to ripening stages of berry development when PA synthesis has stopped.

Recently, the *Arabidopsis ugt80B1* mutant was described as exhibiting a transparent testa phenotype, and corresponded in fact to the *tt15* mutant. In addition to reduced PA levels, the amount of sterol glycoside (SG) and acyl-SG, important lipid membrane components, is affected. TT15 would indirectly affect PA composition through its involvement in modifying tonoplast or trafficking vesicle composition (DeBolt *et al.*, 2009).

The (co)-existence of VT and/or LT model in flavonoid accumulating cells and the precise involvement of the identified actors in these models are still unclear. Finding PA derivatives in vacuolar-like structures in *tt19* mutants would rather suggest a role for this GST in the LT mechanism (Kitamura *et al.*, 2010). However, in this mutant, flavonols were mislocalised in the cytosol instead of inside tapetosomes (Hsieh & Huang, 2007). The link between the PA pathway and vacuole biogenesis also appears to be very tight and complex. Several mutations in the pathway result in abnormal vacuole inside the cell: *tds4* and *tt12* like *aha10* mutants lack the large central vacuole (Abrahams *et al.*, 2002; Baxter *et al.*, 2005). In *tds4* mutants, PA precursors stained with osmium tetroxide, most probably leucocyanidin since *TDS4* encodes LDOX, were clearly located in small vesicles with size ranging from 0.2 to 2 µM (Abrahams *et al.*, 2003). The authors suggested that the formation of PA intermediates acts as a signal to the cell to produce a normal large and central vacuole to store the end products of the pathway (Abrahams *et al.*, 2003). However, events regulating the vacuole formation seem more complex, since cells of mutants affected downstream in the pathway such as *aha10*, containing epicatechin monomers, also failed to exhibit a normal vacuolar shape (Baxter *et al.*, 2005).

7.2.3 PA polymerisation

The polymerisation process remains one of the main black boxes in PA biosynthesis. Several points are still unknown: what is the nature of the extension units, is this reaction catalysed or not, and where does it take place.

7.2.3.1 Nature of the extension units

All of the proposed mechanisms are based on an addition of a nucleophilic flavan-3-ol, through one of its C6 or C8 free positions, onto an activated electrophilic species that serves as the precursor for the upper unit (substituted in C4).

Flavan-3,4-diols and their stereochemistry
The most commonly accepted pathway considers the flavan-3,4-diol (syn. leucoanthocyanidin) as the precursor of the upper unit. In this process, a flavan-3-ol either undergoes condensation (i.e. addition with loss of a water molecule) with the flavan-3,4-diol (Geissman & Yoshimura, 1966) or adds to intermediate compounds formed by dehydration of the flavan-3,4-diol, such as the quinone methide, the flav-3-en-ol or its protonated carbocation (Jacques & Haslam, 1974).

Nevertheless, in most cases within the plant kingdom, the extension units observed are 2,3-*cis* (usually, 2*R*,3*R*). On the other hand, leucoanthocyanidins and their derivatives are thought to be 2,3-*trans* (2*R*,3*S*), assuming that the stereochemistry of their dihydroflavonol precursor is maintained. Indeed, dihydroflavonols are predominantly under the 2*R*,3*S* configuration, although others have been observed in several plants (Grayer & Veitch, 2006). However, the stereochemistry of the 'classical' leucoanthocyanidins (i.e. leucocyanidin, leucopelargonidin and leucodelphinidin) is rarely investigated, due to their instability and the impossibility to isolate them (Xie & Dixon, 2005; Davies & Schwinn, 2006). Taking advantage of its stability, Drewes and Roux (1966) identified both 2,3-*cis*- and -*trans*-teracacidin, a leucoanthocyanidin derivative found in *Acacia auriculiformis*, whereas only the *trans*-configuration could be detected for the corresponding dihydroflavonols. This suggests either the existence of an epimerase activity or the ability of DFR to epimerise its substrate in parallel with its reductase activity. It can be postulated that any reductase of the pathway belonging to the reductase epimerase dehydrogenase family, i.e. DFR, LAR or ANR, exhibits an epimerase activity, in addition to its reductase one. In fact, Gargouri *et al.* (2010) demonstrated that grape ANR could epimerise (+)-catechin to (−)-epicatechin, when its activity was assessed in the reverse condition, i.e. in excess of NADP+. However, after reduction of (+)-taxifolin with an enzymatic extract of Douglas fir cell culture, the formed 3,4-diol was converted to catechin dimer and small amounts of the trimer upon addition of catechin under acid conditions (Stafford & Lester, 1984). No epicatechin-catechin dimer was detected, suggesting that the enzymatic cocktail present in the *in vitro* reaction, able to reduce dihydroflavonol, did not exhibit any additional epimerase activity, or that particular conditions required for its synthesis were not met.

Anthocyanidins
Other mechanisms have been proposed for this reaction. They involve reaction intermediates in which the stereochemistry is lost. Several such intermediates can be envisaged. The formation of the 2*R*,3*R* flavan-3-ol monomers from anthocyanidins, catalysed by ANR, has been recently demonstrated. Similarly, the 2*R*,3*R* extension units may derive from the quinonoidal form of the anthocyanidin, resulting from deprotonation of its flavylium form (Jacques & Haslam, 1974). Addition of flavan-3-ols onto the C4 of anthocyanins has been confirmed to occur in wine. However, this reaction yields A-type instead of B-type PAs (Bishop & Nagel 1984, Remy-Tanneau *et al.*, 2003). Finding plants where LDOX activity is affected and with catechin as PA terminal unit would provide informative arguments to confirm that anthocyanidin can serve as precursor for upper units. Barley mutants *ant1*, *ant2* and *ant5* do not produce anthocyanin due to probable mutation in *LDOX* gene, but contain as much PAs as the wild plant (Jende-Strid, 1993). However, since barley PA elongation units are also catechin units, it cannot be generalised to plant kingdom. Here again, grape PA mixed composition appears as quite pertinent to solve this question.

Quinone methide and flav-3-ene-3-ol
Alternatively, the 2*R*,3*R* quinone methide formed from the dihydroflavonol may undergo tautomeric rearrangement, through the flav-3-ene-3-ol, to its 2*R*,3*S* epimer, that can yield 2*R*,3*S* elongation units (Hemingway & Laks, 1985). Flav-3-ene-3-ols have also been proposed to be the precursors of elongation units (Jacques & Haslam, 1974; Xie & Dixon, 2005). They could be intermediate in the reactions catalysed by ANR (Xie *et al.*, 2004) and LAR (Pfeiffer *et al.*, 2006). *In vitro* reactions involving those enzymes did not lead to production of oligomers (Tanner *et al.*, 2003; Xie *et al.*, 2004). However, we cannot rule out the possibility that adequate conditions for polymerisation were not met.

7.2.3.2 Enzymatic or chemical polymerisation

Whether PA condensation is catalysed by a 'condensing enzyme' *in vivo* is still a mater of debate. *In vitro* formation of polymers from catechin and chemically reduced taxifolin, i.e. leucocyanidin, has been achieved non-enzymatically (Delcour *et al.*, 1983). DP of the produced polymers can be modulated by modulating the starting amount of each molecule. Ectopic expression of *Medicago* or grape *ANR* in tobacco flower led to the production of oligomers (Xie *et al.*, 2003; Bogs *et al.*, 2005). On the other hand, the barley PA mutant *ant26*, containing equivalent amount of catechin as in wild type, but only traces amount of PAs, could rather suggest the existence of a condensing enzyme (Jende-Strid, 1993). However, we cannot rule out the possibility that the mutation affects the existence of correct conditions for condensation of monomers with precursors of elongation units such as suitable pH, appropriate trafficking of the terminal and/or elongation units.

Oxidative polymerisation of flavan-3-ols has also been proposed to yield PA (Weinges *et al.*, 1969a). However, (epi)catechin oxidation generate isomers of procyanidins in which the constitutive units are linked by biphenyl or biphenyl-ether bonds between the B-ring of one unit and the A-ring of the other (Weinges *et al.*, 1969b, Young *et al.*, 1987; Guyot *et al.*, 1996). Activation of the C4 position of (epi)catechin under oxidative conditions appears unlikely unless oxidation of the *ortho*-diphenolic B-ring to the corresponding quinone is impeded. Substitution of the 3′ hydroxyl as encountered in epicatechin 3′ glucoside described as an intermediate in flavan3-ol biosynthesis (Pang *et al.*, 2008) could provide such conditions.

The *Arabidopsis tt10* mutant lacks the brown coloration in its seeds and exhibits larger amounts of extractable PAs, suggesting that the laccase coded by the *TT10* gene oxidises flavan-3-ols to brown unknown pigments (Pourcel *et al.*, 2005). Its involvement in oxidative polymerisation where the 3′ hydroxyl function of epicatechin is protected by glucosylation to prevent random polymerisation has been proposed (Pang *et al.*, 2008). A *N*-terminal peptide signal predicts an apoplast location for TT10 protein (Pourcel *et al.*, 2005). This putative location, to be confirmed by immunolocalisation or visualisation of GFP fusion protein, is not easily reconcilable with the involvement of TT10 in polymerisation of PAs located inside the vacuole. An apoplastic localisation of TT10 would rather fit with its participation to the formation of oxidised complexes with polysaccharides and other cellular material during seed maturation (Marles *et al.*, 2003).

7.2.3.3 Subcellular localisation of polymerisation

A related question is: where does polymerisation take place inside the cell, and subsequently, what are the species transported into the vacuole? Feeding cell culture with [^{14}C]phenylalanine, Stafford *et al.* (1982) noticed that, six hours later, less radioactivity was incorporated in terminal units than in upper units or free monomers, suggesting that upper and lower units arose from different steps of the pathway rather than condensation of similar units, and that polymers were built with terminal units previously stored in a different place than newly synthesised units. Chemical studies indicate that acidic pH would be necessary for polymerisation (Delcour *et al.*, 1983), suggesting an intracellular compartment for this reaction, such as vacuole or vesicle. On the other hand, LAR and ANR were localised in the cytoplasm (Pang *et al.*, 2007). One also will have to answer whether elongation and terminal units are transported via the same mechanism.

7.3 Regulation of the pathway

The analysis of the different PA mutant collections led to the identification of several regulatory proteins. To date, according to our knowledge from *Arabidopsis*, the identified PA regulatory proteins could be classified into six major families: MYB, bHLH, WD40, WRKY, WIP-type zinc finger, and MADS (reviewd by Lepiniec *et al.*, 2006). TT1 and TT16 belong to WIP-type zinc finger and MADS transcription factor families, respectively (Nesi *et al.*, 2002; Sagasser *et al.*, 2002). Both of them are necessary for PA accumulation during seed development but seem to be involved primarily in cellular differentiation to allow PA synthesis. TTG2, a WRKY transcription factor, affects seed and trichome development, and it is supposed to regulate PA accumulation downstream *TTG1* (Johnson *et al.*, 2002). *TTG1*, *TT8* and *TT2* encode for a WD40 protein, a bHLH protein and an R2R3-MYB protein, respectively (Walker *et al.*, 1999; Nesi *et al.*, 2000, 2001). The three corresponding mutants produce seeds without PAs. As the formation of transcription factor (TF) complex is known to be a fundamental process in fine gene regulation of eukaryotes, Baudry and coworkers (2004) demonstrated that TT2, TT8 and TTG1 form a TF complex to activate *BAN* expression. They pointed out that the TT2/TT8 (MYB/bHLH) complex was required for *BAN* activation by recognising and binding directly to *BAN* promoter. TTG1 forms a ternary complex with TT2 and TT8 and enhances their activation ability, in accordance with the common function of WD40 protein, known to facilitate protein–protein interaction (Baudry *et al.*, 2004; Ramsay & Glover, 2005). Negative regulators are also identified in *A. thaliana*, like MYB4, an R2R3-MYB repressor which downregulates gene expression of the cinnamate-4-hydroxylase, *C4H*, a structural enzyme of phenylpropanoid pathway upstream to flavonoid branch (see Fig. 7.3; Jin *et al.*, 2000). Overexpression of a small MYB, MYBL2, was shown to reduce PA content in seed, which is probably due to competition with TT2 and other MYBs activating the pathway (Dubos *et al.*, 2008). Recently, a new dominant *tt* mutant *sk-21d* was identified (Gao *et al.*, 2010). The causal gene is *KAN4*, encoding for a GARP MYB-like protein, initially identified as involved in cellular differentiation (McAbee *et al.*, 2006), whose overexpression leads to a decrease of PA in seed.

Thanks to the development of genomic tools and based on knowledge from *Arabidopsis*, regulators of PA synthesis were recently identified in other species: a WD40 protein for *M. truncatula* (Pang *et al.*, 2009) and MYB factors for *Brassica napus, Lotus japonicus,* poplar and persimmon fruit (Wei *et al.*, 2007; Yoshida *et al.*, 2008; Akagi *et al.*, 2009; Mellway *et al.*, 2009). However, there is no catechin or epigallocatechin, nor galloylation in *Arabidopsis*, *Medicago* and tobacco. Consequently, these plants, although they are commonly used as heterologous transformation systems, may not prove suitable for regulation dissection of more complex PA pathways. Comparably, grapevines is a useful tool to study fine PA regulation since it proceeds larger diversity in PA content and composition inter- and intra-plant

In the past 5 years, several regulatory proteins of phenylpropanoid/PA pathway were identified in grape berries (Deluc *et al.*, 2006, 2008; Bogs *et al.*, 2007; Matus *et al.*, 2008, 2010; Terrier *et al.*, 2009; Hichri *et al.*, 2010). Their expression profile and activation of their putative targets are summarised in Fig. 7.5 and Table 7.2, respectively. MYB5a and MYB5b were identified through MYB conserved sequence by cDNA screening (Deluc *et al.*, 2006, 2008). MYB5a, showing higher expression in skins than in seeds, is transcribed in the early stage of berry development and its transcript abundance decreases to undetectable level after véraison, the onset of ripening (Fig. 7.5). Its close relative, MYB5b, is expressed in both seed and skin throughout berry development with a higher transcript level during ripening (Fig. 7.5). Ectopic expression of either MYB5a or MYB5b in tobacco induces anthocyanin and PA accumulation and both were able to activate several flavonoid synthesis related genes, from upstream phenylpropanoid synthetic genes such as *C4H* and *CHI* to *LDOX*, *ANR* and *LAR* (Table 7.2). MYB5a and MYB5b are suggested as putative regulators of PA synthesis. Their regulatory role in PA synthesis could be direct activation on structural gene promoters and/or indirect regulation through other regulatory protein

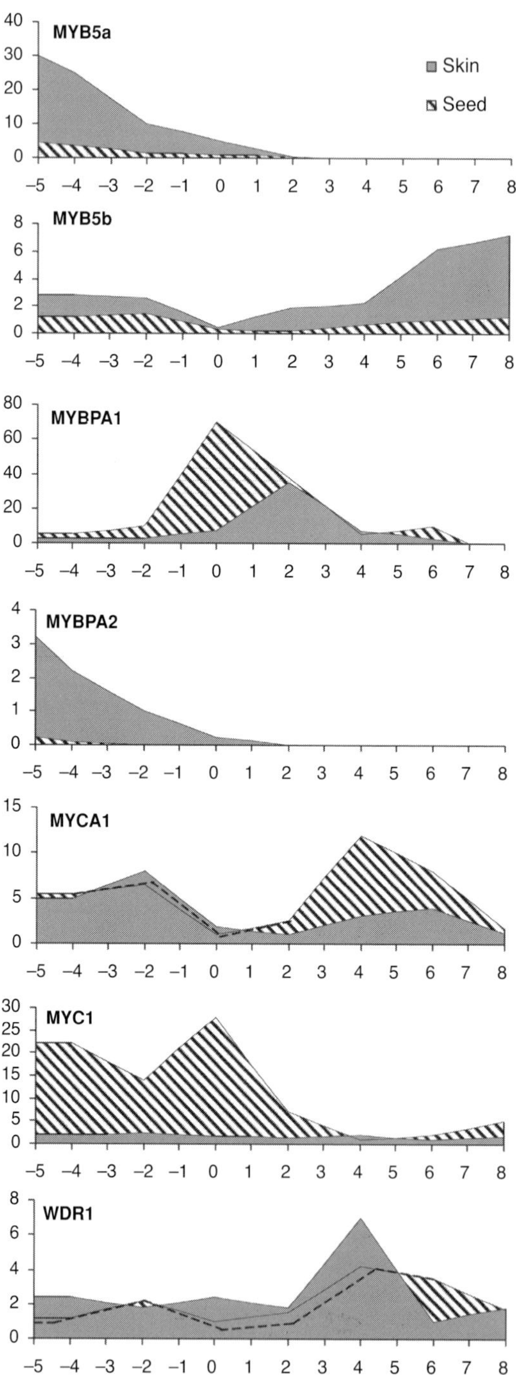

Fig. 7.5   Expression profile of identified grape PA regulators during berry development. Dark grey, the relative expression level of considered regulator in berry skin; hatching, the relative expression level of considered regulator in seed. Numbers on the *x*-coordinate indicate the weeks from véraison (onset of ripening). The scale in *y*-coordinate is in arbitrary units. Dashed line indicates expression level in seed for MYCA1 and WDR1, masked by illustration of skin expression. (Resumed from Bogs *et al.*, 2007; Deluc *et al.*, 2006, 2008; Hichri *et al.*, 2010; Matus *et al.*, 2010; Terrier *et al.*, 2009.)

Table 7.2   Direct and indirect induction of PA genes by grape regulator genes. For synthetic genes of flavonoid pathway, early box genes are those common to all flavonoid derivatives. Late box genes are those common to anthocyanin and PA synthesis and specific structural genes of these two branchs. WDR1 and MYCA1 were discarded from the table because of the absence of available functional data.

		MYB5a	MYB5b	MYBPA1	MYBPA2	MYC1
Phenylpropanoid general pathway	PAL	-	n. a.	*	*	n. a.
	C4H	*	n. a.	-	*	n. a.
	4CL	-	n. a.	*	*	n. a.
Early Box	CHS	*	*	*	*	n. a.
	CHI	*+	*+	+	-	+
	F3H	*	*	*	*	n. a.
	F3'H	n. a.	n. a.	*	*	n. a.
	F3'5'H	+	+	+	n. a.	n. a.
Late Box	DFR	*	*	*	*	n. a.
	LDOX	*+	*+	*+	*	n. a.
PA branch	ANR	+	+	*+	*	+
	LAR	+	+	*+	-	n. a.
Anthocyanin branch	UFGT	-	-	-	-	+

Source: Resumed from Bogs *et al.* (2007), Deluc *et al.* (2006, 2008), Terrier *et al.* (2009) and Hichri *et al.* (2010). −, no significant induction; *, indirect induction concluded from overexpression experiences; +, direct induction by physical activation verified by promoter essays; na, not available.

activation (Deluc *et al.*, 2006, 2008). Two R2R3-MYB, MYBPA1 and MYBPA2, controlling more specifically PA synthesis were identified (Bogs *et al.*, 2007; Terrier *et al.*, 2009). Phylogenetic analysis revealed that MYBPA2 is closer to AtTT2, and they share a conserved *C*-terminal motif with OsMYB3, a likely signature of the MYB subgroup they belong to, while this sequence was not found in MYBPA1 (Bogs *et al.*, 2007; Terrier *et al.*, 2009). This indicated the probable functional homology between MYBPA2 and AtTT2. Expression profile of *MYBPA2* resembles that of *MYB5a*: the transcript level was principally found in skin during early berry developmental stages and undetectable during ripening (Fig. 7.5). This expression profile correlates with the expression profile of PA structural genes in skin, especially *ANR* and *LAR2,* while the expression profile of *MYBPA1* correlates better with PA structural gene expression in seeds (Bogs *et al.*, 2005, 2007; Terrier *et al.*, 2009). Ectopic expression of *MYBPA1* restored the wild-type phenotype in *Arabidopsis tt2* mutant and was able to directly activate the promoters of grape *CHI*, *F3′5′H*, *LDOX*, *LAR1* and *ANR* through transient promoter assay (Bogs *et al.*, 2007). Ectopic expression of *MYBPA1* and *MYBPA2* in grape hairy roots led to PA accumulation and induced expression of diverse genes identified by large-scale transcriptomic screening (Terrier *et al.*, 2009). Among the genes induced both by *MYBPA1* and *MYBPA2*, we could note that the most induced ones were putative flavonoid synthetic genes. Nevertheless, the microarray used in this study covered only 50% of known grape unigenes (Terrier *et al.*, 2009; http://www.ncbi.nlm.nih.gov/unigene). Therefore, this result does not offer an exhaustive view of the whole genome expression induced by MYBPA1 and MYBPA2.

Since the protein regulatory complexes MYB/bHLH/WD40 are known to activate flavonoid synthetic gene expression, efforts of grapevine researchers were also focused on identifying unknown bHLH and WD40 proteins in flavonoid biosynthesis regulation. Recently, two WD40-type and two bHLH-type proteins involved in flavonoid synthesis were identified (Hichri *et al.*, 2010; Matus *et al.*, 2010). Expression patterns of the two WD-40 proteins, WDR1 and WDR2, were similar in grape tissues, and they were expressed both in skin and seed throughout grape berry development with a higher expression level during ripening (Fig. 7.5). However, ectopic expression of WDR1 in wild-type *Arabidopsis* led to anthocyanin over-production in leaves and shoot, while no anthocyanin differences were observed in WDR2 transgenic *Arabidopsis*, despite 61% amino acid identity between these two proteins (Matus *et al.*, 2010). The expression pattern of WDR1 and MYCA1 (a bHLH protein) correlated to the expression patterns of *ANR* and *UFGT*, and were, therefore, suggested as putative regulators of PA and anthocyanin synthesis (Bogs *et al.*, 2005; Matus *et al.*, 2010). However, no PA accumulation difference was reported for WDR1-overexpressed *Arabidopsis* transgenic plants. In addition, since it is known that the expression patterns do not always correlate exactly with metabolite biosynthetic enzyme activities, further studies are necessary to characterise the precise roles of *WDR1* and *MYCA1* in flavonoid regulation. The physical interaction of the other bHLH protein, MYC1, with MYB5a, MYB5b, MYBPA1 and MYBA, the well-known TF specifically promoting anthocyanin synthesis (Kobayashi *et al.*, 2002), was verified by yeast two-hybrid assay (Hichri *et al.*, 2010). Interestingly, the cotransfection of MYC1 and MYBA in grape cell suspension led to accumulation of anthocyanins, while single transfection of MYBA or MYC1 was ineffective (Hichri *et al.*, 2010). This indicates that putative regulator complexes recruit at least MYB and MYC factors in grape flavonoid synthesis, as in *Arabidopsis*. Indeed, in the transient promoter assay of *MYBPA1*, Bogs and coworkers (2007) already showed the need of cointroducing a bHLH factor in order to activate promoters of target genes. In the same study, a higher induction level of *VvANR* promoter compared with the promoter of *AtANR* was observed after transfection of either *AtTT2* or *VvMYBPA1*. This could be an indication that other grape native regulators are involved in promoter control (Bogs *et al.*, 2007). In addition, a sequence study revealed that there are at least eight putative *AtTT2* orthologues in the grape genome (Matus *et al.*, 2008). The actual role of these genes remains to be determined. After a global vision of all regulators in PAs, the next step would be the identification of *cis*-element motifs.

A good way to study a biosynthesis pathway is to use mutant collection when available. Thus, transparent testa *Arabidopsis* mutants and barley ant mutants have been used to study PA biosynthesis (Koornneef, 1981; Jende-Strid, 1993). Recent studies about PA regulation in plant kingdom were principally based on knowledge deriving from *Arabidopsis*. The classical scheme is to firstly identify putative candidate by sequence homology with *Arabidopsis* genes, followed by expression profiling, mutant complementation, promoter assay, etc., in order to characterise gene function. This working scheme based on *a priori* knowledge may be a good tool to gain first insight of a complex regulating network in a plant lacking available genomic tools such as complete genome sequence or mutant banks. However, in a complex regulating network as that for grape PA synthesis, the *a priori* scheme could reveal too partial to overcome all the obstacles concerning spatio-temporary expression variation, diversity of PAs between different organs and tissues, influence of environmental stimulus and coordination between PA synthetic genes. Global approach as eQTL offers another way without *a priori* to characterise regulating factors in a perennial plant as grape. The concept of eQTL first came from Jansen and Nap (2001), who proposed to combine genetic and genomic tools to understand molecular and genetic basis of complex quantitative traits. Briefly, it consists in recording global gene expression patterns in a segregating population and detecting all the significant 'eQTL' ('e' for 'expression') for each measured gene. Global gene expression is often measured by microarray, and recently, by next-generation sequencing (Wang *et al.*, 2009). Alternatively, in a more targeted approach, one can also measure expression level of a set of genes, such as structural genes of pathway of interest, by RT-qPCR when *a priori* knowledge is available (Yamashita *et al.*, 2005). There are two features of eQTL: *cis*-eQTL and *trans*-eQTL. *Cis*-eQTL are proximal QTL, which are located near target genes, probably due to polymorphism in *cis*-element. In contrast, *trans*-eQTL are those located in distant genomic regions, supposed to be regulator proteins of target genes, such as TF. With global eQTL approach, one could identify at once all probable regulatory loci of expressed genes and use these data to construct networks in order to elucidate connexions between genes. There are two predominant ways to construct network. One is to use *a priori* knowledge to analyse the regulatory relation of the pathway of interest; a good example in plant is the glucosinolate biosynthetic pathway in *Arabidopsis* (Kliebenstein, 2009). The *a priori* knowledge is defined by information found in the literature, which cannot be exhaustive. As a complementary approach, one can also use directly eQTL results to construct *a posteriori* networks. This approach already proved successful in yeast and Arabidopsis (Lee *et al.*, 2006; Keurentjes *et al.*, 2007).

Despite recent advance in grapevine PA pathway, many black boxes remain to be elucidated. Concerning the regulation aspect, the first task to be accomplished is the identification of regulatory proteins and *cis*-element motifs recognised by corresponding TF. Secondly, an interesting and difficult question to be answered is how regulators coordinate between each other, and what kind of signal address to these regulators, to activate different expression patterns according to organs and tissues.

Previous strategies linked to the identification of mutants through their seed colour appear ineffective to dissect the trafficking pathway, allowing PA storage, the mechanisms controlling the DP, the level of galloylation or the origin of elongation *trans*- or *cis*-units. Moreover, it precludes the identification of functions carried by several isogenes expressed in the same PA producing cell. Several strategies are currently developed to identify new genes involved in the PA pathway such as transcriptomic screening between plants or organs differing in their PA content. This can be achieved through natural variation, such as between different cultivars of *Diospyros kaki* (Ikegami *et al.*, 2009), through plant treatment, such as spraying ethanol on fruits, which resulted in impairing PA synthesis (Ikegami *et al.*, 2007), or through overexpression of transcription factors controlling the pathway in transgenic plants, leading to PA over-accumulation (Pang *et al.*, 2008; Terrier *et al.*, 2009). Using natural composition variation through classical quantitative genetics or association genetics also appears as an attractive way to identify novel functions in the pathway. The particular composition of grape PAs, including galloylation, mixture of *cis*- and *trans*-elementary units, the contrasting PA profile in seed and skin and between cultivars, is a fascinating phenomena which should be useful to study PA synthesis and regulation. As grape PAs have diverse benefits for plant and human health, the well-established knowledge on their synthetic pathway combined with advance in biotechnology will contribute to variety, food quality and human health improvement.

References

Abrahams, S., Lee, E., Walker, A.R., Tanner, G.J., Larkin, P.J. & Ashton, A.R. (2003) The Arabidopsis TDS4 gene encodes leucoanthocyanidin dioxygenase (LDOX) and is essential for proanthocyanidin synthesis and vacuole development. *The Plant Journal*, **35**, 624–636.

Abrahams, S., Tanner, G.J., Larkin, P.J. & Ashton, A.R. (2002) Identification and biochemical characterization of mutants in the proanthocyanidin pathway in Arabidopsis. *Plant Physiology*, **130**, 561–576.

Achnine, L., Blancaflor, E.B., Rasmussen, S. & Dixon, R.A. (2004) Colocalization of L-phenylalanine ammonia-lyase & cinnamate 4-hydroxylase for metabolic channeling in phenylpropanoid biosynthesis. *Plant Cell*, **16**, 3098–3109.

Ageorges, A., Fernandez, L., Vialet, S., Merdinoglu, D., Terrier, N. & Romieu, C. (2006) Four specific isogenes of the anthocyanin metabolic pathway are systematically co-expressed with the red colour of grape berries. *Plant Science*, **170**, 372–383.

Akagi, T., Ikegami, A., Tsujimoto, T., *et al.* (2009) DkMyb4 Is a Myb Transcription Factor Involved in Proanthocyanidin Biosynthesis in Persimmon Fruit. *Plant Physiology*, **151**, 2028–2045.

Alfenito, M.R., Souer, E., Goodman, C.D., *et al.* (1998) Functional complementation of anthocyanin sequestration in the vacuole by widely divergent glutathione S-transferases. *Plant Cell*, **10**, 1135–1149.

Baur, P.S. & Walkinshaw, C.H. (1974) Fine structure of tannin accumulation in callus cultures of *Pinus elliotti* (slash pine). *Canadian Journal of Botany*, **52**, 615–619.

Baudry, A., Heim, M.A., Dubreucq, B., Caboche, M., Weisshaar, B. & Lepiniec, L. (2004) TT2, TT8, and TTG1 synergistically specify the expression of BANYULS and proanthocyanidin biosynthesis in *Arabidopsis thaliana*. *Plant Journal*, **39**, 366–380.

Baxter, I.R., Young, J.C., Armstrong, G., *et al.* (2005) A plasma membrane H^+-ATPase is required for the formation of proanthocyanidins in the seed coat endothelium of *Arabidopsis thaliana*. *Proceedings of the National Academy of Sciences of the United States of America*, **102**, 2649–2654.

Bindon, K.A., Smith, P.A. & Kennedy, J.A. (2010) Interaction between grape-derived proanthocyanidins and cell wall material. 1. Effect on proanthocyanidin composition and molecular mass. *Journal of Agricultural Food Chemistry*, **58**, 2520–2528.

Bishop, P.B. & Nagel, C.W. (1984) Characterization of the condensation product of malvidin 3,5-diglucoside and catechin. *Journal of Agricultural Food Chemistry*, **32**, 1022–1026.

Bogs, J., Downey, M.O., Harvey, J.S., Ashton, A.R., Tanner, G.J. & Robinson, S.P. (2005) Proanthocyanidin synthesis and expression of genes encoding leucoanthocyanidin reductase and anthocyanidin reductase in developing grape berries and grapevine leaves. *Plant Physiology*, **139**, 652–663.

Bogs, J., Ebadi, A., McDavid, D. & Robinson, S.P. (2006) Identification of the flavonoid hydroxylases from grapevine and their regulation during fruit development. *Plant Physiology*, **140**, 279–291.

Bogs, J., Jaffe, F.W., Takos, A.M., Walker, A.R. & Robinson, S.P. (2007) The grapevine transcription factor VvMYBPA1 regulates proanthocyanidin synthesis during fruit development. *Plant Physiology*, **143**, 1347–1361.

Boss, P.K. & Thomas, M.R. (2002) Association of dwarfism and floral induction with a grape 'green revolution' mutation. *Nature*, **416**, 847–850.

Boukharta, M., Girardin, M. & Metche, M. (1988) Procyanidines galloylées du sarment de vigne (*Vitis vinifera*) – Séparation et identification par chromatographie liquide haute performance et chro-matographie en phase gazeuse. *Journal of Chromatography*, **455**, 406–409.

Braidot, E., Zancani, M., Petrussa, E., *et al.* (2008) Transport and accumulation of flavonoids in grapevine (*Vitis vinifera* L.). *Plant Signal Behaviour*, **3**, 626–632.

Burbulis, I.E. & Winkel-Shirley, B. (1999) Interactions among enzymes of the *Arabidopsis* flavonoid biosynthetic pathway. *Proceedings of the National Academy of Sciences of the United States of America*, **96**, 12929–12934.

Cadot, Y., Minana Castello, M.T. & Chevalier, M. (2006) Flavan-3-ol compositional changes in grape berries (*Vitis vinifera* L. cv Cabernet franc) before véraison, using two complementary analytical approaches, HPLC reversed phase and histochemistry. *Analytica Chimica Acta*, **563**, 65–75.

Castellarin, S.D., Di Gaspero, G., Marconi, R., *et al.* (2006) Colour variation in red grapevines (*Vitis vinifera* L.), genomic organisation, expression of flavonoid 3′-hydroxylase, flavonoid 3′,5′-hydroxylase genes and related metabolite profiling of red cyanidin-/blue delphinidin-based anthocyanins in berry skins. *BMC Genomics*, **7**, 12.

Chafe, S.C. & Durzan, D.J. (1973) Tannin inclusions in cell suspension cultures of white spurce. *Planta*, **113**, 251–262.

Chaïb, J., Torregrosa, L., Mackenzie, D., Corena, P., Bouquet, A. & Thomas, M.R. (2010) The grape microvine – a model system for rapid forward and reverse genetics of grapevines. *The Plant Journal*, **62**, 1083–1092.

Cheynier, V., Moutounet, M. & Sarni-Manchado, P. (1998) Les composés phénoliques. In: *Oenologie Fondements Scientifiques et Technologiques* (ed. C. Flanzy), pp. 123–162. Editions Tec & Doc, Paris.

Cheynier, V. & Rigaud, J. (1986) HPLC separation and characterization of flavonols in the skins of *Vitis vinifera* var. Cinsault. *American Journal of Enology and Viticulture*, **37**, 248–252.

Conn, S., Curtin, C., Bézier, A., Franco, C. & Zhang, W. (2008) Purification, molecular cloning, and characterization of glutathione S-transferases (GSTs) from pigmented *Vitis vinifera* L. cell suspension cultures as putative anthocyanin transport proteins. *Journal of Experimental Botany*, **59**, 3621–3634.

Cutanda-Perez, M., Ageorges, A., Gomez, C., *et al.* (2009) Ectopic expression of VlmybA1 in grapevine activates a narrow set of genes involved in anthocyanin synthesis and transport. *Plant Molecular Biology*, **69**, 633–648.

DaSilva, F.G., Iandolino, A., Al-Kayal, F., *et al.* (2005) Characterizing the grape transcriptome. Analysis of expressed sequence tags from multiple Vitis species and development of a compendium of gene expression during berry development. *Plant Physiology*, **139**, 574–597.

Da Silva Porto, P.A., Laranjinha, J.A. & de Freitas, V.A. (2003) Antioxidant protection of low density lipoprotein by procyanidins: structure/activity relationships. *Biochemical Pharmacology*, **66**, 947–954.

Davies, K.M. & Schwinn, K.E. (2006) Molecular biology and biotechnology of flavonoid biosynthesis. In: *Flavonoids – Chemistry, Biochemistry and Applications* (eds. Ø.M. Andersen & K.R. Markham), pp. 143–219. CRC Press, Boca Raton, FL.

Debeaujon, I., Peeters, A.J., Léon-Kloosterziel, K.M. & Koornneef, M. (2001) The TRANSPARENT TESTA12 gene of Arabidopsis encodes a multidrug secondary transporter-like protein required for flavonoid sequestration in vacuoles of the seed coat endothelium. *Plant Cell*, **13**, 853–871.

DeBolt, S., Scheible, W.R., Schrick, K., *et al.* (2009) Mutations in UDP-Glucose:sterol glucosyltransferase in Arabidopsis cause transparent testa phenotype and suberization defect in seeds. *Plant Physiology*, **151**, 78–87.

Delcour, J.A., Ferreira, D. & Roux, D.G. (1983) Synthesis of condensed tannins. Part 9. The condensation sequence of leucocyanidin with (+)-catechin and with the resultant procyanidins. *Journal of the Chemical Society Perkin Transactions*, **1**, 1711–1717.

Deluc, L., Barrieu, F., Marchive, C., *et al.* (2006) Characterization of a grapevine R2R3-MYB transcription factor that regulates the phenylpropanoid pathway. *Plant Physiology*, **140**, 499–511.

Deluc, L., Bogs, J., Walker, A.R., *et al.* (2008) The transcription factor VvMYB5b contributes to the regulation of anthocyanin and proanthocyanidin biosynthesis in developing grape berries. *Plant Physiology*, **147**, 2041–2053.

Deluc, L.G., Grimplet, J., Wheatley, M.D., *et al.* (2007) Transcriptomic and metabolite analyses of Cabernet Sauvignon grape berry development. *BMC Genomics*, **22**, 429.

Devic, M., Guilleminot, J., Debeaujon, I., *et al.* (1999) The BANYULS gene encodes a DFR-like protein and is a marker of early seed coat development. *The Plant Journal*, **19**, 387–398.

Dixon, R.A., Xie, D.Y. & Sharma, S.B. (2005) Proanthocyanidins – a final frontier in flavonoid research? *New Phytologist*, **165**, 9–28.

Downey, M.O., Harvey, J.S. & Robinson, S.P. (2003) Analysis of tannins in seeds and skins of Syrah grapes throughout berry development. *Australian Journal of Grape and Wine Research*, **9**, 15–27.

Drewes, S.E. & Roux, D.G. (1966) A new flavan-3,4-diol from *Acacia auriculiformis* by paper ionophoresis. *Biochemical Journal*, **98**, 493–500.

Dubos, C., Le Gourrierec, J., Baudry, A., *et al.* (2008) MYBL2 is a new regulator of flavonoid biosynthesis in *Arabidopsis thaliana*. *The Plant Journal*, **55**, 940–953.

Flint-Garcia, S.A., Thornsberry, J.M. & Buckler, E.S. (2003) Structure of linkage disequilibrium in plants. *Annual Review of Plant Biology*, **54**, 357–374.

Forkmann, G., Heller, W. & Grisebach, H. (1980) Anthocyanin biosynthesis in flowers of Matthiola-Incana. Flavanone-3-hyroxylase and flavonoid-3′-hydroxylase. *Zeitschrift Für Naturforschung C – A Journal of Biosciences*, **35**, 691–695.

Fournand, D., Vicens, A., Sidhoum, L., Souquet, J.M., Moutounet, M. & Cheynier, V. (2006) Accumulation and extractability of grape skin tannins and anthocyanins at different advanced physiological stages. *Journal of Agricultural Food Chemistry*, **54**, 7331–7338.

Fournier-Level, A., Le Cunff, L., Gomez, C., *et al.* (2009) Quantitative genetic bases of anthocyanin variation in grape (*Vitis vinifera* L. ssp. sativa) berry: a quantitative trait locus to quantitative trait nucleotide integrated study. *Genetics*, **183**, 1127–1139.

Froemel, S., Devlaming, P., Stotz, G., Wiering, H., Forkmann, G. & Schram, A.W. (1985) Genetic and biochemical studies on the conversion of flavanones to dihydroflavonols in flowers of *Petunia hybrida*. *Theoretical and Applied Genetics*, **70**, 561–568.

Fulcrand, H., Remy, S., Souquet, J.-M., Cheynier, V. & Moutounet, M. (1999) Study of wine tannin oligomers by on-line liquid chromatography electrospray ionisation mass spectrometry. *Journal of Agricultural Food Chemistry*, **47**, 1023–1028.

Gao, P., Li, X., Cui, D., Wu, L., Parkin, I. & Gruber, M.Y. (2010) A new dominant Arabidopsis transparent testa mutant, sk21-D, and modulation of seed flavonoid biosynthesis by KAN4. *Plant Biotechnology Journal*, **8**, 979–993.

Gargouri, M., Chaudière, J., Manigand, C., *et al.* (2010) The epimerase activity of anthocyanidin reductase from Vitis vinifera and its regiospecific hydride transfers. *Biological Chemistry*, **391**, 219–227.

Geissman, T. & Yoshimura, N. (1966) Synthetic proanthocyanidin. *Tetrahedron Letters*, **24**, 2669–2673.

Gomez, C., Terrier, N., Torregrosa, L., *et al.* (2009) *Vinis vinifera* MATE-type proteins act as vacuolar H^+-dependent acylated anthocyanin transporters. *Plant Physiology*, **150**, 402–415.

Grayer, R.J. & Veitch, N.C. (2006) Flavanones and dihydroflavonols. In: *Flavonoids – Chemistry, Biochemistry and Applications* (eds. Ø.M. Andersen & K.R. Markham), pp. 917–1002. CRC Press, Boca Raton, FL.

Grotewold, E. & Davies, K. (2008) Trafficking and sequestration of anthocyanins. *Natural Product Communications*, **3**, 1251–1258.

Guyot, S., Vercauteren, J. & Cheynier, V. (1996) Colourless and yellow dimers resulting from (+)-catechin oxidative coupling catalysed by grape polyphenoloxidase. *Phytochemistry*, **42**, 1279–1288.

Haslam, E. (1980) In vino veritas : oligomeric procyanidins and the ageing of red wines. *Phytochemistry* **19**, 2577–2582.

Heller, W. & Forkmann, G. (1988) Biosynthesis. In: *The Flavonoids – Advances in Research since 1980* (ed. J.B. Harborne), pp. 399–425. Chapman & Hall, London.

Hemingway, R.W. & Laks, P. (1985) Condensed tannins, a proposed route to 2R,3R-(2,3-cis)-proanthocyanidins. *Journal of Chemical Society Chemical Communications*, 746–747.

Hernandez-Jimenez, A., Gomez-Plaza, E., Martinez-Cutilla, A. & Kennedy, J.A. (2009) Grape skin and seed proanthocyanidins from Monastrell X Syrah grapes. *Journal of Agricultural Food Chemistry*, **57**, 10798–10803.

Hichri, I., Heppel, S.C., Pillet, J., *et al.* (2010) The basic helix-loop-helix transcription factor MYC1 is involved in the regulation of the flavonoid biosynthesis pathway in grapevine. *Molecular Plant*, **3**, 509–523.

Hsieh, K. & Huang, A.H. (2007) Tapetosomes in *Brassica* tapetum accumulate endoplasmic reticulum-derived flavonoids and alkanes for delivery to the pollen surface. *Plant Cell*, **19**, 582–596.

Hrazdina, G. & Jensen, R.A. (1992) Spatial organization of enzymes in plant metabolic pathways. *Annual Review of Plant Physiology and Plant Molecular Biology*, **43**, 241–267.

Ikegami, A., Akagi, T., Potter, D., *et al.* (2009) Molecular identification of 1-Cys peroxiredoxin and anthocyanidin/flavonol 3-O-galactosyltransferase from proanthocyanidin-rich young fruits of persimmon (*Diospyros kaki* Thunb.). *Planta*, **230**, 841–855.

Ikegami, A., Eguchi, S., Kitajima, A., Inoue, K. & Yonemori, K. (2007) Identification of genes involved in proanthocyanidin biosynthesis of persimmon (*Diospyros kaki*) fruit. *Plant Science*, **172**, 1037–1047

Jacques, D. & Haslam, E. (1974) Biosynthesis of plant proanthocyanidins. *Journal of Chemical Society Chemical Communication*, 231–232.

Jaillon, O., Aury, J.M., *et al.* (2007) The grapevine genome sequence suggests ancestral hexaploidization in major angiosperm phyla. *Nature*, **449**, 463–467.

Jende-Strid, B. (1993) Genetic control of flavonoid biosynthesis in barley. *Hereditas*, **119**, 187–204.

Jende-Strid, B. & Møller, B.L. (1981) Analysis of proanthocyanidins in wild-type and mutant barley (*Hordeum vulgare* L.). *Carlsberg Research Communication*, **46**, 53–64.

Jeong, S.T., Goto-Yamamoto, N., Hashizume, K. & Esaka, M. (2006) Expression of the flavonoid 3′-hydroxylase and flavonoid 3′,5′-hydroxylase genes and flavonoid composition in grape (*Vitis vinifera*). *Plant Science*, **170**, 61–69.

Jeong, S.T., Goto-Yamamoto, N., Hashizume, K. & Esaka, M. (2008) Expression of multi-copy flavonoid pathway genes coincides with anthocyanin, flavonol and flavan-3-ol accumulation of grapevine. *Vitis*, **47**, 135–140.

Jin, H., Cominelli, E., Bailey, P., *et al.* (2000) Transcriptional repression by AtMYB4 controls production of UV-protecting sunscreens in Arabidopsis. *EMBO Journal*, **19**, 6150–6161.

Johnson, C.S., Kolevski, B. & Smyth, D.R. (2002) *TRANSPARENT TESTA GLABRA2*, a trichome and seed coat development gene of Arabidopsis, encodes a WRKY transcription factor. *Plant Cell*, **14**, 1359–1375.

Kajiya, K., Kumazawa, S. & Nakayama, T. (2001) Steric effects on interaction of tea catechins with lipid bilayers. *Bioscience Biotechnology and Biochemistry*, **65**, 2638–2643.

Kajiya, K., Kumazawa, S. & Nakayama, T. (2002) Effects of external factors on the interaction of tea catechins with lipid bilayers. *Bioscience Biotechnology and Biochemistry*, **66**, 2330–2335.

Kennedy, J.A., Hayasaka, Y., Vidal, S., Waters, E.J. & Jones, G.P. (2001) Composition of grape skin proanthocyanidins at different stages of berry development. *Journal of Agricultural Food Chemistry*, **49**, 5348–5355.

Kennedy, J.A., Matthews, M.A. & Waterhouse, A.L. (2002) Effect of maturity and vine water status on grape skin and wine flavonoids. *American Journal of Enology and Viticulture*, **53**, 268–274.

Keurentjes, J.J.B., Fu, J.Y., Terpstra, I.R., *et al.* (2007) Regulatory network construction in Arabidopsis by using genome-wide gene expression quantitative trait loci. *Proceedings of the National Academy of Sciences of the United States of America*, **104**, 1708–1713.

Kitamura, S., Matsuda, F., Tohge, T., *et al.* (2010) Metabolic profiling and cytological analysis of proanthocyanidins in immature seeds of *Arabidopsis thaliana* flavonoid accumulation mutants. *The Plant Journal*, **62**, 549–559.

Kitamura, S., Shikazono, N. & Tanaka, A. (2004) TRANSPARENT TESTA 19 is involved in the accumulation of both anthocyanins and proanthocyanidins in Arabidopsis. *The Plant Journal*, **37**, 104–114.

Kliebenstein, D. (2009) A quantitative genetics and ecological model system: understanding the aliphatic glucosinolate biosynthetic network via QTLs. *Phytochemistry Reviews*, **8**, 243–254.

Kobayashi, S., Ishimaru, M., Hiraoka, K. & Honda, C. (2002) Myb-related genes of the Kyoho grape (*Vitis labruscana*) regulate anthocyanin biosynthesis. *Planta*, **215**, 924–933.

Koes, R., Verweij, W. & Quattrochio, F. (2005) Flavonoids: a colorful model for the regulation and evolution of biochemical pathways. *Trends in Plant Science*, **10**, 236–242.

Koornneef, M. (1981) The complex syndrome of TTG mutants. *Arabidopsis Information Service*, **18**, 45–51.

Koornneef, M. (1990) Mutations affecting the testa colour in Arabidopsis. *Arabidopsis Information Service*, **27**, 1–4.

Koukol, J. & Conn, E.E. (1961) Metabolism of aromatic compounds in higher plants. 4. Purification and properties of phenylalanine deaminase of *Hordeum vulgare*. *Journal of Biological Chemistry*, **236**, 2692–2710.

Kreuzaler, F. & Hahlbrock, K. (1972) Enzymatic synthesis of aromatic compounds in higher plants: formation of naringenin (5,7,4′-trihydroxyflavanone) from *p*-coumaroyl coenzyme A and malonyl coenzyme A. *FEBS Letters*, **15**, 69–72.

Lee, S.I., Pe'ert, D., Dudley, A.M., Church, G.M. & Koller, D. (2006) Identifying regulatory mechanisms using individual variation reveals key role for chromatin modification. *Proceedings of the National Academy of Sciences of the United States of America*, **103**, 14062–14067.

Lepiniec, L., Debeaujon, I., Routaboul, J.M., *et al.* (2006) Genetics and biochemistry of seed flavonoids. *Annual Review of Plant Biology*, **57**, 405–430.

Lizarraga, D., Lozano, C., Briedé, J.J., *et al.* (2007) The importance of polymerization and galloylation for the antiproliferative properties of procyanidin-rich natural extracts. *FEBS Journal* **274**, 4802–4811.

Lodhi, M.A., Daly, M.J., Ye, G.N., Weeden, N.F. & Reisch, B.I. (1995) A molecular marker based linkage map of Vitis. *Genome*, **38**, 786–794.

Mané, C., Souquet, J.M., Ollé, D., *et al.* (2007) Optimization of simultaneous flavanol, phenolic acid and anthocyanin extraction from grapes using an experimental design; application to the characterization of Champagne grape varieties. *Journal of Agricultural Food Chemistry*, **55**, 7224–7233.

Marinova, K., Pourcel, L., Weder, B., *et al.* (2007) The Arabidopsis MATE transporter TT12 acts as a vacuolar flavonoid/H^+-antiporter active in proanthocyanidin-accumulating cells of the seed coat. *Plant Cell*, **19**, 2023–2038.

Marles, M.A., Ray, H. & Gruber, M.Y. (2003) New perspectives on proanthocyanidin biochemistry and molecular regulation. *Phytochemistry*, **64**, 367–383.

Marrs, K., Alfenito, M., Lloyd, A. & Walbot, V. (1995) A glutathione S-transferase involved in vacuolar transfer encoded by the maize gene Bronze-2. *Nature*, **375**, 397–400.

Matus, J., Poupin, M., Cañon, P., Bordeu, E., Alcalde, J. & Arce-Johnson, P. (2010) Isolation of WDR and bHLH genes related to flavonoid synthesis in grapevine (*Vitis vinifera* L.). *Plant Molecular Biology*, **72**, 607–620.

Matus, J.T., Aquea, F. & Arce-Johnson, P. (2008) Analysis of the grape MYB R2R3 subfamily reveals expanded wine quality-related clades and conserved gene structure organization across *Vitis* and *Arabidopsis* genomes. *BMC Plant Biology*, **22**, 83.

McAbee, J.M., Hill, T.A., Skinner, D.J., *et al.* (2006) *ABERRANT TESTA SHAPE* encodes a KANADI family member, linking polarity determination to separation and growth of Arabidopsis ovule integuments. *The Plant Journal*, **46**, 522–531.

Mellway, R.D., Tran, L.T., Prouse, M.B., Campbell, M.M. & Constanbel, C.P. (2009) The wound-, pathogen-, and ultraviolet B-responsive MYB134 gene encodes an R2R3 MYB transcription factor that regulates proanthocyanidin synthesis in poplar. *Plant Physiology*, **150**, 924–941.

Menting, J.G.T., Scopes, R.K. & Stevenson, T.W. (1994) Characterization of flavonoid 3′,5′-hydroxylase in microsomal membrane-fraction of *Petunia hybrida* flowers. *Plant Physiology*, **106**, 633–642.

Mol, J., Grotewold, E. & Koes, R. (1998) How genes paint flowers and seeds. *Trends in Plant Science*, **3**, 212–217.

Monagas, M., Gomez-Cordoves, C., Bartolomé, B., Laureano & Ricardo da Silva, J. (2003) Monomeric, oligomeric, and polymeric flavan-3-ol composition of wines and grapes from *Vitis vinifera* L. cv graciano, tempranillo, and cabernet sauvignon. *Journal of Agricultural Food Chemistry*, **51**, 6475–6481.

Moustafa, E. & Wong, E. (1967) Purification and properties of chalcone-flavanone isomerase from soybean seed. *Phytochemistry*, **6**, 625–632.

Mueller, L., Goodman, C., Silady, R. & Walbot, V. (2000) AN9, a petunia Glutathione S-Transferase required for anthocyanin sequestration, is a flavonoid-binding protein. *Plant Physiology*, **123**, 1561–1570.

Müller, M.L., Jensen, M. & Taiz, L. (1999) The vacuolar H^+-ATPase of lemon fruits is regulated by variable H^+/ATP coupling and slip. *Journal of Biological Chemistry*, **274**, 10706–10716.

Nesi, N., Debeaujon, I., Jond, C., Pelletier, G., Caboche, M. & Lepiniec, L. (2000) The *TT8* gene encodes a basic helix-loop-helix domain protein required for expression of *DFR* and *BAN* genes in Arabidopsis siliques. *Plant Cell*, **12**, 1863–1878.

Nesi, N., Debeaujon, I., Jond, C., Pelletier, G., Caboche, M. & Lepiniec, L. (2001) The Arabidopsis *TT2* gene encodes an R2R3 MYB domain protein that acts as a key determinant for proanthocyanidin accumulation in developing seed. *Plant Cell*, **13**, 2099–2114.

Nesi, N., Debeaujon, I., Jond, C., *et al.* (2002) The *TRANSPARENT TESTA16* locus encodes the ARABIDOPSIS BSISTER MADS domain protein and is required for proper development and pigmentation of the seed coat. *Plant Cell*, **14**, 2463–2479.

Pang, Y., Peel, G.J., Sharma, S.B., Tang, Y. & Dixon, R.A. (2008) A transcript profiling approach reveals an epicatechin-specific glucosyltransferase expressed in the seed coat of *Medicago truncatula*. *Proceedings of the National Academy of Sciences of the United States of America*, **105**, 14210–14205.

Pang, Y., Peel, G.J., Wright, E., Wang, Z. & Dixon, R.A. (2007) Early steps in proanthocyanidin biosynthesis in the model legume *Medicago truncatula*. *Plant Physiology*, **145**, 601–615.

Pang, Y., Wenger, J.P., Saathoff, K., *et al.* (2009) A WD40 repeat protein from *Medicago truncatula* is necessary for tissue-specific anthocyanin and proanthocyanidin biosynthesis but not for trichome development. *Plant Physiology*, **151**, 1114–1129.

Parham, R.A. & Kaustinen, H.M. (1977) On the site of tannin synthesis in plant cells. *Botanical Gazette*, **138**, 465–467.

Pezza, R.J., Villarreal, M.A., Montich, G.G. & Argaraña, C.E. (2002) Vanadate inhibits the ATPase activity and DNA binding capability of bacterial MutS. A structural model for the vanadate-MutS interaction at the Walker A motif. *Nucleic Acids Research*, **30**, 4700–4708.

Pfeiffer, J., Kuhnel, C., Brandt, J., *et al.* (2006) Biosynthesis of flavan-3-ols by leucoanthocyanidin 4-reductases and anthocyanidin reductases in leaves of grape (*Vitis vinifera* L.), apple (*Malus domestica* Borkh.) and other crops. *Plant Physiology and Biochemistry*, **44**, 323–334.

Pourcel, L., Routaboul, J.M., Kerhoas, L., Caboche, M., Lepiniec, L. & Debeaujon, I. (2005) TRANSPARENT TESTA10 encodes a laccase-like enzyme involved in oxidative polymerization of flavonoids in Arabidopsis seed coat. *Plant Cell*, **17**, 2966–2980.

Prieur, C., Rigaud, J., Cheynier, V. & Moutounet, M. (1994) Oligomeric and polymeric procyanidins from grape seeds. *Phytochemistry*, **36**, 781–784.

Rafalski, J.A. (2010) Association genetics in crop improvement. *Current Opinion in Plant Biology*, **13**, 174–180.

Ramsay, N.A. & Glover, B.J. (2005) MYB-bHLH-WD40 protein complex and the evolution of cellular diversity. *Trends in Plant Science*, **10**, 63–70.

Remy-Tanneau, S., Le Guerneve, C., Meudec, E. & Cheynier, V. (2003) Characterization of a colorless anthocyanin-flavan-3-ol dimer containing both carbon-carbon and ether interflavanoid link-ages by NMR and mass spectrometries. *Journal of Agricultural Food Chemistry*, **51**, 3592–3597.

Ribéreau-Gayon, P. (1964) Les composés phénoliques du raisin et du vin II. Les flavonosides et les anthocyanosides. *Annales de Physiologie Végétale*, **6**, 211–242.

Ricardo da Silva, J.M., Rigaud, J., Cheynier, V., Cheminat, A. & Moutounet, M. (1991) Procyanidin dimers and trimers from grape seeds. *Phytochemistry*, **30**, 1259–1264.

Routaboul, J.M., Kerhoas, L., Debeaujon, I., *et al.* (2006) Flavonoid diversity and biosynthesis in seed of *Arabidopsis thaliana*. *Planta*, **224**, 96–107.

Russell, D.W. & Conn, E.E. (1967) The cinnamic acid 4-hydroxylase of pea seedlings: isolation, activity, cofactor requirements. *Archive of Biochemistry and Biophysics*, **122**, 256–258.

Sagasser, M., Lu, G.-H., Hahlbrock, K. & Weisshaar, B. (2002) A. thaliana TRANSPARENT TESTA 1 is involved in seed coat development and defines the WIP subfamily of plant zinc finger proteins. *Genes and Development*, **16**, 138–149.

Saslowsky, D.E. & Winkel-Shirley, B. (2001) Localization of flavonoid enzymes in Arabidopsis roots. *The Plant Journal*, **27**, 37–48.

Shih, C.H., Chu, H., Tang, L.K., *et al.* (2008) Functional characterization of key structural genes in rice flavonoid biosynthesis. *Planta*, **228**, 1043–1054.

Souquet, J.M., Cheynier, V., Brossaud, F. & Moutounet, M. (1996) Polymeric proanthocyanidins from grape skins. *Phytochemistry*, **43**, 509–512.

Souquet, J.M., Labarbe, B., Le Guernevé, C., Cheynier, V. & Moutounet, M. (2000) Phenolic composition of grape stems. *Journal of Agricultural Food Chemistry*, **48**, 1076–1080.

Souquet, J.M., Veran, F., Mané, C. & Cheynier, V. (2006) Optimization of extraction conditions on phenolic yields from the different parts of grape clusters – quantitative distribution of their proanthocyanidins. *Proceedings of the XXIII International Conference on Polyphenols*, Winnipeg, Manitoba, Canada.

Sparvoli, F., Martin, C., Scienza, A., Gavazzi, G. & Tonelli, C. (1994) Cloning and molecular analysis of structural genes involved in flavonoid and stilbene biosynthesis in grape (*Vitis vinifera* L.) *Plant Molecular Biology*, **24**, 743–755.

Stafford, H.A. (1974) Possible multienzyme complexes regulating the formation of C6-C3 phenolic compounds and lignins in higher plants. *Recent Advances in Phytochemistry*, **8**, 53–79.

Stafford, H.A. & Lester, H.H. (1982) Enzymic and nonenzymic reduction of (+)-dihydroquercetin to its 3,4,-diol. *Plant Physiology*, **70**, 695–698.

Stafford, H.A. & Lester, H.H. (1984) The conversion of (+)-dihydroquercetin and flavan-3,4-cis-diol (leucocyanidin) to (+)-catechin by reductases extracted from cell suspension of Douglas fir. *Plant Physiology*, **76**, 184–186.

Stafford, H.A., Shimamoto, M. & Lester, H.H. (1982) Incorporation of [14C]phenylalanine into flavan-3-ols and procyanidins in cell suspension cultures of Douglas fir. *Plant Physiology*, **69**, 1055–1059.

Suzuki, T., Yamazaki, N., Sada, Y., Oguni, I. & Moriyasu, Y. (2003) Tissue distribution and intracellular localization of catechins in tea leaves. *Bioscience Biotechnology and Biochemistry*, **67**, 2683–2686.

Sze, H. (1985) H^+-translocating ATPases advances using membrane vesicles. *Annual Review of Plant Physiology*, **36**, 175–208.

Tanner, G.J., Francki, K.T., Abrahams, S., Watson, J.M., Larkin, P.J. & Ashton, A.R. (2003) Proanthocyanidin biosynthesis in plants. Purification of legume leucoanthocyanidin reductase and molecular cloning of its cDNA. *Journal of Biological Chemistry*, **278**, 31647–31656.

Terrier, N., Ageorges, A., Abbal, P. & Romieu, C. (2001) Generation of ESTs from grape berry at various developmental stages. *Journal of Plant Physiology*, **158**, 1575–1583.

Terrier, N., Glissant, D., Grimplet, J., *et al.* (2005) Isogene specific oligo arrays reveal multifaceted changes in gene expression during grape berry (*Vitis vinifera* L.) development. *Planta*, **222**, 820–831.

Terrier, N., Torregrosa, L., Ageorges, A., *et al.* (2009) Ectopic expression of VvMybPA2 promotes proanthocyanidin biosynthesis in grapevine and suggests additional targets in the pathway. *Plant Physiology*, **149**, 1028–1041.

Tesnière, C., Torregrosa, L., Pradal, M., *et al.* (2006) Effects of genetic manipulation of alcohol dehydrogenase levels on the response to stress and the synthesis of secondary metabolites in grapevine leaves. *Journal of Experimental Botany*, **57**, 91–99.

Torregrosa, L. & Bouquet, A. (1997) *Agrobacterium tumefaciens* and *Agrobacterium rhizogenes* co-transformation to obtain grapevine hairy roots producing the coat protein of grapevine chrome mosaic nepovirus. *Plant Cell, Tissue and Organ Culture*, **49**, 53–62.

Velasco, R., Zharkikh, A., Troggio, M., *et al.* (2007) A high quality draft consensus sequence of the genome of a heterozygous grapevine variety. *PLoS ONE*, **19**, e1326.

Verriès, C., Guiraud, J.L., Souquet, J.M., Vialet, S., Terrier, N. & Ollé, D. (2008) Validation of an extraction method on whole pericarp of grape berry (*Vitis vinifera* L. cv Syrah) to study biochemical and molecular aspects of flavan-3-ol synthesis during berry development. *Journal of Agricultural Food Chemistry*, **56**, 5896–5904.

Verweij, W., Spelt, C., Di Sansebastiano, G.P., *et al.* (2008) An H$^+$ P-ATPase on the tonoplast determines vacuolar pH and flower colour. *Nature Cell Biology*, **10**, 1456–1462.

Vidal, S., Cartalade, D., Souquet, J., Fulcrand, H. & Cheynier, V. (2002) Changes in proanthocyanidin chain-length in wine-like model solutions. *Journal of Agricultural Food Chemistry*, **50**, 2261–2266.

VonWettstein, D. (2007) From analysis of mutants to genetic engineering. *Annual Review of Plant Biology*, **58**, 1–19.

Walker, A.R., Davison, P.A., Bologesi-Winfield, A.C., *et al.* (1999) The *TRANSPARENT TESTA GLABRA1* locus, which regulates Ttrichome differentiation and anthocyanin biosynthesis in Arabidopsis, encodes a WD40 repeat protein. *Plant Cell*, **11**, 1337–1350.

Wang, Z., Gerstein, M. & Snyder, M. (2009) RNA-Seq: a revolutionary tool for transcriptomics. *Nature Reviews Genetics*, **10**, 57–63.

Wei, Y.-L., Li, J.-N., Lu, J., Tang, Z.-L., Pu, D.-C. & Chai, Y.-R. (2007) Molecular cloning of *Brassica napus* TRANSPARENT TESTA 2 gene family encoding potential MYB regulatory proteins of proanthocyanidin biosynthesis. *Molecular Biology Reports*, **34**, 105–120.

Weinges, K., Bahr, W., Ebert, W., Goritz, K. & Marx, H.D. (1969a) Konstitution, Entstehung und Bedeutung der Flavonoid-Gerbstoffe. *Fortschritte der Chemie Organischer Naturstoffe*, **27**, 158–260.

Weinges, K., Ebert, W., Huthwelker, D., Mattauch, H. & Perner, J. (1969b) Oxydative kupplung von phenolen, II. Konstitution und bildungsmechanismus des dehydro-dicatechins A. *Justus Liebigs Annalen der Chemie*, **726**, 114–124.

Weinges, K. & Piretti, M.V. (1971) Isolierung des $C_{30}H_{26}O_{12}$-procyanidins B1 aus Wientrauben. *Justus Liebigs Annalen der Chemie*, **748**, 218–220.

Welter, L.J., Grando, M.S. & Zyprian, E. (2010) Basics of grapevine genetic analysis. In: *Grapes* (eds. J.M. Martinez-Zapater & A.F. Adam-Blondon), pp. 137–159. Science Publishers, Enfileld, NH.

Winkel-Shirley, B. (2001) Flavonoid biosynthesis. A colorful model for genetics, biochemistry, cell biology, and biotechnology. *Plant Physiology*, **126**, 485–493.

Xie, D.Y. & Dixon, R.A. (2005) Proanthocyanidin biosynthesis – still more questions than answers? *Phytochemistry*, **66**, 2127–2144.

Xie, D.Y., Sharma, S.B. & Dixon, R.A. (2004) Anthocyanidin reductases from *Medicago truncatula* and *Arabidopsis thaliana*. *Archive of Biochemistry and Biophysics*, **422**, 91–102.

Xie, D.Y., Sharma, S.B., Paiva, N.L., Ferreira, D. & Dixon, R.A. (2003) Role of anthocyanidin reductase, encoded by BANYULS in plant flavonoid biosynthesis. *Science*, **299**, 396–399.

Yamashita, S., Wakazono, K., Nomoto, T., Tsujino, Y., Kuramoto, T. & Ushijima, T. (2005) Expression quantitative trait loci analysis of 13 genes in the rat prostate. *Genetics*, **171**, 1231–1238.

Yazaki,, K. (2005) Transporters of secondary metabolites. *Current Opinion in Plant Biology*, **8**, 301–307.

Yoshida, K., Iwasaka, R., Kaneko, T., Sato, S., Tabata, S. & Sakuta, M. (2008) Functional differentiation of *Lotus japonicus* TT2s, R2R3-MYB transcription factors comprising a multigene family. *Plant Cell Physiology*, **49**, 157–169.

Young, D.A., Young, E., Roux, D.G., Brandt, E.V. & Ferreira, D. (1987) Synthesis of (+)-catechin and (+)-mesquitol, conformation of bis (+)-catechins. *Journal of Chemical Society, Perkin Transactions*, **1**, 2345–2351.

Yu, J. & Buckler, E.S. (2006) Genetic association mapping and genome organization of maize. *Current Opinion in Biotechnology*, **17**, 155–160.

Zhang, H., Wang, L., Deroles, S., Bennett, R. & Davies, K. (2006) New insight into the structures and formation of anthocyanic vacuolar inclusions in flower petals. *BMC Plant Biology*, **17**, 29.

Zhao, J. & Dixon, R.A. (2009) MATE transporters facilitate vacuolar uptake of epicatechin 3′-O-glucoside for proanthocyanidin biosynthesis in *Medicago truncatula* and Arabidopsis. *Plant Cell*, **21**, 2323–2340.

Zhao, J. & Dixon, R.A. (2010) The 'ins' and 'outs' of flavonoid transport. *Trends in Plant Science* **15**, 72–80.

Zhao, J., Pang, Y. & Dixon, R.A. (2010) The mysteries of proanthocyanidin transport and polymerization. *Plant Physiology*, **153**, 437–443.

Chapter 8
Phenolic Compounds in Plant Defense and Pathogen Counter-defense Mechanisms

Fouad Daayf, Abdelbasset El Hadrami, Ahmed F. El-Bebany,
Maria A. Henriquez, Zhen Yao, Holly Derksen,
Ismaïl El-Hadrami and Lorne R. Adam

Abstract: Phenolics have been in the center of many discoveries on plant defenses to pathogens, including bacteria, fungi, and viruses. More recently, this fascinating group of metabolites was also pointed out as a target of invading microbes to defeat such defenses. Here, we summarize findings in several host–pathogen interactions involving the oomycete *Phytophthora infestans* and soilborne pathogens such as *Verticillium dahliae*, depicting the role of phenolics in plant defenses against them, and how they target such responses to settle their infection. The defenses described here include the upregulation of genes in the phenylpropanoid pathway, and the accumulation of phenolics such as scopoletin (a coumarin), *p*-coumaric acid methyl ester (a hydroxycinnamic acid derivative), or rutin (a flavonoid) in response to several pathogens. Counter-defense mechanisms developed by these pathogens, such as downregulation of defense genes and degradation of phenolics, are described as well.

Keywords: phytoanticipins; phytoalexins; defense signaling; counter-defense mechanisms; salicylic acid; jasmonic acid; ethylene; proteomics; transcriptomics; metabolomics; cDNA-AFLP

8.1 Introduction

Plants developed the ability to fight pathogens that attack them and cause diseases. Sometimes, diseases turn into epidemics that result into vast damages to crops, and sometimes to humans. There are many examples in history of plant diseases that caused human tragedies, and the most famous is probably the Irish famine that followed potato late blight epidemics

Recent Advances in Polyphenol Research, Volume 3, First Edition. Edited by Véronique Cheynier,
Pascale Sarni-Manchado and Stéphane Quideau.

in the 1840s. More than a million of people died of starvation and about the same number emigrated. Many of the highly devastating epidemics happened some time ago, but the risk for more to occur again is always present. In general, disease control relies on management techniques such as cultural practices (i.e., rotations), the use of resistant lines, chemical pesticides, and sometimes, biological control methods. These methods are ideally used in combination as part of an integrated disease management strategy. However, for these strategies to be successful and sustainable, they need to be based on genuine knowledge of the underlying biology of each specific plant–pathogen interaction. This means understanding the pathogen strategies to attack the host plant, the plant's ability to deploy defenses against such attacks, and the effects of the environment on the interactions.

Until recently, the majority of host–pathogen interactions were described in terms of pathogens attacking their hosts and the hosts responding to the attack. Scientists used this model to test several hypotheses and further discovered most of what we know today about pathogens' strategies to invade their hosts and the plant mechanisms leading plants either to resist or succumb to the invasion. More recently, investigations on host defense failure to stop some pathogens generated turning point questions about the mechanisms of disease susceptibility, and a starting point towards unraveling the strategies used by pathogens to avoid or overcome host defenses. Most scientists at the forefront of discovery in this field have been examining the range of strategies used by pathogens to overcome plant defenses and carry on their invasion. Plants use several pathways to protect their tissues against invaders. In contrast, these invaders may build counter-defense strategies to manipulate the host reactions, making their attack a successful one.

Plant phenolics have been in the center of a myriad of discoveries related to plant defenses to different pathogens, including bacteria, fungi, and viruses (Nicholson & Hammerschmidt, 1992; Bennett & Wallsgrove, 1994; Dixon & Paiva, 1995; Daayf *et al.*, 1997a; Treutter, 2006; El Hadrami & Daayf, 2009). It is not surprising that this fascinating group of metabolites has become one of the pivotal elements in the plant reactions targeted and hijacked by invading microbes to serve their own settling (El Hadrami *et al.*, 2009). Here, we summarize our contributions to this knowledge gained through the investigation of several host–pathogen systems.

8.2 Plant defenses and pathogen counter-defenses

Plant defenses to pathogens can be either structural or metabolic, and within each type, they can be either preformed or induced upon infection. Many of the preformed defenses are structural, such as leaf waxes and lignified cell walls. Waxes act as a water-repellent surface, which reduces the risks for infections, whereas cell walls reinforced with lignin or suberin serve as extra-barriers for the pathogen to overcome before getting to the cells' content. Preformed defenses can also be metabolic, including molecules with antimicrobial properties such as fungitoxic exudates (Walker & Stahmann, 1955), phenolics (El Hadrami *et al.*, 2005), saponins (Osbourn, 1996), and others. The lack of compounds essential for the establishment of the infection may also be considered as a preformed mechanism of defense leading to resistance. Resistance mechanisms may also include lack of recognition between the host and the pathogen (i.e., nonhost resistance), lack of specific receptors on the host membrane to essential virulence factors of the pathogen (i.e., HC-toxin), or lack of other substances essential for the pathogen's own growth or development (Vogel & Somerville, 2000).

When plants are attacked by pathogens, mutual recognition results into complex signaling reactions (Hausladen & Stamler, 1998; Dangl & Jones, 2001) that trigger a series of highly coordinated defense responses (Bennett & Wallsgrove, 1994; Hammond-Kosack & Jones, 1996) including an increase in transcript levels of several genes during either compatible or incompatible interactions (Birch *et al.*, 2003; Ros *et al.*, 2004). Incompatibility means a quick ending of the infection with no further multiplication/sporulation of the pathogen, whereas compatibility permits a more or less establishment of the disease. These reactions culminate into such reactions as cell wall lignification and suberization (Stein *et al.*, 1993; Daayf *et al.*, 1997b), *de novo* synthesis of pathogenesis-related (PR) proteins (van Loon & van Strien, 1999), and accumulation of phytoalexins and phytoanticipins (Daayf *et al.*, 1997b, 1997c, 2000, 2003; Harborne, 1999; Ongena *et al.*, 1999; Dangl & Jones, 2001; El Hassni *et al.*, 2004; El Hadrami & Daayf, 2009). Phenolics or related compounds have been reported as part of most of these reactions.

Both PR proteins and phytoalexins are known to be involved in plant disease resistance (Hammond-Kosack & Jones, 1996). Phytoalexins are plant secondary metabolites, produced *de novo* mainly through the phenylpropanoid and the terpenoid pathways, in response to a variety of stresses (Hammerschmidt, 1999; Wang *et al.*, 2008). Phenylalanine ammonia-lyase (PAL) and hydroxymethyl-glutaryl-CoA reductase (HMGR) catalyze the first steps in these two pathways. Compounds involved in plant defense against pathogens can be either synthesized upon infection or simply released from their nontoxic conjugated forms, themselves either constitutive or induced (Daayf *et al.*, 1997c). Phytoalexins from the same plant families tend to range within the same chemical classes. For example, *Solanaceae* and *Malvaceae*'s phytoalexins belong mostly to the sesquiterpenes (Choi *et al.*, 1992; Lyon *et al.*, 1995; Hammerschmidt, 1999), whereas those from the *Leguminosae* are mainly isoflavonoids and polyacetylenes (Dixon *et al.*, 1992; Hammerschmidt, 1999). However, there are plant species that produce phytoalexins related to more than one chemical class. Phytoalexins are important in disease resistance because they can directly affect the integrity of pathogens' structures (cell membranes, DNA) or functions (oxidative phosphorylation). However, phytoalexins do not provide an absolute protection against microbial infections because many pathogens have evolved mechanisms to protect themselves from these substances (El Hadrami *et al.*, 2009).

PR proteins are grouped into several classes, based on their functions such as glucanases, chitinases, or peroxidases (van Loon & van Strien, 1999). They can destroy fungal cell walls or alter their physiology (Wang *et al.*, 2005, 2006), while other enzymes help detoxify the pathogen toxins (i.e., fusaric acid) (El Hadrami *et al.*, 2005). Several enzymes such as oxidases (i.e., polyphenol oxidases and peroxidases) help generate oxidation products that are toxic to the pathogen (El Hadrami *et al.*, 1997; Daayf *et al.*, 2003; Arfaoui *et al.*, 2007).

The synthesis and accumulation of phytoalexins and PR proteins occurs following a cascade of signal transduction reactions, involving factors such as hydrogen peroxide, nitric oxide, calcium, protein kinases and phosphatases, systemin, ethylene, salicylic, jasmonic, and abscisic acids (Nawrath & Métraux, 1999; Romero-Puertas *et al.*, 2004; Catinot *et al.*,

2008). Many studies have reported cross-talk among these pathways (Pieterse *et al.*, 2001; Kunkel & Brooks, 2002; Nandi *et al.*, 2003; Romero-Puertas *et al.*, 2004). However, most questions remain unanswered regarding how they interact in different crops (host–pathogen interactions).

In this chapter, we mainly focus on the role of phenolic compounds as structural components of defense or as metabolites acting either directly on pathogens or participating to the complex signaling network that monitors the outcome of the host response to each particular pathogen.

The appearance and development of plant diseases are due to inadequate plant defenses (timing, location, and/or intensity), or to pathogen strategies to bypass these defenses (Osbourn, 1996; Bouarab *et al.*, 2002; El Hadrami *et al.*, 2009). Some microbes have evolved highly specialized tactics to suppress plant defenses and induce susceptibility in hosts that normally carry a good level of resistance (Abramovitch & Martin, 2004). For example, the degradation of plant phytoalexins by some pathogens is known as a counter-defense strategy (Weltring *et al.*, 1995; Soby *et al.*, 1996). More interestingly, some products resulting from the degradation of plant defense compounds (Ham *et al.*, 1997) can suppress plant defenses (Bouarab *et al.*, 2002).

8.3 Phenolic-related plant responses to pathogens

Phenolics-based plant defense mechanisms include physical changes such as lignification and suberization of the plant cell walls (Stein *et al.*, 1993), as well as metabolic changes such as *de novo* synthesis of PR proteins (Carr & Klessig, 1989), and biosynthesis and accumulation of phenylpropanoid secondary metabolites (Bennett & Wallsgrove, 1994). Many of the phytoalexins described so far are produced through the phenylpropanoid pathway. Actually, the activation of the latter in plants is one of the initial reactions to attacks by pathogens (Dixon & Paiva, 1995). In addition, this pathway contributes not only to the pool of free metabolites but also to the group of compounds that are integrated into cell wall reinforcement.

Although much work was also described by others on the role of phenolics in plant defenses, we are only focusing in this review on the studies we have been conducting on several plant–pathogen systems.

*8.3.1 Cotton–***Verticillium dahliae**

In the cotton–*Verticillium dahliae* pathosystem, we used susceptible and resistant cotton lines to perform phytochemical, cytological, and histochemical studies where we investigated cotton defense responses to a highly aggressive and defoliating strain of *V. dahliae*. Early cotton tissue responses to this fungal pathogen consisted of reinforcement in structural barriers with polysaccharides, including callose and cellulose. Ultrastructural modifications of parenchyma cells of the vascular tissues were associated with strong production of terpenoids and phenolics. These defense reactions were detected early in roots of the resistant line, 1–4 days after inoculation, while they were seen later in roots of the susceptible

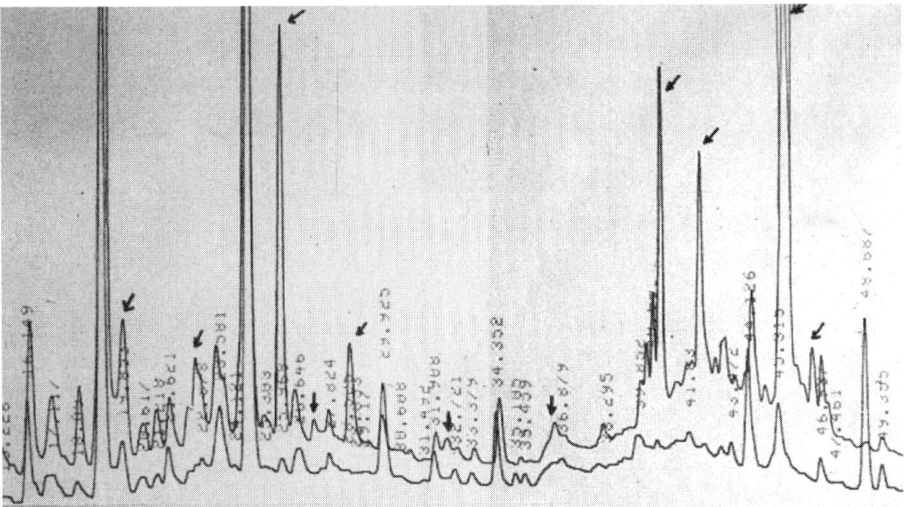

Fig. 8.1 High-performance liquid chromatography profile of cotton tissues inoculated (upper line) or not (lower line) with *Verticillium dahliae*, showing the appearance of new peaks.

line. Regarding the secondary metabolites produced during this interaction, prior to our work, four sesquiterpenes were initially defined as cotton's phytoalexins: hemigossypol (HG), deoxy-hemigossypol (dHG), methoxy-hemigossypol (MHG), and deoxy-methoxy-hemigossypol (dMHG). Then only dHG was recognized as a true phytoalexin based on its timely accumulation in sufficient amounts to be effective. However, dHG alone did not justify the whole cotton lines' resistance. Therefore, we wondered whether other mechanisms complement the activity of dHG. Given the role of phenolics in disease resistance of many plant species, we investigated their involvement in this host–pathogen interaction. As a result, we identified scopoletin as a prominent phenolic compound differentially accumulating in response to *V. dahliae*, in addition to other phenolics including flavan-3-ols, flavan-3,4-diols, and proanthocyanidins. We have also determined the coordinated spatiotemporal accumulation of both sesquiterpenes and phenolics using thin layer chromatography (plate 8.1), high-performance liquid chromatography (HPLC; Fig. 8.1), fluorescence (plate 8.2), and electronic microscopy (Daayf *et al.*, 1997b). Scopoletin seems to be part of the defense system of many plants, and we have also observed its accumulation in potato tubers infected with *Phytophthora infestans* (M.A. Henriquez *et al.*, unpublished data), as well as in plants infiltrated by thaxtomin A, the toxin produced by the actinomycete when attacking Solanaceae (Lerat *et al.*, 2009), in a collaboration with Drs. Bouarab and Beaulieu from the University of Sherbrooke.

8.3.2 *Cucumber*—Sphaerotheca fuliginea

A commercial formulation of plant extracts from the giant knotweed *Reynoutria sachalinensis* (Milsana®) was shown to effectively control cucumber powdery mildew *Sphaerotheca fuliginea* in the greenhouse. The mode of action was unknown, the extract had no direct effects on the fungus, and previous phenolic analysis did not show accumulation of any free phenolics with antifungal activity against *S. fuliginea*. This was in agreement with the claim in the literature that phytoalexins didn't exist in cucurbits. We investigated this interaction, confirmed that there was no direct toxicity on the fungus causing powdery mildew in cucumber, and deduced that Milsana had an indirect mode of action (induced resistance) (Daayf *et al.*, 1995). However, in studying the involvement of phenolics induced by this biological treatment, we considered both free and conjugated phenolics. After hydrolysis of aqueous extracts, we have demonstrated that *p*-coumaric acid methyl ester acts as a phytoalexin in cucumber (Daayf *et al.*, 1997c). This was the first evidence of phytoalexins in cucumber and in the cucurbits family in general. Further analyses to determine to what extent phenolic compounds played a role in resistance against powdery mildew following the prophylactic treatment with Milsana included the assessment of specific phenolic compounds (*p*-coumaric, caffeic, and ferulic acids and *p*-coumaric acid methyl ester) in susceptible (cultivar Mustang) and tolerant (cultivar Flamingo) cucumber cultivars (Daayf *et al.*, 2000). Levels of these phenolics increased in all treatments (with leaf extracts of *R. Sachalinensis*, powdery mildew, or both) except the control, 1 or 2 days after treatment. In the fraction containing free phenolics, from the tested compounds, only ferulic acid showed an increase in cv. Flamingo (partially resistant), and was particularly evident following treatments. The rapid increase in hydroxycinnamic acids in the two cultivars following Milsana treatment, suggested a role of these compounds in disease reduction, especially that they showed antifungal activity when tested against common pathogens of cucumber (*Botrytis cinerea*, *Pythium ultimum*, and *Pythium aphanidermatum*). However, it was noticeable that methyl esters were more fungitoxic than their corresponding free acids.

In further work with cucumber, Fawe *et al.* (1998) reported the accumulation of flavonoid phytoalexins in cucumber in response to powdery mildew when plants were treated with silicon. We have also demonstrated that phenolic compounds play an important role in the defense of this crop against *P. ultimum* and *P. aphanidermatum* (Ongena *et al.*, 2000).

8.3.3 *Chickpea*—Fusarium oxysporum *f. sp.* ciceris

Chickpea fusariosis resulting from the infection by *Fusarium oxysporum* f. sp. *ciceris* (F.o.c.) is a serious vascular disease that reduces yield potential in commonly grown cultivars. Several efforts have been made to reduce the disease severity using biological control methods. In this context, we have reported that pretreatment of chickpea seedlings with Rhizobium isolates before inoculation with F.o.c. enhanced the accumulation of mRNA transcripts involved in the isoflavonoids pathway. These include PAL, CHS, DFR along with peroxydases and polyphenols oxidases involved in their oxidation. In parallel, HPLC analyses revealed a substantial accumulation of the isoflavones formononetin and biochanin A and their glycoside conjugates in chickpea roots inoculated with Rhizobium isolates and/or challenged with F.o.c., as compared with the controls (Arfaoui *et al.*, 2007).

8.3.4 *Potato*–Verticillium dahliae

Potato Verticillium wilt is among the late season diseases that reduce tuber yield and quality. The pathogen is soil- and seed-borne, with highly resilient resting structures and an extensive list of hosts, making some management practices, such as rotations, inefficient. Recently, we have investigated the responses of potato plants to *V. dahliae*. Like in many Solanaceae plant species, the main secondary metabolites known to play a role in potato defenses against pathogens have been shown to be in the sesquiterpenoid pathway. Several sesquiterpenes have been described in potatoes as phytoalexins, and included rishitin, lubimin, phytoburin, solavetivone, and others (Lyon *et al.*, 1995; Yao *et al.*, 1995). In our investigations, we approached this interaction by first establishing a working model, where we used two potato lines with differential levels of susceptibility to *V. dahliae* and two isolates of the latter with different levels of aggressiveness (Alkher *et al.*, 2009; El-Bebany *et al.*, 2010). Using this model, we have been investigating the expression of several potato defense genes in response to *V. dahliae*, based on their involvement in the defense signaling through salicylic or jasmonic acid (SA, JA) pathways. So far, it has not been clear which of these, or any other signaling networks, is the main pathway involved in plant defenses to *V. dahliae*, although the JA pathway was suggested to be the main one involved in tomato responses to this pathogen. Our findings indicate that both pathways are involved to a certain degree and that the success of plant defense mechanisms in counteracting *V. dahliae*'s infection strategies entails a complex coordination of these defense-signaling networks both in time and in space (H. Derksen *et al.*, unpublished data).

In another study, we selected a set of bacteria and plant extracts that are able to reduce potato *Verticillium* wilt and investigated the mechanisms by which successful biocontrol agents were able to reduce the disease impact (Uppal *et al.*, 2008). Our analyses revealed a high induction of the flavonoid rutin in response to effective treatments, *versus* a low accumulation in response to less effective treatments (El Hadrami *et al.*, 2011). Such a variation in rutin content correlated well with variation in disease expression under both controlled and field conditions.

8.3.5 Potato–**Phytophthora infestans**

Potato late blight, caused by *P. infestans*, is historically notorious because it led to the Irish famine in the 1840s. In Canada, like in other parts of the world, the previously predominant population of *P. infestans* (US-1 genotype) has been replaced by new, more aggressive populations (i.e., US-8 genotype) (Daayf & Platt, 2000). This made late blight disease management much more challenging. Last and this year have shown a resurgence of late blight in many locations across North America, with highly aggressive genotypes on tomato and potato. This makes disease management challenging, especially when the full understanding of the host–pathogen interactions is still lacking.

Both *PAL* and *HMG* have been previously shown to be involved in potato resistance to late blight (Dixon *et al.*, 1992; Yoshioka *et al.*, 1996). *PAL* encodes phenylalanine ammonia lyase, which converts phenylalanine to cinnamic acid, an essential step for the synthesis of a wide variety of phenolic compounds such as flavonoids, isoflavonoids, and lignins (Dixon *et al.*, 1992). *HMG* encodes 3-hydroxy-3-methylglutaryl coenzyme A reductase, which is essential for the biosynthesis of sesquiterpene phytoalexins in potato, i.e., rishitin, solavetivone, and lubimin (Choi *et al.*, 1992).

Initially, we had described the differential expression of *PAL1* and *HMG2* using Northern blot analysis in cultivars Russet Burbank (susceptible) and Kennebec (moderately resistant) upon inoculation with isolates from the US-1 and US-8 genotypes of *P. infestans* (Wang *et al.*, 2004, 2008). The differential expression of these genes was assessed both locally at the infection site and remotely in proximal and distal sites.

Further analysis included the use of quantitative real-time RT-PCR as a more accurate tool to confirm the differential accumulation patterns of several genes in response to the two *P. infestans* isolates (Wang *et al.*, 2008). In parallel, we assessed the accumulation of metabolites deriving from the PAL and HMGR pathways, as a result of their induction by infection. Both *PAL-1* and *HMGR-2* were upregulated earlier in Kennebec than in Russet Burbank and in response to US-1 than to US-8. This was confirmed using both Northern blot and qRT-PCR (Wang *et al.*, 2004, 2008). In addition, the upregulation of these genes was systemic, which is in line with the systemic acquired resistance (SAR) previously described in potatoes (Cohen *et al.*, 1993). However, it was the inhibition of accumulation of these transcripts at the local site of infection, in parallel with their systemic accumulation in uninfected remote sites that was more interesting and that we are describing in the subsequent text.

In a later study, we have also analyzed, *in vivo*, the effects of glucans extracted from *P. infestans*, the elicitor eicosapentanoic acid (EPA) and *P. infestans* isolates on the accumulation of phenolic compounds and the expression level of genes belonging to the phenylpropanoid pathway in the susceptible and the partially resistant potato cultivars "Russet Burbank" and "Defender," respectively. Our results showed a correlation among the variation in the disease level, the expression of PAL (*PAL-1*, *PAL-2*), 4-coumarate:coenzyme A ligase (*4CL*) and chalcone synthase (*CHS*) genes, as well as the accumulation of phenolic compounds. Our findings imply that genetic resistance in potato against *P. infestans* is not the result of isolated reactions against the pathogen, other than the combination of different factors that suggest a polygenic trait or horizontal resistance (Henriquez *et al.*, 2009; Henriquez & Daayf, 2010).

Andreu *et al.* (1998) showed that the addition of EPA and glucan to potato tubers reduced the accumulation of phytoalexins, compared with the potato tuber treated only with EPA. In our investigation, we found that when both treatments, EPA and glucans, were applied on the same plants, followed by the inoculation of the high-aggressive *P. infestans* isolate (US-8), over the 80% of the leaf area in the susceptible cultivar was affected by late blight disease, showing a higher disease severity than the same cultivar infected only with US-8. In addition, a suppression of chlorogenic acid was only observed in this treatment as well as in Russet Burbank inoculated with EPA and glucans, indicating a synergistic effect (Henriquez *et al.*, 2009).

8.3.6 Sunflower–Verticillium dahliae

Verticillium wilt is among the high-yield limiting factors in confectionary and oilseed sunflowers. Most grown hybrids are susceptible or carry a single resistance gene that has recently been overcome, rendering conventional management strategies vain. In this pathosystem, we used two highly- and two weakly aggressive isolates of *V. dahliae* to inoculate moderately resistant and susceptible sunflower hybrids. In parallel, we also used VdNEP (*V. dahliae* necrosis- and ethylene-inducing protein), an elicitor from *V. dahliae*, to infiltrate sunflower plants. VdNEP exhibited a dual role in the interaction between sunflower and *V. dahliae*. It induced wilting symptoms such as chlorosis, necrosis, and vascular discoloration, but also acted as an elicitor, by triggering host defense responses, including the hypersensitive cell death in *Nicotiana benthamiana* leaves and sunflower cotyledons. In addition, it activated the production of reactive oxygen species (i.e., H_2O_2) and the accumulation of fluorescent compounds in sunflower leaves, as well as PR genes (*Ha-PR-3* and *Ha-PR-5*), two defensin genes (*Ha-PDF* and *Ha-CUA1*) and genes encoding *Ha-ACO*, *Ha-CHOX*, *Ha-GST,* and *Ha-SCO*, suggesting that multiple signaling pathways are involved in *V. dahliae*–sunflower interaction (Yao *et al.*, 2011). Two SA-related genes (*Ha-PAL* and *Ha-NML1*) were slightly suppressed after infiltration with VdNEP, suggesting a possible involvement of VdNEP in affecting sunflower defenses (Yao *et al.*, 2011).

The expression of *Ha-PAL* and *Ha-NML1* was lower in VdNEP-treated sunflower leaves 24 and 48 hours after infiltration (h.a.i.) as compared with control plants. The levels of these transcripts kept increasing and decreasing in both treatments, but never exceeded those at time T0 (Yao *et al.*, 2011). This suggested that *V. dahliae* may be interfering with sunflower responses to infection and that VdNEP might be involved in such a mechanism.

8.3.7 *Date palm*–Fusarium oxysporum *f. sp.* albedinis

Date palm (*Phoenix dactylifera*) is an important socio-economical crop in sub-Saharan and arid regions. Bayoud, caused by *F.o.* f. sp. *albedinis*, is a devastating vascular fusariosis that is jeopardizing the cultivation of some of the highly- and well-appreciated cultivars due to susceptibility. During the last two decades, an effort to understand the plant–pathogen interactions has led to examination of the role of induced resistance in the control of bayoud (Ramos *et al.*, 1997; El Hadrami *et al.*, 1998; Daayf *et al.*, 2003; El Hassni *et al.*, 2004; J'aiti *et al.*, 2009). We have shown that hydroxycinnamates and their derivatives play an important differential role between susceptible and moderately resistant cultivars (El Hadrami, 1995; El Hadrami *et al.*, 1997; Ramos *et al.*, 1997; Daayf *et al.*, 2003). These studies revealed the potential of priming date palm resistance using synthetic chemicals, biocontrol agents and antagonists or hypoaggressive isolates of *Fusarium* spp. They also reported on qualitative and quantitative changes in terms of the phenolics pool of primed plant *versus* the nonprimed and subjected to inoculation. An accumulation of nonconstitutive hydroxycinnamic acid derivatives was also reported as strong biochemical markers of resistance. One of the specific compounds is a sinapic derivative known as I2 was shown to exhibit phytoalexin-like properties (Ramos *et al.*, 1997; El Hadrami, 2002). A panel of oxidoreductases was also associated with defense responses. These include peroxidases and polyphenoloxidases (El Hassni *et al.*, 2004; J'aiti *et al.*, 2009).

8.3.8 *Canola*–Leptosphaeria maculans

Canola (*a.k.a.* oilseed rape; *Brassica napus*) is the second largest oilseed crop worldwide (Raymer, 2002) and is well known for the quality and health benefits of its oil. It has been also recognized as an important alternative source for biodiesel and other by-products. The most important fungal disease of this crop is blackleg, also called phoma leaf spot and stem canker (Fitt *et al.*, 2006), which is caused mainly by the pathogenic fungus *Leptosphaeria maculans* (Mendes-Pereira *et al.*, 2003). We have reported that *Leptosphaeria biglobosa*, a related species generally causing much milder blackleg symptoms, may cause severe symptoms on canola cotyledons under certain temperature and humidity conditions (El Hadrami & Daayf, 2009). Concurrently to severe symptoms under high relative humidity (RH) conditions, canola plants synthesized less hydroxycinnamates and eventually accumulated less lignin in their cell walls, as compared with plants incubated under ambient RH conditions (El Hadrami & Daayf, 2009). This suggested that such hydroxycinnamates were important in the canola response to blackleg.

In a later investigation, we exploited the fact that *L. biglobosa* and *L. maculans* can co-habit on canola for a limited period of time during the growing season and the competition between the two species to characterize the host responses to isolates from both species used individually or subsequently in the infection. These resulted in the finding that hydroxycinnamates were the major phenolics that accumulate differentially in response to either pre- or coinoculation involving a weakly, then a highly aggressive isolate of the pathogen. Taking into consideration that these hydroxycinnamates are precursors for the synthesis of lignin and phenylamide phytoalexins, we suggested that their detection in response to canola priming with a weakly aggressive isolate might explain the reduction of the disease normally induced by further inoculation with highly aggressive isolates (El Hadrami *et al.*, 2010) and pinpointed to their potential exploitation in further disease management strategies in this crop.

8.3.9 Saskatoons–Entomosporium mespili

Amelanchier alnifolia (Saskatoon berries tree) is an important fruit shrub to the economy of the Canadian prairies and is yearly threatened by Entomosporium leaf and berry spots. The pathogenic fungus *Entomosporium mespili* that causes this disease is difficult to control using conventional methods and yield losses are continuously recorded. Recently, we explored the potential implementation of new strategies relying on the use of resistance inducers such as SA, JA, and Canada milkvetch extract. We showed using two Saskatoon cultivars, Smoky and Martin, that JA and CMV extract enhanced plant defense responses, especially a differential induction of the synthesis/accumulation of PAL (Wolski *et al.*, 2010). In parallel, pretreatment of the leaves with these inducers significantly reduced the disease levels especially when applied to the moderately tolerant cultivar Martin. Analysis of the soluble phenolics pool revealed an accumulation of various derivatives of hydroxycinnamic acids and proanthocyanidins that was in line with the observed disease levels on the tested cultivars. The gathered results in this preliminary investigation allowed us to set the stage for a better understanding of the host–pathogen interactions and determine biomarkers of resistance to be targeted in further investigations.

8.4 Pathogens counter-defense against plants' phenolic-related defenses

Phenolics play a pivotal role in disease resistance in many plant species. However, in parallel, several pathogens have developed adaptive strategies to counteract plant defenses by targeting phenolics. Gathering such information is essential in searching for knowledge-based methods to manage plant diseases.

Counteracting plant defense responses plays an important role in plant–microbe interactions, especially those involving biotrophic and hemi-biotrophic pathogens, such as *P. infestans*, which require living plant tissues to establish a successful infection (Heath, 2000). While plant-defense suppressors have been well studied in both viral and bacterial plant pathogens (He *et al.*, 2006; Zhang *et al.*, 2006), they are not as well understood in fungal and oomycete pathogens. The production of defense suppressors has been reported for many fungi, including *Mycosphaerella pinodes* (Shiraishi *et al.*, 1978) and *P. infestans* (Doke *et al.*, 1979), but much remains to be done to establish a clear understanding of how those and other suppressors function. Suppressors can inhibit phytoalexin production and may be key pathogenicity factors for the fungi/oomycetes producing them. Several fungal species have also evolved mechanisms to detoxify phytoalexins (Shiraishi *et al.*, 1978; Yoshioka *et al.*, 1990). Other mechanisms where suppressors inhibit plant defenses at the gene transcription level have also been under investigation.

8.4.1 **Phytophthora infestans**

Suppression of potato defenses by *P. infestans* has been suggested for several decades and glucans (Andreu *et al.*, 1998), and other molecules (Tian *et al.*, 2004) from the pathogen have been suggested to play such a role.

Using Northern blot (Wang *et al.*, 2004) and real-time RT-PCR (Wang *et al.*, 2008) analyses, we showed that the highly aggressive *P. infestans* US-8 reduced the level of potato defense genes' expression in potato much more than the mildly aggressive US-1 (Wang *et al.*, 2004). These results strongly suggested the suppression of potato *pal1* and *hmgr2* gene expression, leading to a reduction in phytoalexins, as a strategy used by *P. infestans*, especially US-8 genotype, to counteract potato defense mechanisms. Besides the glucans suggested by Andreu *et al.* (1998), other potential counter-defense molecules from *Phytophthora* species have been suggested recently, including glucanase inhibitor proteins and a protease inhibitor (EPI1) produced by *Phytophthora sojae* and *P. infestans*, respectively (Rose *et al.*, 2002; Tian *et al.*, 2005).

8.4.2 **Verticillium dahliae**

PAL, a key enzyme in the phenylpropanoid pathway, is involved in the production of plant phenolic compounds such as salicylic acid (SA). *NML1* (*NIM1*/*NPR1*-like 1) shares high sequence similarity with *NPR1*, which acts downstream of SA in the SAR signal transduction pathway (Cao *et al.*, 1994). SA pretreatment did not enhance resistance of *Arabidopsis* against *V. dahliae* infection compared with the wild-type plants (Veronese *et al.*, 2003). However, the slight downregulation of *Ha-PAL* and *Ha-NML1* may suggest that sunflower resistance against *Verticillium* wilt might not be completely SA-independent. Actually, if PAL and NML are downregulated, this may suggest that they were important in sunflower resistance to *V. dahliae* until the pathogen became able to reduce their activity. This might also explain why JA/ET-related genes are upregulated as a plant response to the pathogen counter-defenses. This said that VdNEP is only one among several other potential effectors that need to be scrutinized individually and as a group to complete the bigger picture of sunflower–*V. dahliae* interaction mechanisms.

It is well known that host root exudates stimulate microsclerotia germination at the precontact early stage of host–*V. dahliae* interactions; in another study, we have used the differential potato–*V. dahliae* model, mentioned earlier, to identify the differentially expressed genes in *V. dahliae* after elicitation by root extracts from either resistant or susceptible potato cultivars (El-Bebany *et al.*, 2011). We identified stress response regulator (*SrrA*) gene differentially expressed in the highly aggressive *V. dahliae* isolate after exposure to either root extracts (El-Bebany *et al.*, 2011). At the proteomics level, we have established and compared the proteomic map (pH 4-7) of the two, highly and weakly aggressive, *V. dahliae* isolates. Thioredoxin and NADH-ubiquinone oxidoreductase 29.9 kDa subunit proteins were differentially expressed in the highly aggressive isolate (El-Bebany *et al.*, 2010). Stress response regulator (*SrrA*), thioredoxin and NADH-ubiquinone oxidoreductase 29.9 kDa subunit have been reported to play a role in stress signal transduction and oxidative burst tolerance in yeast and several pathogenic fungi (Trotter & Grant, 2002; Kim & Kim, 2006; Sellam *et al.*, 2007; Vargas-Perez *et al.*, 2007). Reactive oxygen species (ROS) are known for triggering plant defense responses including production of phytoalexins. That means these genes/proteins may help *V. dahliae* tolerating host's ROS toxicity and/or overcoming plant defense responses and, consequently, suppressing phenolics production (El-Bebany *et al.*, 2011).

Another potential plant defense suppressor, isochorismatase hydrolase, was found differentially expressed in the highly aggressive through the comparative proteomic analysis of the two *V. dahliae* isolates (El-Bebany *et al.*, 2010). SA is biosynthesized in plants via phenylpropanoid and isochorismate pathways (Wildermuth *et al.*, 2001). The role of isochorismatase family in suppressing plant defenses by hydrolyzing isochorismate and interfering SA biosynthesis was hypothesized to occur by several pathogenic fungi (Soanes *et al.*, 2008). This potential suppression in signaling plant defenses may help explain the findings by Yao *et al.* (2011) and H. Derksen *et al.* (unpublished data); however, differences in defense responses from host to another always should be considered.

All together, these genes/proteins may contribute to pathogens' counter-defense against plant defense mechanisms and help pathogens to establish a successful infection. However, more investigations of these factors are underway through generating overexpression and knock-out mutants of these factors in the weakly and highly aggressive isolates, respectively, then profile their expression during *V. dahliae* interactions with an array of host plants.

8.5 Concluding remarks

Phenolics are at the forefront of plant defenses against pathogens. They are involved in many processes that restrict pathogen growth and development. We have shown their contribution in plant defenses against foliar and root pathogens, including biotrophs, necrotrophs, and hemibiotrophs. Such involvement was either directly induced in response to pathogenic attacks or reacting to a third party such as biological control agents. We have also shown phenolics' involvement in reactions in different parts of the plant, including leaves, stems, and roots. Such reactions were either local or systemic. However, it was even more interesting to note that pathogens have evolved strategies that allow them to cope with the toxicity/inhibitory effect of phenolics. Recent attention to these facts allowed the discovery of new mechanisms of plant defense suppression. Several examples illustrated earlier add to the body of knowledge and increase our understanding about plant defenses and pathogen counter-defenses. The mechanisms that have been denoted include the suppression of essential transcripts involved in the biosynthesis of defense-related metabolites, detoxification of phytoalexins, and interference with key signaling regulatory pathways. More studies are still needed to unravel the relationships among these networks that warrant the outcome of such interactions. However, it is clear that the warfare between plants and pathogens will continue to evolve possibly through new mechanisms yet to be discovered.

Acknowledgments

We acknowledge funding from the Natural Sciences and Engineering Research Council of Canada (NSERC), Groupe Polyphénols, Manitoba Agri-Food Research and Development Initiative (ARDI), the University of Manitoba, the Manitoba Rural Adaptation Council (MRAC), the Canola Council of Canada, Keystone Potato Producers Association, Connery Vegetable Producers, Dow AgroSciences Canada, Jeffries Vegetable Growers, Cargill Ltd., the National Sunflower Association of Canada, Jamors Producers, McCain Foods, and the Prairie Fruit Growers Association.

References

Abramovitch, R.B. & Martin, G.B. (2004) Strategies used by bacterial pathogens to suppress plant defenses. *Current Opinion in Plant Biology*, **7**, 356–364.

Alkher, H., El Hadrami, A., Rashid, K.Y., Adam, L.R. & Daayf, F. (2009) Cross-pathogenicity of *Verticillium dahliae* between potato and sunflower. *European Journal of Plant Pathology*, **124**, 505–519.

Andreu, A., Tonón, C., Van Damme, M., Van Damme, M., Huarte, M. & Daleo, G. (1998) Effect of glucans from different races of *Phytophthora infestans* on defense reactions in potato tuber. *European Journal of Plant Pathology*, **104**, 777–783.

Arfaoui, A., El Hadrami, A., Mabrouk, Y., *et al.* (2007) Treatment of chickpea with Rhizobium isolates enhances the expression of phenylpropanoid defense-related genes in response to infection by *Fusarium oxysporum* f. sp. *ciceris*. *Plant Physiology and Biochemistry*, **45**, 470–479.

Bennett, R.N. & Wallsgrove, R.M. (1994) Secondary metabolites in plant defence mechanisms. *New Phytologist*, **127**, 617–633.

Birch, P.R.J., Avrova, A.O., Armstrong, M., *et al.* (2003) The potato-*Phytophthora infestans* interaction transcriptome. *Canadian Journal of Plant Pathology*, **25**, 226–231.

Bouarab, K., Melton, R., Peart, J., Baulcombe, D. & Osbourn, A. (2002) Fungal pathogenesis: a saponin-detoxifying enzyme mediates suppression of plant defences. *Nature*, **418**, 889–892.

Cao, H., Bowling, S.A., Gordon, S. & Dong, X. (1994) Characterization of an Arabidopsis mutant that is nonresponsive to inducers of systemic acquired resistance. *The Plant Cell*, **6**, 1583–1592.

Carr, J.P. & Klessig, D.F. (1989) The pathogenesis-related proteins of plants. In: *Genetic Engineering, Principles and Methods* (ed. J.K. Setlow), pp. 65–89. Plenum Press, New York.

Catinot, J., Buchala, A., Abou-Mansour, E. & Métraux, J.P. (2008) Salicylic acid production in response to biotic and abiotic stress depends on isochorismate in *Nicotiana benthamiana. FEBS Letters*, **582**, 473–478.

Choi, D., Ward, B.L. & Bostock, R.M. (1992) Differential induction and suppression of potato 3-hydroxy-3-methylglutaryl coenzyme A reductase genes in response to *Phytophthora infestans* and to its elicitor arachidonic acid. *Plant Cell*, **4**, 1333–1344.

Cohen, Y., Gisi, U. & Niderman, T. (1993) Local and systemic protection against *Phytophthora infestans* induced in potato and tomato plants by jasmonic acid and jasmonic-methyl-ester. *Phytopathology*, **83**, 1054–1062.

Daayf, F., Bel-Rhlid, R. & Bélanger, R.R. (1997a) Methyl ester of p-coumaric acid, a phytoalexin-like compound from long English cucumber leaves. *Journal of Chemical Ecology*, **23**, 1517–1526.

Daayf, F., El Bellaj, M., El Hassni, M., J'Aiti, F. & El Hadrami, I. (2003) Elicitation of soluble phenolics in date palm *(Phoenix dactylifera)* callus by *Fusarium oxysporum* f. sp. *albedinis* culture medium. *Environmental and Experimental Botany*, **49**, 41–47.

Daayf, F., Nicole, M., Boher, B., Pando-Bahuon, A. & Geiger, J.P. (1997b) Early defense reactions of cotton (*Gossypium sp.*) to *Verticillium dahliae* Kleb. *European Journal of Plant Pathology*, **103**, 125–136.

Daayf, F., Ongena, M., Boulanger, R., El hadrami, I. & Bélanger, R. (2000) Induction of phenolic compounds in two cultivars of cucumber by treatment of treatment of healthy and powdery mildew-infected plants with extracts of *Reynoutria sachalinensis. Journal of Chemical Ecology*, **26**, 1579–1593.

Daayf, F. & Platt, H.W. (2000) Changes in metalaxyl resistance among glucose phosphate isomerase genotypes of *Phytophthora infestans* in Canada during 1997–1998. *American Journal of Potato Research*, **77**, 311–318.

Daayf, F., Schmitt, A. & Bélanger, R.R. (1995) The effects of plant extracts of *Reynoutria sachalinensis* on powdery mildew development and leaf physiology of long English cucumber. *Plant Disease*, **79**, 577–580.

Daayf, F., Schmitt, A. & Bélanger, R.R. (1997c) Evidence of phytoalexins in cucumber leaves infected with powdery mildew following treatment with leaf extracts of *Reynoutria sacchalinensis. Plant Physiology*, **113**, 719–727.

Dangl, J.L. & Jones, J.D.G. (2001) Plant pathogens and integrated defence responses to infection. *Nature*, **411**, 826–833.

Dixon, R.A., Choudhary, A.D., Dalkin, D., *et al.* (1992) Molecular biology of stress induced phenylpropanoid and isoflavonoid biosynthesis in alfalfa. In: *Phenolic Metabolism in Plants* (eds. H.A. Stafford & R.K. Ibrahim), pp. 91–138. Plenum Press, New York.

Dixon, R.A. & Paiva, N.L. (1995) Stress-induced phenylpropanoid metabolism. *Plant Cell*, **7**, 1085–1097.

Doke, N., Garas, N.A. & Kuc, J. (1979) Partial characterization and aspects of the mode of action of a hypersensitivity-inhibiting factor (HIF) isolated from *Phytophthora infestans. Physiological Plant Pathology*, **15**, 117–126.

El-Bebany, A.F., Henriquez, M.A., Badawi, M., Adam, L.R., El Hadrami, A. & Daayf, F. (2011) Induction of putative pathogenicity-related genes in *Verticillium dahliae* in response to elicitation with potato root extracts. *Environmental and Experimental Botany*, **72**, 251–257.

El-Bebany, A.F., Rampitsch, C. & Daayf, F. (2010) Proteomic analysis of the phytopathogenic soilborne fungus *Verticillium dahliae* reveals differential protein expression in isolates that differ in aggressiveness. *Proteomics*, **10**, 289–303.

El Hadrami, A., Adam, L.R. & Daayf, F. (2011) Biocontrol treatments confer protection against *Verticillium dahliae* infection of potato by inducing anti-microbial metabolites. *Molecular Plant-Microbe Interactions*, **24**, 328–335.

El Hadrami, A. & Daayf, F. (2009) Priming canola resistance to blackleg with weakly aggressive isolates leads to an activation of hydroxycinnamates. *Canadian Journal of Plant Pathology*, **31**, 393–406.

El Hadrami, A., El Hadrami, I. & Daayf, F. (2009) Suppression of induced plant defense responses by fungal and oomycete pathogens. In: *Molecular-Plant Microbe Interactions* (eds. K. Bouarab, N. Brisson & F. Daayf), pp. 231–268. CABI, UK.

El Hadrami, A., Fernando, W.G.D. & Daayf, F. (2010) Variations in relative humidity modulate *Leptosphaeria* spp. pathogenicity and interfere with canola mechanisms of defence. *European Journal of Plant Pathology*, **126**, 187–202.

El Hadrami, A., Kone, D. & Lepoivre, P. (2005) Effect of juglone on active oxygen species and antioxidants in susceptible and partial resistant banana cultivars to Black Leaf Streak Disease. *European Journal of Plant Pathology*, **113**, 241–254.

El Hadrami, I. (1995) L'embryogenèse somatique chez *Phoenix dactylifera* L. Quelques facteurs limitants et marqueurs biochimiques. Thèse de Doctorat d'Etat, Faculté des Sciences Semlalia, Université Cadi Ayyad. Marrakech, Morocco.

El Hadrami, I. (2002) Infections racinaires localisées et rôle des dérivés hydroxycinnamiques dans la résistance du Palmier dattier (*Phoenix dactylifera* L.) au *Fusarium oxysporum albedinis*, agent causal du bayoud. *Polyphénols Actualités*, **22**, 19–27.

El Hadrami, I., El Bellaj, M., El Idrissi, A., J'Aiti, F., El Jaafari, S. & Daayf, F. (1998) Biotechnologies végétales et amélioration du Palmier dattier (*Phoenix dactylifera* L.), pivot de l'agriculture oasienne marocaine. *Cahiers Agricultures*, **7**, 463–468.

El Hadrami, I., Ramos, T., El Bellaj, M., El Idrissi Tourane, A. & Macheix, J.J. (1997) A sinapic derivative as induced defence compound of date palm against *Fusarium oxysporum* f. sp. *albedinis*, the agent causing bayoud disease. *Journal of Phytopathology*, **145**, 329–333.

El Hassni, M., J'Aiti, F., Dihazi, A., Baraka, E.A., Daayf, F. & El Hadrami, I. (2004) Enhancement of defence responses against bayoud disease by treatment of date palm seedlings with an hypoaggressive Fusarium oxysporum isolate. *Journal of Phytopathology*, **152**, 182–189.

Fawe, A., Abou-Zaid, M., Menzies, J.G. & Bélanger, R.R. (1998) Silicon-mediated accumulation of flavonoid phytoalexins in cucumber. *Phytopathology*, **88**, 396–401.

Fitt, B.D.L., Brun, H., Barbetti, M.J. & Rimmer, S.R. (2006) World-wide importance of phoma stem canker (*Leptosphaeria maculans* and *L. biglobosa*) on oilseed rape (*Brassica napus*). *European Journal of Plant Pathology*, **114**, 3–15.

Ham, K.S., Wu, S.C., Darvill, A.G. & Albersheim, P. (1997) Fungal pathogens secrete an inhibitor protein that distinguishes isoforms of plant pathogenesis-related endo-beta-1, 3-glucanases. *Plant Journal*, **11**, 169–179.

Hammerschmidt, R. (1999) Phytoalexins: what have we learned after 60 years? *Annual Review of Phytopathology*, **37**, 285–306.

Hammond-Kosack, K.E. & Jones, J.D.G. (1996) Resistance gene-dependent plant defense responses. *Plant Cell*, **8**, 1773–1791.

Harborne, J.B. (1999) The comparative biochemistry of phytoalexin induction in plants. *Biochemical Systematics and Ecology*, **27**, 335–367.

Hausladen, A. & Stamler, J.S. (1998) Nitric oxide in plant immunity. *Proceedings of the National Academy of Sciences of the United States of America*, **95**, 10345–10347.

He, P., Shan, L., Lin, N.C., *et al.* (2006) Specific bacterial suppressors of MAMP signaling upstream of MAPKKK in Arabidopsis innate immunity. *Cell*, **125**, 563–375.

Heath, M.C. (2000) Advances in imaging the cell biology of plant-microbe interactions. *Annual Review of Phytopathology*, **38**, 443–459.

Henriquez, M.A. & Daayf, F. (2010) Identification and cloning of differentially expressed genes involved in the interaction between potato and *Phytophthora infestans* using a subtractive hybridization and cDNA-AFLP combinational approach. *Journal of Integrative Plant Biology*, **52**, 453–467.

Henriquez, M.A., Wolski, E.A., Molina, O.I., Adam,, L.R., Andreu,, A.B. & Daayf, F. (2009) Functional genomics analysis of *Verticillium dahliae* towards identification of its pathogenicity-related genes. *XIV International Congress on Molecular Plant-Microbe Interactions.* Québec City, PQ, July 19–23.

J'Aiti, F., Verdeil, J.L. & El Hadrami, I. (2009) Effect of jasmonic acid on the induction of polyphenoloxidase and peroxidase activities in relation to date palm resistance against *Fusarium oxysporum* f. sp. *albedinis. Physiological and Molecular Plant Pathology*, **74**, 84–90.

Kim, J.H. & Kim, D.H. (2006) Cloning and characterization of a thioredoxin gene, CpTrx1, from the chestnut blight fungus *Cryphonectria parasitica. Journal of Microbiology*, **44**, 556–561.

Kunkel, B.N. & Brooks, D.M. (2002) Cross talk between signaling pathways in pathogen defense. *Current Opinion in Plant Biology*, **5**, 325–331.

Lerat, S., Babana, A., El Oirdi, M., *et al.* (2009) *Streptomyces scabies* and its toxin thaxtomin A induce scopoletin biosynthesis in tobacco *and Arabidopsis thaliana. Plant Cell Reports*, **28**, 1895–1903.

Lyon, G.D., Reglinski, T. & Newton, A.C. (1995) Novel disease control compounds: the potential to immunize plants against infection. *Plant Pathology*, **44**, 407–427.

Mendes-Pereira, E., Balesdent, M.H., Brun, H. & Rouxel, T. (2003) Molecular phylogeny of the *Leptosphaeria maculans – L. biglobosa* species complex. *Mycological Research*, **107**, 1287–1304.

Nandi, A., Kachroo, P., Fukushige, H., Hildebrand, D., Klessig, D.F. & Shah, J. (2003) Ethylene and jasmonic acid signaling pathways affect NPR1-independent expression of defense genes without impacting resistance to *Pseudomonas syringae* and *Peronospora parasitica* in the Arabidopsis ssi1 mutant. *Molecular Plant-Microbe Interactions*, **16**, 588–599.

Nawrath, C. & Métraux, J.P. (1999) Salicylic acid induction-deficient mutants of Arabidopsis express PR-2 and PR-5 and accumulate high levels of camalexin after pathogen inoculation. *Plant Cell*, **11**, 1393–1404.

Nicholson, R.L. & Hammerschmidt, R. (1992) Phenolic compounds and their role in disease resistance. *Annual Review of Phytopathology*, **30**, 369–389

Ongena, M., Daayf, F., Jacques, P., *et al.* (1999) Protection of cucumber against Pythium root rot by fluorescent pseudomonads: predominant role of induced resistance over siderophores and antibiosis. *Plant Pathology*, **48**, 66–76.

Ongena, M., Daayf, F., Jacques, P., *et al.* (2000) Systemic induction of phytoalexins in cucumber in response to treatments with fluorescent pseudomonads. *Plant Pathology*, **49**, 523–530.

Osbourn, A. (1996) Saponins and plant defence – a soap story. *Trends in Plant Science*, **1**, 4–9.

Pieterse, C.M.J., Ton, J. & van Loon, L.C. (2001) Cross-talk between plant defence signaling pathways: boost or burden? *AgBiotechNet,* **3**, ABN 068.

Ramos, T., El Bellaj, M., El Idrissi-Tourane, A., Daayf, F. & El Hadrami, I. (1997) Phenolamides of palm rachis, components of defense reaction of date palm against *Fusarium oxysporum* f. sp. *albedinis*, the causal agent of Bayoud. *Journal of Phytopathology*, **145**, 487–493.

Raymer, P.L. (2002) Canola: an emerging oilseed crop. In: *Trends in New Crops and New Uses* (eds. J. Janick & A. Whipkey), pp. 122–126. ASHS Press, Alexandria, VA.

Romero-Puertas, M.C., Perazzolli, M., Zago, E.D. & Delledonne, M. (2004) Nitric oxide signaling functions in plant-pathogen interactions. *Cellular Microbiology*, **6**, 795–803.

Ros, B., Thümmler, F. & Wenzel, G. (2004) Analysis of differentially expressed genes in a susceptible and moderately resistant potato cultivar upon *Phytophthora infestans* infection. *Molecular Plant Pathology*, **5**, 191–201.

Rose, J.K.C., Ham, K.-S., Darvill, A.G. & Albersheim, P. (2002) Molecular cloning and characterization of glucanase inhibitor proteins: coevolution of a counterdefense mechanism by plant pathogens. *Plant Cell*, **14**, 1329–1345.

Sellam, A., Dongo, A., Guillemette, T., Hudhomme, P. & Simoneau, P. (2007) Transcriptional responses to exposure to the brassicaceous defence metabolites camalexin and allylisothiocyanate in the necrotrophic fungus *Alternaria brassicicola. Molecular Plant Pathology*, **8**, 195–208.

Shiraishi, T., Oku, H., Yamashita, M. & Ouchi, S. (1978) Elicitor and suppressor of pisatin induction in spore germination fluid of pea pathogen, *Mycosphaerella pinodes*. *Annals of the Phytopathological Society of Japan*, **44**, 659–665.

Soanes, D.M., Alam, I., Cornell, M., *et al.* (2008) Comparative genome analysis of filamentous fungi reveals gene family expansions associated with fungal pathogenesis. *PLoS ONE*, **3**, e2300.

Soby, S., Caldera, S., Bates, R. & VanEtten, H. (1996) Detoxification of the phytoalexins maackiain and medicarpin by fungal pathogens of alfalfa. *Phytochemistry*, **41**, 759–765.

Stein, B.D., Klomparens, K.L. & Hammerschmidt, R. (1993) Histochemistry and ultrastructure of the induced resistance response of cucumber plants to *Colletotrichum lagenarium*. *Journal of Phytopathology*, **137**, 177–188.

Tian, M., Benedetti, B. & Kamoun, S. (2005) A second Kazal-like protease inhibitor from *Phytophthora infestans* inhibits and interacts with the apoplastic pathogenesis-related protease P69B of tomato. *Plant Physiology*, **38**, 1785–1793.

Tian, M., Huitema, E., da Cunha, L., Torto-Alalibo, T. & Kamoun, S. (2004) A Kazal-like extracellular serine protease inhibitor from *Phytophthora infestans* targets the tomato pathogenesis-related protease P69B. *The Journal of Biological Chemistry*, **279**, 26370–26377.

Treutter, D. (2006) Significance of flavonoids in plant resistance: a review. *Environmental Chemistry Letters*, **4**, 147–157.

Trotter, E.W. & Grant, C.M. (2002) Thioredoxins are required for protection against a reductive stress in the yeast *Saccharomyces cerevisiae*. *Molecular Microbiology*, **46**, 869–878.

Uppal, A.K., El Hadrami, A., Adam, L.R., Tenuta,, M. & Daayf,, F. (2008) Biological control of potato Verticillium wilt under controlled and field conditions using selected bacterial antagonists and plant extracts. *Biological Control*, **44**, 90–100.

van Loon, L.C. & van Strien, E.A. (1999) The families of pathogenesis-related proteins, their activities, and comparative analysis of PR-1 type proteins. *Physiological and Molecular Plant Pathology*, **55**, 85–97

Vargas-Perez, I., Sanchez, O., Kawasaki, L., Georgellis, D. & Aguirre, J. (2007) Response regulators SrrA and SskA are central components of a phosphorelay system involved in stress signal transduction and asexual sporulation in *Aspergillus nidulans*. *Eukaryotic Cell*, **6**, 1570–1583.

Veronese, P., Narasimhan, M.L., Stevenson, R.A., *et al.* (2003) Identification of a locus controlling Verticillium disease symptom response in Arabidopsis thaliana. *Plant Journal*, **35**, 574–587.

Vogel, J. & Somerville, S. (2000) Isolation and characterization of powdery mildew-resistant *Arabidopsis* mutants. *Proceedings of the National Academy of Sciences of the United States of America of the United States of America*, **97**, 1897–1902.

Walker, J.C. & Stahmann, M.A. (1955) Chemical nature of disease resistance in plants. *Annual Review of Plant Physiology*, **6**, 351–366.

Wang, X., El Hadrami, A., Adam, L.R. & Daayf, F. (2004) US-1 and US-8 genotypes of *Phytophthora infestans* differentially affect local, proximal and distal gene expression of phenylalanine ammonia-lyase and 3-hydroxy, 3-methylglutaryl CoA reductase in potato leaves. *Physiological and Molecular Plant Pathology*, **65**, 157–167.

Wang, X., El Hadrami, A., Adam, L.R. & Daayf, F. (2005) Genes encoding pathogenesis-related proteins PR-2, PR-3 and PR-9, are differentially regulated in potato leaves inoculated with isolates from US-1 and US-8 genotypes of *Phytophthora infestans* (Mont.) de Bary. *Physiological and Molecular Plant Pathology*, **67**, 49–56.

Wang, X., El Hadrami, A., Adam, L.R. & Daayf, F. (2006) Local and distal gene expression of pr-1 and pr-5 in potato leaves inoculated with isolates from the old (US-1) and the new (US-8) genotypes of *Phytophthora infestans* (Mont.) de Bary. *Environmental and Experimental Botany*, **57**, 70–79.

Wang, X., El Hadrami, A., Adam,, L.R. & Daayf,, F. (2008) Differential activation and suppression of potato defence responses by *Phytophthora infestans* isolates representing US-1 and US-8 genotypes. *Plant Pathology*, **57**, 1026–1037.

Weltring, K.-M., Schaub, H.-P. & Barz, W. (1995) Metabolism of pisatin stereoisomers by *Ascochyta rabiei* strains transformed with the pisatin demethylase gene *of Nectria haematococca* MP VI. *Molecular Plant-Microbe Interaction*, **8**, 499–505.

Wildermuth, M.C., Dewdney, J., Wu, G. & Ausubel, F.M., (2001) Isochorismate synthase is required to synthesize salicylic acid for plant defence. *Nature*, **414**, 562–565.

Wolski, E.A., Henriquez, M.A., Adam, L.R., *et al.* (2010) Induction of defense genes and secondary metabolites in saskatoons (*Amelanchier alnifolia* Nutt.) in response to *Entomosporium mespili* using jasmonic acid and Canada milk vetch extracts. *Environmental and Experimental Botany*, **68**, 273–282.

Yao, K., De Luca, V. & Brisson, N. (1995) Creation of a Metabolic Sink for Tryptophan Alters the Phenylpropanoid Pathway and the Susceptibility of Potato to *Phytophthora infestans*. *Plant Cell*, **7**, 1787–1799.

Yao, Z., Rashid, K.Y., Adam, L.R. & Daayf, F. (2011) *Verticillium dahliae*'s VdNEP acts both as a plant defense elicitor and a pathogenicity factor in the interaction with *Helianthus annuus*. *Canadian Journal of Plant Pathology*, **33**, 375–388.

Yoshioka, H., Miyabe, M., Hayakawa, Y. & Doke, N. (1996) Expression of genes for phenylalanine ammonia-lyase and 3-hydroxy-3-methylglutaryl CoA reductase in aged potato tubers infected with *Phytophthora infestans*. *Plant Cell Physiology*, **37**, 81–90.

Yoshioka, H., Shiraishi, T., Yamada, T., Ichinose, Y. & Oku, H. (1990) Suppression of pisatin production and ATPase activity in pea plasma membranes by orthovanadate, verapamil and a suppressor from *Mycosphaerella pinodes*. *Plant and Cell Physiology*, **31**, 1139–1146.

Zhang, X., Yuan, Y.R., Pei, Y., Lin, S.S., Tuschl, T., Patel, D.J. & Chua, N.H. (2006) *Cucumber mosaic virus*-encoded 2b suppressor inhibits *Arabidopsis* Argonaute1 cleavage activity to counter plant defense. *Genes and Development*, **20**, 3255–3268.

Chapter 9
Absorption and Metabolism of Dietary Chlorogenic Acids and Procyanidins

Gary Williamson and Angelique Stalmach

Abstract: Chlorogenic acids and procyanidins are commonly consumed polyphenols. Chlorogenic acids are present in the diet in many foods including coffee and fruits, and procyanidins are found in many fruits and in cocoa. They are poorly absorbed in their intact forms, but their catabolites, after metabolism by colonic microbiota, are very efficiently absorbed. The catabolites include dihydrocinnamic and other phenolic acids, which appear in the plasma later than the intact parent compounds. These catabolites have additional biological activities, which may even exceed that of the intact parent compounds *in vivo*.

Keywords: polyphenol; procyanidin; microbiota; chlorogenic acid; phenolic acid; catabolism; colon; bioavailability; metabolism

9.1 Introduction

Although many biological activities have been proposed for polyphenols, phenolic acids and tannins (PPT), efficacy *in vivo* depends on their bioavailability. Understanding the pathways of metabolism of polyphenols can lead to new insights into bioefficacy and identification of potential molecular candidates responsible for the observed effects. There are several ways to measure and report the bioavailability of PPT. Animal and cellular studies give mechanistic information, but the only way to determine the amount absorbed relevant for human nutrition is to measure absorption and excretion in humans. Measuring bioavailability of PPT in humans is most commonly by single-dose post-prandial measurements in blood or urine from healthy volunteers. After consumption of a single dose of PPT, either pure or in a food, the concentration in blood with time and the amount excreted in urine are measured. The rate and extent of small intestinal absorption and pre-systemic metabolism are the most important parameters controlling the early phases of the plasma concentration time curve, whereas the colonic microbiota are important for metabolism into lower molecular

Recent Advances in Polyphenol Research, Volume 3, First Edition. Edited by Véronique Cheynier, Pascale Sarni-Manchado and Stéphane Quideau.

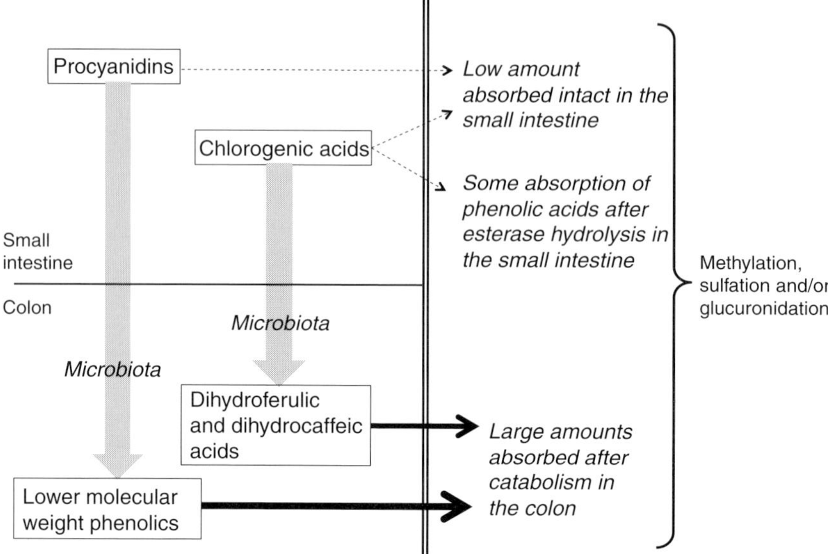

Fig. 9.1   Relationship between metabolism of procyanidins and chlorogenic acids.

weight compounds which are then absorbed. The latter leads to different compounds in plasma compared to those that were consumed. This review examines recent advances in understanding the absorption and metabolism of two classes of PPT, the chlorogenic acids and the procyanidins. Both classes are absorbed to only a very small extent in the intact form, but catabolites are extensively absorbed after microbial transformation in the colon (Fig. 9.1).

9.2 Procyanidins

Proanthocyanidins are found in fruits, bark, leaves and seeds of many plants, and in foods such as fruits and berries, nuts, beans, some cereals, beverages such as wine and beer, and at high levels in cocoa and dark chocolate (Gu *et al.*, 2004). The procyanidins, possessing 3′,4′-dihydroxyl groups on the B-ring, are the most abundant proanthocyanidins in plants, consisting of oligomers of two or more (epi)catechin units. The monomeric units of proanthocyanidins are most commonly linked by C–C bonds generally between the C(4)-position of the 'upper' flavonoid and the C(8)-position of the 'lower' flavonoid as in procyanidin B1–B4 (Pascual-Teresa *et al.*, 2000). A good example of a food source of procyanidins is cocoa, which, apart from monomeric (–)-epicatechin, contains dimeric to decameric (or larger) oligomers (Hammerstone *et al.*, 1999). Commonly consumed food, such as dark chocolate, apples, red wine and cranberry juice, contain large amounts of procyanidins (Hammerstone *et al.*, 2000).

Consumed procyanidins pass through the digestive tract and at best trace amounts may be absorbed intact in the small intestine (minor pathway) but most reach the colon where prior to absorption they are degraded to simpler phenolic structures by the gut microbiota (major pathway).

9.2.1 *Gut lumen stability of procyanidins*

Procyanidins bind to salivary proteins, which may lead to pH-dependent protein precipitation (Bacon & Rhodes, 2000; de Freitas & Mateus, 2001). The intra-flavanol linkage and the structural monomeric units comprising procyanidins strongly influence this precipitation, whereas the degree of polymerisation has a much smaller effect. Oligomeric cocoa procyanidins at pH 2 were degraded into monomeric and dimeric forms. The trimeric to hexameric forms were more rapidly degraded than the dimers, and this degradation was pepsin independent (Spencer *et al.*, 2000). Procyanidin dimers B2 and B5 are unstable at pH 1.8 (Zhu *et al.*, 2000). However, neither of these *in vitro* studies took into account the presence of protein and other food-derived compounds in the stomach, and the buffering capacity of food which increases the pH after the stomach has received food. In humans, *in vivo,* after consumption of a cocoa beverage, the stomach pH increased to 5.4, and did not return to pH 2 until 45 minutes later when the stomach was empty. Direct stomach sampling established that the procyanidins are stable in the stomach (Rios *et al.*, 2002) and that most procyanidins pass through the stomach and reach the small intestine intact without any degradation. The jejunal pH can reach 8.5, where procyanidins are expected to degrade, but procyanidins appear to be stable in the small intestinal lumen due to the influence of protein and other food constituents (Serra *et al.*, 2009). Procyanidins may also be stabilised by the presence of fat in the intestinal lumen (Ortega *et al.*, 2009). Ileostomy subjects are deprived of a functional colon, surgically removed for medical reasons, and upon complete surgical recovery, ileostomists are as healthy as individuals with an intact colon with similar dietary habits (Kennedy *et al.*, 1982). The stability of procyanidins from apple juice in ileostomists was examined. From a dose of ~10 mg procyanidin B2 or ~5 mg procyanidin B1 in the juice, nothing remained in the ileostomy fluid, whereas 17% of the dose of (—)-epicatechin (~15 mg) remained (Kahle *et al.*, 2005). When apple juice was consumed by ileostomists, low levels of B1 and B4 disappeared from the ileal fluid, but there was an 88% recovery of oligomeric procyanidins (Kahle *et al.*, 2007). It is not always clear if the compounds are absorbed or are partly unstable in the lumen in these experiments.

9.2.2 *Absorption of intact procyanidins from the small intestine*

Procyanidin dimer B2 is absorbed intact in humans but only to a very limited extent: after consumption of ~256 mg procyanidin B2 in cocoa, the peak plasma concentration of unconjugated procyanidin B2 at 2 hours was 41 $\pm$ 4 nM which is ~100-fold less than (–)-epicatechin (323 mg total monomers gave 5.9 μM (–)-epicatechin in plasma) (Holt *et al.*, 2002). Procyanidin B1 is present in plasma in humans after consumption of grape seed extract (Sano *et al.*, 2003). Procyanidin B2 is present at low levels in human urine after consumption of cocoa (Urpi-Sarda *et al.*, 2009).

Isolated compounds were given to rats, and subsequently, pure dimer was absorbed into plasma less well than trimer, tetramer and pentamer fractions (Shoji *et al.*, 2006). A pure and single dose of 20 mg procyanidin B3 administration to rats gave no detectable intact compound in urine or plasma. In addition, no monomeric catechins were found after B3 administration, even though catechins and 3′-methyl catechins were readily detected in plasma after administration of catechin alone (Donovan *et al.*, 2002). On oral administration of procyanidin dimer B2 to rats, only a low amount (0.34%) of the provided dose was detected as intact procyanidin B2 in urine (Baba *et al.*, 2002).

Grape seed extract is high in procyanidins, and after administration of grape seed polyphenol extract containing procyanidin B1 and B2, no intact procyanidins were detected in serum or urine (Nakamura & Tonogai, 2003). After administration of grape seed extract to rats by gavage, 0.5% of the original dose of B1, 0.3% of B2, 1% of B3, 0.4% of B4 and 4.3% of trimer C2 were in urine, but with no procyanidins detectable in plasma. Substantial amounts were found in the stomach of rats after feeding, by gavage, up to several hours after administration, with very little in the duodenum, jejunum and ileum, but some in the caecum and in the colon (Tsang *et al.*, 2005).

We can conclude that the absorption of intact procyanidins is very low, close to the limit of quantification by most liquid chromatography–mass spectrometric (LC–MS) methods. Hence, some studies have reported no procyanidin, and others very low levels, indicating that transport across the small intestine occurs, but only to a very limited extent.

9.2.3 Mechanism of absorption across small intestine

The permeation of ^{14}C-labelled (+)-catechin, procyanidin dimer, trimer and polymers across Caco-2 cells indicated that there was very little difference in permeation between monomer, dimer and trimer from the apical to the basolateral side, but the permeability of the polymers was tenfold lower (Deprez *et al.*, 2001). The radiolabelled metabolites were not identified. In the rat intestinal perfusion model, procyanidin dimers A1, A2 and B2 were absorbed without conjugation or methylation, but one or two orders of magnitude lower than (−)-epicatechin. The latter was partly methylated and 100% conjugated. The presence of tetrameric procyanidin enhanced the absorption of B2, but not A1. 'A'-type trimers were not absorbed, and procyanidins B2 and A2 were not excreted in the bile (Appeldoorn *et al.*, 2009b). Whether the permeation of procyanidins is by paracellular or transcellular diffusion, or by active or facilitated transport, is uncertain at the present time, but it is clear that the rate is intrinsically low. This may be due to slow transport or rapid efflux.

9.2.4 Absorption from the colon after microbial metabolism

The main site for absorption of procyanidins is the colon, but only after metabolism to other compounds by the microbiota. A radiolabelled mixture of procyanidins from *Vitis vinifera*, which consisted mainly of (+)-catechin, (−)-epicatechin and procyanidin dimers B1, B2, B3 and B4, was fed orally to rats. One-third of the label was found in the urine, and the percentage decreased with increasing dose. At the higher doses, about 50% was in the faeces, with 4–19% in expired air depending on the dose. Most of the radioactivity in urine was associated with low molecular weight compounds, such as hippuric acid, ethyl catechol and 3-(3′-hydroxyphenyl)propionic acid, in faeces as ethyl catechol, and in bile as vanillic acid (3-hydroxy-4-methoxybenzoic acid) and 3-(3′,-hydroxyphenyl)propionic acid. The existence of these metabolites was dependent on the presence of colonic microbiota, but distribution in tissues was only measured after intravenous injection (Harmand & Blanquet, 1978). Later studies extended these findings, and all agree on the importance of the colonic microbiota. After cocoa consumption in humans, 3-(3′-hydroxyphenyl)propionic acid, 3′-hydroxyphenylacetic acid, 3′,4′-dihydroxyphenylacetic acid, 3-hydroxybenzoic acid and ferulic acid were present in urine, with highest amounts between 9 and 48 hours, indicating that these were microbial metabolites (Rios *et al.*, 2003). 5-(3′,4′-Dihydroxyphenyl)-γ-valerolactone and 5-(3′-methoxy-4′-hydroxyphenyl)-γ-valerolactone were detected in human urine after soluble cocoa powder consumption together with procyanidin B2, hippuric acid, ferulic acid and 3-hydroxyphenylacetic acid (Urpi-Sarda *et al.*, 2009).

After feeding [^{14}C]-procyanidin B2 to rats, ∼82% of the ^{14}C label was recovered in urine. Blood concentration of total [^{14}C] reached a maximum at ∼6 hours after ingestion of [^{14}C]-procyanidin B2. The terminal half-lives were similar after intravenous or oral administration, but total clearance and the apparent volumes of distribution was eightfold larger after oral dosing. This shows that most of the parent compound administered orally is degraded by the gut microbiota prior to absorption into the blood and that these microbial metabolites have a distribution in the body different from the compounds circulating after the intravenous dose (Stoupi *et al.*, 2010b). When procyanidins B3, trimer C2 and polymer were fed to rats, no parent compound or monomeric catechins were found in urine, but when catechin monomers were fed, 26% of the dose was excreted in urine including the 3′-methylated form. Sixteen metabolites of microbial origin were described, accounting for 11% of the catechin monomers but less for procyanidins (procyanidin B3 yield of 6.5%, trimer C2 only 0.7%). The main urinary products of the procyanidin B3 dimer were 3-(3′-hydroxyphenyl)propionic acid, hippuric acid, 4-hydroxyhippuric acid, 4-hydroxybenzoic acid and 3-hydroxy-4-methoxybenzoic acid (Gonthier *et al.*, 2003a).

Experiments have also been performed on isolated microbiota from faeces *in vitro*. Incubation of pure procyanidin B2 with human colonic microbiota gave a range of metabolites, which appeared in a time-dependent manner. Approximately 10% of procyanidin B2 was converted to epicatechin by scission of the interflavan bond. Catabolism favoured removal of the 4′-hydroxyl rather than the 3′-hydroxyl group, and both β-oxidation and α-oxidation occurred. The main metabolite, which peaked at 24 hours, was 3-(3′-hydroxyphenyl)propionic acid in the *in vitro* incubation. Consistent with this, this compound was also found in human urine after consumption of procyanidin-rich cocoa (Rios *et al.*, 2003). Many intermediates have also been partially identified, and many of these are 'dimeric' catabolites (molecular weight greater than 290), which produced MS fragment ions characteristic of flavan-3-ols and/or proanthocyanidins. One catabolite was identified tentatively as either 6 or 8-hydroxy-procyanidin B2, some were reduced in at least one of the epicatechin units, and some contained an apparently unmodified epicatechin unit. These 'dimeric' catabolites were detected up to 9 hours after the start of the incubation and together accounted for ∼20% of the substrate (Stoupi *et al.*, 2010a). Metabolism of radiolabelled procyanidin polymers (average degree of polymerisation ∼6) by human colonic microbiota *in vitro* showed that they were almost totally degraded after 48 hours. The main metabolites were 3-(3′-hydroxyphenyl)propionic acid, 3- and 4-hydroxyphenylacetic acid, 5-(3′-hydroxyphenyl)valeric acid and phenylpropionic acid (Deprez *et al.*, 2000). A mixture of procyanidin dimers were metabolised by human microbiota into low molecular weight compounds. The main products were 3,4-dihydroxyphenylacetic acid and 5-(3′,4′-dihydroxyphenyl)-γ-valerolactone. Other metabolites detected were 3- and 4-hydroxyphenylacetic acid, 3-hydroxyphenylpropionic acid, phenylvaleric acids, monohydroxylated phenylvalerolactone and 1-(3′,4′-dihydroxyphenyl)-3-(2″,4″,6″-trihydroxyphenyl)propan-2-ol. There were also some differences between microbial metabolism of procyanidin and monomeric substrates (Appeldoorn *et al.*, 2009a).

In summary, the absorption of lower molecular weight metabolites of procyanidins is the major metabolic pathway, and >80% of the dose can be considered 'bioavailable' by this route. This contrasts with the very low bioavailability of intact procyanidins (in the small intestine). After consumption of procyanidins, the flavanols are mainly converted to C_6–C_2 and C_6–C_3-dihydroxy forms (Williamson & Clifford, 2010). It is important to note that the C_6–C_5 intermediates seem unique to flavan-3-ols (catechins and procyanidins).

9.3 Chlorogenic acids and hydroxycinnamates

Chlorogenic acids are a family of esters formed between *trans*-hydroxycinnamic ('phenolic') acids and quinic acid, and the main classes are caffeoylquinic acids (CQA), CQA lactones, feruloylquinic acids (FQA), *p*-coumaroylquinic acids and dicaffeoylquinic acids (diCQA) (Clifford, 1999). More recently, other minor cinnamoyl conjugates have been found in the green unroasted bean, with the presence of cinnamoyl-amino acids (Clifford & Knight, 2004), as well as dimethoxycinnamoylquinic acids, caffeoyl-dimethoxycinnamoylquinic acids, feruloyl-dimethoxycinnamoylquinic acids and diferuloylquinic acids, *p*-coumaroyl-caffeoylquinic acids, *p*-coumaroyl-feruloylquinic acids, *p*-coumaroyl-dimethoxycinnamoylquinic acids and di-*p*-coumaroylquinic acids (Clifford *et al.*, 2006a, 2006b).

9.3.1 Transport of chlorogenic acids

Intact chlorogenic acids are not efficiently absorbed and transported across the small intestinal epithelium (Spencer *et al.*, 1999; Azuma *et al.*, 2000; Konishi & Kobayashi, 2004; Konishi *et al.*, 2006; Lafay *et al.*, 2006), but are mainly metabolised in the colon after hydrolysis of the quinic moiety (Plumb *et al.*, 1999; Couteau *et al.*, 2001; Gonthier *et al.*, 2006). However, 5-CQA can be absorbed to a limited extent in the upper gastrointestinal tract: <0.1 μM was present in the portal vein and abdominal artery of rats following gastric infusion of 2.25 μmol of 5-CQA (Konishi *et al.*, 2006). The stomach may also be a site for absorption since 16% of the dose infused in rats was absorbed from the gastric lumen, and recovered intact at concentrations of 3.3 and 1.6 μM in the gastric vein and aorta, respectively (Lafay *et al.*, 2006).

A number of studies have used the ileostomy model to measure the intestinal absorption of dietary polyphenols. Upon ingestion of 385 μmol of chlorogenic acids contained in instant coffee, 446 μmol contained in apple juice and 2.8 mmol of 5-CQA by ileostomy volunteers, the recoveries obtained in ileal effluents ranged from 3.6% for CQA lactones to 10.8–46% for the *p*-CoQAs, 10.2–67% for the CQAs, 46% for the diCQAs and 77% for the FQAs (Olthof *et al.*, 2001; Kahle *et al.*, 2005; Stalmach *et al.*, 2010). A wide variation in the levels recovered in ileal effluent is observed between the different studies, but taken as total chlorogenic acids, about one-third was absorbed in the stomach and/or small intestine (Olthof *et al.*, 2001; Stalmach *et al.*, 2010), with chlorogenic acids being mostly stable in the gastrointestinal tract (Takenaka *et al.*, 2000; Olthof *et al.*, 2001; Rechner *et al.*, 2001b). The remaining two-thirds of chlorogenic acids pass through to the colon where the colonic bacteria carry out further metabolic processes (Olthof *et al.*, 2003; Gonthier *et al.*, 2003b, 2006).

The presence of a low amount of intact 5-CQA in plasma and urine following oral ingestion of coffee (Ito *et al.*, 2005; Azzini *et al.*, 2007; Stalmach *et al.*, 2009), and pure compounds in humans and rats (Olthof *et al.*, 2001; Gonthier *et al.*, 2003b), suggests that the bioavailability of quinic esters of hydroxycinnamic acids is low, and these compounds are able to cross the gastrointestinal epithelium in their intact form to a limited extent.

Following the intake of chlorogenic acids from various foods, a wide variety of metabolites have been identified in plasma and urine. These were mainly sulphated, methylated and glucuronidated hydroxycinnamic acids, as well as glycine conjugates, hydrogenated hydroxycinnamic acids and lower molecular weight phenolic acids, but also intact chlorogenic acids. The wide range of metabolites identified reflects the extensive and complex metabolism of chlorogenic acids upon absorption. Analysis of the various metabolites has been carried with and without β-glucuronidase, sulphatase and esterase treatment together with analysis by LC–MS.

9.3.2 Chlorogenic acid absorption in humans

The C_{max} of free and conjugated hydroxycinnamic acids are generally sub-μM, reaching the circulation rapidly under an hour following oral administration. Urinary excretion accounted for up to 29% of intake (Stalmach *et al.*, 2009), indicating that coffee chlorogenic acids and their products of metabolism are relatively highly bioavailable (Manach *et al.*, 2005). The pharmacokinetic profile of intact chlorogenic acids in plasma is very variable depending on the laboratory making the measurements because of the lack of standards, absence of standard methods and the specificity of deconjugating enzymes. It ranged from as little as 2 nM of 5-CQA following a single serving of coffee (Stalmach *et al.*, 2009) to 5.9 μM of 5-CQA following a single serving of green coffee extract (Farah *et al.*, 2008), a huge difference in values. C_{max} concentrations were reached under 4 hours following intake indicating an absorption in the upper gastrointestinal tract, although a large inter-individual variation could be observed following intake of the green coffee extract (Farah *et al.*, 2008). Urinary excretion of intact chlorogenic acids ranged from 0.29% for 5-CQA to 4.9% for FQAs (Olthof *et al.*, 2001, 2003; Ito *et al.*, 2005; Stalmach *et al.*, 2009). Chlorogenic acids may be differentially bioavailable depending on the nature of the compound ingested. The intake of 5-CQA contained in coffee beverage, green coffee extract or artichoke heads resulted in a wide variation in the pharmacokinetic profile, with C_{max} values ranging from 2.2 nM to 5.9 µM, area under the curve (AUC) values of 4.1 to 17.9 nmol h L^{-1}, and T_{max} values of 0.7 to 3.3 hours (Azzini *et al.*, 2007; Monteiro *et al.*, 2007; Farah *et al.*, 2008; Stalmach *et al.*, 2009). Apart from the potential matrix effect in which 5-CQA was ingested as well as the dose consumed (from 119 to 1068 µmol), it appeared that the group of Farah and Monteiro (Monteiro *et al.*, 2007; Farah *et al.*, 2008) reported a much higher bioavailability of 5-CQA in circulation compared to Stalmach *et al.* (2009) and Azzini *et al.* (2007) with C_{max} values expressed as a percentage of intake of ~0.01% following instant coffee and artichoke consumption, compared to 0.9–14.8% after brewed coffee or green coffee extract. Similarly, Farah *et al.* (2008) reported a high C_{max} value for the diCQAs of 6.6 µM following an intake of ca. 43 µmol, corresponding to ~46% of intake. Neither of the other groups reported the presence of diCQAs in plasma, although Azzini *et al.* (2007) did not mention this in their analysis. The other main difference in the pharmacokinetic profile of intact chlorogenic acids observed between the groups is that of the FQAs which were detected in the plasma of volunteers following the consumption of instant coffee (Stalmach *et al.*, 2009), but not following the intake of a green coffee extract (Farah *et al.*, 2008), despite similar levels ingested. The differences observed between the groups may arise from the matrix in which chlorogenic acids were ingested (beverages or food *vs* extract) as well as the methods of analysis used (enzymatic treatment *vs* use of LC–MS in selective ion monitoring mode for the detection and quantification of metabolites). Although the bioavailability of intact chlorogenic acids requires further investigation, current and past studies reporting the absorption and metabolism of these compounds in humans and animal models do not generally report a high bioavailability and Farah *et al.* (2008) and Monteiro *et al.* (2007) are the first group reporting C_{max} values in the micromolar range associated with intact chlorogenic acids. These discrepancies need to be resolved if the field is to progress, perhaps by inter-laboratory standardisation.

During their passage through the gastrointestinal tract and into the body, there is extensive metabolism of chlorogenic acids following their ingestion. The 0–24-hour excretion of the parent compounds and their metabolites in humans corresponded to 0.29% of intact 5-CQA (Olthof *et al.*, 2001) and to up to 29% of total chlorogenic acids ingested (Stalmach *et al.*, 2009). Within 1 hour of intake, as a result of absorption in the small intestine, low plasma C_{max} values were obtained with unmetabolised chlorogenic acids as well as conjugated hydroxycinnamic acids (Nardini *et al.*, 2002; Wittemer *et al.*, 2005; Azzini *et al.*, 2007; Renouf *et al.*, 2010; Stalmach *et al.*, 2009). More substantial absorption occurred, from the colon, with T_{max} values of 5–10 hours being attained for free and conjugated dihydroferulic and dihydrocaffeic acids (Wittemer *et al.*, 2005; Azzini *et al.*, 2007; Renouf *et al.*, 2010; Stalmach *et al.*, 2009). This differential in the metabolism has been reported in studies examining the bioavailability of chlorogenic acids from artichoke heads and extracts (Rechner *et al.*, 2001a; Wittemer *et al.*, 2005; Azzini *et al.*, 2007), as well as from coffee beverages (Renouf *et al.*, 2010; Stalmach *et al.*, 2009). Metabolites found in plasma and urine of human volunteers following consumption of coffee and artichoke extracts are of similar nature. Free and conjugated dihydrocaffeic and dihydroferulic acids reached their maximum concentration within 5–8 hours , ranging from 41 to 325 nM for free and conjugated dihydrocaffeic acid and from 111 to 550 nM of dihydroferulic acid (Wittemer *et al.*, 2005; Azzini *et al.*, 2007; Renouf *et al.*, 2010; Stalmach *et al.*, 2009). This biphasic profile of absorption results, in part, from a metabolism taking place at different sites of the gastrointestinal tract.

9.3.3 Chlorogenic acid metabolism

Although some studies have reported a lack of hydrolysis of chlorogenic acids taking place in the upper gastrointestinal tract and/or liver (Plumb *et al.*, 1999; Azuma *et al.*, 2000; Olthof *et al.*, 2001), the rapid circulation of metabolites as determined by the short T_{max} values suggests the presence of intracellular esterases capable of releasing the hydroxycinnamic acids prior to absorption and conjugation. Some authors (Andreasen *et al.*, 2001; Kern *et al.*, 2003) reported the presence of intracellular cinnamoyl esterases located in the small intestine, and capable of hydrolysing methyl esters of hydroxycinnamic acids. The main site of hydrolysis of chlorogenic acids and other esters of free and bound hydroxycinnamic acids is the colon where these compounds are subject to microbial metabolism (Buchanan *et al.*, 1996; Kroon *et al.*, 1997; Andreasen *et al.*, 2001; Couteau *et al.*, 2001).

Regarding the metabolism of intact chlorogenic acids, rat liver and small intestine were reported to carry out various metabolic processes such as methylation and glucuronidation, catalysed by catechol-*O*-methyl transferase and uridine diphosphate glucuronosyl transferase, respectively (Yang *et al.*, 2005). In the plasma of rats, 1,5-diCQA was methylated and glucuronidated, and glucuronides of 5-CQA were in the urine following intravenous injection (Choudhury *et al.*, 1999; Yang *et al.*, 2005, 2006). Following coffee intake, traces of sulphated CQAs and their lactone derivatives were detected in the urine of humans (Stalmach *et al.*, 2009, 2010).

Following uptake in the upper gastrointestinal tract, metabolites of chlorogenic acids are present in ileal effluent of ileostomists given coffee, predominantly as sulphates and glucuronides of CQA, CQA lactones and FQAs (Stalmach *et al.*, 2010). The majority of conjugated metabolites recovered in ileal fluid comprised sulphates (77%), with only 7% as glucuronides. The human small intestine epithelium is a site for phase II enzyme reactions (Lin *et al.*, 1999), and there is enteric recycling of metabolites (Chen *et al.*, 2003; Poquet *et al.*, 2008a). The presence of ferulic and isoferulic acids was reported in plasma and urine samples of rats and humans following the intake of CQAs (Graefe & Veit, 1999; Azuma *et al.*, 2000; Rechner *et al.*, 2001a; Wittemer *et al.*, 2005; Lafay *et al.*, 2006; Azzini *et al.*, 2007), suggesting an extensive methylation of the caffeic moiety, taking place either in the small intestine (Kern *et al.*, 2003) or liver (Moridani *et al.*, 2002; Mateos *et al.*, 2006). This metabolic process appears to occur preferentially in the C(4)-position of the phenyl ring (Lafay *et al.*, 2006; Poquet *et al.*, 2008b). Similarly, methylation of intact chlorogenic acids was reported in rats with methylated diCQAs identified in plasma, urine and bile after oral administration (Yang *et al.*, 2005). This extensive methylation pathway could explain the recovery of FQAs but little or no CQAs in blood and urine of humans after coffee consumption (Stalmach *et al.*, 2009, 2010). Besides methylation, sulphation of hydroxycinnamic acids is the preferential pathway of metabolism in humans, with a regioselectivity preference for sulphation of the 3-hydroxyl of caffeic and dihydrocaffeic acids (Poquet *et al.*, 2008b; Stalmach *et al.*, 2009; Wong *et al.*, 2010). Some authors (Yang *et al.*, 2005, 2006) investigated the extensive metabolism of 1,5-diCQA upon oral and intravenous administration to rats and reported the production of methylated and methyl-glucuronidated isomers in plasma and urine. Similarly, there was an isomerisation in the human gastrointestinal tract after the consumption of 5- and 4-CQAs in apple juice (Kahle *et al.*, 2007), with 1- and 3-CQAs recovered in the ileostomy fluid. The liver has also been identified as a site of isomerisation following incubation of 5-CQA with HepG2 cells (Mateos *et al.*, 2006).

The recent data on chlorogenic acids and on procyanidins illustrate the complex pathways involved in their metabolism. Importantly, both classes of compound are absorbed intact only to a very small extent, and it is the catabolites that are absorbed after extensive metabolism by the microbiota in the colon. These reactions give rise to many different compounds *in vivo*, which can contribute to the biological activity of these phenolic antioxidants, and in some cases, the activity of the microbial catabolites could exceed that of the parent compounds. In this way, an understanding of the bioavailability helps to target *in vitro* measurements of the biological activity of chlorogenic acids and procyanidins.

References

Andreasen, M.F., Kroon, P.A., Williamson, G. & Garcia-Conesa, M.T. (2001) Esterase activity able to hydrolyze dietary antioxidant hydroxycinnamates is distributed along the intestine of mammals. *Journal of Agricultural and Food Chemistry*, **49**, 5679–5684.

Appeldoorn, M.M., Vincken, J.P., Aura, A.M., Hollman, P.C. & Gruppen, H. (2009a) Procyanidin dimers are metabolized by human microbiota with 2-(3,4-dihydroxyphenyl) acetic acid and 5-(3,4-dihydroxyphenyl)-gamma-valerolactone as the major metabolites. *Journal of Agricultural and Food Chemistry*, **57**, 1084–1092.

Appeldoorn, M.M., Vincken, J.P., Gruppen, H. & Hollman, P.C. (2009b) Procyanidin dimers A1, A2, and B2 are absorbed without conjugation or methylation from the small intestine of rats. *Journal of Nutrition*, **139**, 1469–1473.

Azuma, K., Ippoushi, K., Nakayama, M., Ito, H., Higashio, H. & Terao, J. (2000) Absorption of chlorogenic acid and caffeic acid in rats after oral administration. *Journal of Agricultural and Food Chemistry*, **48**, 5496–5500.

Azzini, E., Bugianesi, R., Romano, F., *et al.* (2007) Absorption and metabolism of bioactive molecules after oral consumption of cooked edible heads of *Cynara scolymus* L. (cultivar Violetto di Provenza) in human subjects: a pilot study. *British Journal of Nutrition*, **97**, 963–969.

Baba, S., Osakabe, N., Natsume, M. & Terao, J. (2002) Absorption and urinary excretion of procyanidin B2 [epicatechin-(4 beta-8)-epicatechin] in rats. *Free Radical Biology and Medicine*, **33**, 142–148.

Bacon, J.R. & Rhodes, M.J. (2000) Binding affinity of hydrolyzable tannins to parotid saliva and to proline-rich proteins derived from it. *Journal of Agricultural and Food Chemistry*, **48**, 838–843.

Buchanan, C.J., Wallace, G., Fry, S.C. & Eastwood, M.A. (1996) In vivo release of ^{14}C-labelled phenolic groups from intact dietary spinach cell walls during passage through the rat intestine. *Journal of the Science of Food and Agriculture*, **71**, 459–469.

Chen, J., Lin, H. & Hu, M. (2003) Metabolism of flavonoids via enteric recycling: role of intestinal disposition. *Journal of Pharmacology and Experimental Therapeutics*, **304**, 1228–1235.

Choudhury, R., Srai, S.K., Debnam, E. & Rice-Evans, C.A. (1999) Urinary excretion of hydroxycinnamates and flavonoids after oral and intravenous administration. *Free Radical Biology and Medicine*, **27**, 278–286.

Clifford, M.N. (1999) Chlorogenic acids and other cinnamates—nature, occurrence and dietary burden. *Journal of the Science of Food and Agriculture*, **79**, 362–372.

Clifford, M.N. & Knight, S. (2004) The cinnamoyl-amino acid conjugates of green robusta coffee beans. *Food Chemistry*, **87**, 457–463.

Clifford, M.N., Knight, S., Surucu, B. & Kuhnert, N. (2006a). Characterization by LC-MS(n) of four new classes of chlorogenic acids in green coffee beans: dimethoxycinnamoylquinic acids, diferuloylquinic acids, caffeoyl-dimethoxycinnamoylquinic acids, and feruloyl-dimethoxycinnamoylquinic acids. *Journal of Agricultural and Food Chemistry*, **54**, 1957–1969.

Clifford, M.N., Marks, S., Knight, S. & Kuhnert, N. (2006b) Characterization by LC-MSn of four new classes of p-coumaric acid-containing diacyl chlorogenic acids in green coffee beans. *Journal of Agricultural and Food Chemistry*, **54**, 4095–4101.

Couteau, D., McCartney, A.L., Gibson, G.R., Williamson, G. & Faulds, C.B. (2001) Isolation and characterization of human colonic bacteria able to hydrolyse chlorogenic acid. *Journal of Applied Microbiology*, **90**, 873–881.

de Freitas, V.. & Mateus, N. (2001) Structural features of procyanidin interactions with salivary proteins. *Journal of Agricultural and Food Chemistry*, **49**, 940–945.

Deprez, S., Brezillon, C., Rabot, S., *et al.* (2000) Polymeric proanthocyanidins are catabolized by human colonic microflora into low-molecular-weight phenolic acids. *Journal of Nutrition*, **130**, 2733–2738.

Deprez, S., Mila, I., Huneau, J.F., Tome, D. & Scalbert, A. (2001) Transport of proanthocyanidin dimer, trimer, and polymer across monolayers of human intestinal epithelial Caco-2 cells. *Antioxidants and Redox Signalling*, **3**, 957–967.

Donovan, J.L., Manach, C., Rios, L., Scalbert, A. & Remesy, C. (2002) Procyanidins are not bioavailable in rats fed a single meal of containing a grape seed extract or the procyanidin dimer B3. *Journal of Nutrition*, **87**, 299–306.

Farah, A., Monteiro, M., Donangelo, C.M. & Lafay, S. (2008) Chlorogenic acids from green coffee extract are highly bioavailable in humans. *Journal of Nutrition* **138**, 2309–2315.

Gonthier, M.P., Donovan, J.L., Texier, O., Felgines, C., Remesy, C. & Scalbert, A. (2003a) Metabolism of dietary procyanidins in rats. *Free Radical Biology and Medicine*, **35**, 837–844.

Gonthier, M.P., Remesy, C., Scalbert, A., *et al.* (2006) Microbial metabolism of caffeic acid and its esters chlorogenic and caftaric acids by human faecal microbiota in vitro. *Biomedicine and Pharmacotherapeutics*, **60**, 536–540.

Gonthier, M.P., Verny, M.A., Besson, C., Remesy, C. & Scalbert, A. (2003b) Chlorogenic acid bioavailability largely depends on its metabolism by the gut microflora in rats. *Journal of Nutrition* **133**, 1853–1859.

Graefe, E.U. & Veit, M. (1999) Urinary metabolites of flavonoids and hydroxycinnamic acids in humans after application of a crude extract from Equisetum arvense. *Phytomedicine*, **6**, 239–246.

Gu, L., Kelm, M.A., Hammerstone, J.F., *et al.* (2004) Concentrations of proanthocyanidins in common foods and estimations of normal consumption. *Journal of Nutrition*, **134**, 613–617.

Hammerstone, J.F., Lazarus, S.A., Mitchell, A.E., Rucker, R. & Schmitz, H.H. (1999) Identification of procyanidins in cocoa (*Theobroma cacao*) and chocolate using high-performance liquid chromatography/mass spectrometry. *Journal of Agricultural and Food Chemistry*, **47**, 490–496.

Hammerstone, J.F., Lazarus, S.A. & Schmitz, H.H. (2000) Procyanidin content and variation in some commonly consumed foods. *Journal of Nutrition*, **130**, 2086s–2092s.

Harmand, M.F. & Blanquet, P. (1978) The fate of total flavanolic oligomers (oft) extracted from *Vitis vinifera* l. in the rat. *European Journal of Drug Metabolism and Pharmacology*, **3**, 15–30.

Holt, R.R., Lazarus, S.A., Sullards, M.C., *et al.* (2002) Procyanidin dimer B2 [epicatechin-(4 beta-8)-epicatechin] in human plasma after the consumption of a flavanol-rich cocoa. *American Journal of Clinical Nutrition*, **76**, 798–804.

Ito, H., Gonthier, M.P., Manach, C., *et al.* (2005) Polyphenol levels in human urine after intake of six different polyphenol-rich beverages. *British Journal of Nutrition*, **94**, 500–509.

Kahle, K., Huemmer, W., Kempf, M., Scheppach, W., Erk, T. & Richling, E. (2007) Polyphenols are intensively metabolized in the human gastrointestinal tract after apple juice consumption. *Journal of Agricultural and Food Chemistry*, **55**, 10605–10614.

Kahle, K., Kraus, M., Scheppach, W. & Richling, E. (2005) Colonic availability of apple polyphenols—A study in ileostomy subjects. *Molecular Nutrition and Food Research*, **49**, 1143–1150.

Kennedy, H.J., Lee, E.C., Claridge, G. & Truelove, S.C. (1982) The health of subjects living with a permanent ileostomy. *Quarterly Journal of Medicine*, **203**, 341–357.

Kern, S.M., Bennett, R.N., Needs, P.W., Mellon, F.A., Kroon, P.A. & Garcia-Conesa, M.T. (2003) Characterization of metabolites of hydroxycinnamates in the in vitro model of human small intestinal epithelium caco-2 cells. *Journal of Agricultural and Food Chemistry*, **51**, 7884–7891.

Konishi, Y. & Kobayashi, S. (2004) Transepithelial transport of chlorogenic acid, caffeic acid, and their colonic metabolites in intestinal caco-2 cell monolayers. *Journal of Agricultural and Food Chemistry*, **52**, 2518–2526.

Konishi, Y., Zhao, Z. & Shimizu, M. (2006) Phenolic acids are absorbed from the rat stomach with different absorption rates. *Journal of Agricultural and Food Chemistry*, **54**, 7539–7543.

Kroon, P.A., Faulds, C.B., Ryden, P., Robertson, J.A. & Williamson, G. (1997) Release of bound ferulic acid from fiber in the human colon. *Journal of Agricultural and Food Chemistry*, **45**, 661–667.

Lafay, S., Gil-Izquierdo, A., Manach, C., Morand, C., Besson, C. & Scalbert, A. (2006) Chlorogenic acid is absorbed in its intact form in the stomach of rats. *Journal of Nutrition*, **136**, 1192–1197.

Lin, J.H., Chiba, M. & Baillie, T.A. (1999) Is the role of the small intestine in first-pass metabolism overemphasized? *Pharmacological Reviews*, **52**, 135–158.

Manach, C., Williamson, G., Morand, C., Scalbert, A. & Remesy, C. (2005) Bioavailability and bioefficacy of polyphenols in humans. I. Review of 97 bioavailability studies. *American Journal of Clinical Nutrition*, **81**, 230S–242S.

Mateos, R., Goya, L. & Bravo, L. (2006) Uptake and metabolism of hydroxycinnamic acids (chlorogenic, caffeic, and ferulic acids) by HepG2 cells as a model of the human liver. *Journal of Agricultural and Food Chemistry*, **54**, 8724–8732.

Monteiro, M., Farah, A., Perrone, D., Trugo, L.C. & Donangelo, C. (2007) Chlorogenic acid compounds from coffee are differentially absorbed and metabolized in humans. *Journal of Nutrition*, **137**, 2196–2201.

Moridani, M.Y., Scobie, H. & O'Brien, P.J. (2002) Metabolism of caffeic acid by isolated rat hepatocytes and subcellular fractions. *Toxicology Letters*, **133**, 141–151.

Nakamura, Y. & Tonogai, Y. (2003) Metabolism of grape seed polyphenol in the rat. *Journal of Agricultural and Food Chemistry*, **51**, 7215–7225.

Nardini, M., Cirillo, E., Natella, F. & Scaccini, C. (2002) Absorption of phenolic acids in humans after coffee consumption. *Journal of Agricultural and Food Chemistry*, **50**, 5735–5741.

Olthof, M.R., Hollman, P.C., Buijsman, M.N., van Amelsvoort, J.M., & Katan, M.B. (2003) Chlorogenic acid, quercetin-3-rutinoside and black tea phenols are extensively metabolized in humans. *Journal of Nutrition*, **133**, 1806–1814.

Olthof, M.R., Hollman, P.C. & Katan, M.B. (2001) Chlorogenic acid and caffeic acid are absorbed in humans. *Journal of Nutrition*, **131**, 66–71.

Ortega, N., Reguant, J., Romero, M.P., Macia, A. & Motilva, M.J. (2009) Effect of fat content on the digestibility and bioaccessibility of cocoa polyphenol by an in vitro digestion model. *Journal of Agricultural and Food Chemistry*, **57**, 5743–5749.

Pascual-Teresa, S., Santos-Buelga, C. & Rivas-Gonzalo, J.C. (2000) Quantitative analysis of flavan-3-ols in Spanish foodstuffs and beverages. *Journal of Agricultural and Food Chemistry*, **48**, 5331–5337.

Plumb, G.W., Garcia Conesa, M.T., Kroon, P.A., Rhodes, M., Ridley, S. & Williamson, G. (1999) Metabolism of chlorogenic acid by human plasma, liver, intestine and gut microflora. *Journal of the Science of Food and Agriculture*, **79**, 390–392.

Poquet, L., Clifford, M.N. & Williamson, G. (2008a) Investigation of the metabolic fate of dihydrocaffeic acid. *Biochemical Pharmacology*, **75**, 1218–1229.

Poquet, L., Clifford, M.N. & Williamson, G. (2008b) Transport and metabolism of ferulic acid through the colonic epithelium. *Drug Metabolism and Disposition*, **36**, 190–197.

Rechner, A.R., Pannala, A.S. & Rice-Evans, C.A. (2001a) Caffeic acid derivatives in artichoke extract are metabolised to phenolic acids in vivo. *Free Radical Research*, **35**, 195–202.

Rechner, A.R., Spencer, J.P., Kuhnle, G., Hahn, U. & Rice-Evans, C.A. (2001b) Novel biomarkers of the metabolism of caffeic acid derivatives in vivo. *Free Radical Biology and Medicine*, **30**, 1213–1222.

Renouf, M., Guy, P.A., Marmet, C., *et al.* (2010) Measurement of caffeic and ferulic acid equivalents in plasma after coffee consumption: small intestine and colon are key sites for coffee metabolism. *Molecular Nutrition and Food Research*, **54**, 760–766.

Rios, L., Bennett, R.N., Lazarus, S.A., Remesy, C., Scalbert, A. & Williamson, G. (2002) Cocoa proanthocyanidins are stable during gastric transit in humans. *American Journal of Clinical Nutrition*, **76**, 1106–1110.

Rios, L.Y., Gonthier, M.P., Remesy, C., *et al.* (2003) Chocolate intake increases urinary excretion of polyphenol-derived phenolic acids in healthy human subjects. *American Journal of Clinical Nutrition*, **77**, 912–918.

Sano, A., Yamakoshi, J., Tokutake, S., Tobe, K., Kubota, Y. & Kikuchi, M. (2003) Procyanidin B1 is detected in human serum after intake of proanthocyanidin-rich grape seed extract. *Bioscience, Biotechnology and Biochemistry*, **67**, 1140–1143.

Serra, A., Macia, A., Romero, M.P., *et al.* (2009) Bioavailability of procyanidin dimers and trimers and matrix food effects in in vitro and in vivo models. *British Journal of Nutrition*, **103**, 944–952.

Shoji, T., Masumoto, S., Moriichi, N., *et al.* (2006) Apple procyanidin oligomers absorption in rats after oral administration: analysis of procyanidins in plasma using the porter method and high-performance liquid chromatography/tandem mass spectrometry. *Journal of Agricultural and Food Chemistry*, **54**, 884–892.

Spencer, J.P., Chowrimootoo, G., Choudhury, R., Debnam, E.S., Srai, S.K. & Rice-Evans, C. (1999) The small intestine can both absorb and glucuronidate luminal flavonoids. *FEBS Letters*, **458**, 224–230.

Spencer, J.P.E., Chaudry, F., Pannala, A.S., Srai, S.K., Debnam, E. & Rice-Evans, C.A. (2000) Decomposition of cocoa procyanidins in the gastric milieu. *Biochemical and Biophysical Research Communications*, **272**, 236–241.

Stalmach, A., Mullen, W., Barron, D., *et al.* (2009) Metabolite profiling of hydroxycinnamate derivatives in plasma and urine after the ingestion of coffee by humans: identification of biomarkers of coffee consumption. *Drug Metabolism and Disposition*, **37**, 1749–1758.

Stalmach, A., Steiling, H., Williamson, G. & Crozier, A. (2010) Bioavailability of chlorogenic acids following acute ingestion of coffee by humans with an ileostomy. *Archives of Biochemistry and Biophysics*, **501**, 98–105.

Stoupi, S., Williamson, G., Drynan, J.W., Barron, D. & Clifford, M.N. (2010a) Procyanidin B2 catabolism by human fecal microflora: Partial characterization of 'dimeric' intermediates. *Archives of Biochemistry and Biophysics*, **501**, 73–78.

Stoupi, S., Williamson, G., Viton, F., *et al.* (2010b) In vivo bioavailability, absorption, excretion, and pharmacokinetics of [^{14}C]procyanidin B2 in male rats. *Drug Metabolism and Disposition*, **38**, 287–291.

Takenaka, M., Nagata, T. & Yoshida, M. (2000) Stability and bioavailability of antioxidants in garland (*Chrysanthemum coronarium* L.). *Bioscience, Biotechnology and Biochemistry*, **64**, 2689–2691.

Tsang, C., Auger, C., Mullen, W., *et al.* (2005) The absorption, metabolism and excretion of flavan-3-ols and procyanidins following the ingestion of a grape seed extract by rats. *British Journal of Nutrition*, **94**, 170–181.

Urpi-Sarda, M., Monagas, M., Khan, N., *et al.* (2009) Epicatechin, procyanidins, and phenolic microbial metabolites after cocoa intake in humans and rats. *Analytical and Bioanalytical Chemistry*, **394**, 1545–1556.

Williamson, G. & Clifford, M.N. (2010) Colonic metabolites of berry polyphenols: the missing link to biological activity? *British Journal of Nutrition*, **104**, s48–s66.

Wittemer, S.M., Ploch, M., Windeck, T., *et al.* (2005) Bioavailability and pharmacokinetics of caffeoylquinic acids and flavonoids after oral administration of Artichoke leaf extracts in humans. *Phytomedicine*, **12**, 28–38.

Wong, C.C., Meinl, W., Glatt, H.R., *et al.* (2010) In vitro and in vivo conjugation of dietary hydroxycinnamic acids by UDP-glucuronosyltransferases and sulfotransferases in humans. *Journal of Nutritional Biochemistry*, **21**, 1060–1068.

Yang, B., Meng, Z., Dong, J., *et al.* (2005) Metabolic profile of 1,5-dicaffeoylquinic acid in rats, an in vivo and in vitro study. *Drug Metabolism and Disposition*, **33**, 930–936.

Yang, B., Meng, Z.Y., Yan, L.P., *et al.* (2006) Pharmacokinetics and metabolism of 1,5-dicaffeoylquinic acid in rats following a single intravenous administration. *Journal of Pharmacology and Biomedical Analysis*, **40**, 417–422.

Zhu, M., Chen, Y. & Li, R.C. (2000) Oral absorption and bioavailability of tea catechins. *Planta Medica*, **66**, 444–447.

Chapter 10

Extra-Virgin Olive Oil—Healthful Properties of Its Phenolic Constituents

Francesco Visioli and Elena Bernardini

Abstract: Olive oil, the primary oil source of the Mediterranean diet, differs significantly in composition from dietary lipids that are consumed by other populations. In particular, extra-virgin olive oil is rich in phenolic compounds that provide aroma and taste. As of now, the cardioprotective activities of olive phenols are quite established: approximately 20 human trials describe the superiority of phenol-rich olive oil to other vegetable oils or sources of fat. Two major fields remain to be investigated, namely, the neurodegenerative disease and the cancer. This article critically reviews the available evidence and the additional research that is still needed in this area.

Keywords: olive oil; polyphenols; hydroxytyrosol; cardiovascular disease; xenobiotics; neurodegeneration; Mediterranean diet

10.1 Introduction

Several epidemiological studies have shown that the incidence of coronary heart disease (CHD) and certain cancers, for example, breast and colon cancers, is low in the Mediterranean basin (Trichopoulou & Critselis, 2004). In addition to a possible contribution of genetics, it has been suggested that this effect is largely due to the protective dietary habits of this area (Trichopoulou *et al.*, 1995; Estruch *et al.*, 2006). The traditional Mediterranean diet, rich in fruit, vegetables, fish, and whole grain, is, in fact, thought to promote good health and longevity (Manios *et al.*, 2006; Visioli *et al.*, 2006). Olive oil, the primary oil source of this diet, differs significantly in composition from dietary lipids that are consumed by other populations (Visioli & Galli, 2000). In particular, extra-virgin olive oil (EVOO) is rich in phenolic compounds (Fig. 10.1), which other vegetable oils do not contain (Boskou, 2000). The formulation of an antioxidant/atherosclerosis hypothesis stimulated experimental and epidemiological studies on the possible role of antioxidants, including olive oil phenolics, in the protection from CHD observed in the Mediterranean area (Visioli *et al.*,

Recent Advances in Polyphenol Research, Volume 3, First Edition. Edited by Véronique Cheynier, Pascale Sarni-Manchado and Stéphane Quideau.

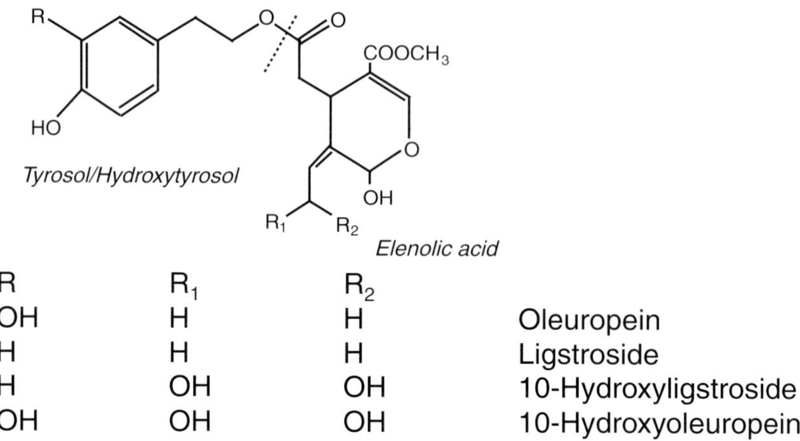

R	R₁	R₂	
OH	H	H	Oleuropein
H	H	H	Ligstroside
H	OH	OH	10-Hydroxyligstroside
OH	OH	OH	10-Hydroxyoleuropein

Fig. 10.1 Chemical structures of the most important aglycones found in olives, extra-virgin olive oil, and olive mill waste waters.

2005). Included among the minor constituents of virgin olive oil are vitamins such as α- and γ-tocopherols (around 200 ppm) and β-carotene, phytosterols, pigments, terpenic acids, flavonoids, squalene, and a number of phenolic compounds, usually grouped under the rubric "polyphenols" (Boskou, 2000).

It needs to be underlined that there is considerable difference between "olive oil" and "EVOO." Whereas the former is nearly devoid of phenolic compounds, the latter contains up to 1 g/kg of such molecules (Visioli & Galli, 1998). The consumer is often confused and should, conversely, be advised to prefer extra-virgin oils, for which there is evidence of healthful activities (as reviewed in the subsequent text). For the sake of brevity, we will just refer to "olive oil": readers beware that only those of high quality play a role in maintaining cardiovascular health. Indeed, this is important from a preventive medicine viewpoint: if the studies reviewed in this chapter are confirmed, the consumers should be informed that the best olive oil to be bought is that rich in phenolic compounds. The trained consumer can then choose the best product based on its sensory characteristics. In brief, olive oils rich in polyphenols and, hence, endowed with biological activities are those with a pungent aroma and a bitter taste (Fig. 10.2). Also, the food industry often stirs consumers away from bitter food items. In addition, the bitter taste is an acquired one and the choice of bitter olive oils is not based on instinct. The converse is true: most olive oils found in the market possess smooth tastes because they are, for the most part, of low quality. In synthesis, the "average" consumer develops the habit of purchasing smooth- or sweet-tasting oils and does not distinguish between the aromas conveyed by the different cultivars. In a way, this is ironic, given that olive oil is very similar to wine in that it is derived from a fruit with minimal processing. As such, not only different cultivars but also different soils, climates, weather conditions, etc. strongly influence to organoleptic characteristics of the final product. While it is fashionable to distinguish the various nuances that grapes provide to wine, i.e., a chardonnay is very different from a pinot, most self-proclaimed gourmets do not apply the same line of reasoning to olive oil. The outcome is that high-quality olive oils are often overlooked, and consequently, their potential health benefits are not exploited.

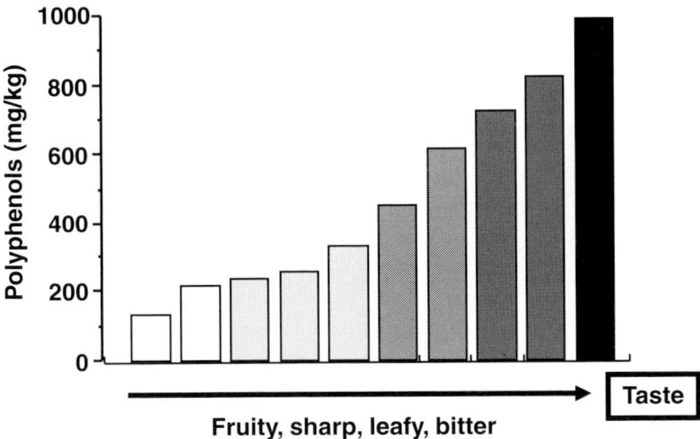

Fig. 10.2   (Poly)phenols provide the peculiar taste of extra-virgin olive oil. This is important from a health viewpoint, in that the consumer should be trained and informed on how to choose high-quality olive oils based on their organoleptic attributes.

In this chapter, we review the current scientific evidence of a healthful role of olive phenols, with particular emphasis on hydroxytyrosol and related molecules.

10.2 Epidemiological studies

Research on the healthful properties of olive phenols stems from epidemiological observations that link preferential olive oil intake with lower incidence of cardiovascular mortality (Trichopoulou *et al.*, 1995). This link has been observed in all Mediterranean countries and has been particularly underlined in Albania (Gjonca & Bobak, 1997). Whether this association is causal or casual has not been resolved (and, probably, never will be). However, from a nutritional point of view, the choice of a phenol-rich olive oil contributes to the dietary intake of biologically active compounds in quantities that have been correlated with a reduced risk of developing CHD (Hertog *et al.*, 1993). Indeed, the use of EVOO as the principal source of dietary oil instead of animal fat, in addition to providing a considerable amount of oleic acid, provides an intake of bioactive compounds with potential healthful effects, as described in the preceding text (Lopez-Miranda *et al.*, 2010). It also appears that the intake and interaction of several "micronutrients" provided by a healthy diet, such as that in use in the Mediterranean area during the mid-1940s, is likely to be the link that affords protection from such pathologies (Ames, 2004; Visioli & Hagen, 2007). In turn, the answer to the current debate on the efficacy of antioxidant supplements is likely to be found in the adoption of a Mediterranean-style diet, in which the abundance of bioactive, functional compounds provided by fruits, vegetables, wine, and olive oil grants a higher protection toward reactive oxygen species (ROS)-induced diseases.

10.3 *In vitro* studies on olive oil's phenolics

The lower incidence of CHD observed in the Mediterranean area (Trichopoulou & Critselis, 2004) lead to the hypothesis that olive oil phenolics exert a protective effect with respect to chemically induced oxidation of human low-density lipoprotein (LDL), which is one of the initial steps in the onset of atherosclerosis (Miller *et al.*, 2003; Stocker & Keaney, 2004) (though the degree of its contribution is yet to be ascertained).

There are, to date, several *in vitro* studies that address the mechanisms underlying the purported protective activities of olive phenols (Covas *et al.*, 2009). It should be noted that these studies provide interesting insight on the cellular and molecular actions of such compounds, but cannot prove any *in vivo* beneficial effect. Nonetheless, the data that originate from *in vitro* studies often foster and build the bases for more elaborate *in vivo* investigations. Finally, it must be underlined that some of the methodologies that were initially employed to assess the biological, namely, antioxidant, activities of olive phenols are now outdated. One notable example is that of the *in vitro* oxidation of human LDL, which likely does not translate in similar *in vivo* actions. When such experiments were started, however, that model was a widely accepted proxy of human pathology (Salami *et al.*, 1995). These days, it can represent a useful substrate to test lipid peroxidation in a complex lipidic and proteic environment such as that of human LDL.

Another noteworthy observation is that the initial *in vitro* studies on olive oil phenolics have been performed with high concentrations of, e.g., oleuropein or hydroxytyrosol. Unfortunately, several groups still publish *in vitro* studies that report biological activities of olive phenols while employing concentrations in excess of 100 mM. Following some studies on olive phenolics' absorption and metabolism (see Section 10.4), this research should be labeled as "irrelevant" to *in vivo* physiology and dietary recommendations. Research on cellular and molecular mechanisms should, of course, be encouraged, but the difference between a mere exercise of biochemistry and investigations that shed molecular light on *in vivo* observations should always be kept in mind.

A brief summary of the first study that was performed to address the question of whether olive phenols could modulate physio-pathological pathways has been given in the subsequent text. Indeed, results obtained on human LDL demonstrate that catechol-like compounds present in virgin olive oil inhibit the formation of lipid oxidation products in a dose-dependent manner and are effective at a concentration lower than that of pure tyrosol, a phenolic component of the oil, and of probucol, used as reference compounds. This effect is probably due to the synergistic action of hydroxytyrosol, oleuropein aglycones, and some flavonoids, such as quercetin, luteolin, and apigenin present in the virgin olive oil extract in minute amounts.

Pure hydroxytyrosol (HT) and oleuropein both potently and dose dependently inhibit copper sulfate-induced oxidation of LDL at concentrations of $10-6$ to $10-4$ M (Visioli & Galli, 1994; Visioli *et al.*, 1995a). This method to assess antioxidant activity as related to the cardiovascular system is now outdated, as it is rather crude and not representative of *in vivo* human situations. However, chemically oxidized LDL provides information on the antioxidant activity of molecules toward lipid substrates. The free radical scavenging activities of hydroxytyrosol and oleuropein have been further confirmed (Visioli & Galli, 1998; Visioli *et al.*, 1998a) by the use of metal-independent oxidative systems and stable free radicals, such as DPPH (Visioli & Galli, 1998), in a series of experiments that demonstrated both a strong metal-chelation and a free-radical scavenging action. In addition, changes in electrophoretic mobility of apo B are also prevented by the phenols. As far as the mechanism of action of olive oil phenolics is concerned, it is well known that the antioxidant properties of *ortho*-diphenols are related to hydrogen donation, which is their ability to improve radical stability by forming an intramolecular hydrogen bond between the free hydrogens of their hydroxyl group and their phenoxyl radicals (Bors & Michel, 2002; Butkovic *et al.*, 2004). Although specific investigations of olive oil phenols are yet to be carried out, studies performed on the structure–activity relationship of flavonoids indicated that the degree of antioxidant activity is strictly related to the number of hydroxyl substitutions (Bors & Michel, 2002; Butkovic *et al.*, 2004). In brief, the most potent (poly)phenolic antioxidants are those richest in *ortho*-diphenolic substituents, and it is not correct to herald the total polyphenolic content of a food item or of an extract without knowing the proportion of *ortho*-diphenols that it contains.

While most *in vitro* studies are focusing on the cardiovascular system (it's worth reminding that cardiovascular pathologies are the first cause of death in the Western world), increasing attention is being paid to cancer and its causes. The mutagenic properties of oxidatively damaged DNA suggest that antioxidants might have protective activity toward tumor formation. Low concentrations of hydroxytyrosol (50 μM) are able to scavenge peroxynitrite and, therefore, prevent $ONOO^-$-dependent DNA damage and tyrosine nitration (Deiana *et al.*, 1999); also, in a model of copper-induced DNA damage, the pro-oxidant activities of hydroxytyrosol (due to its copper-reducing properties) become evident at nonphysiological concentrations (>500 μM) and are 40-fold weaker than those of the widely employed reducing agent ascorbate (Deiana *et al.*, 1999). Interestingly, two human intervention studies (Weinbrenner *et al.*, 2004; Salvini *et al.*, 2006) confirmed these data *in vivo*, indicating that EVOO might decrease DNA damage, hence, lessening cancer risk (more on this is described in the subsequent text).

The immune system plays roles in the development of cardiovascular pathologies, even though the extent and precise nature of its contribution is yet to be ascertained. Oleuropein increases the functional activity of immune-competent cells (macrophages), as demonstrated by a significant increase (58.7 ± 4.6%) in the lipopolysaccharide (LPS)-induced production of nitric oxide, a bactericidal and cytostatic agent (Visioli *et al.*, 1998b). This increase is consequent to a direct tonic effect of oleuropein on the inducible form of the enzyme nitric oxide synthase (iNOS), as demonstrated by Western blot analysis of cell homogenates and by coincubation of LPS-challenged cells with the iNOS inhibitor L-nitromethylarginine methylester (Visioli *et al.*, 1998b).

A correlation between inflammation and cardiovascular diseases has long been established, namely, after the publication of the controversial (at that time) Russel Ross' review (Ross, 1999); monocyte/macrophages and NF-kB play a pivotal role. The effects of an EVOO extract, particularly rich in phenolic compounds, were investigated on NF-kB translocation in monocytes and monocyte-derived macrophages (MDM) isolated from healthy volunteers. In a concentration-dependent manner, the EVOO extract inhibited p50 and p65 NF-kB translocation in both unstimulated and phorbol-myristate acetate (PMA)-challenged cells, being particularly effective on the p50 subunit (Brunelleschi *et al.*, 2007). Interestingly, this effect occurred at concentrations found in human plasma after nutritional ingestion of virgin olive oil and was quantitatively similar to that exerted by ciglitazone, a PPAR-γ ligand. However, the EVOO did not affect PPAR-γ expression in monocytes and MDM. These data provide further evidence of the beneficial effects of EVOO by indicating its ability to inhibit NF-kB activation in human monocyte/macrophages.

10.4 *In vivo* studies

The first step toward demonstrating *in vivo* effects of olive oil phenolics was to assess their bioavailability. In fact, experimental evidence that flavonoids and phenolic compounds are absorbed from the diet is accumulating (Williamson & Manach, 2005; Manach *et al.*, 2009). In the year 2000, Visioli *et al.* demonstrated that olive oil phenolics are dose dependently absorbed in humans and that they are excreted in the urine, mainly as glucuronide conjugates (Caruso *et al.*, 2001); it is noteworthy that increasing amounts of phenolics administered with olive oil stimulated the rate of conjugation with glucuronide (Visioli *et al.*, 2000b). These data add to the growing experimental evidence that indicates absorption and urinary disposition of polyphenols, including those from EVOO (Vissers *et al.*, 2002, 2004), in humans (Fig. 10.3). Associated with the first bioavailability study there was the first demonstration of an *in vivo* antioxidant activity, as shown by reduced F_2-isoprostane excretions by healthy volunteers (Visioli *et al.*, 2000a). The mechanisms for absorption are currently being investigated in models of colon and stomach environments (Corona *et al.*, 2006).

Further studies elucidated the metabolic pathways of hydroxytyrosol and oleuropein, which form elevated quantities of homovanillyl alcohol and homovanilic acid (Caruso *et al.*,

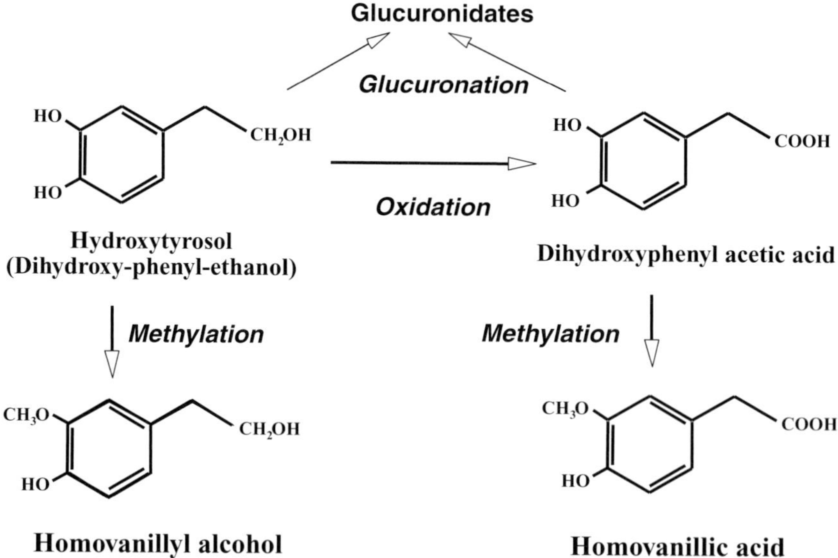

Fig. 10.3 Human metabolism of olive phenolics. Olive phenols are dose dependently absorbed by humans, and they are then extensively conjugated. The major proportion is recovered in urine as glucoronide conjugates. It is noteworthy that homovanillyl alcohol is a marker of extra-virgin olive oil consumption: its urinary concentrations depend on intake.

2001; Miro-Casas *et al.*, 2003). The most complete study in this area is from Miro-Casas and collaborators, who developed a method to quantify hydroxytyrosol and its metabolites in plasma (Miro-Casas *et al.*, 2003). In brief, absorption of hydroxytyrosol in nearly complete and its plasma half-life is 2.43 hours (Miro-Casas *et al.*, 2003). It is noteworthy that HT exists in the brain as an endogenous catabolite of catecholic neurotransmitters, such as dopamine and norepinephrine, but its presence in urine has never, until recently, been described. On the other hand, the formation of homovanillic alcohol (HVAlc), the *O*-methylated derivative of HT, was reported by Manna *et al.* (1999) in human Caco-2 cell incubated with HT. We also reported the urinary excretion of HVAlc in large excess over its basal excretion (57 ± 3 μg excreted in 24 hours, means ± SD, $n = 6$) and described the substrate-induced enhancement of HVA formation, also a product of catecholamines metabolism, in addition to its basal urinary excretion (1660 ± 350 μg excreted in 24 hours, means ± SD, $n = 6$) (Caruso *et al.*, 2001). Indeed, the results reported suggest that HT increases the basal excretion of HVA, even at the low doses of phenols administered. Future investigations will adopt commercially available virgin olive oils, thus allowing the further elucidation of the *in vivo* kinetics of olive oil phenolics in usual consumption quantities. It is noteworthy that, after intake, hydroxytyrosol associates with lipoproteins (Bonanome *et al.*, 2000; Gonzalez-Santiago *et al.*, 2010), which confers this molecule the potential to decrease lipoprotein oxidizability. Finally, the metabolism of hydroxytyrosol differs greatly between rats and humans and depends on the matrix with which hydroxytyrosol is administered (Visioli *et al.*, 2003).

In terms of biological activities, Covas *et al.* recently reviewed approximately 15 human intervention studies, the vast majority of which indicate that EVOO (rich in phenols) is superior to seed oils and olive oil devoid of phenols in modulating selected surrogate markers of cardiovascular disease (Covas, 2007). It is of note that while some human experiments have been performed using doses of olive oil that do not approximate habitual consumption, several others did employ more realistic quantities (Covas, 2007). One example is an investigation of the effects of olive oil phenols on postprandial events. Bogani *et al.* (2007) evaluated the effects of moderate, real-life doses of two olive oils, differing only in their phenolic content, on some *in vivo* indexes of oxidative stress (plasma antioxidant capacity and urinary hydrogen peroxide levels) in a postprandial setting. Moreover, the authors assessed whether phenolic compounds influence a few arachidonic acid metabolites involved in the atherosclerotic processes, such as leukotriene B$_4$ (LTB$_4$) and thromboxane B$_2$ (TXB$_2$). Six subjects in each group received the three oils (30 mL/day of olive oil (OO), corn oil, or EVOO, distributed among meals) in a Latin square design. The results (Fig. 10.4) demonstrate that EVOO is capable of reducing the postprandial events that associate with inflammation and oxidative stress (Bogani *et al.*, 2007).

Other studies investigated the effects of olive phenols on specific risk markers. The effect of EVOO on platelet aggregation and plasma concentrations of homocysteine (Hcy) redox forms, in relation to the phenolic compounds' concentration, was also investigated in rats (Priora *et al.*, 2008). Three olive oil samples with similar fatty acid but different phenolic compound concentrations were used: refined olive oil (RF) with traces of phenolic compounds (control oil), native EVOO with low phenolic compounds concentration (LC), and EVOO with high phenolic compounds concentration (HC) enriching LC with its own

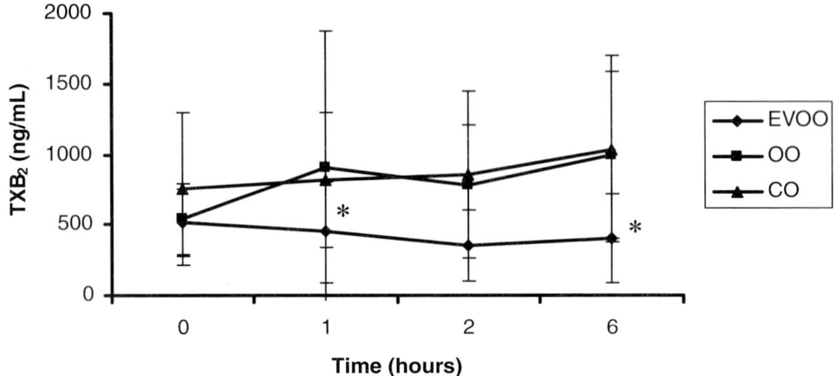

Fig. 10.4   Thromboxane B_2 (TXB$_2$, µg/mL serum) production by whole blood drawn at the indicated times (days) from volunteers who were administered either refined or extra-virgin olive oil (EVOO) in a cross-over fashion. Data are from Bogani *et al.* (2007). Blood was incubated at 37°C for 1 hour and serum was separated by centrifugation. TXB$_2$ was measured by a commercial kit. $p < 0.05$ with respect to T_0; $^*p < 0.05$ with respect to WO; $^\#p < 0.05$ with respect to T_{49}.

phenolic compounds. Oil samples were administered to rats by gavage (1.25 mL/kg body weight) using two experimental designs: acute (24-hour food deprivation and killed 1 hour after oil administration) and subacute (12-day treatment, a daily dose of oil for 12 days, and killed after 24 hours of food deprivation).

Platelet aggregability is a proxy of thrombogenic potential and an important marker of cardiovascular risk. Platelet aggregation was induced by ADP (*ex vivo* tests) and a reduction in platelet reactivity occurred in cells from rats given LC in the subacute study and in cells from rats administered HC in both studies, as indicated by an increase in the agonist half maximal effective concentration. HC inhibited platelet aggregation induced by low ADP doses (reversible aggregation) in cells of rats in both the acute and subacute studies, whereas LC had this effect only in the subacute experiment. Moreover, in rats administered HC in both experiments, the plasma concentration of free reduced Hcy (rHcy) was lower and Hcy bound to protein by disulfide bonds (bHcy) was greater than in RF-treated rats. bHcy was also greater in rats given LC than in RF-treated rats in the subacute experiment. Plasma free-oxidized Hcy was greater in rats given LC and HC than in those administered RF only in the subacute experiment. These results, which add to the body of *in vivo*, namely, animal, literature (Deiana *et al.*, 2007; Fki *et al.*, 2007) show that phenolic compounds in EVOO inhibit platelet aggregation and reduce the plasma rHcy concentration, effects that may be associated with cardiovascular protection (Assanelli *et al.*, 2004; Priora *et al.*, 2008).

The most complete human study to date is the Eurolive study, led by the group of Covas (Covas *et al.*, 2006). In this important trial, the authors enrolled 200 subjects (healthy volunteers) and evaluated the effects of three different oils, namely, with low, medium, and high polyphenol content, on plasma lipids and on circulating markers of oxidative stress. The results show that olive phenols afford a modest, but significant protection toward oxidative stress, as evaluated *in vivo* (Table 10.1). Given the large number of subjects involved, these data likely apply to the overall population. One comment on this study is warranted, even

Table 10.1 Effects of olive oils with low, medium, or high polyphenol content on markers of oxidative stress, in healthy volunteers.

Variable	Low pp	Medium pp	High pp	P
CD (mmol/mol cho)	2.61	2.55	2.37	0.011
Hydroxy FA (nmol/L)	179	176	157	0.038
Ox LDL (U/L)	48	47	46	0.014
F_2-isoprostanes (mmol/L)	28.3	28.7	28.1	0.34

Source: Data are from Covas *et al.* (2006).
CD, conjugated dienes; FA, fatty acids; LDL, low-density lipoprotein; pp, polyphenols.

more so within the context of this book. The changes that were recorded were of low magnitude. The same reasoning applies to the Bogani *et al.* study outlined in the preceding text. It should be remembered that olive oil (and, for that matter, any other food item) is not a drug. The changes its components induce in the human body are necessarily small. However, such changes occur through an entire lifetime. In turn, the intake of salubrious foods modifies risk factors in a way that brings about positive health consequences long term. While it is difficult to discriminate the effects of a single food item and to single them out of a complex lifestyle and associated diet, the notion that bioactive components of food exert healthful activities is gaining scientific evidence.

At the same time, the difference between food (including olive oil) and medicines calls for moderation when the "medicinal" properties of individual food items are heralded. It is not scientifically correct to suggest that one single food item, be it olive oil or green tea or tomato, can cure or prevent disease in a remarkable way. The correct notion should be to select foods whose components have proven, albeit limited in magnitude, biological activities and build a balanced diet round them, to decrease cardiovascular and cancer risk.

10.5 Olive oil and cancer

Epidemiological evidence has shown an inverse correlation between olive oil consumption and cancer in different sites (Martin-Moreno *et al.*, 1994; La Vecchia *et al.*, 1995; Braga *et al.*, 1998; Franceschi *et al.*, 1999; Norrish *et al.*, 2000; Hodge *et al.*, 2004), notably versus breast cancer (Martin-Moreno *et al.*, 1994; La Vecchia *et al.*, 1995; Gerber, 1997; Garcia-Segovia *et al.*, 2006). Some animal studies also demonstrated a protective activity against chemically induced carcinogenesis, such as dimethylbenz[a]anthracene-induced mammary tumors and azoxymethane-induced colon carcinoma (Fabiani *et al.*, 2009). Furthermore, olive oil may protect from UV-induced skin cancer (Ichihashi *et al.*, 2000) and it reduces the incidence of spontaneous appearance of liver tumors in mice (Thuy *et al.*, 2001).

More recently, some intervention studies have investigated the DNA protective potential of olive oil phenols with conflicting results. Weinbrenner *et al.* (2004) found a decreased amount of 8-oxo-7,8-dihydro-2′-deoxyguanosine in mitochondrial DNA of mononuclear cells and in urine after short-term consumption of olive oil with a linear trend significantly correlated to the content of phenols. Similarly, Salvini *et al.* (2006) showed a 30% reduction of oxidative DNA damage in peripheral blood lymphocytes during intervention on post-menopausal women with virgin olive oil containing high amounts of phenols. On the other hand, no significant effect was detected on urinary excretion of etheno-DNA adducts after consumption of phenol-rich olive oil (Hillestrom *et al.*, 2006). Furthermore, it was shown that the urinary excretion of oxidation products of guanine was not modified after intake of olive oil with low, medium, and high phenolic content (Machowetz *et al.*, 2007). In turn, olive oil phenolics may possess anticarcinogenic properties, never clearly demonstrated *in vivo* (Hillestrom *et al.*, 2006).

10.6 Potential mechanisms of action of olive phenols—to be elucidated

To counteract the potentially noxious effects of xenobiotics, higher vertebrates, including humans, have developed a battery of genes encoding phase II and antioxidant enzyme expression. These include the expression of various superoxide dismutase (SOD) isoforms, catalase, glutathione peroxidase, glutathione reductase, various glutathione-*S*-transferase (GST) isoforms, NAD(P)H: quinone oxidoreductase 1 (NQO1), and heme oxygenase (HO)-1, which can exert cytoprotective, antioxidant, and anti-inflammatory effects. One mechanism by which cells respond to oxidative injury is through the antioxidant response element (ARE), a cis-acting enhancer sequence that regulates the transcription of various cytoprotective genes. Upon toxic injury, the transcription factor nuclear factor erythroid 2-related factor 2 (Nrf2) translocates to the nucleus and dimerizes with small Maf proteins to form a transactivation complex that, consequently, binds to the ARE. Nrf2-induced ARE activation coordinates the expression of many genes involved in combating oxidative stress and toxicity in a broad range of tissues and cell types (Fig. 10.5).

One of the hallmarks of illness is a reduced capacity of cellular homeostatic mechanisms that protect the body against a variety of oxidative, toxicological, and pathological insults. A notable example is the age-related decline in hepatic glutathione (GSH) levels. This leads to a loss of antixenobiotic (phase II) enzymes, which detoxify the organs by eliminating most of the insults mentioned in the preceding text. Due to the central role of GSH in cellular protective mechanisms, the induction of enzymes required for its synthesis represents a key adaptive response to oxidative injury. Sometimes, however, when levels of oxidative stress become elevated, GSH and the enzymes from which it is synthesized do not concomitantly increase but actually decline in many tissues. This lack of a cellular compensatory response to loss in GSH and the existence of a pro-oxidant state in aging cells suggest that the coordination of cellular antioxidant defenses may be altered with age and with illness. A potential loss in transcriptional regulation associated with illness and/or aging may be indicative of a global decline in phase II defense systems.

A previous study showed that the administration of olive mill wastewater (OMWW) increases total circulating GSH concentrations in healthy human volunteers (Visioli *et al.*, 2009). These data suggest that OMWW and its components, e.g., hydroxytyrosol, are able to stimulate phase II enzymes and, consequently, GSH synthesis and the detoxification

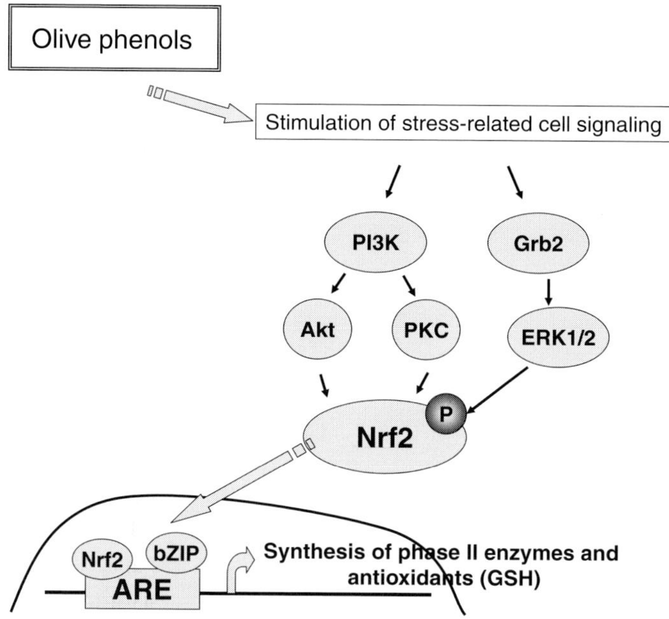

Fig. 10.5 Proposed mechanism of action of olive phenols, which stimulate the endogenous response to produce antioxidant and detoxifying molecules. This scheme is proposed on the basis of the results of Visioli *et al.* (2009).

pathways. Indeed, two recent papers confirm this hypothesis, at least *in vitro* (Martin *et al.*, 2010; Zhu *et al.*, 2010).

In summary, the observed effects of OMWW on glutathione levels might be governed by the ARE-mediated increase in phase II enzyme, namely, GCL and GSH synthetase expression, in turn, leading to enhance glutathione synthesis. If confirmed, the potential applications of these preliminary findings are manifold and could extend to patients with reduced circulating glutathione levels, for example, patients on chronic hemodialysis, patients suffering from Alzheimer or HIV infection. Furthermore, elderly people may potentially benefit from an increase of phase II enzyme activity and of glutathione levels.

Because numerous and diverse compounds activate phase II detoxification pathways and many similar compounds are known to be present in OMWW, our hypothesis is that it will be an effective adjunct to boosting antioxidant and detoxification pathways in otherwise susceptible populations.

10.7 Focus on hydroxytyrosol

Hydroxytyrosol is peculiar to olives (and, hence, to olive oil) and is being exploited as a potential supplement or preservative to be employed in the nutraceutical, cosmeceutical, and food industry. It is noteworthy that the number of publications on hydroxytyrosol surpassed that on oleuropein in the 2000 (Fig. 10.6). This indicates increasing interest for this molecule, which can be currently synthesized (as opposed to oleuropein) and for

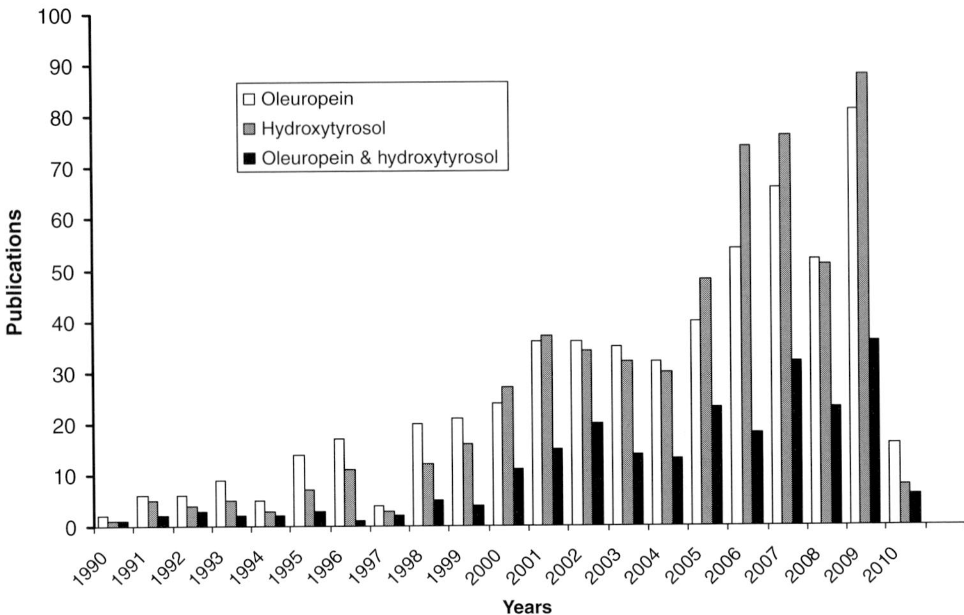

Fig. 10.6   Year-by-year (from January 1, 1985 to December 31, 2010) analysis of publications on oleuropein, hydroxytyrosol, or both combined. Data were obtained by searching Web of Science: Science citation index 1985–2010 (Thomson Reuters). Research strings were as follows: TS = (Hydroxytyrosol) Databases = SCI-EXPANDED Timespan = All Years; Topic = (Oleuropein) Databases = SCI-EXPANDED Timespan = All Years; TS = (Hydroxytyrosol AND Oleuropein) Databases = SCI-EXPANDED Timespan = All Years.

which industrial exploitation is envisaged. Remarkable dates are 1994 (first report of the antioxidant properties of oleuropein in an *in vitro* model of human LDL oxidation; Visioli & Galli, 1994); 1995 (exploration of the antioxidant activities of hydroxytyrosol; Visioli *et al.*, 1995a); 2000 (first demonstration of the human absorption of olive oil phenolics; Visioli *et al.*, 2000b); 2006 (publication of the Eurolive study; Estruch *et al.*, 2006); 2007 (first *in vivo* evidence of neuroprotection afforded by hydroxytyrosol; Schaffer *et al.*, 2007). As expected, most of the publications come from Italy and Spain, two countries where olive oil production is elevated and where research is being actively pursued (Fig. 10.7).

The biological activities of hydroxytyrosol can be summarized as follows:

(1) *Antioxidant activity*: Hydroxytyrosol is a potent inhibitor of metal-induced oxidation of LDL. In addition, metal-independent oxidation is also significantly retarded by hydroxytyrosol. The antioxidant activities of hydroxytyrosol, which has been proven to be more effective than BHT or vitamin E, were further confirmed, by the use of stable free radicals, such as DPPH, Also, hydroxytyrosol is a scavenger of superoxide anions generated by either human polymorphonuclear cells or by the xanthine/xanthine oxidase system. Furthermore, a scavenging effect of hydroxytyrosol was demonstrated with respect to hypochlorous acid, a potent oxidant produced *in vivo* at the site of inflammation and a major component of chlorine-based bleaches that can often come into contact with food during manufacturing. Antioxidant activities have also been demonstrated

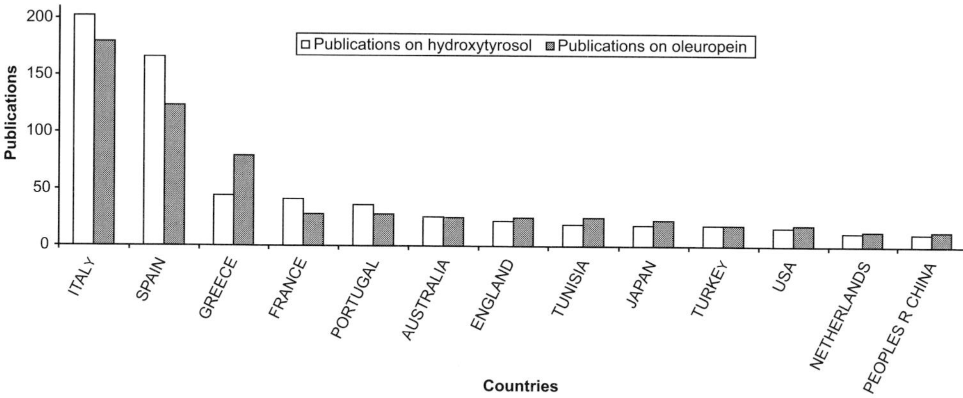

Fig. 10.7   Per-country analysis of publications on oleuropein and hydroxytyrosol. Data were obtained by searching Web of Science, Science citation index 1985–2010 (Thomson Reuters). Research strings were as follows: TS = (Hydroxytyrosol) Databases = SCI-EXPANDED Timespan = All Years; Topic = (Oleuropein) Databases = SCI-EXPANDED Timespan = All Years. Only countries with a number of publications >10 on both topics are reported.

versus DNA damage, hydrogen peroxide-induced insult to red blood cells. These results have been discussed in the preceding text and can be found in several reviews (e.g., Perez-Jimenez *et al.*, 2005). The signaling pathways involved in the biological activities of hydroxytyrosol are being elucidated (Corona *et al.*, 2009; Incani *et al.*, 2010) and will likely be the subject of several future investigations.

(2) *Interference with enzymes*: Hydroxytyrosol is able to modulate several enzymatic activities linked to cardiovascular disease. Among them, inhibition of platelet aggregation and pro-inflammatory enzymes such as 5-lipoxygenase, and stimulation of the inducible form of nitric oxide synthase have been demonstrated *in vitro*. In *in vitro* models, hydroxytyrosol is not able to upregulate the activity of the endothelial form of nitric oxide synthase, leaving its role in modulation of vasomotion unresolved. While the majority of data have been obtained *in vitro*, several experiments have been performed in laboratory animals. In addition, there are approximately 15 human experiments that compared olive oil with EVOO (which, however, contains phenols other than hydroxytyrosol) (Covas *et al.*, 2009). Finally, hydroxytyrosol and related olive phenols have been tested, as supplements, in humans. The most notable result is the inhibition of thromboxane B_2 production by whole blood, suggesting antithrombotic activity *in vivo* (Visioli & Galli, 2003; Leger *et al.*, 2005).

Animal experiments confirm, *in vivo*, most of the evidence obtained *in vitro*. In particular, hydroxytyrosol retains its antioxidant activity once ingested (though the human metabolic pathway has been elucidated and shows extensive glucuronidation and subsequent urinary excretion) (Visioli *et al.*, 2001), protects from second-hand smoke-induced oxidative damage (Visioli *et al.*, 2000d), inhibits platelet aggregation (Priora *et al.*, 2008), increases brain cell resistance to oxidation and mitochondrial membrane potential (Schaffer *et al.*, 2007). Human experiments initially showed that hydroxytyrosol from

EVOO is absorbed and excreted in the urine (Visioli *et al.*, 2000b). Further experiments confirmed hydroxytyrosol's anti-inflammatory (Bitler *et al.*, 2005) and antithrombotic potential (Visioli & Galli, 2003; Leger *et al.*, 2005) and its ability to ameliorate osteoarthritis (Bitler *et al.*, 2007). As a caveat, such experiments have been performed with mixtures of olive phenols in which hydroxytyrosol was the most active ingredient, but not the exclusive one. Synergy with other olive phenols cannot, at present, be excluded.

The safety profile of hydroxytyrosol appears to be excellent (Babich & Visioli, 2003): no untoward effects have been demonstrated even at very high doses (D'Angelo *et al.*, 2001; Christian *et al.*, 2004; Soni *et al.*, 2006). Hence, hydroxytyrosol has been granted a GRAS status. However, one study showed increased development of atherosclerosis in apoE knock out mice administered hydroxytyrosol (Acin *et al.*, 2006). This is in contrast with a rabbit study in which hydroxytyrosol exerted antiatherosclerotic effects (Gonzalez-Santiago *et al.*, 2006). It is worth mentioning that, in another rabbit study, the administration of resveratrol (a molecule for which human studies are lacking even though the lay press heralds it as a panacea) increased arterial thickness without affecting lipid metabolism or overall antioxidant status (Wilson *et al.*, 1996). In synthesis, hydroxytyrosol is currently one of the most actively investigated natural phenols and is endowed with interesting pharmacological activities, many of which have been demonstrated *in vivo*. Given its excellent safety profile, future availability as human supplement might be conceivable (Granados-Principal *et al.*, 2010).

It is noteworthy that the large number of patents on hydroxytyrosol, oleuropein and other olive phenols is continually increasing (Tables 10.2 and 10.3). Several industrial applications are being envisioned, although—to the best of our knowledge—there is currently only one supplement based on OMWW that is being sold in the American and Asian markets. However, due to the continual production of data in various areas of human health, other products might be launched in the near future in the fields of cosmetology or food science. Finally, new derivatives of hydroxytyrosol, namely, its methylesters, are being developed and tested for potential future applications (Tofani *et al.*, 2010).

It must be underlined that hydroxytyrosol is not the exclusive responsible for the healthful effects of olive phenols. Other molecules—as outlined in this article—contribute to the effects and, at present, their individual role is difficult to single out.

10.8 Olive mill waste water as a source of olive phenols

Another interesting and emerging facet of olive oil minor components' biology is the recent development of their recovery from waste waters for nutraceutical purposes (Agalias *et al.*, 2007; Obied *et al.*, 2008). It is noteworthy that the olive paste is continuously hosed with lukewarm water during the milling, a process that is called malaxation (Visioli *et al.*, 1995b). The resulting "waste water" (OMWW) is produced in extremely large quantities (~800,000 tons/year in Italy) and, despite the fact that it contains a considerable amount of phenols (more than 1% w/v), is currently disposed of. More than a decade ago, Visioli *et al.* demonstrated that waste water extracts have powerful (in the ppm range) *in vitro* antioxidant

Table 10.2   Patents on oleuropein.

PN: WO2010082600   A1   20100722   [WO201082600]
AP: 2010WO-JP50338 20100114
TI: AGENT FOR MAINTAINING HEALTHY STATE OF PERIODONTAL TISSUE COMPRISING
       OLEUROPEIN AND DEGRADED PRODUCT THEREOF
IN: TANIGUCHI MASAYO; KOBAYASHI ETSUKO; YASUDA TAKAKO; OGAWA SHOKO
PR: 2009JP-0006851 20090115

PN: WO2010070183   A1   20100624   [WO201070183]
AP: 2009WO-ES70598 20091217
TI: PHARMACEUTICAL COMPOSITION COMPRISING OLEUROPEIN FOR UTILIZATION IN
       ANGIOGENESIS AND VASCULOGENESIS INDUCTION
IN: QUESADA GOMEZ JOSE MANUEL; SANTIAGO MORA RAQUEL MARIA; CASADO DIAZ
       ANTONIO
PR: 2009ES-0000009 20081218

PN: JP2009227616   A   20091008   [JP2009227616]
AP: 2008JP-0075967 20080324
TI: OSTEOGENESIS PROMOTER CONTAINING OLEUROPEIN AND/OR HYDROXYTYROSOL AS
       ACTIVE INGREDIENT
IN: HAGIWARA YOSHIMI; MIYAZAKI HITOSHI
PR: 2008JP-0075967 20080324

PN: JP2009191012   A   20090827   [JP2009191012]
AP: 2008JP-0033153 20080214
TI: INFERTILITY TREATING AGENT CONTAINING OLEUROPEIN, OLEUROPEIN DERIVATIVE OR
       HYDROXYTYROSOL AS ACTIVE COMPONENT
IN: MIYAZAKI HITOSHI; ISODA HIROKO; WAKABAYASHI YUKI; MOKTAR ZARROUK
PR: 2008JP-0033153 20080214

PN: HR20070332   A2   20090228   [HR20070332]
AP: 2007HR-0000332 20070720
      HR20070332   B3   20100731   [HR20070332]
TI: PRECESS FOR PRODUCING OLEUROPEIN EXTRACT USING ULTRASOUND EXTRACTION
IN: GROSS SRECKO
PR: 2007HR-0000332 20070720

PN: ITMI20080514   A1   20090928   [IT2008MI0514]   WO2009118380
      A1   20091001   [WO2009118380]
AP: 2008IT-MI00514 20080327   2009WO-EP53596 20090326
TI: USE OF OLEUROPEIN AND DERIVATIVES IN THE TREATMENT OF TYPE 2 DIABETES
       MELLITUS AND PATHOLOGIES ASSOCIATED WITH PROTEIN AGGREGATION PHENOMENA
IN: BERTI ANDREA; STEFANI MASSIMO; RIGACCI STEFANIA
PR: 2008IT-MI00514 20080327

PN: WO2008136037   A2   20081113   [WO2008136037]   WO2008136037
      A3   20081224   [WO2008136037]   EP2235032   A2   20101006   [EP2235032]
AP: 2008WO-IT00303 20080505 2008EP-0763854 20080505
TI: CHEMICAL-CATALYTIC METHOD FOR THE PERACYLATION OF OLEUROPEIN AND ITS
       PRODUCTS OF HYDROLYSIS
IN: PROCOPIO ANTONIO; SINDONA GIOVANNI; COSTA NICOLA; GASPARI MARCO; NARDI
       MONICA
PR: 2007IT-MI00903 20070504 2007IT-MI00904 20070504 2008WO-IT00303 20080505

PN: KR20080081636   A   20080910   [KR20080081636]   WO2008108817
      A1   20080912   [WO2008108817]   AU2007202276
      A1   20080925   [AU2007202276]   KR100878549   B1   20090114   [KR100878549]

(Continued)

Table 10.2   (*Continued*)

AP: 2007KR-0021940 20070306 2007WO-US20953 20070928 2007AU-0202276 20070521
TI: NUTRITIONAL SUPPLEMENT WITH OLEUROPEIN
IN: HALL RICHARD; KWON JAMES
PR: 2007KR-0021940 20070306

PN: KR100834444   B1   20080609   [KR100834444]
AP: 2007KR-0004199 20070115
TI: PHARMACEUTICAL COMPOSITION COMPRISING OLEUROPEIN ISOLATED FROM FRAXINUS
     RHYNCHOPHYLLA FOR PREVENTION AND TREATING TOXOPLASMOSIS
IN: KIM YOUN CHUL; BAK HYEON; JIANG JING HUA
PR: 2007KR-0004199 20070115

PN: CN101003557   A   20070725   [CN101003557]   CN100473656
     C   20090401   [CN100473656C]
AP: 2006CN-0098344 20061212
TI: Method for preparing extractive of olive leaves rich in oleuropein in high purity
IN: WANG CHENGZHANG GAO
PR: 2006CN-0098344 20061212

PN: KR20080026017   A   20080324   [KR20080026017]
AP: 2007KR-0028060 20070322
TI: THE COMPOSITION WITH OLIVE LEAF EXTRACT WHICH IS CONTAINED AN OLEUROPEIN
     AS MAJOR COMPONENT AND THE MANUFACTURING METHOD OF THE SAME
IN: JAMES KWON
PR: 2006KR-0090527 20060919

PN: EP1795201   A1   20070613   [EP1795201]
AP: 2006EP-0386039 20061204
TI: Isolation of oleuropein from the leaves of olive tree
IN: KEFALAS PANAGIOTIS
PR: 2005GR-0100601 20051209

PN: WO2007051829   A1   20070510   [WO200751829]   ES2274721
     A1   20070516   [ES2274721]   ES2274721   B1   20080601   [ES2274721]   EP1954293
     A1   20080813   [EP1954293]   US2008227728   A1   20080918   [US20080227728]
AP: 2006WO-EP68053 20061103 2005ES-0002712 20051103 2006EP-0829930 20061103 2006US-0092549
     20061103
TI: OLEUROPEIN FOR THE TREATMENT OF CONDITIONS ASSOCIATED TO THE PERIPHERAL
     VASCULAR DISEASE
IN: CARVAJAL MARTIN LUIS; SEVILLA TIRADO FRANCISCO JAVIE; AZNAR ANTONANZAS
     YOLANDA
PR: 2005ES-0002712 20051103 2006WO-EP68053 20061103

PN: KR20050091672   A   20050915   [KR20050091672]
AP: 2005KR-0078709 20050826
TI: CHEWING GUM COMPOSITION OF CONTAINING OLEUROPEIN
IN: JUN YOUNG SOO
PR: 2005KR-0078709 20050826

PN: US2005103711   A1   20050519   [US20050103711]
AP: 2003US-0714967 20031118
TI: Isolation of oleuropein aglycon from olive vegetation water
IN: EMMONS WAYNE; GUTTERSEN CONNIE
PR: 2003US-0714967 20031118

PN: EP1489135   A1   20041222   [EP1489135]   WO2004113431
     A1   20041229   [WO2004113431]
AP: 2003EP-0076877 20030617 2004WO-EP06575 20040617

Table 10.2   (*Continued*)

TI: Use of secoiridoids, preferably oleuropein, for cross-linking biopolymers
IN: DRIEHUIS FRANK; FLORIS THEODORUS A G; DE KRUIF CORNELIS G
PR: 2003EP-0076877 20030617

PN: FR2853549 A1 20041015 [FR2853549]   WO2004091591
     A2 20041028 [WO200491591]   CA2521967
     A1 20041028 [CA2521967]   WO2004091591
     A3 20041125 [WO200491591]   EP1617836
     A2 20060125 [EP1617836]   US2006193931
     A1 20060831 [US20060193931]   FR2853549 B1 20071109 [FR2853549]
AP: 2003FR-0004584 20030411 2004WO-FR50156 20040409 2004CA-2521967 20040409 2004EP-0742843
     20040409 2005US-0552723 20051011
TI: NUTRITIONAL OR THERAPEUTIC COMPOSITION CONTAINING THE COMPOUND
     OLEUROPEINE OR ONE OF THE DERIVATIVES THEREOF
IN: COXAM VERONIQUE; SKALTSOUNIS LEANDROS; PUEL CAROLINE; MAZUR ANDRE
PR: 2003FR-0004584 20030411 2004WO-FR50156 20040409

PN: JP2003335693 A 20031125 [JP2003335693]   JP4405135 B2 20100127 [JP4405135]
AP: 2002JP-0142915 20020517
TI: DRIED OLIVE LEAF HAVING HIGH OLEUROPEIN CONTENT AND EXTRACT OF THE LEAF
IN: HATAKE SHUICHI; NAKAMURA HIROMICHI
PR: 2002JP-0142915 20020517

PN: WO02094193 A1 20021128 [WO200294193]   CA2447231
     A1 20021128 [CA2447231]   US2003004117
     A1 20030102 [US20030004117]   US6632798
     B2 20031014 [US6632798]   US2004048808
     A1 20040311 [US20040048808]   EP1397105
     A1 20040317 [EP1397105]   IL158899 D0 20040512 [IL-158899]
     ZA200308763 A 20040526 [ZA200308763]   BR0209922
     A 20040727 [BR200209922]   CN1531435 A 20040922 [CN1531435]   JP2005508856
     T 20050407 [JP2005508856] MXPA03010528 A 20050419 [MX2003PA010528]   EP1397105
     B1 20060726 [EP1397105]   AT333882
     T 20060815 [ATE333882]   DE60213407 D1 20060907 [DE60213407]   DK1397105
     T3 20061127 [DK1397105T]   PT1397105
     E 20061229 [PT1397105]   ES2266512 T3 20070301 [ES2266512]   CN1315479
     C 20070516 [CN1315479C]   AU2002311985
     B2 20080410 [AU2002311985]   AU2002311985 B9 20080710 [AU2002311985]
AP: 2002WO-US16191 20020522 2002CA-2447231 20020522 2002US-0153003 20020522
     2003US-0657414 20030908 2002EP-0739332 20020522 2002IL-0158899 20020522 2003ZA-0008763
     20031111 2002BR-0009922 20020522 2002CN-0812125 20020522 2002JP-0590914 20020522
     2003MX-PA10528 20031117 2002AT-0739332 20020522 2002DE-6013407 20020522 2002DK-0739332
     20020522 2002PT-0739332 20020522 2002ES-0739332 20020522
     DE60213407 T2 20070823 [DE60213407]   2002AU-0311985 20020522
TI: COMPOSITIONS FOR INHIBITING ANGIOGENESIS
IN: HAMDI HAMDI K; TAVIS JEFFREY H; CASTELLON RAQUEL
PR: 2001US-P292947 20010523 2002US-0153003 20020522 2002WO-US16191 20020522
     2003US-0657414 20030908

PN: JP2002128678 A 20020509 [JP2002128678]
AP: 2000JP-0316811 20001017
TI: METHOD FOR PRODUCING EXTRACT COMPOSITION CONTAINING OLEUROPEIN
IN: HATAKE SHUICHI
PR: 2000JP-0316811 20001017

(*Continued*)

Table 10.2   (*Continued*)

PN:	WO200183031 A2 20011108 [WO200183031] AU5931101 A 20011112 [AU200159311] WO200183031 A3 20020411 [WO200183031] US6440465 B1 20020827 [US6440465] NO20025263 D0 20021101 [NO200205263] US2002164386 A1 20021107 [US20020164386] NO20025263 A 20021219 [NO200205263] EP1278574 A2 20030129 [EP1278574] IL152592 D0 20030529 [IL-152592] CZ20023854 A3 20030618 [CZ200203854] MXPA02010799 A 20040906 [MX2002PA010799] NZ522709 A 20050729 [NZ-522709] EP1278574 B1 20081029 [EP1278574] AT412417 T 20081115 [ATE412417] DE60136348 D1 20081211 [DE60136348] IL152592 A 20081229 [IL-152592] US7670620 B2 20100302 [US7670620]
AP:	2001WO-US14011 20010501 2001AU-0059311 20010501 2000US-0562400 20000501 2002NO-0005263 20021101 2002US-0185388 20020628 2001EP-0932815 20010501 2001IL-0152592 20010501 2002CZ-0003854 20010501 2002MX-PA10799 20021031 2001NZ-0522709 20010501 2001AT-0932815 20010501 2001DE-6036348 20010501
TI:	TOPICAL COMPOSITION FOR THE TREATMENT OF PSORIASIS AND RELATED SKIN DISORDERS
IN:	MEISNER LORRAINE F
PR:	2000US-0562400 20000501 2001WO-US14011 20010501 2002US-0185388 20020628

Source: Data were obtained by searching the QPAT database (Questel©). Time limits were January 1990 to November 2010.

activity (Visioli *et al.*, 1995b; Visioli *et al.*, 1999); thus, OMWW could be recovered and employed as a cheap source of natural antioxidants. Indeed, animal experiments (Visioli *et al.*, 2000d, 2001; Fki *et al.*, 2007) and a couple of human studies (Visioli & Galli, 2003; Leger *et al.*, 2005) confirmed that waste waters are a source of bioactive phenols and dietary supplements derived from OMWW are already available in the market. The latest of such studies showed that OMWW increases glutathione levels in healthy volunteers (Visioli *et al.*, 2009). This activity can be explained by the induction of phase II enzymes mentioned in the preceding text. If these findings are replicated, they might provide the bases to explain, at least in part, the chemopreventive actions attributed to olive oil, in that activation of phase II enzymes and subsequent detoxification of noxious compounds would lower the risk of mutagenesis.

Recently, Schaffer *et al.* addressed the question of whether hydroxytyrosol, as a component of OMWW, could modulate neuronal activities (Schaffer *et al.*, 2007). The results, obtained in mice, demonstrated that feeding HT-rich OMWW extract could improve neuronal function (as assessed by mitochondrial membrane potential) and provide antioxidant protection. While much has to be done to prove or disprove the neuroprotective activities of olive phenols, these data provide partial explanation for the observed higher cognitive capacity of subjects who consume high proportion of olive oil as predominant source of fat (Solfrizzi *et al.*, 2008, 2010).

10.9 Conclusions

As of 2011, the cardioprotective activities of olive phenols are quite established, although the extent and precise nature of such actions needs further clarification. It must be underlined

Table 10.3 Patents on olive mill waste water.

PN: GR1006942   B   20100827   [GR1006942]
AP: 2009GR-0100480 20090907
TI: INTEGRATED OPERATIONAL SYSTEM FOR THE MANAGEMENT OF OLIVE WASTE IN OIL FACTORIES
IN: TZANETOS EMMANOUIL; GOMOS EFSTRATIOS DIMITRIOU
PR: 2009GR-0100480 20090907

PN: WO2010086480   A1   20100805   [WO201086480]
AP: 2010WO-ES00042 20100127
TI: METHOD FOR THE PREPARATION OF PRODUCTS HAVING A HIGH TRITERPENE CONTENT AND RESULTING PRODUCTS
IN: GARCIA-GRANADOS LOPEZ DE HIERRO ANDRES; PARRA SANCHEZ ANDRES; MARTINEZ RODRIGUEZ ANTONIO; RIVAS SANCHEZ FRANCISCO
PR: 2009ES-0000346 20090127 2009ES-0001156 20090428

PN: WO2009060461   A2   20090514   [WO200960461] WO2009060461
     A3   20091126   [WO200960461]
AP: 2008WO-IL01477 20081110
TI: METHOD AND APPARATUS FOR PRODUCING FUEL GAS FROM BIOMASS
IN: BILDER ZAKHARI; GELFMAN VLADIMIR; VLADISLAWSKY YURI
PR: 2007US-P987040 20071110

PN: WO2008082343   A1   20080710   [WO200882343]   EP2102110
     A1   20090923   [EP2102110]   US2010240769   A1   20100923   [US20100240769]
AP: 2007WO-SE01177 20070228 2007EP-0852160 20070228 2007US-0521689 20070228
TI: OLIVE WASTE RECOVERY
IN: TORNBERG EVA; GALANAKIS CHARIS
PR: 2006SE-0002835 20061229 2007WO-SE01177 20070228

PN: EP1852495   A1   20071107   [EP1852495]   ZA200703861   A   20080827   [ZA200703861]
AP: 2007EP-0251846 20070502 2007ZA-0003861 20070514
TI: Method of manufacturing a combustible material
IN: VASILLOS PROKOPOS
PR: 2006ZA-0003447 20060502 2007ZA-0003861 20070514

PN: IL172991   D0   20060611   [IL-172991]   WO2007077553
     A2   20070712   [WO200777553]   EP1979447   A2   20081015   [EP1979447]   US2009031619
     A1   20090205   [US20090031619] WO2007077553   A3   20090409   [WO200777553]
AP: 2006IL-0172991 20060105 2006WO-IL01471 20061221 2006EP-0821654 20061221 2006US-0087564
     20061221
TI: SOLID FUEL
IN: MAMAN ELYAKIM; SEGNER YUVAL
PR: 2006IL-0172991 20060105 2006WO-IL01471 20061221

PN: US2007077279   A1   20070405   [US20070077279]   WO2007039262
     A1   20070412   [WO200739262]   EP1928268   A1   20080611   [EP1928268] KR20080052617
     A   20080611   [KR20080052617] IN0930/DELNP/2008   A   20080627   [IN2008DN00930]
     CN101277618   A   20081001   [CN101277618] JP2009510003   T   20090312   [JP2009510003]
AP: 2006US-0540955 20061002 2006WO-EP09527 20061002 2006EP-0792349 20061002 2008KR-7007555
     20080328 2008IN-DN00930 20080201 2006CN-80036429 20061002 2008JP-0532685 20061002
TI: NOVEL COMPOSITIONS CONTAINING POLYPHENOLS
IN: SCHWEIKERT LONI; STEINKE PETER
PR: 2005US-P721993 20050930 2006US-0540955 20061002 2006WO-EP09527 20061002

FAN: 20090111892350

(Continued)

Table 10.3   (*Continued*)

PN: AU2003900226 D0 20030206 [AU2003900226] WO2004064978
     A1 20040805 [WO200464978]
AP: 2003AU-0900226 20030121 2004WO-AU00061 20040121
TI: A FILTER SYSTEM
IN: LOBBAN SARAH ELIZABETH CHENERY; LOBBAN MARK RICHARD
PR: 2003AU-0900226 20030121

PN: ES2188340 A1 20030616 [ES2188340] ES2188340 B1 20041016 [ES2188340]
AP: 2000ES-0002635 20001102
TI: Improvements in the object of main patent #9301902 relative to a : Automatic stations for the purification
     of purin, olive waste and residual water".
IN: SENTIS CROS JAUME
PR: 2000ES-0002635 20001102

PN: ES2144359 A1 20000601 [ES2144359] ES2144359 B1 20010116 [ES2144359]
AP: 1998ES-0000416 19980226
TI: Biotechnological method for recovering the oil retained in the moist olive waste left after pressing
PA: FIESTAS ROS DE URSINOS JOSE AN
PR: 1998ES-0000416 19980226

PN: ES2103206 A1 19970816 [ES2103206] ES2103206 B1 19980401 [ES2103206]
AP: 1995ES-0002528 19951227
TI: Industrial process for the treatment, recycling and conversion of olive juice and olive waste into pure
     organic fertilizers
PA: HIDALGO CICUENDEZ ARTURO
PR: 1995ES-0002528 19951227

PN: ES2092444 A1 19961116 [ES2092444] ES2092444 B1 19970701 [ES2092444]
AP: 1995ES-0000186 19950201
TI: Process for obtaining alternative electrical power through the use of olive juice.
IN: GOMEZ CASTELLOTE FRANCISCO
PR: 1995ES-0000186 19950201

PN: ITMI931692 D0 19930729 [IT93MI1692] PT101515 A 19950131 [PT-101515] GR94100253
     A 19950331 [GR94100253] IT1264823 B1 19961010 [IT1264823] ES2112125
     A1 19980316 [ES2112125] PT101515 B 19991130 [PT-101515] ES2112125
     B1 20000116 [ES2112125]
AP: 1993IT-MI01692 19930728 1994PT-0101515 19940506 1994GR-0100253 19940527 1994ES-0000799
     19940418
TI: Simplified process for the continuous extraction of oil from the first and second pressings of olives and the
     like
IN: PIERALISI GENNARO
PR: 1993IT-MI01692 19930728

PN: ES2061405 A2 19941201 [ES2061405] ES2061405 R 19950301 [ES2061405] ES2061405
     B1 19951101 [ES2061405]
AP: 1993ES-0000944 19930504
TI: Production of polyhydroxyalkanoates (PHAs) by Azotobacter chroococcum PHA
IN: GONZALEZ LOPEZ JESUS; MARTINEZ TOLEDO M VICTORIA; SALMERON MIRON
     VICTORIANO
PR: 1993ES-0000944 19930504

PN: ES2059283 A1 19941101 [ES2059283] ES2059283 B1 19950516 [ES2059283]
AP: 1993ES-0000988 19930429
TI: Procedure and installation for the treatment of dreg wines, olive waste water and other similar industrial
     waste
IN: CUADRA RUIZ JOSE
PR: 1993ES-0000988 19930429

Table 10.3   (*Continued*)

PN: ES2056745   A1   19941001   [ES2056745] ES2056745   B1   19950401   [ES2056745]
AP: 1993ES-0000490 19930310
TI: Procedure for obtaining mannitol and derived products from the branches and leaves of the olive tree and olive waste water and olive fruit stalks
IN: GARCIA GRANADOS LOPEZ DE HIERR
PR: 1993ES-0000490 19930310

PN: ES2051242   A1   19940601   [ES2051242] ES2051242   B1   19941201   [ES2051242]
AP: 1992ES-0002386 19921125
TI: System for the utilization of olive waste water in the stabilization of soils
IN: SORIANO CARRILLO JESUS; FERNANDEZ DEL CAMPO Y CUEVAS J
PR: 1992ES-0002386 19921125

PN: IL100392   D0   19920906   [IL-100392]
AP: 1991IL-0100392 19911217
TI: PREPARATION OF FOOD BY THE USE OF OLIVE-WASTE DERIVED COMBUSTION MATERIAL
PA: IMAD SALAIMI
PR: 1991IL-0100392 19911217

PN: ES2019830   A6   19910701   [ES2019830]
AP: 1990ES-0001189 19900426
TI: Procedure for treatment of olive waste water for the purification thereof and obtaining byproducts with agronomic or industrial uses
IN: NO CONSTA
PR: 1990ES-0001189 19900426

PN: ES2021191   A6   19911016   [ES2021191]
AP: 1990ES-0000486 19900219
TI: Plant for the full purification of olive juice
IN: PEREZ AMER JORGE PABLO; GARCIA VINAO AGUSTIN F; BALLESTER DIAZ LORENZO
PR: 1990ES-0000486 19900219
PN: ES2018637   A6   19910416   [ES2018637]
AP: 1989ES-0001806 19890418
TI: Improvements made to the method for obtaining biosurfactants in a substrate of refined oils and fat.
IN: MANRESA PRESAS ANGELES; MERCADE GIL M ELENA; ROBERT SAMPIETRO MARTA; GUINEA SANCHEZ JESUS; BOSCH VERDEROL M PILAR; PARRA JUEZ JOSE LUIS; ESPUNY TOMAS M JOSE
PR: 1989ES-0001806 19890418

PN: IT8822835   D0   19881202   [IT8822835] GR89100788   A   19910315   [GR89100788] IT1227676
     B   19910423   [IT1227676] ES2019015   A6   19910516   [ES2019015] GR1000829
     B   19930125   [GR1000829] AR246498   A1   19940831   [AR-246498]
AP: 1988IT-0022835 19881202 1989GR-0100788 19891128 1990ES-0000067 19891201 1989AR-0315579 19891130
TI: PROCESS FOR PURIFYING THE VEGETATION LIQUORS PRODUCED BY OIL PRESSES
IN: ROBERTIELLO ANDREA; GALVAGNO MAURO; PENNA GINO DELLA
PR: 1988IT-0022835 19881202

PN: ES2007071   A6   19890601   [ES2007071]
AP: 1987ES-0002000 19870707
TI: Basidiomycetes mushroom culture
PA: KIRCH ERNST JURGEN
PR: 1987ES-0002000 19870707

Source: Data were obtained by searching the QPAT database (Questel©). Time limits were January 1990 to November 2010.

that the field of olive phenols and health, as related to the cardiovascular system, is very advanced (far more than that of, e.g., green tea or red wine). Approximately 20 human trials describe the superiority of phenol-rich olive oil to other vegetable oils or sources of fat.

Two major fields remain to be investigated, namely, that of neurodegenerative disease and that of cancer. Epidemiological evidence is suggestive, but experimental studies are lacking; hence, we cannot draw firm conclusions on the activities of hydroxytyrosol and other phenolics.

It must be also put forward that quality is essential to olive oil's salubrious activities. Only high-quality olive oils, i.e., those whose phenolic fraction is abundant and proportionally rich in *ortho*-diphenols, might exert biological activities. In brief, either olive oil is of high quality and, alas, expensive, or its actions on human physiology are likely to be limited, given that monounsaturated fatty acids are being considered as neutral or as mildly beneficial.

In summary, there is enough experimental evidence to recommend the production and the consumption of high-quality EVOO as the fat of choice and as part of a balanced diet and a healthful lifestyle.

Acknowledgments

Throughout the years, work in FV's laboratory has been supported by various grants from Carapelli s.p.a. (a collaboration fostered by Dr. Alissa Mattei) and the EU (namely, grant FAIR CT 97 3097 "Natural antioxidant from olive oil processing waste waters"). Several students, researchers, and collaborators provided invaluable help. Last, but not least, such work has been performed under the guidance of Professor Claudio Galli, University of Milan.

References

Acin, S., Navarro, M.A., Arbones-Mainar, J.M., *et al.* (2006) Hydroxytyrosol administration enhances atherosclerotic lesion development in Apo E deficient mice. *Journal of Biochemistry (Tokyo)*, **140**, 383–391.

Agalias, A., Magiatis, P., Skaltsounis, A.L., *et al.* (2007) A new process for the management of olive oil mill waste water and recovery of natural antioxidants. *Journal of Agricultural and Food Chemistry*, **55**, 2671–2676.

Ames, B.N. (2004) A role for supplements in optimizing health: the metabolic tune-up. *Archives of Biochemistry and Biophysics*, **423**, 227–234.

Assanelli, D., Bonanome, A., Pezzini, A., *et al.* (2004) Folic acid and vitamin E supplementation effects on homocysteinemia, endothelial function and plasma antioxidant capacity in young myocardial-infarction patients. *Pharmacological Research*, **49**, 79–84.

Babich, H. & Visioli, F. (2003) In vitro cytotoxicity to human cells in culture of some phenolics from olive oil. *Farmaco*, **58**, 403–407.

Bitler, C.M., Matt, K., Irving, M., *et al.* (2007) Olive extract supplement decreases pain and improves daily activities in adults with osteoarthritis and decreases plasma homocysteine in those with rheumatoid arthritis. *Nutrition Research*, **27**, 470–477.

Bitler, C.M., Viale, T.M., Damaj, B. & Crea, R. (2005) Hydrolyzed olive vegetation water in mice has anti-inflammatory activity. *Journal of Nutrition*, **135**, 1475–1479.

Bogani, P., Galli, C., Villa, M. & Visioli, F. (2007) Postprandial anti-inflammatory and antioxidant effects of extra virgin olive oil. *Atherosclerosis*, **190**, 181–186.

Bonanome, A., Pagnan, A., Caruso, D., *et al.* (2000) Evidence of postprandial absorption of olive oil phenols in humans. *Nutrition, Metabolism & Cardiovascular Diseases*, **10**, 111–120.

Bors, W. & Michel, C. (2002) Chemistry of the antioxidant effect of polyphenols. *Annals of the New York Academy of Sciences*, **957**, 57–69.

Boskou, D. (2000) Olive oil. *World Review of Nutrition & Dietetcis*, **87**, 56–77.

Braga, C., La Vecchia, C., Franceschi, S., *et al.* (1998) Olive oil, other seasoning fats, and the risk of colorectal carcinoma. *Cancer*, **82**, 448–453.

Brunelleschi, S., Bardelli, C., Amoruso, A., *et al.* (2007) Minor polar compounds extra-virgin olive oil extract (MPC-OOE) inhibits NF-kappa B translocation in human monocyte/macrophages. *Pharmacological Research*, **56**, 542–549.

Butkovic, V., Klasinc, L. & Bors, W. (2004) Kinetic study of flavonoid reactions with stable radicals. *Journal of Agricultural and Food Chemistry*, **52**, 2816–2820.

Caruso, D., Visioli, F., Patelli, R., Galli, C. & Galli, G. (2001) Urinary excretion of olive oil phenols and their metabolites in humans. *Metabolism*, **50**, 1426–1428.

Christian, M.S., Sharper, V.A., Hoberman, A.M., *et al.* (2004) The toxicity profile of hydrolyzed aqueous olive pulp extract. *Drug and Chemical Toxicology*, **27**, 309–330.

Corona, G., Deiana, M., Incani, A., Vauzour, D., Dessi, M.A. & Spencer, J.P. (2009) Hydroxytyrosol inhibits the proliferation of human colon adenocarcinoma cells through inhibition of ERK1/2 and cyclin D1. *Molecular Nutrition & Food Research*, **53**, 897–903.

Corona, G., Tzounis, X., Assunta, D.M., *et al.* (2006) The fate of olive oil polyphenols in the gastrointestinal tract: implications of gastric and colonic microflora-dependent biotransformation. *Free Radical Research*, **40**, 647–658.

Covas, M.I. (2007) Olive oil and the cardiovascular system. *Pharmacological Research*, **55**, 175–186.

Covas, M.I., Konstantinidou, V. & Fito, M. (2009) Olive oil and cardiovascular health. *Journal of Cardiovascular Pharmacology*, **54**, 477–482.

Covas, M.I., Nyyssonen, K., Poulsen, H.E., *et al.* (2006) The effect of polyphenols in olive oil on heart disease risk factors: a randomized trial. *Annals of Internal Medicine*, **145**, 333–341.

D'Angelo, S., Manna, C., Migliardi, V., *et al.* (2001) Pharmacokinetics and metabolism of hydroxytyrosol, a natural antioxidant from olive oil. *Drug Metabolism and Disposition*, **29**, 1492–1498.

Deiana, M., Aruoma, O.I., Bianchi, M.L., *et al.* (1999) Inhibition of peroxynitrite dependent DNA base modification and tyrosine nitration by the extra virgin olive oil-derived antioxidant hydroxytyrosol. *Free Radical Biology & Medicine*, **26**, 762–769.

Deiana, M., Rosa, A., Corona, G., *et al.* (2007) Protective effect of olive oil minor polar components against oxidative damage in rats treated with ferric-nitrilotriacetate. *Food and Chemical Toxicology*, **45**, 2434–2440.

Estruch, R., Martinez-Gonzalez, M.A., Corella, D., *et al.* (2006) Effects of a Mediterranean-style diet on cardiovascular risk factors: a randomized trial. *Annals of Internal Medicine*, **145**, 1–11.

Fabiani, R., Fuccelli, R., Pieravanti, F., De Bartolomeo, A. & Morozzi, G. (2009) Production of hydrogen peroxide is responsible for the induction of apoptosis by hydroxytyrosol on HL60 cells. *Molecular Nutrition & Food Research*, **53**, 887–896.

Fki, I., Sahnoun, Z. & Sayadi, S. (2007) Hypocholesterolemic effects of phenolic extracts and purified hydroxytyrosol recovered from olive mill wastewater in rats fed a cholesterol-rich diet. *Journal of Agricultural and Food Chemistry*, **55**, 624–631.

Franceschi, S., Favero, A., Conti, E., *et al.* (1999) Food groups, oils and butter, and cancer of the oral cavity and pharynx. *British Journal of Cancer*, **80**, 614–620.

Garcia-Segovia, P., Sanchez-Villegas, A., Doreste, J., Santana, F. & Serra-Majem, L. (2006) Olive oil consumption and risk of breast cancer in the Canary Islands: a population-based case-control study. *Public Health Nutrition*, **9**, 163–167.

Gerber, M. (1997) Olive oil, monounsaturated fatty acids and cancer. *Cancer Letters*, **114**, 91–92.

Gjonca, A. & Bobak, M. (1997) Albanian paradox, another example of protective effect of Mediterranean lifestyle? *The Lancet*, **350**, 1815–1817.

Gonzalez-Santiago, M., Fonolla, J. & Lopez-Huertas, E. (2010) Human absorption of a supplement containing purified hydroxytyrosol, a natural antioxidant from olive oil, and evidence for its transient association with low-density lipoproteins. *Pharmacological Research*, **61**, 364–370.

Gonzalez-Santiago, M., Martin-Bautista, E., Carrero, J.J., *et al.* (2006) One-month administration of hydroxytyrosol, a phenolic antioxidant present in olive oil, to hyperlipemic rabbits improves blood lipid profile, antioxidant status and reduces atherosclerosis development. *Atherosclerosis*, **188**, 35–42.

Granados-Principal, S., Quiles, J.L., Ramirez-Tortosa, C.L., Sanchez-Rovira, P. & Ramirez-Tortosa, M.C. (2010) Hydroxytyrosol: from laboratory investigations to future clinical trials. *Nutrition Reviews*, **68**, 191–206.

Hertog, M.G., Feskens, E.J., Hollman, P.C., Katan, M.B. & Kromhout, D. (1993) Dietary antioxidant flavonoids and risk of coronary heart disease: the Zutphen Elderly Study. *The Lancet*, **342**, 1007–1011.

Hillestrom, P.R., Covas, M.I. & Poulsen, H.E. (2006) Effect of dietary virgin olive oil on urinary excretion of etheno-DNA adducts. *Free Radical Biology & Medicine*, **41**, 1133–1138.

Hodge, A.M., English, D.R., McCredie, M.R., *et al.* (2004) Foods, nutrients and prostate cancer. *Cancer Causes & Control*, **15**, 11–20.

Ichihashi, M., Ahmed, N.U., Budiyanto, A., *et al.* (2000) Preventive effect of antioxidant on ultraviolet-induced skin cancer in mice. *Journal of Dermatological Science*, **23**(Suppl. 1), S45–S50.

Incani, A., Deiana, M., Corona, G., *et al.* (2010) Involvement of ERK, Akt and JNK signalling in H_2O_2-induced cell injury and protection by hydroxytyrosol and its metabolite homovanillic alcohol. *Molecular Nutrition & Food Research*, **6**, 788–796.

La Vecchia, C., Negri, E., Franceschi, S., Decarli, A., Giacosa, A. & Lipworth, L. (1995) Olive oil, other dietary fats, and the risk of breast cancer (Italy). *Cancer Causes & Control*, **6**, 545–550.

Leger, C.L., Carbonneau, M.A., Michel, F., *et al.* (2005) A thromboxane effect of a hydroxytyrosol-rich olive oil wastewater extract in patients with uncomplicated type I diabetes. *European Journal of Clinical Nutrition*, **59**, 727–730.

Lopez-Miranda, J., Perez-Jimenez, F., Ros, E., *et al.* (2010) Olive oil and health: summary of the II international conference on olive oil and health consensus report, Jaen and Cordoba (Spain) 2008. *Nutrition, Metabolism & Cardiovascular Diseases*, **20**, 284–294.

Machowetz, A., Poulsen, H.E., Gruendel, S., *et al.* (2007) Effect of olive oils on biomarkers of oxidative DNA stress in Northern and Southern Europeans. *FASEB Journal*, **21**, 45–52.

Manach, C., Hubert, J., Llorach, R. & Scalbert, A. (2009) The complex links between dietary phytochemicals and human health deciphered by metabolomics. *Molecular Nutrition & Food Research*, **53**, 1303–1315.

Manios, Y., Detopoulou, V., Visioli, F. & Galli, C. (2006) Mediterranean diet as a nutrition education and dietary guide: misconceptions and the neglected role of locally consumed foods and wild green plants. *Forum of Nutrition*, **59**, 154–170.

Manna, C., Della, R.F., Cucciolla, V., *et al.* (1999) Biological effects of hydroxytyrosol, a polyphenol from olive oil endowed with antioxidant activity. *Advances in Experimental Medicine and Biology*, **472**, 115–130.

Martin, M.A., Ramos, S., Granado-Serrano, A.B., *et al.* (2010) Hydroxytyrosol induces antioxidant/detoxificant enzymes and Nrf2 translocation via extracellular regulated kinases and phosphatidylinositol-3-kinase/protein kinase B pathways in HepG2 cells. *Molecular Nutrition & Food Research*, **54**, 956–66.

Martin-Moreno, J.M., Willett, W.C., Gorgojo, L., *et al.* (1994) Dietary fat, olive oil intake and breast cancer risk. *International Journal of Cancer*, **58**, 774–780.

Miller, Y.I., Chang, M.K., Binder, C.J., Shaw, P.X. & Witztum, J.L. (2003) Oxidized low density lipoprotein and innate immune receptors. *Current Opinion in Lipidology*, **14**, 437–445.

Miro-Casas, E., Covas, M.I., Farre, M., *et al.* (2003) Hydroxytyrosol disposition in humans. *Clinical Chemistry*, **49**, 945–952.

Norrish, A.E., Jackson, R.T., Sharpe, S.J. & Skeaff, C.M. (2000) Men who consume vegetable oils rich in monounsaturated fat: their dietary patterns and risk of prostate cancer (New Zealand). *Cancer Causes & Control*, **11**, 609–615.

Obied, H.K., Bedgood, D., Mailer, R., Prenzler, P.D. & Robards, K. (2008) Impact of cultivar, harvesting time, and seasonal variation on the content of biophenols in olive mill waste. *Journal of Agricultural and Food Chemistry*, **56**, 8851–8858.

Perez-Jimenez, F., Alvarez, d.C., Badimon, L., *et al.* (2005) International conference on the healthy effect of virgin olive oil. *European Journal of Clinical Investigation*, **35**, 421–424.

Priora, R., Summa, D., Frosali, S., *et al.* (2008) Administration of minor polar compound-enriched extra virgin olive oil decreases platelet aggregation and the plasma concentration of reduced homocysteine in rats. *Journal of Nutrition*, **138**, 36–41.

Ross, R. (1999) Atherosclerosis—an inflammatory disease. *New England Journal of Medicine*, **340**, 115–126.

Salami, M., Galli, C., De Angelis, L. & Visioli, F. (1995) Formation of F2-isoprostanes in oxidized low density lipoprotein: inhibitory effect of hydroxytyrosol. *Pharmacological Research*, **31**, 275–279.

Salvini, S., Sera, F., Caruso, D., *et al.* (2006) Daily consumption of a high-phenol extra-virgin olive oil reduces oxidative DNA damage in postmenopausal women. *British Journal of Nutrition*, **95**, 742–751.

Schaffer, S., Podstawa, M., Visioli, F., Bogani, P., Muller, W.E. & Eckert, G.P. (2007) Hydroxytyrosol-rich olive mill wastewater extract protects brain cells in vitro and ex vivo. *Journal of Agricultural and Food Chemistry*, **55**, 5043–5049.

Solfrizzi, V., Capurso, C., D'Introno, A., *et al.* (2008) Dietary fatty acids, age-related cognitive decline, and mild cognitive impairment. *Journal of Nutrition, Health & Aging*, **12**, 382–386.

Solfrizzi, V., Frisardi, V., Capurso, C., *et al.* (2010) Dietary fatty acids in dementia and predementia syndromes: epidemiological evidence and possible underlying mechanisms. *Ageing Research Reviews*, **9**, 184–199.

Soni, M.G., Burdock, G.A., Christian, M.S., Bitler, C.M. & Crea, R. (2006) Safety assessment of aqueous olive pulp extract as an antioxidant or antimicrobial agent in foods. *Food and Chemical Toxicology*, **44**, 903–915.

Stocker, R. & Keaney, J.F., Jr. (2004) Role of oxidative modifications in atherosclerosis. *Physiological Reviews*, **84**, 1381–1478.

Thuy, N.T., He, P. & Takeuchi, H. (2001) Comparative effect of dietary olive, safflower, and linseed oils on spontaneous liver tumorigenesis in C3H/He mice. *Journal of Nutritional Science and Vitaminology (Tokyo)*, **47**, 363–366.

Tofani, D., Balducci, V., Gasperi, T., Incerpi, S. & Gambacorta, A. (2010) Fatty acid hydroxytyrosyl esters: structure/antioxidant activity relationship by ABTS and in cell-culture DCF assays. *Journal of Agricultural and Food Chemistry*, **58**, 5292–5299.

Trichopoulou, A. & Critselis, E. (2004) Mediterranean diet and longevity. *European Journal of Cancer Prevention*, **13**, 453–456.

Trichopoulou, A., Katsouyanni, K., Stuver, S., *et al.* (1995) Consumption of olive oil and specific food groups in relation to breast cancer risk in Greece. *Journal of the National Cancer Institute*, **87**, 110–116.

Visioli, F., Bellomo, G. & Galli, C. (1998a) Free radical-scavenging properties of olive oil polyphenols. *Biochemical and Biophysical Research Communications*, **247**, 60–64.

Visioli, F., Bellomo, G., Montedoro, G. & Galli, C. (1995a) Low density lipoprotein oxidation is inhibited in vitro by olive oil constituents. *Atherosclerosis*, **117**, 25–32.

Visioli, F., Bellosta, S. & Galli, C. (1998b) Oleuropein, the bitter principle of olives, enhances nitric oxide production by mouse macrophages. *Life Sciences*, **62**, 541–546.

Visioli, F., Bogani, P., Grande, S., Detopoulou, V., Manios, Y. & Galli, C. (2006) Local food and cardioprotection: the role of phytochemicals. *Forum of Nutrition*, **59**, 116–129.

Visioli, F., Bogani, P., Grande, S. & Galli, C. (2005) Mediterranean food and health: building human evidence. *Journal of Physiology and Pharmacology*, **56**(Suppl. 1), 37–49.

Visioli, F., Caruso, D., Galli, C., Viappiani, S., Galli, G. & Sala, A. (2000a) Olive oils rich in natural catecholic phenols decrease isoprostane excretion in humans. *Biochemical and Biophysical Research Communications*, **278**, 797–799.

Visioli, F., Caruso, D., Plasmati, E., *et al.* (2001) Hydroxytyrosol, as a component of olive mill waste water, is dose-dependently absorbed and increases the antioxidant capacity of rat plasma. *Free Radical Research*, **34**, 301–305.

Visioli, F. & Galli, C. (1994) Oleuropein protects low density lipoprotein from oxidation. *Life Sciences*, **55**, 1965–1971.

Visioli, F. & Galli, C. (1998) The effect of minor constituents of olive oil on cardiovascular disease: new findings. *Nutrition Reviews*, **56**, 142–147.

Visioli, F. & Galli, C. (2000) Olive oil: more than just oleic acid. *American Journal of Clinical Nutrition*, **72**, 853.

Visioli, F. & Galli, C. (2003) Olives and their production waste products as sources of bioactive compounds. *Current Topics in Nutrition Research*, **1**, 85–89.

Visioli, F., Galli, C., Bornet, F., *et al.* (2000b) Olive oil phenolics are dose-dependently absorbed in humans. *FEBS Letters*, **468**, 159–160.

Visioli, F., Galli, C., Grande, S., *et al.* (2003) Hydroxytyrosol excretion differs between rats and humans and depends on the vehicle of administration. *Journal of Nutrition*, **133**, 2612–2615.

Visioli, F., Galli, C., Plasmati, E., *et al.* (2000d) Olive phenol hydroxytyrosol prevents passive smoking-induced oxidative stress. *Circulation*, **102**, 2169–2171.

Visioli, F. & Hagen, T.M. (2007) Nutritional strategies for healthy cardiovascular aging: focus on micronutrients. *Pharmacological Research*, **55**, 199–206.

Visioli, F., Romani, A., Mulinacci, N., *et al.* (1999) Antioxidant and other biological activities of olive mill waste waters. *Journal of Agricultural and Food Chemistry*, **47**, 3397–3401.

Visioli, F., Vinceri, F.F. & Galli, C. (1995b) Waste waters' from olive oil production are rich in natural antioxidants. *Experientia*, **51**, 32–34.

Visioli, F., Wolfram, R., Richard, D., Abdullah, M.I. & Crea, R. (2009) Olive Phenolics Increase Glutathione Levels in Healthy Volunteers. *Journal of Agricultural and Food Chemistry*, **57**, 1793–1796.

Vissers, M.N., Zock, P.L. & Katan, M.B. (2004) Bioavailability and antioxidant effects of olive oil phenols in humans: a review. *European Journal of Clinical Nutrition*, **58**, 955–965.

Vissers, M.N., Zock, P.L., Roodenburg, A.J., Leenen, R. & Katan, M.B. (2002) Olive oil phenols are absorbed in humans. *Journal of Nutrition*, **132**, 409–417.

Weinbrenner, T., Fito, M., de la Torre, R., *et al.* (2004) Olive oils high in phenolic compounds modulate oxidative/antioxidative status in men. *Journal of Nutrition*, **134**, 2314–2321.

Williamson, G. & Manach, C. (2005) Bioavailability and bioefficacy of polyphenols in humans. II. Review of 93 intervention studies. *American Journal of Clinical Nutrition*, **81**(Suppl. 1), 243S–255S.

Wilson, T., Knight, T.J., Beitz, D.C., Lewis, D.S. & Engen, R.L. (1996) Resveratrol promotes atherosclerosis in hypercholesterolemic rabbits. *Life Sciences*, **59**, L15–L21.

Zhu, L., Liu, Z., Feng, Z., *et al.* (2010) Hydroxytyrosol protects against oxidative damage by simultaneous activation of mitochondrial biogenesis and phase II detoxifying enzyme systems in retinal pigment epithelial cells. *The Journal of Nutritional Biochemistry*, **21**, 1089–1098.

Chapter 11
Analysis and Characterisation of Flavonoid Phase II Metabolites

Celestino Santos-Buelga, Susana González-Manzano, Montserrat Dueñas and Ana M. González-Paramás

Abstract: Most dietary flavonoids are little bioavailable and in the human organism are largely metabolised to different conjugates (sulphates, glucuronides and methylethers) that further appear in the bloodstream. Knowledge about identity, activity and tissue distribution of these metabolites is still incomplete, and robust analytical methods are required for their precise quantification at the levels to which they are found in biological fluids and tissues. On the other hand, pure metabolites are required for activity and mechanistic studies as well as to be used as analytical standards. In this chapter, available methods for the preparation of flavonoid metabolites (namely, of quercetin and catechins), the procedures for their extraction, purification and cleanup in animal and human fluids and tissues, and the employment of liquid chromatography–mass spectrometry (LC–MS^n), as an essential tool for their characterisation and analysis in biological samples, are revised.

Keywords: Flavonoid metabolism; flavonoid metabolites; sulphates; glucuronides; methylated metabolites; synthesis; sample preparation; analysis in biological fluids; identification.

11.1 Introduction

Flavonoids are a major class of plant phenolics that are widely distributed in the human diet and have been related to health promotion. Indeed, a range of biological activities has been demonstrated for different flavonoids in *in vitro* models and animal assays. However, many of those studies have been carried out with the compounds in the form that they are found in plants and food, while the nature of flavonoids in biological samples is different. Most flavonoids, except flavanols (i.e. catechins and proanthocyanidins), occur in foodstuffs as

Recent Advances in Polyphenol Research, Volume 3, First Edition. Edited by Véronique Cheynier, Pascale Sarni-Manchado and Stéphane Quideau.

glycosides, which are little bioavailable and in the human organism are largely metabolised by phase II enzymes to different conjugates (sulphates, glucuronides and methylethers) that further appear in the bloodstream. Circulating metabolites would be those able to reach biological targets and are, therefore, crucial to explain the health effects associated to the intake of dietary flavonoids.

Conjugated metabolites are likely to possess different biological properties than do their precursors. Indeed, the number and position of the free hydroxyl groups in the flavonoid structure are known to be important for the antioxidant capacity (Rice-Evans *et al.*, 1996; Cao *et al.*, 1997), and thus differences in this and other types of biological activities would be expected depending on the type of conjugating residues and the position at which they are located on the flavonoid structure (Day *et al.*, 2000). As reviewed by Williamson *et al.* (2005), it appears that for biological effects in general, the position of conjugation is more important that the nature of the conjugation. Therefore, *in vitro* studies aiming to elucidate the mechanisms by which flavonoids exert their effects must consider metabolites rather than commercially available glycosylated compounds as found in foodstuffs (Kroon *et al.*, 2004). However, far less is known about the activity of the metabolites than about their precursors, and aspects like their distribution, the forms able to reach particular cellular or molecular targets, tissue accumulation or further transformations are yet to be established. In these circumstances, the ability to detect, identify and determine the metabolites is crucial to understand absorption, metabolism, kinetics or biological effects of flavonoids. One of the main difficulties that have prevented to progress in all these studies is the lack of commercial conjugated metabolites. Sufficient amounts of pure compounds are required for activity and mechanistic studies, as well as to be used as standards for optimising methods for their analysis and identification in biological samples.

In the following sections, after an outline about flavonoid metabolism, the methods for the preparation of flavonoid metabolites and the procedures for their extraction and analysis in animal and human fluids and tissues will be revised. The review will be mainly focused on catechins and quercetin (Fig. 11.1), as refers to our experience and as key examples of major dietary flavonoids.

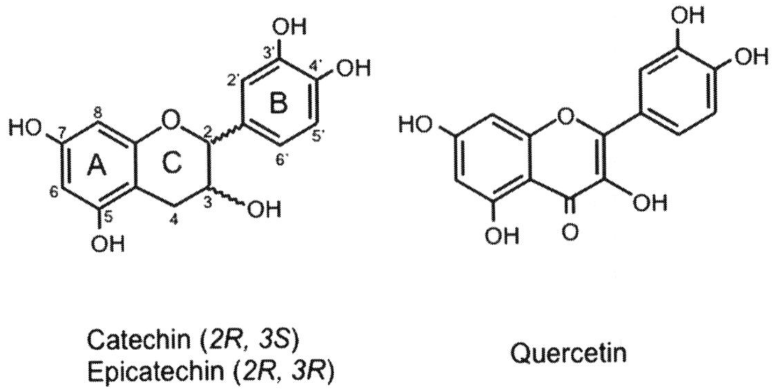

Fig. 11.1 Structures of catechin, epicatechin and quercetin.

11.2 Flavonoid metabolism

Absorption and bioavailability of dietary flavonoids vary widely from one compound to another depending on their chemical structure and are also influenced by the food matrix and individual characteristics. Most flavonoids are poorly absorbed from the intestine and highly metabolised. Isoflavones seem to be the best-absorbed dietary flavonoids; catechins, flavanones and flavonol glycosides are intermediate, whereas proanthocyanidins, catechin gallates and anthocyanins would be the worst absorbed (Kroon *et al.*, 2004; Williamson *et al.*, 2005).

Flavonoids, with the exception of flavan-3-ols (i.e. catechins, proanthocyanidins), mostly occur in plants and foodstuffs as glycosides and, in general, the first step in their metabolism is likely to be deglycosylation before absorption in the small intestine. Hydrolysis of some flavonoid glycosides might already occur in the oral cavity, as both saliva and oral microbiota show ß-glucosidase activity (Requena *et al.*, 2010). *In vitro* studies using cell cultures have demonstrated that flavonol-3-*O*-glycosides can be hydrolysed by the action of mouth bacteria and/or cytosolic enzymes from epithelial cells giving rise to the corresponding aglycones (Walle *et al.*, 2005). The human oral cavity has an abundant microbiota dominated by anaerobic bacteria and some studies have indicated that the same genera can be found in oral and colonic samples (Maukonen *et al.*, 2008). However, there are virtually no studies focusing on the extent of polyphenol/flavonoid transformation in the oral cavity. A *Streptococcus milleri* strain isolated from the oral cavity was able to deglycosylate rutin into quercetin (Parisis & Pritchard, 1983).

The mechanism most usually assumed for flavonoid deglycosylation is hydrolysis by lactase phloridizin hydrolase (LPH) in the brush-border of the small intestine epithelial cells (Day *et al.*, 2003, 2000; Day & Williamson, 2001; Németh *et al.*, 2003). Resulting aglycones would further enter the enterocite by passive diffusion. It has also been suggested that some particular flavonoids like quercetin-4′-*O*-glucoside could be actively transported into epithelial cells by the active sodium-dependent glucose transporter SGLT1 and be hydrolysed in the cell by a cytosolic β-glucosidase (CBG) (Gee *et al.*, 2000; Day *et al.*, 2003). Results obtained by other authors do not support this latter mechanism, but rather suggest that SGLT1 might not transport flavonoids (Kottra & Daniel, 2007).

During transfer across the enterocite, the most absorbed flavonoids undergo conjugation reactions, i.e. *O*-methylation, glucuronidation and sulphation, through the respective action of catechol-*O*-methyltransferase (COMT), UDP-glucuronosyltransferase (UGT) and sulphotransferase. The human intestine UGT's appear to be especially effective in conjugating 3′,4′ catechol units (Boersma *et al.*, 2002). Further conversions by phase II enzymes take place in the liver, where enterohepatic transport in the bile would also occur and some metabolites recycled back to the small intestine. It has been suggested that whilst methylation and glucuronidation are produced in both intestinal cells and liver, sulphation would mainly take place in the liver (Scalbert & Williamson, 2000; Rechner *et al.*, 2002). According to Vaidyanathan and Walle (2002), in catechins metabolism, sulphation would be the major pathway in both the human liver and intestine. Nevertheless, large differences in phase II metabolism of flavonoids exist both among species and individuals, as well as depending on the type of compound, as enzyme isoforms may exhibit their selectivity toward different flavonoids and position preferences.

Except for particular flavonoids like some catechins, such as epigallocatechin-3,*O*-gallate (Hollman *et al.*, 1997) or certain isoflavones (Williamson, 2002), aglycones are not found in the blood but the phase II metabolites, glucuronides, sulphates and methylated derivatives, are the predominant circulating forms. Using differential hydrolyses, Morand *et al.* (2001) concluded after feeding rats with flavonoid-enriched diets, quercetin was present in rat plasma mostly in the form of glucurono–sulpho conjugates (91.5%) and glucuronides (8.5%), whereas catechin was present as glucurono–sulpho conjugates (68%), sulphates (22%) and glucuronides (8%). However, no methylated derivatives that should represent a relevant part of the metabolites were analysed by those authors.

Different cells, including liver, blood and kidney cells also contain enzymes capable of deconjugating the glucuronidated and sulphated forms of flavonoids (Sperker *et al.*, 1997; O'Leary *et al.*, 2001, 2003). Deconjugation is also produced in inflammation situations, as neutrophils and macrophages express β-glucuronidase activity (O'Leary *et al.*, 2001; Shimoi *et al.*, 2001; Kawai *et al.*, 2008). Thus, flavonoid aglycones and/or their methylated derivatives may be expected to be released in some target sites, contributing to the *in vivo* effects of dietary flavonoids.

Cytotoxic and mutagenic effects have also been described for quercetin and other catechol-containing flavonoids in *in vitro* studies, which have been attributed to the pro-oxidant activity of electrophilic quinone-type products resulting from flavonoid oxidation (MacGregor & Jurd, 1978; Metodiewa *et al.*, 1999). The 3′- and 4′-hydroxyl groups are important targets for phase II metabolism and conjugation of these hydroxyl groups has been shown to considerably attenuate the cellular implications of the pro-oxidant activity of quercetin (Boersma *et al.*, 2002; van der Woude *et al.*, 2006) that is thought to be involved in its genotoxic effects observed *in vitro* (van der Hoeven *et al.*, 1984). Another reaction that can counteract the pro-oxidant effects of the electrophilic flavonoid quinones is the conjugation with glutathione (GSH) that is present in significant levels in most tissues. Different *in vitro* studies have shown that quercetin and other catechol-containing flavonoids, such as catechin, taxifolin, luteolin and fisetin are capable of reacting with GSH to generate mono and diglutathionyl adducts in the presence of chemically or enzymatically induced oxidative stress (Galati *et al.*, 2001; Moridani *et al.*, 2001; Awad *et al.*, 2003; Hong & Mitchell, 2006). This reaction could occur either spontaneously or catalysed by the phase II enzymes glutathione S-transferases (GSTs) (Hong & Mitchell, 2006). The formation of glutathione adducts of phase II methylated derivatives of quercetin (i.e. isorhamnetin and tamarixetin) has also been shown *in vitro* (van der Woude *et al.*, 2006). The *in vivo* formation of glutathionyl conjugates of flavonoids has not been as explored as other conjugation reactions, although the presence in human urine of glutathione derivatives of hydroxyphenylacetic acids generated by the colonic microbial degradation of quercetin was reported by Hong and Mitchell (2006).

Quite a relevant fraction of dietary flavonoids is not absorbed in the small intestine and, together with the metabolites returned to the intestinal lumen via entero-hepatic circulation, reaches the large intestine, where compounds are subjected to the action of the colonic microflora. Intestinal bacteria show diverse deglycosylating activities, thus releasing aglycones that might be absorbed in small extent and, more probably, degraded to simpler phenolic derivatives (Aura *et al.*, 2002; Aura, 2008). Degradation of flavonoid aglycones by colonic microflora involves ring-C cleavage and reactions affecting functional groups, such as dehydroxylation, demethylation or decarboxylation (Winter *et al.*, 1989; Aura, 2008). Various hydroxylated aromatic compounds derived from the ring-A (i.e. phloroglucinol, 3,4-dihydroxybenzaldehyde or 3,4-dihydroxytoluene), and phenolic acids derived from the ring-B have been reported as relevant products of the colonic transformation of flavonoids (Aura *et al.*, 2002; Rechner *et al.*, 2002; Aura, 2008; Selma *et al.*, 2009). Colonic metabolites might also play a key role on the biological activity of dietary flavonoids.

11.3 Preparation of metabolites

In recent years, considerable progress has been made in identifying the nature of the circulating metabolites of some major dietary flavonoids, mostly favoured by the development of electrochemical, spectroscopic and spectrometric detection techniques coupled to HPLC. In the case of quercetin, around 20 different conjugated metabolites have been described (Hong & Mitchell, 2004; van der Woude *et al.*, 2004), with quercetin-3-*O*-glucuronide, quercetin-3'-*O*-sulphate and isorhamnetin-3-*O*-glucuronide indicated as being the major circulating forms in humans (Day *et al.*, 2001; Mullen *et al.*, 2004, 2006; Williamson *et al.*, 2005). Regarding catechins, methylated derivatives at positions 3' and 4' of the ring-B, and glucuronide conjugates at positions 5 or 7 of the ring-A and 3' of the ring-B have been reported in human plasma (Donovan *et al.*, 1999a; Natsume *et al.*, 2003). Nevertheless, even for these quite studied flavonoids, there are still many unresolved analytical questions. Among others, the actual identities of some metabolites are still uncertain (i.e. concrete positions of the conjugating groups) and robust analytical methods are required for the precise quantification at the levels to which they appear in animal fluids and organs.

Although relevant circulating flavonoid metabolites were identified, the lack of commercial standards is still an important bottleneck for the study of these compounds in biological samples. Using LC–MS techniques it is possible to establish the type and number of conjugating residues in a given metabolite, and based on fragmentation patterns to conclude whether they are located on the A-, B- or C- rings (Dueñas *et al.*, 2008; Gonzalez-Manzano *et al.*, 2009). However, no distinction can usually be made between conjugation at, e.g. positions 3'- and 4'- of the ring-B, or 5- and 7- on the ring-A; only access to fully identified conjugates would allow a clear distinction between these position isomers (Barron, 2008).

The availability of pure metabolite standards is thus required not only for correct identification and quantification, but also to check their stability, ability to be deconjugated by particular enzymes, as well as for studies on biological activity, kinetics, distribution, etc. Different approaches have been proposed for the preparation of flavonoid conjugates, including isolation from natural sources and different types of enzymatic and chemical syntheses.

11.3.1 Isolation from plant sources

Isolation from biological fluids after consumption of the flavonoid is not a suitable alternative, due to the very low levels of the metabolites that can be found in these samples, either obtained from humans or animal assays. Different flavonoid glucuronides, sulphates and methylated derivatives can also be found as components in plants, from which they could be isolated. Although the common presence of other coexisting polyphenols make the isolation tedious and low yields are commonly obtained, the availability of these sources and the usual low difficulty of this approach make this possibility attractive for the isolation of some metabolites at (semi)preparative scale.

The presence of quercetin-3-*O*-glucuronide (Q3GA), the main quercetin glucuronide in humans, has been reported in a variety of plants (Williamson *et al.*, 2005). This compound is the majority flavonol in green beans, with contents that can reach up to 15 mg/kg of fresh beans (Price *et al.*, 1998), although levels largely vary depending on the variety that has to be carefully considered before starting the isolation. In our laboratory, Q3GA has been successfully obtained from this plant source using extraction with 70% methanol followed by fractionation on polyamide and further isolation by semi-preparative HPLC (Dueñas *et al.*, 2008). Even though the yields are low and the procedure is time-consuming, the flavonoid composition of the green beans is not too complex, which makes the isolation relatively easy. This is particularly interesting since syntheses of glucuronides, either chemical or enzymatic, usually result in poor yields and/or complex mixtures of compounds (Barron *et al.*, 2003).

Flavonoid sulphates are commonly present in plants as complex mixtures and their structural variation is wider than that of flavonoid glucuronides, with mono-sulphates usually coexisting with diversely substituted sulphated derivatives and compounds containing both sulphate and sugar substituents (Barron & Ibrahim, 1988). This diversity linked to the low stability of the sulphate ester linkage makes their isolation difficult. On the other hand, plant sulphated flavonoids may differ appreciably from those derived from mammalian metabolism, in which only mono-sulphated conjugates have been reported and the identified metabolites are not sulphated at position 3 as usually happens in plant derivatives (Williamson *et al.*, 2005). Species of the genus *Flaveria* are rich in flavonol sulphates (Mariel-Agnese *et al.*, 1999), and in our laboratory, leaves of *Flaveria bidentis* have been used for isolation of quercetin- and isorhamnetin-3-*O*-sulphate (Dueñas *et al.*, 2010). With this aim, leaves extracts were fractioned on Sephadex LH-20 using a gradient of 10–70% aqueous ethanol and compounds further isolated by semi-preparative HPLC. The yield of the process was very low, since the plant contains a variety of flavonol sulphates, including various polysulphate derivatives.

11.3.2 Enzymatic synthesis of metabolites

Methods for the synthesis of quercetin glucuronides using different enzymes, either commercial UDP-glucuronosyl transferases or purified microsomal preparations from rat, pig or humans have been reported by Day *et al.* (2000), Boersma *et al.* (2002) or Plumb *et al.* (2003). Glucuronidation and sulphation of epicatechin (EC) have also been assayed by Vaidyanathan & Walle (2002) using human and rat enzymes as well as recombinant UDP-glucuronosyltransferase and sulphotransferase isoforms. Sulphation of flavonoids, including quercetin and catechin, has also been carried out using an arylsulphotransferase purified from human intestinal bacteria (Koizumi *et al.*, 1990, 1991), and plant sulpho-transferases obtained from *Flaveria* species (Varin *et al.*, 1987). Depending on the origin and type of enzyme preparations, metabolites may differ greatly from those with human signification. Furthermore, relevant differences are observed in the ability of the distinct preparations to produce conjugated metabolites. Thus, whereas rat liver microsomes efficiently glucuronidated EC, no evidence of glucuronidation by human liver, small intestinal and colon microsomes was found. On the contrary, sulphation was produced by human liver and intestinal enzymes, whilst rat enzymes were considerably less efficient (Vaidyanathan & Walle, 2002). In general, all enzymatic methods produce variable mixtures of metabolites, including poly-conjugated products, requiring high purification to obtain pure compounds and leading to poor yields.

A method using post-lysosomal fractions of pig liver based on that proposed by Plumb *et al.* (2003) has been used in our laboratory for the preparation of quercetin and isorhamnetin glucuronides (Dueñas *et al.*, 2008). Optimised conditions allowed us in obtaining yields up to 19% for quercetin 4′-*O*-glucuronide and 7% for the quercetin 3′-*O*-glucuronide. However, very low amounts of Q3GA were obtained, as well as of isorhamnetin glucuronides, which not exceeded 1% for the majority metabolite (isorhamnetin 4′-*O*-glucuronide). A relevant observation was that the yield in the synthesis of glucuronides is highly affected by the microsomal preparations used. Freezing and further defrosting of the liver notably reduces the yields, especially after a long storage in cold. Thus, it is very important to use recently obtained fresh liver and even in that case great variability was obtained depending on the animal. Better results were obtained when the synthesis was carried out in water (pH 8.3) instead in HEPES buffer (pH 5.5 or 7.2) as proposed in the original method; this change improved yields and made the process much cheaper, which is relevant, taking into account that it is necessary to use important volumes of buffer when sufficient quantities of metabolites have to be prepared.

11.3.3 Chemical synthesis

A diversity of flavonoid conjugates can be prepared by chemical synthesis. When the parent flavonoid is commercially available, like in the cases of quercetin and catechins, hemisynthesis is the preferred option, making the process easier and more advantageous than total synthesis. Comprehensive reviews on the subject have been published by Barron *et al.* (2003) and Barron (2008), and some efficient protocols for the hemisynthesis of different quercetin glucuronides and/or sulphates have also been reported by Alluis and Dangles (2001), Bouktaib *et al.* (2002), Jones *et al.* (2005) and Needs and Kroon (2006). The synthesis of flavonoid sulphates is usually less favourable than that of glucuronides, due to the incomplete regioselectivity of the sulphation reaction, which leads to the formation of complex mixtures and difficult separation of the products (Williamson *et al.*, 2005). This, linked to the instability of the sulphated conjugates, makes purification critical with usual losses of compounds during isolation.

Methylated derivatives of quercetin, such as isorhamnetin ($3'$-O-methylquercetin) and tamarixetin ($4'$-O-methylquercetin), and of other flavonoids are widespread in plants and are also commercially available and, therefore, their preparation is of little interest. However, methyl ethers of catechins, as well as sulphate and glucuronide conjugates, are not known as plant products, and chemical or enzymatic syntheses represent their sole possible source (Williamson *et al.*, 2005). Methods for the preparation of methylated derivatives of catechins have been described by Donovan *et al.* (1999b) and Cren-Olive *et al.* (2002), that were further adapted in our laboratory (Gonzalez-Manzano *et al.*, 2009). The methylethers of (epi)catechin were synthesised by reaction of catechin or epicatechin with methyl iodide in the presence of potassium carbonate. The majority formation of the $3'$- and $4'$-O-methylethers of (epi)catechin was observed, with minor formation of the methylated derivatives at 5- and 7- positions, and no production of the 3-O-methylether, indicating that the ring-B is the preferential site for methylation of catechins.

Contrary to quercetin, there was a lack of suitable procedures for the preparation of catechin glucuronides and sulphates. Methods for the hemisynthesis of (epi)catechin glucuronides and sulphates have been recently optimised in our laboratory (Gonzalez-Manzano *et al.*, 2009). Glucuronides were synthesised based on the Koenigs-Knorr using acetobrom-α-D-glucuronic acid methyl ester in the presence of potassium carbonate in acetone as glucuronidation reagent (Stachulski & Jenkins, 1998). In the conditions employed, a mixture of (epi)catechin triacetylglucuronic methyl esters is produced; acetyl and methyl esters moieties bound to the glucuronide have to be removed by treatment with sodium methylate and water, respectively, and the (epi)catechin monoglucuronides separated by fractionation on Sephadex LH-20 and semi-preparative HPLC. The method allows obtaining reasonable yields (around 30–40%) of the (epi)catechin $3'$- and $4'$-O-glucuronides, whilst low yields are obtained of the 5- and 7-O-glucuronides ($<10\%$) and hardly 3-glucuronide is produced.

The method used for the synthesis of (epi)catechin sulphates was based on the one described by Jones *et al.* (2005) for the preparation of quercetin sulphates. Reaction of the catechins with a ten-fold molar excess of sulphur trioxide-N-triethylamine complex in dioxan gave a mixture of products in which (epi)catechin mono- and di-sulphates are predominant. Mono-sulphates were separated by fractionation on a Sephadex LH-20 column and, if necessary, individual compounds further purified by semi-preparative HPLC. As for the glucuronides, the conjugation of the hydroxyls of the ring-B was favoured, and hardly formation of the derivative at 3-position was observed. Even though the procedure is time-consuming due to a complex mixture of sulphates is produced and the final yield of individual compounds is not high (around 2–4% for the $3'$- and $4'$-sulphates), this method allowed us obtaining the mono-sulphates on (epi)catechin in sufficient amounts for their use as analytical standards and in *in vitro* assays (Gonzalez-Manzano *et al.*, 2009). A similar reaction using triethylamine–sulphur trioxide complex in dimethylacetamide had been previously employed by Gunnarsson and Desai (2002), although in that case no mono-sulphates were produced but only the penta-sulphated ester was obtained.

11.3.4 Purification of metabolites

A critical step in the preparation of metabolites is the purification of the mixtures of compounds formed upon their (hemi)synthesis. Usually, reversed-phase HPLC is employed for compound isolation, requiring gradient elution between an aqueous acidified polar solvent and a less polar organic solvent, more often acetonitrile or methanol, for the efficient separation of the flavonoid metabolites (Section 11.6.3). However, the presence of an acid may cause some problems of stability of the conjugated flavonoid metabolites. Whereas methylated derivatives are usually stable, some acid-catalysed cleavage of glucuronide and sulphate conjugates might take place, especially during further evaporation of the fractions obtained from the HPLC separation (Needs & Kroon, 2006). Even though the cleavage of the compounds was not total, the appearance of the aglycone in the extracts would require further purification, complicating the process and reducing the final yields. Indeed, relevant losses of product can be produced in this step. In our laboratory, previous to RP–HPLC separation, a pre-purification of the initial reaction mixtures on a Sephadex LH-20 column using water:ethanol gradients is usually performed. Simpler mixtures of two or three metabolites are obtained from this fractionation easier to separate on RP–HPLC, and sometimes even pure compounds are yielded, thus avoiding HPLC purification. As no acid is used in this stage, the acid-catalysed cleavage of the conjugating moieties is prevented.

The HPLC separation of mono-substituted conjugated metabolites, either glucuronides, sulphates or methylated derivatives, is not always easy; since structures only differ in the position of the substituting moiety, their elution characteristics are similar, especially in the case of the compounds substituted on the same flavonoid ring. According to Needs and Kroon (2006), C-18 reverse phase columns, and gradients of 0.1% aqueous TFA and acetonitrile are usually effective for preparative scale purification of glucuronides. Those authors also indicate that the use of sodium salts of the glucuronides can be used to increase the loading capacity of the system, as they are more water-soluble and could be incorporated in more concentrated aqueous solution. Nevertheless, this would give two peaks for each glucuronide, one early peak corresponding to the sodium form, and a later one to the acid form, which would make the approach suited to simple mixtures only. Aqueous TFA–acetonitrile gradients are not equally suitable for the separation of the mono-sulphated derivatives that, presumably due to their low pKa, show quite similar retention behaviour, producing usual overlapping and variable elution order. In order to overcome this problem, Needs and Kroon (2006) suggested substituting TFA by 50 mM aqueous ammonium acetate. In our experience, gradients using a diluted solution of aqueous acetic acid and methanol as solvents also provide reasonable separations of either quercetin or catechin conjugates, including sulphated derivatives. In addition, the lower boiling point of the methanol (64.7°C at standard pressure) compared to acetonitrile (81.6°C) further facilitates the evaporation of the solvent. Addition of water to the fractions obtained from HPLC separation prior to evaporation also helps to prevent glucuronide and sulphate cleavage, as it reduces the impact of acid concentration, especially if the extracts are not evaporated to total dryness. Once the organic solvent is eliminated, the resulting aqueous extracts can be submitted to lyophilisation.

According to Needs & Kroon (2006), glucuronides in their acid forms are indefinitely stable in solution in 50% aqueous methanol at –20°C, although filtration before use is required since some material precipitates during refrigeration, which leads to compound losses. Such losses are not produced when stored in 100% methanolic solution, but under these conditions the glucuronic acid moieties are prone to self-catalysed methyl esterification. Those authors also suggest converting glucuronides to their sodium salts by titration of the solutions in 50% methanol with aqueous sodium hydroxide to pH 6.0; the solids obtained after evaporation would be stable at –20°C. The salt forms are soluble in water, which is highly convenient for use in cell culture studies. Similarly, sulphates would be stable as solid sodium salts at –20°C, or in solution in aqueous methanol or methanol. Nevertheless, in our experience, sulphates can only be stored for a few weeks in either form; the formation of the aglycone is rapidly observed, especially when they are kept in partially aqueous solutions, but also in solid form. As far as we are aware, storage of the freeze-dried compounds is the most suitable form for keeping pure metabolites, especially if combined with an inert atmosphere and low temperature, and even in that case the stability of the sulphated derivatives is not too high. In contrast, methylated derivatives are quite stable, although, as for the aglycones, they may suffer oxidation, especially in aqueous solutions as the pH, temperature and ionic strength increase (Dangles *et al.*, 1999).

11.4 Characterisation of flavonoid metabolites

The structural characterisation of phase II metabolites of flavonoids involves the determination of the nature of the aglycone, the type and number of conjugated residues and the positions at which they are located. As for plant flavonoids, spectroscopic techniques can be used with this purpose, namely ultraviolet visible spectrophotometry, mass spectrometry (MS) and nuclear magnetic resonance (NMR). The coupling (hyphenation) of these techniques to high performance liquid chromatography (HPLC) and other separation methods, like capillary electrophoresis, has become a fundamental approach for the identification and/or confirmation of the identity of target and unknown chemical compounds, without requiring previous compound isolation or even high purity of sample.

11.4.1 UV spectra

Absorption spectra of flavonoids show characteristic maximum wavelengths around 240–290 nm (Band II) and in the 300–550 nm range (Band I) (Mabry *et al.*, 1970). The presence of different functional groups, such as hydroxy / methoxy groups, and glycosyl and acyl residues produce some changes in the maxima and spectra shape that can be used for tentative identification of the compound. Patterns for the identification of flavonoids based on their UV-vis spectra and shifts in the maximum wavelengths of absorption (λmax) produced by different reagents had already been given in the 1970s (Mabry *et al.*, 1970; Harborne, 1973; Markham, 1982). Nevertheless, for this identification, the isolation of the compound is previously required, so as to prepare solutions in the adequate solvent media. HPLC coupled to diode array detection (HPLC–DAD) allows obtaining on-line

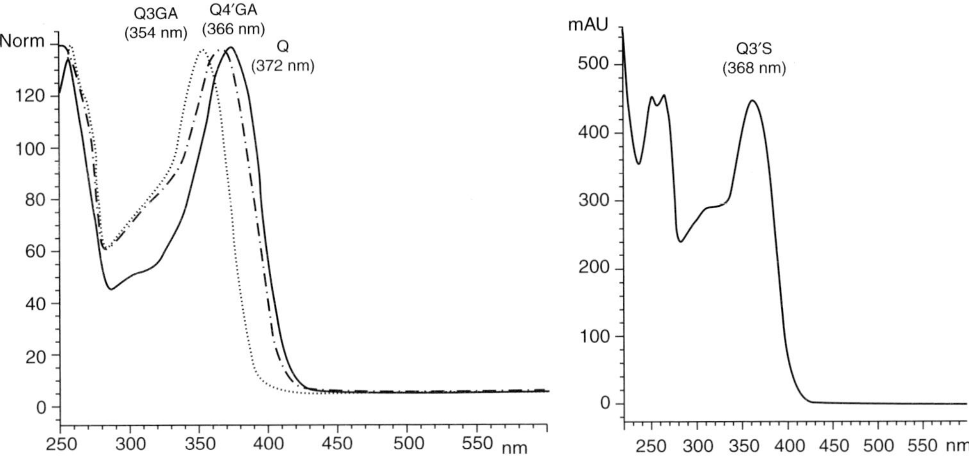

Fig. 11.2   UV spectra and maximum wavelengths of absorption of quercetin (Q) and some quercetin conjugates (GA, glucuronide; S, sulphate) obtained by HPLC–DAD.

UV-vis spectra that, although not as concluding as the information extracted from the isolated pure compounds, can be used as a screening tool for classification of compounds (i.e. flavonoid group and/or type of aglycone) and even orientate about their identity (e.g. type and location of substituents), especially in combination with their chromatographic retention characteristics (Santos-Buelga *et al.*, 2003).

Day *et al.* (2000) observed that conjugation of the hydroxyl group at position 3 of quercetin causes a Band I hypsochromic shift of around 12–17 nm, whereas a shorter hypsochromic shift (3–5 nm) occurred when conjugation was at position 4′, and no spectrum modification was produced for substitutions at 7- and 3′- positions. In agreement with Day's observations, in our laboratory, a large hypsochromic shift of λ max in Band I in relation to quercetin (18 nm) was found in the spectrum of Q3GA obtained online with the DAD; however, small hypsochromic shifts (4–6 nm) in relation to the aglycone was found for the quercetin conjugation substituted in 3′- or 4′-positions, with no relevant difference according to the position of substitution (Fig. 11.2). It should be noted that absorption spectra reported by Day *et al.* (2000) were obtained in methanol, whereas in our case they were recorded online in the solvent composition existing at their time of elution. In the case of catechins, small hypsochromic shifts in maximum wavelength (3–5 nm) with regard to the aglycone were found for compounds substituted at position 4′-, whereas hardly or no modification was observed for the conjugates at 5-, 7- and 3′-positions, either glucuronides, sulphates or methylated derivatives.

In a recent paper published by Singh *et al.* (2010), the UV spectra of a range of monoglucuronidated flavones and flavonols were recorded with the DAD and some interesting observations about the position of conjugation were obtained based on the shifts produced in λ max. It was found that glucuronidation of the 3- and 4′-hydroxyl groups resulted in hypsochromic shifts of λ max in band I of 13–30 nm and 5–10 nm, respectively, whereas glucuronidation of the 5-OH caused a band II λ max hypsochromic shift of 5–10 nm.

Glucuronidation of the 7-OH group did not cause any λ max change in band I or II. However, the extent of the conclusions obtained by those authors was limited by the fact that no compounds substituted at 3′-OH and/or containing more than one substituent on ring-B (like quercetin or catechins) were checked.

11.4.2 Nuclear magnetic resonance

When the compound is available in pure state, NMR techniques are of choice for the structural identification of flavonoid metabolites. However, the low amounts at which metabolites occur in human and animal fluids make their isolation unsuitable in view to their NMR analysis. Thus, the preparation by synthesis or other suitable methods is required to obtain sufficient amounts of the compounds so that their identity can be established by NMR.

NMR data of different quercetin and catechin metabolites have been obtained by various authors (e.g. Barron & Ibrahim, 1987; Donovan *et al.*, 1999b; Day *et al.*, 2001; Moon *et al.*, 2001; Boersma *et al.*, 2002; Natsume *et al.*, 2003; Jones *et al.*, 2005; Needs & Kroon, 2006; Smolarz *et al.*, 2008; Dueñas *et al.*, 2010). A key point is to establish the precise position of substitution of the conjugating residue on the ring-B, either 3′- or 4′-. Recently, ^{1}H, ^{13}C NMR data and HMBC correlations have been obtained for quercetin 3′-*O*-sulphate and quercetin 4′-*O*-sulphate in our laboratory (non-published results). The assignment of the location of the sulphate group at the 3′ position was based on the upfield and downfield displacements of protons and carbons with respect to quercetin. In the ^{1}H–NMR spectrum representative upfield shifts of the H-2′ (8.19 ppm) and H-6′ (8.00 ppm), which are in position *ortho* and *para* with respect to sulphate group, compared to quercetin, where H-2′ and H-6′ appeared at 7.72 and 7.61 ppm, respectively, are observed. Further confirmation of the presence of the sulphate group at position 3′ was obtained from the carbon shifts observed in the ^{13}C–NMR spectrum. A large upfield shift of 4.9 ppm was produced in the C-3′ in relation to quercetin, while C-2′ and C-4′ and C-6′ underwent significant downfield shifts (–7.9, –3.9 and –5.8 ppm, respectively). Similar observations were made by other authors (Barron & Ibrahim, 1987; Jones *et al.*, 2005; Needs & Kroon, 2006). However, opposite to what happened for the 3′-*O*-sulphate, the 4′-*O*-sulphate derivative displayed neither upfield nor downfield displacement for protons and carbons with respect to quercetin. This could be explained by the sulphate group at 4′ is in *para* position in relation to the ring-C heterocycle that would not modify the magnetic field of the molecule with respect to the unconjugated quercetin. The monosulphate nature of the compound was checked by HPLC–DAD–ESI/MS (pseudomolecular ions [M-H]$^{-}$ at *m/z* 381) after the NMR analysis, so as to confirm that it had not been hydrolysed during the analysis. Similar observations were also made for the catechin 4′-*O*-sulphate, supporting the idea that the presence of a unique sulphate residue at position 4′ does not produce upfield displacement in the carbons bearing the sulphate group nor downfield shifts of the carbons in position *ortho* and *para*. As far as we are aware, this is the first time that data for the 4′-*O*-sulphates of these compounds have been reported.

Hyphenation of liquid chromatography to NMR spectroscopy (LC–NMR) emerged in the mid-1990s as a potentially powerful technique for the separation and structural elucidation of unknown compounds in mixtures. However, this technique still presents some drawbacks, like problems for solvent suppression, differences in chemical shifts compared to usual off-line NMR analysis, lack of reference on-line NMR spectra or relatively low sensitivity, thus requiring more amount of sample and higher solvent volumes, which affects the chromatographic resolution and separation leading to peak broadening (Silva Elipe, 2003; Wolfender *et al.*, 2001). Despite the advances made concerning the improvement in NMR flow probe technology (e.g. miniaturisation of the flow-cells and cryogenically cooled NMR probes), introduction of new techniques for solvent suppression or coupling to an on-line solid-phase extraction unit for concentrating and focusing an analyte peak prior to its spectra (LC–SPE–NMR) that have contributed to improve the sensitivity, this hyphenated technique has not yet become popular, among others due to the high price of the equipments. The technique also offers the possibility of performing experiments in stop-flow mode, helping to reduce the amount of sample required for the analysis and increasing the detection limit; this mode also gives the possibility to 2D NMR correlation experiments (COSY, NOESY, HMBC, etc), since the sample can remain inside the flow cell even for days. Indeed, LC–NMR is not available to many laboratories, and actually few papers have been published regarding the analysis of flavonoids by this technique (Queiroz *et al.*, 2002, 2005; Andrade *et al.*, 2002; Le Gall *et al.*, 2003; de Rijke *et al.*, 2004; Exarchou *et al.*, 2006), all of them referring to plant composition and none dealing with body metabolites.

11.4.3 Mass spectrometry

Relevant information for compound identification, in some cases quite definite, can be obtained from their mass spectra. When coupled to HPLC, atmospheric-pressure ionisation (API) techniques are usually employed, and especially electrospray ionisation (ESI), as both polarity and molecular weight of flavonoids match well the requirements of this interface. Flavonoid metabolites can be detected in both positive or negative ion mode. The main fragmentation patterns of flavonoids are apparently independent of the actual ionisation mode or the types of analysers applied (e.g. triple quadrupole, hybrid quadrupole-time of flight or ion trap), although significant differences could exist in the relative abundances of fragment ions. Therefore, methods based on detecting the presence or absence of distinctive fragment ions are usually preferred over techniques relying only on relative intensity changes observed for the isomeric compounds (Vukics & Guttman, 2010).

Intact glucuronides are best observed under ESI conditions, although in some instances, APCI or atmospheric pressure photo ionisation (APPI) can also be employed. The masses of the $[M+H]^+$ and $[M-H]^-$ ions are 176 Da higher than those of the non-conjugated metabolites. In positive-ion mode, a neutral loss of 176 u leading to an abundant fragment $[M+H-176]^+$ is usually observed. In negative-ion mode, an abundant $[M-H-176]^-$ and a usually less abundant ion at *m/z* 175 (deprotonated glucuronide moiety) are generally found. In addition, secondary fragment ions at *m/z* 113 (loss of CO_2 and water from *m/z* 175), and a less intense ion at *m/z* 85 (extrusion of CO from *m/z* 113) appear (Levsen *et al.*, 2005).

The presence of a sulphate group increases the *m/z* value of the quasi-molecular ion of the flavonoid by 80 u. Sulphate conjugates can only be ionised intact using the ESI method, preferably in negative-ion mode. The negative-ion MS/MS spectra of the $[M-H]^-$ quasi-molecular ion of sulphate esters are dominated by the loss of SO_3 (-80 Da), leading to the ion $[M-H-SO_3]^-$ and a usually less abundant $[M-H-H_2SO_4]^-$ ion (loss of 98 Da). Similarly, under positive-ion conditions, the $[M+H-SO_3]^+$ ion (loss of 80 Da) is formed. In some reports, other fragment ions were observed at very low abundance, i.e. the ion $[M+H-HSO_3{}^\bullet]^+$ under positive-ion formation, and the ion $SO_4{}^{-\bullet}$ under negative-ion formation (Levsen *et al.*, 2005).

Methylated derivatives of flavonoids are usually ionised in positive-ion mode with ESI, but also the APCI and APPI methods are well suited. Methylation, increases the $[M+H]^+$ ion mass by 14 Da. In general, in positive mode, the CH_3–O bond is not cleaved and, therefore, the conjugation is not always evident from the MS/MS spectra (Levsen *et al.*, 2005). Nevertheless, in negative ion mode the loss of the methyl group from the deprotonated molecule leading to $[M-H-15]^-$ ions has been indicated to constitute a characteristic fragmentation, which is usually accompanied by $[M-H-15-28]^-$ and $[M-H-15-29]^-$ fragment ions, corresponding to the losses of CO (28 u) and HCO (29 u) (Justesen, 2001). Lower intensity $[M-H-15]^-$ ions seem to be produced when flavonoids are methylated on the ring-B than on rings A or C (Markham, 1982).

The methylation site may be deduced based on the analysis of the product ion spectra obtained in MS^n, since fragments that include the methyl group increase their m/z values by 14. However, if a phenyl ring contains two or more hydroxyl groups, fragmentation data usually do not allow differentiating between the possible methylation sites. Thus, in the case of (epi)catechin and quercetin, it is possible to know whether the methylation is located in the rings A or B but not the precise position (i.e. 3′- / 4′-, or 5- / 7-) and only additional NMR data would allow for an unambiguous assignment of the exact methylation site (Cren-Olive *et al.*, 2000). In a recent paper by Kuhnert *et al.* (2010), the regioisomers ferulic and isoferulic acids that only differ in the relative positions of substitution of a hydroxy and a methoxy group on the aromatic ring, were distinguished based on differences in the relative intensities of the fragments observed in MS/MS. The reasons for such differences were explained as due to the preferential losses of CO_2 followed by loss of CH_3, in the case of ferulic acid, and of CH_3 followed by rapid loss of CO_2, in the case of isoferulic acid. It would be worth checking if similar behaviour is produced for flavonoids differing in the location of hydroxy and methoxy groups in the ring-B, such as 3′-*O*- and 4′-*O*-methylcatechins, or tamarixetin and isorhamnetin.

The assignment of the conjugation site based on the analysis of the product ions in the MS^n spectra is less feasible in the case of the glucuronide and sulphate derivatives, since those groups are usually cleaved. Nevertheless, some conjugated residues could remain attached to the aglycone yielding minority fragments that could be used as diagnostic fragments for deciding the ring of location, as shown by Gonzalez-Manzano *et al.* (2009) for glucuronidated and sulphated catechins.

The identity of the aglycone can be deduced from the m/z value of the major fragment resulting after separation of conjugated moieties, either in positive or in negative ion modes. In the case of aglycones with the same m/z values, further MS^3 fragmentation and comparison with standards or literature data can be required. Major fragmentation pathways for flavonoids are shown in Fig. 11.3.

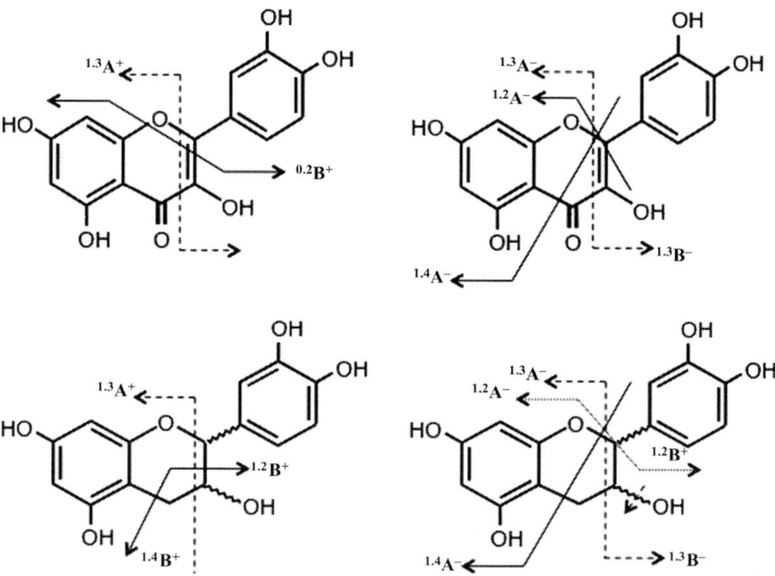

Fig. 11.3   Major fragmentation pathways and product ions obtained for quercetin and catechins in positive and negative ion mode. The fragmentation nomenclature proposed by Ma *et al.* (1997) is used. The labels ^{ij}A and ^{ij}B refer to the fragment containing A- and B-ring, respectively, and the superscripts i and j indicate the C-ring bonds that have been broken.

More information on identification of flavonoid metabolites using different LC–MS equipments can be found in many papers published over the last decade that can be consulted for further information (e.g. Qu *et al.*, 2001; Wittig *et al.*, 2001; Natsume *et al.*, 2003, Mullen *et al.*, 2003, 2004; Hong & Mitchell, 2004; van der Woude *et al.*, 2004; Levsen *et al.*, 2005; Wang & Morris, 2005; Dueñas *et al.*, 2008; Kawai *et al.*, 2008; Santos *et al.*, 2008).

11.5 Extraction and preparation

Extraction of flavonoids from biological matrices is crucial to both their qualitative and quantitative analysis. Methods employed for flavonoid extraction from plants and food are not usually directly applicable to biological fluids and organs, due to the very low levels of metabolites and the complexity and variability of the biological matrices. The samples most usually employed are urine and blood (plasma or serum), although other fluids (bile, milk), tissues and organs (liver, kidney, brain, lung or endothelium) or faeces may also be analytical targets. These samples are normally rich in fat and proteins, and also contain a diversity of compounds that may interfere in the extraction and/or further in the analysis of the target metabolites. Different matrices have distinct composition, and different metabolites also show differences in solubility, therefore, different extraction techniques may also be required in each case. Since the knowledge about flavonoid metabolites is still incomplete, the conditions have to be carefully optimised to ensure that all metabolites are efficiently extracted, so that no information is lost. As for plant and food analysis, extraction procedures for biological matrices should be validated, in particular when dealing with quantitative analysis. Only fully validated methods allow controversial data be critically assessed.

An outline of the procedures for flavonoid extraction and separation from biological matrices is given in the subsequent text. For more detailed information, the reviews by Day & Morgan (2003), Prasain *et al.* (2004), de Rijke *et al.* (2006), Xing *et al.* (2007), Novakova & Vlckova (2009) or Marchi *et al.* (2010) can also be consulted.

11.5.1 Sample preparation and storage

Animal fluids and tissues are rich in components like proteins, carbohydrates, salts and lipids that may affect extraction, purification and analysis of flavonoid metabolites. In most cases, a pre-treatment to eliminate these products is required. Protein precipitation, solvent extraction and solid phase extraction (SPE) are procedures commonly used for sample preparation. The addition of an internal standard (IS) previous to sample extraction is also usual to correct for unknown losses during extraction, especially when quantitative determination has to be done. Apigenin, taxifolin or dihydroflavone are examples of compounds employed as IS in the analysis of flavonoids in biological samples.

Solvent-induced protein precipitation is the sample preparation technique most usually employed for the analysis of flavonoid metabolites in plasma and other physiological fluids. Samples are usually filtered and/or centrifuged before and especially after solvent treatment in order to separate the resultant protein precipitates. This is a fast and simple procedure applicable to both hydrophilic and hydrophobic compounds that can be automated or semiautomated with the use of multiple-well plates. As the procedure does not involve actual extraction, the chance of workup losses during sample preparation is reduced (Prasain *et al.*, 2004). However, the extracts (supernatants) are relatively unclean as they might still contain a significant amount of un-precipitated plasma components. Therefore, selectivity is usually low and can induce analyte coprecipitation or mass spectrometry signal suppression, so that further purification by SPE may be required (Novakova & Vlckova, 2009). Miscible organic solvents (acetonitrile, methanol) have been most used for deproteinisation, as they give better results than classical precipitation with trichloacetic acid or other acidic compounds that may affect the glucuronidated and sulphated conjugated forms. In the case of urine, some authors avoid protein precipitation and reduce the pre-treatment to centrifugation or filtration to remove suspended particles; since electrolytes and other hydrophilic components that could interfere with detection of the flavonoids usually elute in the first gradient steps in the HPLC separation, before the elution of the target analytes has begun, they would not constitute a problem for metabolite detection (Xing *et al.*, 2007). Nevertheless, as indicated for plasma samples, if only protein precipitation or simple pretreatment are used, components remaining in the sample may cause interferences when analyses are performed by LC–MS, and therefore it is not unusual to combine them with SPE.

Solid samples (i.e. animal and human tissues and organs) are usually first homogenised, which may be preceded by freezing with liquid nitrogen or freeze-drying. Frozen samples should be kept at $-70°C$, and storage in airtight vials at $-20°C$ is adequate for freeze-dried samples. For their extraction, they can be homogenised in the presence of the solvent in a suitable device.

In storage of samples, care needs to be taken in respect of possible endogenous enzyme activity and stability of the flavonoids and their conjugates. Quercetin, for example, is particularly prone to oxidation when the pH exceeds 7.4, which may occur during storage of plasma in a freezer resulting in losses and degradation compounds present as sample artefacts. Quercetin-3-conjugates are more resistant to oxidation, which may also lead to an imbalance in calculation of metabolite concentrations. Quercetin appears to be more stable bound to proteins, as the proteins may afford some protection to oxidation reactions. Antioxidants, such as ascorbic acid, are often added both for storage and during preparation to decrease losses of polyphenols by oxidation (Day & Morgan, 2003).

11.5.2 Hydrolysis

In animal fluids and tissues, a relevant fraction of the flavonoids is in the form of β-glucuronides and/or sulphates, which are more polar than aglycones and methylated derivatives, and, therefore, have different extraction requirements. A common strategy, especially in early studies, was to submit the samples to previous hydrolysis and further extract them with usual solvents for flavonoid aglycones. Both enzymatic and acid hydrolysis of samples can be employed. Extraction in methanol with simultaneous hydrolysis with 2N HCl has been used for the analysis of flavonol conjugates (glucuronides and sulphates) in plasma and urine (Hollman, 2001; Morand *et al.*, 2001; Arts *et al.*, 2003). Obviously, in these circumstances, no information about the identity of the conjugated metabolites is obtained. On the other hand, the use of harsh acid hydrolysis conditions may result in losses of polyphenol through degradation and derivatisation (Day & Morgan, 2003). The use of mild hydrolysis conditions with organic acids, such as 10% acetic acid, might offer some structural information, based on the identification of intermediate products from the partial cleavage of the conjugating moieties; for instance, sulphate ester bonds are cleaved faster than glycosidic bonds (Barron & Ibrahim, 1988).

Enzymatic hydrolyses can be performed with β-glucuronidase and sulphatase separately or in mixture. Preparations containing both types of enzymes from *Helix pomatia* or purified fractions of individual enzymes from this or other sources can be employed (Day & Morgan, 2003). The use of enzymes has to be carefully monitored, since purity and activity may vary from batch to batch and depending on the biological material to be analysed, and they might also contain contaminants. To ensure that the enzymes are active in the incubates, known amounts of synthetic substrates can be added to the mixture, such as phenolphthalein glucuronide or 4-methylumbelliferone sulphate (Xing *et al.*, 2007), or ^{13}C-labeled flavonoid conjugates (Setchell *et al.*, 2003). On the other hand, commercially available sulphatases are usually contaminated with glucosidase and uronidase activities (Barron & Ibrahim, 1988), and β-glucuronidase preparations may also contain sulphatase or β-glucosidase activity (Day & Morgan, 2003). Despite these facts, many authors do not assess the activity and purity of the enzyme preparations and sometimes the source of the enzyme(s) used is not even provided. When relevant native enzymes are likely to be present in the cells or tissues, then care needs to be taken that they do not interfere in the processes, by deactivating (which may be the case with the extraction solvents) or inhibiting them with specific inhibitors. Those enzymes would remain active when tissue samples are homogenised in buffer before hydrolysis and solvent extraction (Day & Morgan, 2003).

Information about the type of conjugates may be obtained according to their sensitivity to the enzymes used (glucuronidases or sulphatases) when the sample is analysed before and after the hydrolysis. Sequential hydrolysis can also be made treating the samples first with β-glucuronidase and later with arylsulphatase. The aglycone concentrations found after the successive hydrolysis of samples can be used to estimate the percentages of the different types of conjugates (glucurono-, sulpho- and glucurono–sulpho conjugates). Organ homogenates can be treated in the same way as biological fluids, but higher amounts of the enzymes should be used (Morand *et al.*, 2001).

The assessment of the nature of a conjugate based on peak disappearance in the chromatographic analysis after enzymatic treatment has to be taken with care (Barron, 2008). In addition to possible contamination with other enzymes present in the preparations or the tissue sample, some compounds have shown resistance to enzymatic hydrolysis, as it is the case of flavonol 3-sulphates that are not hydrolysed with aryl sulphatase (Barron & Ibrahim, 1988). In plant tissues, several flavonoids have been identified where the sulphate group is attached to the uronic acid of a glucuronyl moiety and not directly to the hydroxyls of the aglycone (Barron *et al.*, 1998). It is possible, though it has not been demonstrated as yet, that these conjugates might be produced during mammalian metabolism and if so treatment with β-glucuronidase would release aglycone initially present as sulphate (Day & Morgan, 2003).

11.5.3 Solvent extraction

Different organic solvents, such as ethyl acetate, methanol, ethanol, acetone or diethyl ether have been used for extraction of metabolites from biological samples. However, the different types of metabolites have different solubility characteristics and, therefore, the type of solvent used is critical for an efficient extraction. It is, thus, imperative to characterise solvent extraction efficiencies for the compounds of interest as well as for the matrix to be extracted (Day & Morgan, 2003).

Ethyl acetate has been largely employed for the extraction of flavonoids in biofluids after acidification (Xing *et al.*, 2007). By acidifying the media, the phenol–phenolate equilibrium shifts towards the less polar phenyl form, thus facilitating extraction with non-polar solvents. Ethyl acetate has the advantages of not being water miscible and having a relatively low boiling point, which make it easily removable. However, it is not adequate for polar metabolites like sulphates and glucuronides (and, if present, other glycosides). In general, extraction with pure organic solvents suits better for lipophylic compounds, like aglycones or their methylated derivatives, although some solvents with a more polar character like acetonitrile, ethanol or methanol might also suit extraction of polar metabolites. Nevertheless, for the extraction of these latter, mixtures of water with the organic solvent are more usual (e.g. 70% aqueous methanol). Acidification also increases the ability to extract glucuronides (and other glycosides), especially when protic polar solvents like methanol and ethanol are used. Soft acidic conditions must be used to prevent hydrolysis of the conjugating residues, especially in the case of sulphates that are particularly sensible to acidification. Acidification is even necessary for the extraction of some particular flavonoids like anthocyanins that are structurally dependent on the pH of the medium, which modifies their characteristics of solubility and is cause of instability. In general, inorganic acids should be avoided, and acidification with acetic or trifluoroacetic acids is preferred. Nevertheless, attention must also be taken since formation of artefacts due to sugar esterification by organic acids might happen during extraction, as observed during extraction of some flavonoid glycosides from food and plant materials. More detailed information on solvents and procedures for polyphenol extraction from biological fluids and tissues can be found in a review by Day and Morgan (2003).

11.5.4 *Solid phase extraction*

Solid phase extraction (SPE) is the most popular sample preparation method used in routine bio-analytical laboratories for sample clean-up and enrichment. It presents several advantages over conventional solvent extraction, such as high recovery, more effective concentration, use of less organic solvent, short time for sample preparation, no foaming or emulsion problems, ease of operation and possibility of automation (Kataoka, 2003).

A large variety of SPE sorbents are commercially available in the form of cartridges, discs, well plates or disposable pipettes, covering a wide range of hydrophilicities and specificities. Conventional SPE cartridges are commonly used, as they are readily available and easy to handle by using vacuum or positive-pressure manifold. However, they still have some disadvantages. Thus, poor batch-to-batch reproducibility sometimes exists, the flow-rate is not easy to control, relatively large volumes of organic solvents are required and they are manufactured for single use, which makes the procedure expensive. SPE discs are an interesting alternative as they use minimal solvent volume and evaporation to dryness and reconstitution are no longer necessary because elution can be performed directly by the mobile phase (Novakova & Vlckova, 2009). SPE can be automated through the use of 96-well plates, facilitating high-throughput analyses of biological samples, and it can also be adapted to on-line HPLC processing. The on-line SPE technique requires the use of program controlled switch valves and column re-configurations, but once optimised it offers speed, lower detection limits and better reproducibility than the off-line method (Xing *et al.*, 2007).

Solid phases of relevance to flavonoids include hydrophobic, ion exchange and function-alised sorbents. The choice depends on the analytes and their physical–chemical properties, which should define the interactions with the sorbent, as well as on the kind of sample matrix and interactions with both the sorbent and the analyte (Novakova & Vlckova, 2009). Silica-bonded reversed-phase sorbents, mostly containing octadecyl (C18) functional groups, have been one of the most frequent materials used for analyte concentration prior to the analysis of flavonoids in plasma and urine (e.g. Liu *et al.*, 1995; Unno *et al.*, 2005; Ding *et al.*, 2006; Lai *et al.*, 2007). Nevertheless, these types of sorbents are hydrophophobic and, therefore, they suit better for aglycones than for more polar metabolites. Bond Elut[®] C18 cartridges were found by Erlund *et al.* (1999, 2000) to provide the best results for the extraction of quercetin from plasma. After testing various solvents, those authors found that the highest recoveries were obtained using an acidic mixture of toluene and dichloromethane (80:20, v/v). Amberlite® XAD-2, another hydrophobic sorbent, consisting of a cross-linked polystyrene copolymer, has also been applied to the extraction of flavonol aglycones from urine (Watson & Pitt, 1998). When using these types of non-polar sorbents, sample solution and solvents are usually slightly acidified to prevent ionisation of the phenolics, which could greatly reduce their retention. Polyamide, a hydrophilic polymeric sorbent, was proposed by Day and Morgan (2003) as a good alternative for extraction of sulphated polyphenols from biological fluids. These metabolites can be strongly bound by polyamide, unwanted material washed away and the sulphates eluted using aqueous methanol at alkaline pH.

More recently, polymeric sorbents incorporating a cross-linked agent (normally, divinyl-benzene, DVB) and different functional groups have been developed that provide better hydrophilic–lipophilic balance and increase the retention of polar compounds. Strata™-X, a surface-modified styrene divinylbenzene polymer from Phenomenex, and Oasis® HLB, a macro-porous poly(N-vinylpyrrolidonedivinylbenzene) from Waters, are two examples of widely used reversed-phase sorbents that may suit both the parent aglycones and more polar metabolites. Other recently developed SPE materials are mixed-mode cartridges that incorporate reversed-phase sorbents and ion-exchange resins, like Oasis MCX (cation-exchange) Oasis MAX (anion-exchange). Examples of applications of these types of cartridges to the recovery of flavonoid and polyphenol metabolites from plasma and urine can be found in the papers of Sheng and Feng (2006), Mata-Bilbao *et al.* (2007), Medina-Remon *et al.* (2009) or Xu *et al.* (2009).

Molecularly imprinted polymers (MIPs), synthetic polymers with highly specific recognition ability for target molecules, have also recently been developed as a selective approach for solid phase extraction. Even though the molecular structures of the flavonoids are not ideal for molecular imprinting, as the presence of several OH groups in the polyphenolic structure could be source of hydrogen bonding and nonlocalised electrostatic interactions between the OH groups (Stalikas, 2007), some attempts have been made in this direction. Thus, Xie *et al.* (2001) used quercetin as template for the preparation of MIPs for the selective extraction of flavonoids from hydrolyzates of *Ginkgo biloba* leaves, and Theodoridis *et al.* (2006) employed rutin and quercetin as the template molecules with similar aim. In both cases, the obtained polymers showed good affinity to bind the target flavonoids and structural analogues, although non-specific high retention of other non-related compounds was also observed. As far as we know, MIPs have not been yet explored for the extraction of flavonoid metabolites from biological samples.

Another recent trend in sample preparation is the use of aptamers, sequences of oligonucleotides able to bind to target molecules with high affinity and specificity, although no applications have been yet reported to the selective isolation of flavonoids.

11.6 Analysis of metabolites in biological samples

Many works have been published dealing with the analysis of flavonoids in plants and food, but the study of these compounds in biological samples shows some particularities, and usual methods employed in plant analysis could not meet the requirements of sensitivity, resolution and specificity for their analysis. The different nature of the matrices and the type of compounds present, and the much lower concentrations of the analytes when compared to plants and food make sample preparation and extraction particularly challenging and critical for suitable detection and accurate quantification. The circulating concentrations of both native and metabolic forms of polyphenols are in the nanomolar to low micromolar range (Spencer *et al.*, 2008; Loke *et al.*, 2009). Maximum plasma concentrations of flavonoids do not usually exceed 1 μM for normal food intakes, with a peak attained around 1–2 hours after ingestion (Kroon *et al.*, 2004).

A first decision to be made in the analysis of metabolites is whether or not to submit the samples to previous hydrolysis with ß-glucuronidases and/or sulphatases so that only aglycones (and, if present, methylated forms) are analysed (see the preceding text). Released aglycones are further easily identified/quantified by comparison with authentic standards. This strategy increases the possibilities of quantification, as individual metabolites may be under the detection or quantification limits. Nevertheless, this is a decision to be carefully considered since hydrolysis techniques are not always efficient and, whatever the procedure used, relevant information is lost about the precise identity of the metabolites and their levels may be also underestimated. The nature of the enzyme and the conditions used for hydrolysis may cause some compounds be destroyed or not be efficiently released. Nowadays, the increasing knowledge on the nature of flavanoid metabolites and the availability of LC–MS equipments in many laboratories have reduced the need of using hydrolysis techniques, and individual analysis of metabolites has become more usual. Nevertheless, hydrolysis may still be undertaken if the aglycone alone is the source of interest or there is not interest or possibility to monitor the individual metabolites or the sensitivity of the analytical method requires the concentration effect given by hydrolysis (Day & Morgan, 2003).

11.6.1 Quantification of total polyphenol metabolites

The classical method to analyse total phenolic compounds is the Folin-Ciocalteu (FC) method. A recent application of this method has been developed for the analysis of total phenolics in urine as a way for the estimation of the intake of dietary polyphenols (Roura *et al.*, 2006). Conditions were optimised to avoid interferences of other substances that are known to react with the FC reagent, such as sulfur dioxide, ascorbic acid, sugar, aromatic amines, organic acids or Fe(II), which were removed using a single solid phase extraction procedure. Oasis® MAX cartridges were found to provide the best results for the extraction of urine polyphenols in a large range of polarity. The method was further optimised to adapt it to the use of 96-well microplates, thus allowing the simultaneous analysis of a large number of urine samples (Medina-Remon *et al.*, 2009). The suitability of this methodological approach to be used in epidemiological and clinical studies was assessed by applying it to the analysis of urine samples from two trials, a large prospective, randomised, crossover assay with different intervention periods, and one cross-sectional study with a free-living population of 60 volunteers. The obtained results confirmed that total phenolics in urine as determined by this methodology correlated with polyphenol intake and can be used as a marker of the consumption and bioavailability of these compounds.

11.6.2 Analysis of individual metabolites

HPLC has become the technique of choice for the analysis of individual flavonoids and phenolic compounds. Other techniques like gas chromatography (GC) or capillary electrophoresis (CE) have also been used for the separation of flavonoids, although they are less versatile and have had less applicability in their analysis.

Flavonoids do not meet requirements of volatility to be directly analysed by GC and need derivatisation, usually the formation of trimethylsilylated (TMS) derivatives. An additional drawback of the GC techniques is the difficulty to analyse flavonoid glycosides, which have to be normally hydrolysed to the aglycones prior to derivatisation. Furthermore, compounds with various hydroxyl substituents may yield several derivatives making quantification difficult. ESI–MS in the selected ion-monitoring mode (SIM) is often used for detection. The molecular ion, $[M+H]^+$, and fragments formed by the loss of CH_3 and/or CO and retro-Diels-Alder (RDA) cleavage are typically used for detection (de Rijke *et al.*, 2006). Despite GC cannot outweigh the advantages of the LC techniques for the analysis of flavonoids, it has been used by some authors for their determination in biological matrices. Thus, GC–MS was employed to determine catechin, epicatechin and their 3′- and 4′-*O*-methylated metabolites, after TMS derivatisation, in human plasma (Luthria *et al.*, 1997; Donovan *et al.*, 1999a) and urine (Donovan *et al.*, 2002a). Similarly, different flavonoids, including quercetin, catechins and anthocyanins, as well as other phenolic compounds like phenolic acids have also been determined using GC–MS in human plasma and urine after consumption of berries and berry juices (Rechner *et al.*, 2002; Kay *et al.*, 2004; Zhang & Zuo, 2004). This technique has also shown useful for the analysis of low molecular weight compounds such as simple phenols, phenolic and benzoic acids, which can be produced in the flavonoid degradation by the colonic microflora (Aura, 2008; Grün *et al.*, 2008; van Dorsten *et al.*, 2010).

Capillary electrophoresis (CE) has only been occasionally used for the separation and analysis of flavonoids and most studies refer to plant materials and plant-derived food, although some recent papers have also been published dealing with applications of CE to the analysis of flavonoids in biological fluids. Chiral capillary electrophoresis has been employed by El-Hady *et al.* (2008) for the separation and quantification of catechin and epicatechin in the urine of human volunteers following tea consumption, and capillary electrophoresis-frontal analysis has been used to study the binding behaviour of different flavonoids to human serum proteins (Diniz *et al.*, 2008; Knjazeva & Kaljurand, 2010). Nevertheless, it seems unreliable that CE can displace HPLC for the analysis of flavonoids. The repeatability of retention/migration times still is better in LC than in CE, the limited consumption of sample and solvents does not appear to have much impact and there is no dramatic difference of run times between both techniques (de Rijke *et al.*, 2006), especially if compared with more recent LC approaches like UPLC.

11.6.3 High performance liquid chromatography

HPLC is a versatile technique that enables the separation of mixtures of compounds with different physical–chemical characteristics, allowing the simultaneous analysis of hydrophobic flavonoid aglycones and hydrophilic flavonoid derivatives, as well as other possible related compounds like their degradation products. Furthermore, the possibility of coupling to different detection devices (e.g. UV-vis, PDA, ESI–MS, fluorimetry, electrochemical or NMR analysers) further increases the possibilities of detection, identification and quantification even at low levels of analytes and in the presence of other interfering and coeluting components. Numerous papers have been published dealing with the HPLC analysis of flavonoids and metabolites. In general, laboratories tend to develop their own methods according to their particular research needs, although there are some usual patterns in the analytical conditions used.

Reversed-phase (RP) columns (commonly, C8- or C18-bonded silica phases) are most often used, although more polar sorbents, as Silica, Sephadex or Polyamide, can also be employed. Elution is usually performed at room temperature, but moderately higher temperatures up to 40°C are sometimes recommended to reduce the time of analysis and because thermostated columns give more repeatable elution times (Stalikas, 2007). Both isocratic and gradient elution can be applied for the separation of flavonoids depending on the number and type of the analytes and the nature of the matrix. Gradient elution is generally performed using binary systems between an aqueous acidified solvent and an organic modifier. Acetonitrile and methanol are the most commonly used organic solvents. Acetonitrile normally leads to better resolution in a shorter analysis time than methanol and, generally, gives sharper peak shapes, resulting in a higher plate number. Acetonitrile is also preferred to methanol when electrochemical detection is used, as it produces lower baseline noise (Rehova *et al.*, 2004). By contrast, methanol is preferable to acetonitrile when used with (semi)preparative purposes, due to its higher volatility. Also, methanol offers the possibility of using higher percentages in the mobile phase, which could protect the HPLC columns. Flavonoids contain ionisable hydroxyl groups and the use on an acid modifier is important to suppress ionisation and prevent the interactions of these groups with residual traces of metals in the stationary phase that are detrimental to peak shape. Thus, keeping a low pH helps prevent peak tailing and improve the resolution and reproducibility of the retention characteristics. The recommended pH for elution is in the range 2–4. Aqueous acidified solvents such as acetic, formic, trifluoroacetic (TFA), phosphoric, and most rarely perchloric acid, as well as citrate, ammonium acetate, ammonium formate and phosphate buffers at low pH have been employed. However, these saline solutions have become less popular due to the dreaded contamination of the ion sources in MS detection (de Rijke *et al.*, 2006). In flavonoid analysis, the most common additives are acetic acid, formic acid, ammonium–acetate and ammonium–formate. Better resolutions are usually obtained with formic acid, although it may cause problems of corrosion of the sources when relatively high percentages are used. Acidification of the organic solvent so that the percentage of acid remains constant during elution is also a common strategy that helps reducing peak widening (Santos-Buelga *et al.*, 2003).

When RP–HPLC is coupled to MS ion suppression effects may be produced due to solvent and sample interferences, which can reduce the ionisation of analytes. Matrix effects can be produced in both ESI and APCI, but ESI is much more susceptible than APCI (Xing *et al.*, 2007). Sample preparations may reduce (clean-up) or magnify (pre-concentration) matrix effects and eluents have also great influence on the ionisation efficiency in the

MS sources. Thus, the selection of mobile phase is a critical factor in achieving not only good chromatographic separation but also appropriate ionisation. Low-surface tension and a low-dielectric constant of the solvent promote ion evaporation, which favours the ionisation process. For instance, TFA is not considered a good solvent for MS detection as it suppresses the ionisation due to ion-pairing and surface tension effects (de Rijke *et al.*, 2006). In negative ion (NI) mode, ammonium formate and ammonium chloride are often used as modifiers to improve the shape peak and to enhance the MS response of analytes. Rauha *et al.* (2001) found that the optimal ionisation conditions were achieved in NI mode using ammonium acetate buffer at pH 4.0 as the aqueous component of the LC solvent system, whereas 0.4% formic acid (pH 2.3) gave the best results in positive ion mode, both in ESI and APCI.

11.6.4 Detection systems

Traditional methods for HPLC analysis of phenolic compounds have mainly been based on UV detection, especially through the use of diode-array detectors (DAD) that allow obtaining the UV-visible spectra of the peaks, constituting a useful tool for compound classification and peak purity assessment. However, this kind of detection usually lacks the sensitivity and selectivity required when dealing with the analysis of flavonoid metabolites in plasma and other bio-samples. Thus, concentration and cleanup procedures are usually required, and even in that case the enrichment attainable might not be sufficient; furthermore, possible losses of less stable analytes may also be produced. Consequently, other detection techniques with lower sensitivity limits have to be used, like fluorimetry (FL), electrochemical (ECD), and mass spectrometric (MS) detection. As an example, the detection limits for the analysis of quercetin in plasma obtained by different authors using different detection systems is shown in the Table 11.1.

Fluorometry detection (FL) has not been much used in flavonoid analysis due to the number of flavonoids that exhibit native fluorescence is limited. Only isoflavones, flavonoids with a free hydroxyl group at 3-position (e.g. flavonol aglycones and catechins) and some methoxylated flavones, as well as other phenolic compounds like coumarins and hydroxycinnamic acids show fluorescence. In those cases, FL may be useful for detection as it minimises background interference and increases sensitivity (Liu *et al.*, 2008). Native

Table 11.1 Detection limits for analysis of quercetin in plasma obtained by different detection systems.

Method[a]	Calculated detection limits (ng/mL)	Reference
HPLC–UV	14	Lai *et al.* (2007)
HPLC–FL[b]	2	Hollman (2001)
HPLC–ECD	0.63	Erlund *et al.* (1999)
LC–MS	0.10	Wang and Morris (2005)
Micro HPLC–ECD	0.06	Jin *et al.* (2004)

[a]Analytical conditions may be different in each case.
[b]After post-column derivatisation with Al^{3+}.

fluorescence has been used as a tool for tracing quercetin in cells and tissues (Nifli *et al.*, 2007), as well as for the HPLC analysis of 3′,4′,5′-trimethoxyflavonol, a potential cancer chemo-preventive compound, in plasma and tissues of mice (Britton *et al.*, 2009). HPLC–FL has also been used for selective detection and quantification of catechins and their methylated metabolites in plasma and other animal samples (Carando *et al.*, 1998; Donovan *et al.*, 1999b; Gossai & Lau-Cam, 2006), reaching limits of detection around 2–5 ng/mL; the fluorescence was measured at 310 nm (emission) with excitation at 280 nm.

Metal ions may bind to flavonoids through the hydroxyls at 3- or 5- positions and 4-keto group, and/or hydroxyls at 3′- and 4′-, producing fluorescent complexes. Thus, complexation with metal ions can be employed to enhance flavonoid fluorescence and increase sensitivity for their detection. An HPLC method for the analysis of flavonols using FL detection after post-column derivatisation with Al^{3+} was described by Hollman *et al.* (1996), which was further applied to pharmacokinetic and absorption studies of flavonols in humans (Hollman *et al.*, 1997; de Vries *et al.*, 1998). The fluorescence of the quercetin–Al complex was measured at 485 nm using an excitation wavelength of 422 nm. In that post-column system, only compounds that contain both a 3-hydroxyl and a 4-keto oxygen had sufficient fluorescence intensity for sensitive detection; compounds lacking a free 3-hydroxyl group, like flavonol metabolites substituted at that position, escape this detection method, and for their analysis deconjugation to the aglycone is previously required. In the conditions optimised for the method, a limit of detection of 2 ng/mL for quercetin in plasma was achieved (Hollman, 2001). HPLC–FL using derivatisation with aluminium was also applied by Paulke *et al.* (2006) to the quantification of quercetin and its methylated metabolites, tamarixetin and isorhamnetin, in rat brain.

HPLC coupled to electrochemical detection (HPLC–ECD) has also been used for the analysis of flavonoids in plants and food and other biological matrices. When performed with care, this is an extremely sensitive technique, with typical detection limits about 10 to 100-fold higher than UV detection, and ten-fold higher than fluorescence detection, depending on the compound studied. The linear response range of an electrochemical detector is approximately 10^5 versus 10^4 for UV absorption (Manach, 2003). The sensitivity of the technique can be further increased using HPLC micro-columns, as reported by Jin *et al.* (2004).

Gradient elution is generally considered as not suitable for electrochemical detection, because the mobile phase changes during the gradient cause a great baseline drift. Nevertheless, the newest HPLC equipments coupled to multielectrode coulometric detection (HPLC–CoulArray) incorporate a software containing a gradient-correcting algorithm that removes the baseline drift and allows the use of gradient elution without restriction. Thus, chromatographic conditions do not differ from those used with UV detection, except that mobile phases must contain an electrolyte to facilitate the current flux, typically 30–100 mM sodium or lithium phosphate. The pH must also be carefully controlled since it influences the current/voltage curve of the compounds, and thus their optimal oxidation potential (Manach, 2003).

Electroactivity of a compound is mainly dependent on the presence of electroactive functional groups such as phenolic hydroxyls or amino groups. Most flavonoids contain various phenolic groups, which makes them suitable for electrochemical detection. Differences in optimal potential between compounds depend on their relative ability to adapt to the loose of electrons, by delocalising charge via resonance or inductive charge stabilisation. For most analytes, the antioxidant efficiency could be related to the value of the first maximum, which corresponds to the lowest energy required to donate an electron. The oxidation potentials of the flavonoids mostly increase on a scale roughly based on the hydroxyl substitutions of the ring-B (Morand *et al.*, 2001). Nevertheless, for the flavonoids there is no always a linear relation between the anti-oxidant activity and the electrochemical signal, since other structural parameters may also play a role. For instance, glycosylation of the 3-OH group decreases the anti-oxidant activity in flavonols, but not their electrochemical behaviour (de Rijke *et al.*, 2006).

Flavonoids present various waves of oxidation corresponding to different moieties capable of undergoing oxidation. Usually, only the lowest wave of oxidation is used for quantification, but with multielectrode detection systems, the electrochemical behaviour of a compound across the array of the applied potentials can be used with qualitative purposes and peak purity assessment. In this detection system, compounds react on several electrodes according to their redox properties giving characteristic response profiles, which linked to the retention time can be used for their identification (Morand *et al.*, 2001). A comprehensive review on the fundamentals and applications of HPLC–CoulArray to flavonoid analysis in biological matrices was published by Manach (2003).

Over the last 15 years, HPLC–ECD including CoulArray detection has been successfully employed by several authors for the analysis of quercetin and catechins and their metabolites in plasma and other biomatrices (see Manach *et al.*, 2005; Jones *et al.*, 1998; Piskula & Terao, 1998a,1998b; Erlund *et al.*, 1999; Lee *et al.*, 2000; Donovan *et al.*, 2002b; Azuma *et al.*, 2003; Chu *et al.*, 2004; Jin *et al.*, 2004; Bolarinwa & Linseisen, 2005).

Liquid chromatography coupled to mass spectrometry (LC–MS) has become the reference technique for the detection and quantification of target flavonoid metabolites and the structural characterisation of unknowns in biological matrices, as demonstrated by the large list of works published on the subject over the last decade (e.g. Wittig *et al.*, 2001; Mullen *et al.*, 2004, 2010; Wang & Morris, 2005; Lan *et al.*, 2007; Mennen *et al.*, 2008; Serra *et al.*, 2009; Zhang *et al.*, 2010, and many others).

Single-stage MS in combination with diode array detection and the help of standards and reference data can be used for the detection and identification of flavonoids. Nevertheless, this technique presents important limitations (interferences of matrix components, formation of adducts, weak fragmentation of some analytes, variability in the relative abundance of fragment ions) that notably limit its applications in metabolite analysis. Tandem–MS offers much greater versatility, selectivity and capability to obtain structural information than single-stage MS. Patterns for the identification of compounds based on their MS fragmentation behaviour have been summarised earlier and relevant reviews on the subject also exist that can be consulted for further information (Ma *et al.*, 1997; Cuyckens & Claeys, 2004; Prasain *et al.*, 2004; Levsen *et al.*, 2005; Justino *et al.*, 2009; Vukics & Guttman, 2010). In tandem–MS, different screening strategies, such as full-scan, neutral-loss, precursor-ion and product-ion scan modes, and different types of ion screening techniques (i.e. single ion monitoring, SIM; selected reaction monitoring, SRM, and multiple reaction monitoring, MRM) can be used for selective detection of compounds and quantification purposes.

SRM offers great selectivity for compound detection, although it should be taken carefully since some compounds can be detected at the same SRM transitions. MRM screening is preferred for quantification, whilst full-scan mode may not always fulfil the requirements of sensitivity needed for metabolic analyses. When using the adequate approach, limits of quantification up to around 0.01 ng/mL can be achieved in LC–MS/MS, which is often sufficient in the quantitative analysis of metabolites (Kostiainen *et al.*, 2003).

Since several instrumental parameters have drastic influence on the ionisation efficiency, tandem MS conditions need to be optimised according to the experimental design and the type of equipment. The optimisation of the MS parameters includes the adjustment of interface parameters such as the ionisation voltage in ESI and the discharge-needle current in APCI, respectively, and the pressure of the spraying/nebulising potential. When using triple quadrupole MS/MS, the collision energy and the pressure of collision gas are other MS parameters to be optimised (Xing *et al.*, 2007). Conditions should be determined experimentally by evaluating the sensitivity and the fragmentation of each analyte. When available, the compound(s) can be directly infused into the source in the mobile phase by a syringe pump.

Triple quadrupole (TQ), ion traps (IT) and hybrid quadrupole-time of flight (Q-TOF) are the most common mass spectrometers used. The main advantage of the IT equipments is the possibility to perform MS^n experiments, which is of great interest for the structural identification; whereas TOF spectrometers can provide high-resolution analysis and the elemental compositions of metabolites with a mass accuracy better than 10 ppm. Q-TOF spectrometers also provide high sensitivity for the determination of metabolites, although they have a relatively poor dynamic linear range for quantitative analysis, compared to that of quadrupole instruments (Xing *et al.*, 2007). TQ instruments have proven useful in group-specific detection of metabolites using some specific marker ions. Transitions from the parent molecular ion to characteristic product ions, like some neutral diagnostic losses and specific retro Diels-Alder fragments can be employed for selective detection of compounds. For instance, characteristic ions of *m/z* 301 (parent ion) and *m/z* 151 (product ion) are usual for quercetin in negative ion mode (Lan *et al.*, 2007). In the case of catechins, fragment at *m/z* 139 is the one preferable in the positive ion mode. This fragment is formed after cleavage of two bonds in the ring-C and can originate from ring-A of catechins (parent ion at *m/z* 291) and both rings A and B of gallocatechins (*m/z* at 307). Another useful discriminating fragment is that at *m/z* 123 released from ring-B of catechins and that is not produced in the case of gallocatechins, due to the presence of the additional hydroxyl in position 3′. In negative ion mode, a characteristic fragment at *m/z* 125, derived from the ring-A, is observed in the MS^2 spectra of all (gallo)catechins, whereas a minor ion at *m/z* at 109 (*m/z* 125-OH) is only produced for catechins and not for gallocatechins (Spacil *et al.*, 2010). Glucuronides and sulphates can be selectively detected by neutral loss scan of 176 and 80 u, respectively, and product ions derived from sulphates (*m/z* at 80 and 97) and glucuronides (*m/z* at 175 and 113) can be used for their detection in the precursor ion mode.

Other detection techniques have also been used by different authors for the analysis of flavonoid metabolites in biological matrices. Radiocounting coupled to HPLC was used by Mullen *et al.* (2002, 2008) in bioavailability studies of quercetin in rats. Animals were fed [2-^{14}C]quercetin-4′-*O*-glucoside and radiolabeled metabolites were measured in body fluids and tissues using an online radioactivity detector. A method using chemiluminescence detection was applied to the analysis of tea catechins in rat and human plasma and tissues (Nakagawa & Miyazawa, 1997a; 1997b). The method involved a post-column oxidation by successive reaction with solutions of acetaldehyde in a buffer containing horseradish peroxidase and hydrogen peroxide. The reaction only occurs with flavonoids possessing vicinal hydroxyl groups and, therefore, could not be used for metabolites substituted in the ring-B.

11.6.5 *Trends in the chromatographic analysis of flavonoid metabolites*

In the last few years, *ultra-high performance liquid chromatography* (UHPLC) has emerged as a suitable alternative to conventional HPLC. This recently developed technique offers higher separation efficiency through the use of columns with small particle size (<2 μm) and diameter (typically, 2.1 or 1 mm) and optimised instrumentation capable of operating at high pressures up to 100 MPa (15,000 psi). This increases dramatically the number of theoretical plates and provides faster separations maintaining or improving resolution and sensitivity. A large variety of sub-2-microne analytical columns are currently available, most of them made of porous particles, based on both hybrid and silica materials with different degrees of polarity (e.g. C8, C18, Phenyl, Cyano, etc.), as reviewed by Novakova and Vlckova (2009). In recent times, many papers using UHPLC coupled to tandem mass spectrometry have been published dealing with the analysis of flavonoid metabolites in samples from biological assays and the number of applications must be expected to increase in further years (see Wang *et al.*, 2008; Serra *et al.*, 2009; Suárez *et al.*, 2009; Wang *et al.*, 2009; Marti *et al.*, 2010; Su *et al.*, 2010; Zhang *et al.*, 2010).

Another alternative to conventional HPLC is *high temperature liquid chromatography* (HTLC), in which standard columns are combined with temperatures higher than 60°C. In these conditions the viscosity of the solvents is decreased reducing the backpressure and allowing an increase in the velocity of phase mobile, which results in faster and more efficient separations. Another advantage of using high temperature is that it allows exploring unusual organic modifiers such as isopropanol for alternative selectivity and faster analysis. Combination of UHPLC with high temperature (up to 90°C) has also been explored by Plumb *et al.* (2007). These approaches have, however, some drawbacks that have reduced their applications, such as the limited availability of stable high temperature-resistant packing materials, and the potential degradation of thermally unstable compounds. Furthermore, a particular set-up is required for, respectively, heating and then cooling the mobile phase before and after travel through the chromatographic column (Novakova & Vlckova, 2009). In addition, care must be taken since the order of elution of the analytes can change and the production of narrow peak widths may compromise the quality of data obtained from detectors such as high-resolution mass spectrometers.

The use of monolith columns has also become popular in past few years. These columns can accept high flow-rates (up to 10 mL/min) in conventional column lengths without generating high backpressures, and require short times for column equilibration when a mobile phase gradient is used. However, the high flow-rates applied induce a high consumption of solvents and make these columns not fully compatible with MS detection. These facts linked to the short number of commercially stationary phases available (C8, C18, plain silica only) and their lower chemical stability compared to porous sorbents have reduced the applications of monolith columns. As far as we know, none of these techniques has been yet applied to the analysis of flavonoids and metabolites in biological samples.

Acknowledgements

The GIP–USAL is financially supported by the Spanish *Ministerio de Ciencia e Innovación* through the projects AGL2007-66108-C04-02 and AGL2009-12001, and the Consolider-Ingenio 2010 Programme (CSD2007-00063).

References

Alluis, B. & Dangles, O. (2001) Quercetin (2-(3,4-dihydroxyphenyl)-3,5,7-trihydroxy-4H-1-benzopyran-4-one) glycosides and sulfates: Chemical synthesis, complexation, and antioxidant properties. *Helvetica Chimica Acta*, **84**, 1133–1156.

Andrade, F.D.P., Santos, L.C., Datchler, M., Albert, K. & Vilegas, W. (2002) Use of on-line liquid chromatography-nuclear magnetic resonance spectroscopy for the rapid investigation of flavonoids from. *Sorocea bomplandii Journal of Chromatography A*, **953**, 287–291.

Arts, I.C.W., Venema, D.P. & Hollman, P.C.H. (2003) Quantitative determination of flavonols in plant foods and biological fluids. In: *Methods in Polyphenol Analysis* (eds. C. Santos-Buelga & G. Williamson), pp. 214–227. The Royal Society of Chemistry, Cambridge.

Aura, A.M. (2008) Microbial metabolism of dietary phenolic compounds in the colon. *Phytochemistry Reviews*, **7**, 407–429.

Aura, A.M., O'Leary, K.A., Williamson, G., *et al.* (2002) Quercetin derivatives are deconjugated and converted to hydroxyphenylacetic acids but not methylated by human fecal flora in vitro. *Journal of Agricultural and Food Chemistry*, **50**, 1725–1730.

Awad, H.M., Boersma, M.G., Boeren, S., Van Bladeren, P.J., Vervoort, J. & Rietjens, I.M. (2003) Quenching of quercetina quinone/quinone methides by different thiolate scavengers: Stability and reversibility of conjugate formation. *Chemical Research in Toxicology*, **16**, 822–831.

Azuma, K., Ipposhui, K., Ito, H., Horie, H. & Terao, J. (2003) Enhancing effect of lipids and emulsifiers on the accumulation of quercetin metabolites in blood plasma after the short-term ingestion of onion by rats. *Bioscience Biotechnology and Biochemistry*, **67**, 2548–2555.

Barron, D. (2008) Recent advances in the chemical synthesis and biological activity of phenolic metabolites. In: *Recent Advances in Polyphenol Research* (eds. F. Daayf & V. Lattanzio), vol. 1, pp. 317–358. Wiley-Blackwell, Oxford.

Barron, D., Cren-Olive, C. & Needs, P. (2003) Chemical synthesis of flavonoid conjugates. In: *Methods in Polyphenol Analysis* (eds. C. Santos-Buelga & G. Williamson), pp. 187–213. The Royal Society of Chemistry, Cambridge.

Barron, D. & Ibrahim, R.K. (1987) Quercetin and patuletin 3,3'-disulphates from. *Flaveria chloraefolia Phytochemistry*, **26**, 1181–1184.

Barron, D. & Ibrahim, R.K. (1988) Synthesis of flavonoid sulfates. 2. The use of aryl sulfatase in the synthesis of flavonol-3-sulfates. *Zeitschrift für Naturforschung C-A Journal of Biosciences*, **43**, 625–630.

Barron, D., Varin, L., Ibrahim, R.K., Harborne, J.B. & Williams, C.A. (1998) Sulphated flavonoids. An update. *Phytochemistry*, **27**, 2375–2395.

Boersma, M.G., van der Woude, H., Bogaards, J., *et al.* (2002) Regioselectivity of phase II metabolism of luteolin and quercetin by UDP-glucuronosyl transferases. *Chemical Research in Toxicology*, **15**, 662–670.

Bolarinwa, A. & Linseisen, J. (2005) Validated application of a new high-performance liquid chromatographic method for the determination of selected flavonoids and phenolic acids in human plasma using electrochemical detection. *Journal of Chromatography B: Analytical Technologies in the Biomedical and Life Sciences*, **823**, 143–151.

Bouktaib, M., Atmani, A. & Rolando, C. (2002) Regio- and stereoselective synthesis of the major metabolite of quercetin, quercetin-3-*O*-beta-D-glucuronide. *Tetrahedron Letters*, **43**, 6263–6266.

Britton, R.G., Fong, I., Saad, S., *et al.* (2009) Synthesis of the flavonoid 3′,4′,5′-trimethoxyflavonol and its determination in plasma and tissues of mice by HPLC with fluorescence detection. *Journal of Chromatography B*, **877**, 939–942.

Cao, G., Sofic, E. & Prior, R.L. (1997) Antioxidant and pro oxidant behavior of flavonoids: Structure-activity relationships. *Free Radical Biology and Medicine*, **22**, 749–760.

Carando, S., Teissedre, P.L. & Cabanis, J.C. (1998) Comparison of (1)-catechin determination in human plasma by high-performance liquid chromatography with two types of detection: fluorescence and ultraviolet. *Journal of Chromatography B*, **707**, 195–201

Chu, K.O., Wang, C.C., Chu, C.Y., Rogers, M.S., Choy, K.W. & Peng, C.P. (2004) Determination of catechins and catechins gallates in tissues by liquid chromatography with coulometric array detection and selective solid phase extraction. *Journal of Chromatography B: Analytical Technologies in the Biomedical and Life Sciences*, **810**, 187–195.

Cren-Olive, C., Deprez, S., Lebrun, S., Coddeville, B. & Rolando, C. (2000) Characterization of methylation site of monomethylflavan-3-ols by liquid chromatography/electrospray ionization tandem mass spectrometry. *Rapid Communications in Mass Spectrometry*, **14**, 2312–2319.

Cren-Olive, C., Lebrun, S. & Rolando, C. (2002) An efficient synthesis of the four mono methylated isomers of (+)-catechin including the major metabolites and of some dimethylated and trimethylated analogues through selective protection of the catechol ring. *Journal of the Chemical Society, Perkin Transactions 1*, **6**, 821–830.

Cuyckens, H. & Claeys, M. (2004) Mass spectrometry in the structural analysis of flavonoids. *Journal of Mass Spectrometry*, **39**, 1–15.

Dangles, O., Fargeix, G. & Dufour, C. (1999) One-electron oxidation of quercetin and quercetin derivatives in protic and non protic media. *Journal of the Chemical Society, Perkin Transactions 2*, **7**, 1387–1395.

Day, A.J., Bao, Y.P., Morgan, M.R.A. & Williamson, G. (2000) Conjugation position of quercetin glucuronides and effect on biological activity. *Free Radical Biology and Medicine*, **29**, 1234–1243.

Day, A.J., Gee, J.M., Dupont, M.S., Johnson, I.T. & Williamson, G. (2003) Absorption of quercetin-3-glucoside and quercetin-4′-glucoside in the rat small intestine: the role of lactase phlorizin hydrolase and the sodium-dependent glucose transporter. *Biochemical Pharmacology*, **65**, 1199–1206.

Day, A.J., Mellon, F., Barron, D., Sarrazin, G., Michael, R.A.M. & Williamson, G. (2001) Human metabolism of dietary flavonoids: Identification of plasma metabolites of quercetin. *Free Radical Research*, **35**, 941–952

Day, A.J. & Morgan, M.R.A. (2003) Methods of polyphenol extraction from biological fluids and tissues. In: *Methods in Polyphenol Analysis* (eds. C. Santos-Buelga & G. Williamson), pp. 17–47. The Royal Society of Chemistry, Cambridge.

Day, A.J. & Williamson, G. (2001) Biomarkers for exposure to dietary flavonoids: a review of the current evidence for identification of quercetin glycosides in plasma. *British Journal of Nutrition*, **86**, 105–110.

de Rijke, E., de Kanter, F., Ariese, F., Brinkman, U.A.T. & Gooijer, C. (2004) Liquid chromatography coupled to nuclear magnetic resonance spectroscopy for the identification of isoflavone glucoside malonates in *T. pratense*. L. leaves. *Journal of Separation Science*, **27**, 1061–1070.

de Rijke, E. , Out, P., Niessen, W.N.A., Ariese, F., Gooijer, C. & Brinkman, U.A.T. (2006) Analytical separation and detection methods for flavonoids. *Journal of Chromatography A*, **1112**, 31–63.

de Vries, J.H., Hollman, P.C.H., Meyboom, S., *et al.* (1998) Plasma concentrations and urinary excretion of the antioxidant flavonols quercetin and kaempferol as biomarkers for dietary intake. *American Journal of Clinical Nutrition*, **68**, 60–65.

Ding, S.J., Dudley, E., Chen, L.J., *et al.* (2006) Determination of active components of *Ginkgo biloba* in human urine by capillary high-performance liquid chromatography/mass spectrometry with on-line column-switching purification. *Rapid Communications in Mass Spectrometry*, **20**, 3619–3624.

Diniz, A., Escuder-Gilabert, L., Lopes, N.P., Villanueva-Camañas, R.M., Sagrado, S. & Medina-Hernández, M.J. (2008) Characterization of interactions between polyphenolic compounds and human serum proteins by capillary electrophoresis. *Analytical and Bioanalytical Chemistry*, **391**, 625–632.

Donovan, J.L., Bell, J.R., Kasim-Karakas, S., *et al.* (1999a) Catechin is present as metabolites in human plasma after consumption of red wine. *Journal of Nutrition*, **129**, 1662–1668.

Donovan, J.L., Kasim-Karakas, S., German, J.B. & Waterhouse, A.L. (2002a) Urinary excretion of catechin metabolites by human subjects after red wine consumption. *British Journal of Nutrition*, **87**, 31–37.

Donovan, J.L., Luthria, D.L., Stremple, P. & Waterhouse, A.L. (1999b) Analysis of (+)-catechin, (−)-epicatechin and their 3′ and 4′ *O*-methylated analogs: A comparison of sensitive methods. *Journal of Chromatogaphy B*, **726**, 277–283.

Donovan, J.L., Manach, C., Rios, L., Morand, C., Scalbert, A., Remesy, C. (2002b) Procyanidins are not bioavailable in rats fed a single meal containing a grapeseed extract or the procyanidin dimer B3. *British Journal of Nutrition*, **87**, 299–306.

Dueñas, M., Mingo-Chornet, H., Perez-Alonso, J.J., Di Paola-Naranjo, R., Gonzalez-Paramas, A.M. & Santos-Buelga, C. (2008) Preparation of quercetin glucuronides and characterization by HPLC-DAD-ESI/MS. *European Food Research and Technology*, **227**, 1069–1076.

Dueñas, M., Surco-Laos, F., Gonzalez-Manzano, S., Gonzalez-Paramas, A.M. & Santos-Buelga, C. (2010) Antioxidant Properties of major metabolites of quercetin. *European Food Research and Technology*, **232**, 103-111.

El-Hady, D.A., Gotti, R. & El-Maali, N.A. (2008) Selective determination of catechin, epicatechin and ascorbic acid in human urine using chiral capillary electrophoresis. *Journal of Separation Science*, **31**, 2252–2259.

Erlund, I., Alfthan, G., Siren, H., Ariniemi, K. & Aro, A. (1999) Validated method for the quantitation of quercetin from human plasma using high-performance liquid chromatography with electrochemical detection. *Journal of Chromatography B*, **727**, 179–189.

Erlund, I., Kosonen, T., Alfthan, G., *et al.* (2000) Pharmacokinteics of quercetin from quercetin aglycone and rutin in healthy volunteers. *European Journal of Clinical Pharmacology*, **56**, 545–553.

Exarchou, V., Fiamegos, Y.C., van Beek, T.A., Nanos, C. & Vervoort, J. (2006) Hyphenated chromatographic techniques for the rapid screening and identification of antioxidants in methanolic extracts of pharmaceutically used plants. *Journal of Chromatography A*, **1112**, 293–302.

Galati, G., Moridani, M.Y., Chan, T.S. & O'Brien, P.J. (2001) Peroxidative metabolism of apigenin and naringenin versus luteolin and quercetin: glutathione oxidation and conjugation. *Free Radical Biology and Medicine*, **30**, 370–382.

Gee, J.M., DuPont, S.M., Day, A.J., *et al.* (2000) Intestinal transport of quercetin glycosides in rats involves both deglycoslylation and interaction with the hexose transport pathway. *Journal of Nutrition*, **130**, 2765–2771.

Gonzalez-Manzano, S., Gonzalez-Paramas, A.M., Santos-Buelga, C. & Dueñas, M. (2009) Preparation and characterization of catechin sulfates, glucuronides, and methylethers with metabolic interest. *Journal of Agricultural and Food Chemistry*, **57**, 1231–1238.

Gossai, D. & Lau-Cam, C.A. (2006) Simple HPLC method, with fluorometric detection, for studying the oral absorption of monomeric catechins in a small animal model. *Journal of Liquid Chromatography & Related Technologies*, **29**, 2571–2586.

Grün, C., van Dorsten, F.A., Jacobs, D.M., *et al.* (2008) GC-MS methods for metabolic profiling of microbial fermentation products of dietary polyphenols in human and *in vitro* intervention studies. *Journal of Chromatography B*, **871**, 212–219.

Gunnarsson, G.T. & Desai, U.R. (2002) Interaction of designed sulfated flavanoids with antithrombin: Lessons on the design of organic activators. *Journal of Medicinal Chemistry*, **45**, 4460–4470.

Harborne, J.B. (1973) *Phytochemical Methods*. Chapman & Hall, London.

Hollman, P.C.H. (2001) Determination of flavonols in body fluids. *Methods in Enzymology*, **335**, 97–103.

Hollman, P.C.H., Van Trijp, J.M., Buysman, M.N. (1996) Fluorescence detection of flavonols in HPLC by post-column chelation with aluminum. *Analytical Chemistry*, **68**, 3511–3515.

Hollman, P.C.H., Van Trijp, J.M., Buysman, M.N., *et al.* (1997) Relative bioavailability of the antioxidant quercetin from various foods in man. *FEBS Letters*, **418**, 152–156.

Hong, J. & Mitchell, A.E. (2006) Identification of glutathione-related quercetin metabolites in humans, *Chemical Research in Toxicology*, **19**, 1525–1532.

Hong, Y. & Mitchell, A.E. (2004) Metabolic profiling of flavonol metabolites in human urine by liquid chromatography and tandem mass spectrometry. *Journal of Agricultural and Food Chemistry*, **52**, 6794–6801.

Jones, D.J.L., Jukes-Jones, R., Verschoyle, R.D., Farmer, P.B. & Gescher, A.A. (2005) A synthetic approach to the generation of quercetin sulfates and the detection of quercetin 3′-*O*-sulfate as a urinary metabolite in the rat. *Bioorganic Medicinal Chemistry*, **13**, 6727–6731.

Jones, D.J.L., Lim, C.K., Ferry, D.R. & Gescher, A.A. (1998) Determination of quercetin in human plasma by HPLC with spectrophotometric or electrochemical detection. *Biomedical Chromatography*, **12**, 232–235.

Justesen, U. (2001) Collision-induced fragmentation of deprotonated methoxylated flavonoids, obtained by electrospray ionization mass spectrometry. *Journal of Mass Spectrometry*, **36**, 169–178.

Justino, G.C., Borges, C.M. & Florêncio, M.H. (2009) Electrospray ionization tandem mass spectrometry fragmentation of protonated flavone and flavonol aglycones: a re-examination. *Rapid Communications in Mass Spectrometry*, **23**, 237–248.

Kataoka, H. (2003) New trends in sample preparation for clinical and pharmaceutical analysis. *Trends in Analytical Chemistry*, **22**, 232–244.

Kawai, Y., Ishisaka, A., Saito, S., *et al.* (2008) Immunochemical detection of flavonoid glycosides: Development, specificity, and application of novel monoclonal antibodies. *Archives of Biochemistry and Biophysics*, **476**, 124–132.

Kay, C.D., Mazza, G., Holub, B.J. & Wang, J. (2004) Anthocyanin metabolites in human urine and serum. *British Journal of Nutrition*, **91**, 933–942.

Knjazeva, T. & Kaljurand, M. (2010) Capillary electrophoresis frontal analysis for the study of flavonoid interactions with human serum albumin. *Analytical and Bioanalytical Chemistry*, **397**, 2211–2219.

Koizumi, M., Akao, T., Katoda, S., Kikuchim, T., Okuda, T. & Kobashi, K. (1991) Enzymatic sulfation of polyphenols related to tannins by arylsulfotransferase. *Chemical & Pharmaceutical Bulletin*, **39**, 2638–2643.

Koizumi, M., Shimizu, M. & Kobashi, K. (1990) Enzymatic sulfation of quercetin by arylsulfotransferase from human intestinal bacterium. *Chemical & Pharmaceutical Bulletin*, **38**, 794–796.

Kostiainen, R., Kotiaho, T., Kuuranne, T. & Auriola, S. (2003) Liquid chromatography/atmospheric pressure ionization–mass spectrometry in drug metabolism studies. *Journal of Mass Spectrometry*, **38**, 357–372.

Kottra, G. & Daniel, H. (2007) Flavonoid glycosides are not transported by the human Na^+/glucose transporter when expressed in *Xenopus laevis* oocytes, but effectively inhibit electrogenic glucose uptake. *Journal of Pharmacology and Experimental Therapeutics*, **322**, 829–835.

Kroon, P.A., Clifford, M.N., Crozier, A., *et al.* (2004) How should we assess the effects of exposure to dietary polyphenols in vitro? *American Journal of Clinical Nutrition*, **80**, 15–21.

Kuhnert, N., Jaiswal, R., Febi Matei, M., Sovdat, T. & Deshpande, S. (2010) How to distinguish between feruloyl quinic acids and isoferuloyl quinic acids by liquid chromatography/tandem mass spectrometry. *Rapid Communications in Mass Spectrometry*, **24**, 1575–1582.

Lai, X.Y., Zhao, Y.Y., Liang, H., *et al.* (2007) SPE-HPLC method for the determination of four flavonols in rat plasma and urine after oral administration of *Abelmoschus manihot* extract. *Journal of Chromatography B-Analytical Technologies in the Biomedical and Life Sciences*, **852**, 108–114.

Lan, K., Jiang, X. & He, J. (2007) Quantitative determination of isorhamnetin, quercetin and kaempferol in rat plasma by liquid chromatography with electrospray ionization tandem mass spectrometry and its application to the pharmacokinetic study of isorhamnetin. *Rapid Communications in Mass Spectrometry*, **21**, 112–120.

Jin, D., Hakamata, H., Takahashi, K., *et al.* (2004). Determination of quercetin in human plasma after ingestion of commercial canned green tea by semi-micro HPLC with electrochemical detection. *Biomedical Chromatography*, **18**, 662–666.

Le Gall, G., DuPont, M.S., Mellon, F.A., *et al.* (2003) Characterization and content of flavonoid glycosides in genetically modified tomato (*Lycopersicon esculentum*) fruits. *Journal of Agricultural and Food Chemistry*, **51**, 2438–2446.

Lee, M.J., Prabhu, S., Meng, X., Li, C. & Yang, C.S. (2000) An improved method for the determination of green and black tea polyphenols in biomatrices by high-performance liquid chromatography with coulometric array detection. *Analytical Biochemistry*, **279**, 164–169.

Levsen, K., Schiebel, H.M., Behnke, B., *et al.* (2005) Structure elucidation of phase II metabolites by tandem mass spectrometry: an overview. *Journal of Chromatography A*, **1067**, 55–72.

Liu, B., Anderson, D., Ferry, D.R., Seymour, L.W., de Takats, P.G. & Kerr, D.J. (1995) Determination of quercetin in human plasma using reversed phase high-performance liquid chromatography. *Journal of Chromatography B*, **666**, 149–155.

Liu, E.H., Qi, L.W., Cao, J., Li, P., Li, C.Y. & Peng, Y.B. (2008) Advances of modern chromatographic and electrophoretic methods in separation and analysis of flavonoids. *Molecules*, **13**, 2521–2544.

Loke, W.M., Jenner, A.M., Proudfoot, J.M., *et al.* (2009) A Metabolite profiling approach to identify biomarkers of flavonoid intake in humans. *Journal of Nutrition*, **139**, 2309–2314.

Luthria, D.I., Jones, A.D., Donovan, J.L. & Waterhouse, A.L. (1997) GC-MS determination of catechin and epicatechin levels in human plasma. *Journal of High Resolution Chromatogaphy*, **20**, 621–623.

Ma, Y.L., Li, Q.M., Van den Heuvel, H. & Claeys, M. (1997) Characterization of flavone and flavonol aglycones by collision-induced dissociation tandem mass spectrometry. *Rapid Communications in Mass Spectrometry*, **11**, 1357–1364.

Mabry, T.J., Markham, K.R. & Thomas, M.B. (1970) *The Systematic Identification of Flavonoids*. Springer-Verlag, New York.

MacGregor, J.T. & Jurd, L. (1978) Mutagenicity of plant flavonoids: structural requirements for mutagenic activity in *Salmonella typhimurium*. *Mutation Research*, **54**, 297–309.

Manach, C. (2003) The use of HPLC with coulometric detection in the analysis of flavonoids in complex matrices. In: *Methods in Polyphenol Analysis* (eds. C. Santos-Buelga & G. Williamson), pp. 63–91. The Royal Society of Chemistry, Cambridge.

Manach, C., Williamson, G., Morand, C., Scalbert, A. & Remesy, C. (2005) Bioavailability and bioefficacy of polyphenols in humans. I. Review of 97 bioavailability studies. *American Journal of Clinical Nutrition*, **81**, 230S–242S.

Marchi, I., Viette, V., Badoud, F., *et al.* (2010) Characterization and classification of matrix effects in biological samples analyses. *Journal of Chromatography A*, **1217**, 4071–4078.

Mariel-Agnese, A., Nuñez-Montoya, S., Ariza-Espinar, L. & Cabrera, J.L. (1999) Chemotaxonomic features in Argentinean species of *Flaveria* (*Compositae*). *Biochemical Systematics and Ecology*, **27**, 739–742

Markham, K.R. (1982) *Techniques of Flavonoid Identification*. Academic Press, London.

Marti, M.P., Pantaleon, A., Rozek, A., *et al.* (2010) Rapid analysis of procyanidins and anthocyanins in plasma by microelution SPE and ultra-HPLC. *Journal of Separation Science*, **33**, 2841–2853.

Mata-Bilbao, M.L., Andres-Lacueva, C., Roura, E., Jauregui, O., Torre, C. & Lamuela-Raventos, R.M. (2007) A new LC/MS/MS rapid and sensitive method for the determination of green tea catechins and their metabolites in biological samples. *Journal of Agricultural and Food Chemistry*, **55**, 8857–8863.

Maukonen, J., Motto, J., Suihko, M.L. & Saarela, M. (2008) Intra-individual diversity and similarity of salivary and faecal microbiota. *Journal of Medical Microbiology*, **57**, 1560–1568.

Medina-Remon, A., Barrionuevo-González, A., Zamora-Ros, R., *et al.* (2009) Rapid Folin–Ciocalteu method using microtiter 96-well plate cartridges for solid phase extraction to assess urinary total phenolic compounds, as a biomarker of total polyphenols intake. *Analytica Chimica Acta*, **634**, 54–60.

Mennen, L.I., Sapinho, D., Ito, H., Galan, P., Hercberg, S. & Scalbert, A. (2008) Urinary excretion of 13 dietary flavonoids and phenolic acids in free-living healthy subjects – variability and possible use as biomarkers of polyphenol intake. *European Journal of Clinical Nutrition*, **62**, 519–525.

Metodiewa, D., Jaiswal, A.K., Cenas, N., Dickancaite, E. & Segura-Aguilar, J. (1999) Quercetin may act as a cytotoxic pro oxidant after its metabolic activation to semiquinone and quinoidal product. *Free Radical Biology and Medicine*, **26**, 107–116.

Moon, J.H., Tsushida, T., Nakahara, K. & Terao, J. (2001) Identification of quercetin 3-O-beta-D-glucuronide as an antioxidative metabolite in rat plasma after oral administration of quercetin. *Free Radical Biology and Medicine*, **30**, 1274–1285.

Morand, C., Manach, C., Donovan, J. & Remesy, C. (2001) Preparation and characterization of flavonoid metabolites present in biological samples. *Methods in Enzymology*, **335**, 115–121.

Moridani, M.Y., Scobie, H., Salehi, P. & O'Brien, P.J. (2001) Catechin metabolism: Glutathione conjugate formation catalyzed by tyrosinase, peroxidase, and cytochrome p450. *Chemical Research in Toxicology*, **14**, 841–848.

Mullen, W., Boitier, A., Stewart, A.J. & Crozier, A. (2004) Flavonoid metabolites in human plasma and urine after the consumption of red onions: analysis by liquid chromatography with photodiode array and full scan tandem mass spectrometric detection. *Journal of Chromatography A*, **1058**, 163–168

Mullen, W., Borges, G., Lean, M.E., Roberts, S.A. & Crozier, A. (2010) Identification of metabolites in human plasma and urine after consumption of a polyphenol-rich juice drink. *Journal of Agricultural and Food Chemistry*, **58**, 2586–2595.

Mullen, W., Edwards, C.A. & Crozier, A. (2006) Absorption, excretion and metabolite profiling of methyl-, glucuronyl-, glucosyl- and sulpho-conjugates of quercetin in human plasma and urine after ingestion of onions. *British Journal of Nutrition*, **96**, 107–116.

Mullen, W., Graft, B.A., Caldwell, S.T., *et al.* (2002) Determination of flavonol metabolites in plasma and tissues of rats by HPLC-radiocounting and tandem mass spectrometry following oral ingestion of [2-^{14}C]quercetin-4′-glucoside. *Journal of Agricultural and Food Chemistry*, **50**, 6902–6909.

Mullen, W., Hartley, R.C. & Crozier, A. (2003) Detection and identification of C-14-labelled flavonol metabolites by high-performance liquid chromatography-radiocounting and tandem mass spectrometry. *Journal of Chromatography A*, **1007**, 21–29.

Mullen, W., Rouanet, J.M., Auger, C., *et al.* (2008) Bioavailability of [2-^{14}C]quercetin-4′-glucoside in rats. *Journal of Agricultural and Food Chemistry*, **56**, 12127–12137.

Nakagawa, K. & Miyazawa, T. (1997a) Chemiluminescence-high performance liquid chromatographic determination of tea catechin, (–)-epigallocatechin-3-gallate, at picomole levels in rat and human plasma. *Analytical Biochemistry*, **248**, 41–49.

Nakagawa, K. & Miyazawa, T. (1997b) Absorption and distribution of tea catechin, (-)-epigallocatechin-3-gallate, in the rat. *Journal of Nutritional Science and Vitaminology*, **43**, 679–684.

Natsume, M., Osakabe, N., Oyama, M., *et al.* (2003) Structures of (–)-epicatechin glucuronide identified from plasma and urine after oral ingestion of (–)-epicatechin: differences between human and rat. *Free Radical Biology and Medicine*, **34**, 840–849.

Needs, P.W. & Kroon, P.A. (2006) Convenient syntheses of metabolically important quercetin glucuronides and sulfates. *Tetrahedron*, **62**, 6862–6868.

Németh, K., Plumb, G.W., Berrin, J.G., *et al.* (2003) Deglycosylation by small intestinal epithelial cell β-glucosidases is a critical step in the absorption and metabolism of dietary flavonoid glycosides in humans. *European Journal of Nutrition*, **42**, 29–42.

Nifli, A.F., Theodoropoulos, P.A., Munier, S., *et al.* (2007) Quercetin exhibits a specific fluorescence in cellular milieu: a valuable tool for the study of its intracellular distribution. *Journal of Agricultural and Food Chemistry*, **55**, 2873–2878.

Novakova, L. & Vlckova, H. (2009) A review of current trends and advances in modern bio-analytical methods: chromatography and sample preparation. *Analytica Chimica Acta*, **656**, 8–35.

O'Leary, K.A., Day, A.J., Needs, P.W., Mellon, F.A., O'Brien, N.M. & Williamson, G. (2003) Metabolism of quercetin-7- and quercetin-3-glucuronides by an in vitro hepatic model: the role of human β-glucuronidase, sulfotransferase, catechol-O-methyltransferase and multi-resistant protein 2 (MRP2) in flavonoid metabolism, *Biochemical Pharmacology*, **65**, 479–491.

O'Leary, K.A., Day, A.J., Needs, P.W., Sly, W.S., O'Brien, N.M. & Williamson, G. (2001) Flavonoid glucuronides are substrates for human liver β-glucuronidase. *FEBS Letters*, **503**, 103–106.

Parisis, D.M. & Pritchard, E.T. (1983) Activation of rutin by human oral bacterial isolates to the carcinogen-mutagen quercetin. *Archives of Oral Biology*, **28**, 583–590.

Paulke, A., Schubert-Zsilavecz, M. & Wurglics, M. (2006) Determination of St. John's wort flavonoid-metabolites in rat brain through high performance liquid chromatography coupled with fluorescence detection. *Journal of Chromatography B*, **832**, 109–113.

Piskula, M.K. & Terao, J. (1998a) Accumulation of (−)-epicatechin metabolites in rat plasma after oral administration and distribution of conjugated enzymes in rat tissues. *Journal of Nutrition*, **128**, 1172–1178.

Piskula, M.K. & Terao, J. (1998b) Quercetin's solubility affects its accumulation in rat plasma after oral administration. *Journal of Agricultural and Food Chemistry*, **46**, 4313–4317.

Plumb, G.W., O'Leary, K., Day, A.J. & Williamson, G. (2003) Enzymatic synthesis of quercetin glucosides and glucuronides. In: *Methods in Polyphenol Analysis* (eds. C. Santos-Buelga & G. Williamson), pp. 177–185. The Royal Society of Chemistry, Cambridge.

Plumb, R., Mazzeo, J.R., Grumbach, E.S., *et al.* (2007) The application of small porous particles, high temperatures, and high pressures to generate very high resolution LC and LC/MS separations. *Journal of Separation Science*, **27**, 1345–1359.

Prasain, J.K., Wang, C.C. & Barnes, S. (2004) Mass spectrometric methods for the determination of flavonoids in biological samples. *Free Radical Biology and Medicine*, **37**, 1324–1350.

Price, K.R., Colquhoun, I.J., Barnes, K.A. & Rhodes, M.J.C. (1998) Composition and content of flavonol glycosides in green beans and their fate during processing. *Journal of Agricultural and Food Chemistry*, **46**, 4898–4903.

Qu, J., Wang, Y.M., Luo, G.A. & Wu, Z.P. (2001) Identification and determination of glucuronides and their aglycones in *Erigeron breviscapus* by liquid chromatography-tandem mass spectrometry. *Journal of Chromatography A*, **928**, 155–162.

Queiroz, E.F., Ioset, J.R., Ndjoko, K., Guntern, A., Foggin, C.M. & Hostettmann, K. (2005) On-line identification of the bioactive compounds from *Blumea gariepina* by HPLC-UV-MS and HPLC-UV-NMR combined with HPLC-micro-fractionation. *Phytochemical Analysis*, **16**, 166–174.

Queiroz, E.F., Wolfender, J.L., Atindehou, K.K., Traore, D. & Hostettmann, K. (2002) On-line identification of the antifungal constituents of *Erythrina vogelii* by liquid chromatography with tandem mass spectrometry, ultraviolet absorbance detection and nuclear magnetic resonance spectrometry combined with liquid chromatographic micro-fractionation. *Journal of Chromatography A*, **974**, 123–134.

Rauha, J.P., Vuorela, H. & Kostiainen, R. (2001) Effect of eluent on the ionization efficiency of flavonoids by ion spray, atmospheric pressure chemical ionization, and atmospheric pressure photoionization mass spectrometry. *Journal of Mass Spectrometry*, **36**, 1269–1280.

Rechner, A.R., Kuhnle, G., Hu, H.L., *et al.* (2002) The metabolism of dietary polyphenols and the relevance to circulating levels of conjugated metabolites. *Free Radical Research*, **36**, 1229–1241.

Rehova, L., Skerikova, V. & Jandera, P. (2004) Optimisation of gradient HPLC analysis of phenolic compounds and flavonoids in beer using a CoulArray detector. *Journal of Separation Science*, **27**, 1345–1359.

Requena, T., Monagas, M., Pozo-Bayón, M.A., *et al.* (2010) Perspectives of the potential implications of wine polyphenols on human oral and gut microbiota. *Trends in Food Science and Technology*, **21**, 332–344.

Rice-Evans, C.A., Miller, N.J. & Paganga, G. (1996) Structure-antioxidant activity relationships of flavonoids and phenolic acids. *Free Radical Biology and Medicine*, **20**, 933–956.

Roura, E., Andres-Lacueva, C., Estruch, R. & Lamuela-Raventos, R.M. (2006) Total polyphenol intake estimated by a modified Folin–Ciocalteu assay of urine. *Clinical Chemistry*, **52**, 749–752.

Santos, M.R., Rodríguez-Gómez, M.J., Justino, G.C., Charro, N., Florencio, M.H. & Mira, L. (2008) Influence of the metabolic profile on the in vivo antioxidant activity of quercetin under a low dosage oral regimen in rats. *British Journal of Pharmacology*, **153**, 1750–1761.

Santos-Buelga, C., García-Viguera, C. & Tomás-Barberán, F.A. (2003) On-line identification of flavonoids by HPLC coupled to diode array detection. In: *Methods in Polyphenol Analysis* (eds. C. Santos-Buelga & G. Williamson), pp. 92–124. The Royal Society of Chemistry, Cambridge.

Scalbert, A. & Williamson, G. (2000) Dietary intake and bioavailability of polyphenols. *Journal of Nutrition*, **130**, 2073S–2085S.

Selma, M.V., Espin, J.C. & Tomas-Barberan, F.A. (2009) Interaction between phenolics and gut microbiota: role in human health. *Journal of Agricultural and Food Chemistry*, **57**, 6485–6501.

Serra, A., Macia, A., Romero, M.P., *et al.* (2009) Determination of procyanidins and their metabolites in plasma samples by improved liquid chromatography–tandem mass spectrometry. *Journal of Chromatography B*, **877**, 1169–1176.

Setchell, K.D., Faughnana, M.S., Avades, T., *et al.* (2003) Comparing the pharmacokinetics of daidzein and genistein with the use of C-13-labeled tracers in premenopausal women. *American Journal of Clinical Nutrition*, **77**, 411–419.

Sheng, Y. & Feng, F. (2006) Sensitive liquid chromatography–tandem mass spectrometry method for the determination of scutellarin in human plasma: Application to a pharmacokinetic study. *Journal of Chromatography B*, **830**, 1–5.

Shimoi, K., Saka, N., Nozawa, R., *et al.* (2001) Deglucuronidation of a flavonoid, luteolin monoglucuronide, during inflammation. *Drug Metabolism and Disposition*, **29**, 1521–1524.

Silva Elipe, M.V. (2003) Advantages and disadvantages of nuclear magnetic resonance spectroscopy as a hyphenated technique. *Analytica Chimica Acta*, **497**, 1–25.

Singh, R., Wu, B., Tang, L., *et al.* (2010). Identification of the position of mono-O-glucuronide of flavones and flavonols by analyzing shift in online UV spectrum (λmax) generated from an online diode array detector. *Journal of Agricultural and Food Chemistry*, **58**, 9384–9395.

Smolarz, H.D., Budzianowski, J., Bogucka-Kocka, A., Kocki, J. & Mendyk, E. (2008) Flavonoid glucuronides with anti-leukaemic activity from *Polygonum amphibium* L. *Phytochemical Analysis*, **19**, 506–513.

Spacil, Z., Novakova, L. & Solich, P. (2010) Comparison of positive and negative ion detection of tea catechins using tandem mass spectrometry and ultra high performance liquid chromatography. *Food Chemistry*, **123**, 535–541.

Spencer, J.P.E., El Mohsen, M.M.A., Minihane, A.M. & Mathers, J.C. (2008) Biomarkers of the intake of dietary polyphenols: strengths, limitations and application in nutrition research. *British Journal of Nutrition*, **99**, 12–22.

Sperker, B., Mürdther, T.E., Schick, M., Eckhardt, K., Bosslet, K. & Kroemer, H.K. (1997) Interindividual variability in expression and activity of human β-glucuronidase in liver and kidney: Consequences for drug metabolism, *Journal of Pharmacology and Experimental Therapeutics*, **281**, 914–920.

Stachulski, A.V. & Jenkins, G.N. (1998) The synthesis of O-glucuronides. *Natural Products Reports*, **15**, 173–186.

Stalikas, C.D. (2007) Extraction, separation, and detection methods for phenolic acids and flavonoids. *Journal of Separation Science*, **30**, 3268–3295.

Su, S., Guo, J., Duan, J., *et al.* (2010) Ultra-performance liquid chromatography–tandem mass spectrometry analysis of the bioactive components and their metabolites of Shaofu Zhuyu decoction active extract in rat plasma. *Journal of Chromatography B*, **878**, 355–362.

Suárez, M., Romero, M.P., Macia, A., *et al.* (2009) Improved method for identifying and quantifying olive oil phenolic compounds and their metabolites in human plasma by microelution solid-phase extraction plate and liquid chromatography–tandem mass spectrometry. *Journal of Chromatography B*, **877**, 4097–4106.

Theodoridis, G., Lasakova, M., Skerikova, V., Tegou, A., Giantsiou, N. & Jandera, P. (2006) Molecular imprinting of natural flavonoid antioxidants: Application in solid-phase extraction for the sample pretreatment of natural products prior to HPLC analysis. *Journal of Separation Science*, **29**, 2310–2321.

Unno, T., Sagesake, Y.M. & Kakuda, T. (2005) Analysis of tea catechins in human plasma by high-performance liquid chromatography with solid-phase extraction. *Journal of Agricultural and Food Chemistry*, **53**, 9885–9889.

Vaidyanathan, J.B. & Walle, T. (2002) Glucuronidation and sulfation of the tea flavonoid epicatechin by the human and rat enzymes. *Drug Metabolim and Disposition*, **30**, 897–903.

van der Hoeven, J.C.M., Brugeeman, L.M. & Debets, F.M.H. (1984) Genotoxicity of quercetin in cultured mammalian cells, *Mutation Research/Genetic Toxicology*, **136**, 9–21.

van der Woude, H., Boersman, M.M.G., Alink, G.M., Vervoort, J. & Rietjens, I.M.C.M. (2006) Consequences of quercetin methylation for its covalent glutathione and DNA adduct formation, *Chemico-Biological Interactions*, **160**, 193–203.

van der Woude, H., Boersman, M.M.G., Vervoort, J. & Rietjens, I.M.C.M. (2004) Identification of 14 quercetin phase II mono- and mixed conjugates and their formation by rat and human phase II in vitro model systems, *Chemical Research in Toxicology*, **17**, 1520–1530.

van Dorsten, F.A., Grün, C., van Velzen, E.J.J., Jacobs, D.M., Draijer, R. & van Duynhove, J.P.M. (2010) The metabolic fate of red wine and grape juice polyphenols in humans assessed by metabolomics. *Molecular Nutrition & Food Research*, **54**, 897–908.

Varin, L., Barron, D. & Ibrahim, R.K. (1987) Enzymatic assay for flavonoid sulfotransferase. *Analytical Biochemistry*, **161**, 176–180.

Vukics, V. & Guttman, A. (2010) Structural characterization of flavonoid glycosides by multi-stage mass spectrometry. *Mass Spectrometry Reviews*, **29**, 1–16.

Walle, T., Browning, A.M., Steed, L.L., Reed, S.G. & Walle, K. (2005) Flavonoid glucosides are hydrolyzed and thus activated in the oral cavity in humans. *Journal of Nutrition*, **135**, 48–52.

Wang, L. & Morris, M.E. (2005) Liquid chromatography–tandem mass spectroscopy assay for quercetin and conjugated quercetin metabolites in human plasma and urine. *Journal of Chromatography B*, **821**, 194–201.

Wang, X., Sun, W., Sun, H., *et al.* (2008) Analysis of the constituents in the rat plasma after oral administration of Yin Chen Hao Tang by UPLC/Q-TOF-MS/MS. *Journal of Pharmaceutical and Biomedical Analysis*, **46**, 477–490.

Wang, Y., Yao, Y., An, R., You, L. & Wang, X. (2009) Simultaneous determination of puerarin, daidzein, baicalin, wogonoside and liquiritin of GegenQinlian decoction in rat plasma by ultra-performance liquid chromatography–mass spectrometry. *Journal of Chromatography B*, **877**, 1820–1826.

Watson, D.G. & Pitt, A.R. (1998) Analysis of flavonoids in tablets and urine by gas chromatography/mass spectrometry and liquid chromatography/mass spectrometry. *Rapid Communications in Mass Spectrometry*, **12**, 153–156.

Williamson, G. (2002) The use of flavonoid aglycones in *in vitro* systems to test biological activities based on bioavailability data, is this a valid approach? *Phytochemistry Reviews*, **1**, 215–222.

Williamson, G., Barron, D., Shimoi, K. & Terao, J. (2005) *In vitro* biological properties of flavonoid conjugates found *in vivo*. *Free Radical Research*, **39**, 457–469.

Winter, J., Moore, L.H., Dowell, V.R. & Bokkenheuser, V.D. (1989) C-ring cleavage of flavonoids by human intestinal bacteria, *Applied and Environmental Microbiology*, **55**, 1203–1208.

Wittig, J., Herderich, M., Graefe, E.U. & Veit, M. (2001) Identification of quercetin glucuronides in human plasma by high performance liquid chromatography–tandem mass spectrometry. *Journal of Chromatography B*, **753**, 237–243.

Wolfender, J.L., Kdjoko, K. & Hostettmann, K. (2001) The potential of LC-NMR in phytochemical analysis. *Phytochemical Analysis*, **12**, 2–22.

Xie, J., Zhu, L., Luo, H., Zhou, L., Li, C. & Xu, X. (2001) Direct extraction of specific pharmacophoric flavonoids from ginkgo leaves using a molecularly imprinted polymer for quercetin. *Journal of Chromatography A*, **934**, 1–11.

Xing, J., Chunfeng, S. & Hongxiang, L. (2007) Recent applications of liquid chromatography–mass spectrometry in natural products bioanalysis. *Journal of Pharmaceutical and Biomedical Analysis*, **44**, 368–378.

Xu, F.Q., Guan, H.S., Li, G.Q. & Liu, H.B. (2009) LC method for analysis of three flavonols in rat plasma and urine after oral administration of *Polygonum aviculare* extract. *Chromatographia*, **69**, 1251–1258.

Zhang, K. & Zuo, Y. (2004) GC-MS Determination of flavonoids and phenolic and benzoic acids in human plasma after consumption of cranberry juice. *Journal of Agricultural and Food Chemistry*, **52**, 222–227.

Zhang, W., Xu, M., Yu, C., Zhang, G. & Tang, X. (2010) Simultaneous determination of vitexin-4′-*O*-glucoside, vitexin-2′-*O*-rhamnoside, rutin and vitexin from hawthorn leaves flavonoids in rat plasma by UPLC–ESI-MS/MS. *Journal of Chromatography B*, **878**, 1837–1844.

Chapter 12
High-speed Countercurrent Chromatography in the Separation of Polyphenols

Andrew Marston

Abstract: Countercurrent chromatography is a versatile version of partition chromatography that avoids the use of solid supports. The most widely used variant of this technique, high-speed countercurrent chromatography (HSCCC), utilises the intriguing hydrodynamic behaviour of two immiscible solvents in a fast-rotating coiled tube to achieve excellent separations. At the same time, there is no irreversible adsorption and total recovery of the injected sample. This is a great advantage for the isolation of polyphenols, which suffer from tailing and sample loss on chromatography with solid supports. Consequently, HSCCC is ideal for separation of this class of natural products.

Keywords: countercurrent chromatography; preparative separation; purification methods; liquid–liquid partition; polyphenols

12.1 Foreword

High-speed countercurrent chromatography (HSCCC) is an all-liquid technique that relies on the partitioning of a sample between two immiscible phases of a given solvent system. As such, it does not suffer from the disadvantages of chromatography on solid supports, such as irreversible adsorption. Consequently, it is an ideal method for the separation of polyphenols, a class of compounds that contain acidic hydroxyl groups and produce extensive tailing during classical column chromatography over solid stationary phases.

The aim of this chapter is not to give an exhaustive review of polyphenol separations by HSCCC, but to provide some examples of the application of countercurrent chromatography, together with an idea of the different approaches and the potential of the method. Further

Recent Advances in Polyphenol Research, Volume 3, First Edition. Edited by Véronique Cheynier, Pascale Sarni-Manchado and Stéphane Quideau.

applications can be found in published compilations (Hostettmann *et al.*, 1998; Ito, 2005; Marston and Hostettmann, 2006).

12.2 High-speed countercurrent chromatography

Countercurrent chromatography (CCC) is a versatile version of partition chromatography that totally eliminates the use of solid supports. The method basically involves liquid–liquid partition of a sample between two immiscible liquid phases, as found in a simple separating funnel experiment. It has its origins in experiments on instruments constructed by Y. Ito at the National Institutes of Health in Bethesda, Maryland, USA, during the early 1970s (Ito, 1981; Mandava & Ito, 1988). Over the years, the method has been radically improved in terms of resolution, separation time and sample loading capacity, and the term high-speed countercurrent chromatography (HSCCC) has now been adopted for modern hydrodynamic instruments. These consist of a coil of PTFE or Tefzel tubing wrapped around a drum (bobbin) (Fig. 12.1). A variable gravity field is produced by a double axis gyratory motion and there is seal-free introduction of solvent onto the column. HSCCC utilises the intriguing hydrodynamic behaviour of the two immiscible solvents in the rotating coiled tube (Conway, 1990). One of the most important experimental parameters (β) is defined as the ratio of the coil radius (r) to the orbital radius of revolution (R). Almost 100% of the column space is used for the mixing of the two immiscible phases in the hydrodynamic system.

In parallel, hydrostatic instruments have been developed. They use a centrifugal force generated by a single-axis centrifuge to hold a stationary phase in place. A constant gravity field is produced by single axis rotation. Rotating seals are necessary for supply of solvent. Mixing of phases is achieved through a series of channels and ducts etched into inert plates. These instruments are called "centrifugal partition chromatographs" (CPC). If the coil is filled with stationary phase of a biphasic solvent system and then the other phase pumped through the coil at a suitable speed, a point is reached at which no further displacement of the stationary phase occurs and the apparatus contains approximately 50% of each of the two

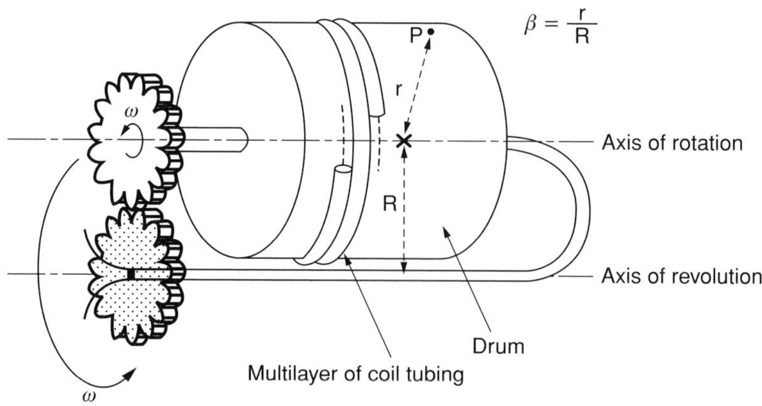

Fig. 12.1   The rotating coil, equipped with planetary gear, of a HSCCC instrument.

phases. Steady pumping-in of mobile phase results in elution of mobile phase alone. This basic system uses only 50% of the efficient column space for actual mixing of the two phases.

The two different techniques, CCC and CPC, have been collectively referred to as "countercurrent separation" (CS) (Pauli *et al.*, 2008). The terminology is confusing because the generally accepted expression "countercurrent chromatography" actually involves the passage of one phase (the mobile phase) of a biphasic solvent system through the non-mobile (stationary phase) second component of the solvent system.

As HSCCC is an all-liquid separation method, it has the following advantages over other chromatographic techniques:

- No irreversible adsorption of the sample
- Quantitative recovery of the introduced sample
- Greatly reduced risk of sample denaturation
- Low solvent consumption
- Favourable economics

Although the efficiency (as represented by the number of theoretical plates) cannot match that of HPLC, the high selectivity and the high stationary to mobile phase ratio more than compensate. In HPLC, around 20% of the volume of the column is stationary (bonded) phase around the silica support, available for interaction with solute, while in CCC the figure for stationary phase content can be as high as 80%.

The essence of a successful HSCCC separation is the correct choice of chromatography solvent. One of the advantages of the technique is that if a solvent is not suitable and separation is not achieved, the sample can be quantitatively recovered. However, this is time consuming and it is preferable to start with the right solvent. Certain guidelines can be followed for this purpose. A suitable two-phase solvent system has to fulfil the following conditions:

(a) To ensure a satisfactory retention of the stationary phase, the settling time of the solvent system should be shorter than 30 seconds.
(b) For efficient separation, the partition coefficient (K) of the target compounds should lie in the range $0.5 < K < 2.0$, and the separation factor between any two components ($\alpha = k_2/k_1$, $k_2 > k_1$) should be greater than 1.5. A smaller K value results in a loss of peak resolution, while a larger value produces excessive band broadening.
(c) In order to avoid wastage of solvent, it is recommended that the two-phase system under consideration produces similar volumes of each phase.

Some methods for the choice of solvent system have been published by Camacho-Frias and Foucault (1996).

Numerous examples of solvent systems used in countercurrent chromatography can be found in the specialist literature, e.g. Abbott and Kleiman (1991), Conway (1990), and Hostettmann *et al.* (1998). These can be tried as a first attempt to find a suitable solvent system. Chloroform–methanol–water or less polar *n*-hexane–ethyl acetate–methanol–water mixtures are good starting points and by modifying the relative proportions of each constituent solvent, it is possible to obtain the required distribution of sample between the two phases. This is the strategy adopted by Oka *et al.* (1991), who describe a systematic approach using the two aforementioned-mentioned systems. In the case of the hexane-containing system, *n*-butanol can be added if the sample is more hydrophilic. In total, Oka describes 27 two-phase solvent systems, giving a broad polarity range. Most of the solvent systems provide close to 1:1 volume ratios of the two phases, together with reasonable settling times. The search for a suitable solvent system can be initiated with *n*-hexane–ethyl acetate–methanol–water 1:1:1:1 or chloroform–methanol–water 10:3:7. A similar approach has been adopted by Margraff (personal communication cited in Camacho-Frias & Foucault, 1996), starting with a mixture of *n*-heptane–ethyl acetate–methanol–water 1:1:1:1 and ending with ethyl acetate–water 1:1 for polar compounds or *n*-heptane–methanol 1:1 for less polar samples.

Ternary (or phase) diagrams provide an excellent means to search for HSCCC solvents. Ternary diagrams suitable for HSCCC applications have been collected together (see, for example, Foucault, 1994). These give the exact compositions of both the mobile and stationary phases and allow the desired quantity of each phase to be prepared independently.

Thin-layer chromatography can be used as a very rapid way of guiding method development. A small amount of the sample is thoroughly mixed in a vial with equal volumes of the upper and lower phases of the chosen solvent system. The same volume (several microlitres) of each phase is applied to a TLC plate, which is then allowed to migrate in the organic (or less polar) phase of the two-phase solvent system. An optimal system gives an equal distribution between the two phases and R_f values of the sample components between 0.2 and 0.5. If the sample is found almost exclusively in one phase, the solvent system is obviously of no use. It should be noted that the R_f value gives only an approximate indication of the appropriateness of a particular solvent because TLC involves both partition and adsorption mechanisms, whereas HSCCC is based on purely liquid–liquid partition phenomena. The TLC method has been refined to make it more predictive and this approach is known as the GUESS ("Generally useful estimate of solvent systems") guide (Friesen & Pauli, 2005).

High-performance liquid chromatography is the most frequently employed and most accurate method for the investigation of solvent systems. In the same way as the TLC method, the solute is partitioned between the two immiscible phases and the respective concentrations in each phase are measured by reversed-phase HPLC. The partition coefficient of each component can be calculated from the detector response following injection of a solution of the compounds in one phase before and after extraction with the other phase of the solvent system. This procedure is of value for natural product mixtures in which the identities of the individual components are not known.

The most direct and accurate method for the choice of solvent system is to run a trial separation on an analytical HSCCC column. The small volume of the coil (ca. 10–40 mL) means that separation is rapid (generally less than 1 hour) and a few milligrams of sample require little solvent consumption for completion of the experiment. An example is the separation of a flavonoid fraction from seeds of *Oroxylum indicum* (Bignoniaceae) whereby trials on a 4.6 mL (0.76 mm i.d. tubing) coil were used to find a suitable solvent system and analytical runs were completed in 6–20 minutes (Chen *et al.*, 2005). Systematic searches for HSCCC solvents using the Oka and Margraff approaches are not adequate for charged analytes such as alkaloids because an additional adjustment of pH and ionic strength is required. Because of its speed, analytical HSCCC is ideal for the trial and error search for the most suitable solvent system in these cases.

Although the number of manufacturers of HSCCC equipment is limited, different sizes of instrument are produced and these allow the separation of sample quantities varying from milligrams up to 100 g. Custom-made instruments for larger samples have been produced for industrial use but these are few.

For polyphenols, UV is obviously the detection method of choice but it is possible to interface HSCCC equipment with mass spectrometry and evaporative light-scattering detection (ELSD).

12.3 Separations of polyphenols

The acidic hydroxyl groups in polyphenols often cause irreversible adsorption on solid stationary phases (such as silica gel and polyamide) during column chromatographic procedures. Countercurrent chromatography is a technique that avoids this problem and it would seem to be ideally suited to the separation of this class of compounds. Furthermore, polyphenols often occur as polar glycosides and these are compatible with the aqueous solvent systems favoured in CCC.

Centrifugal countercurrent chromatography has a different selectivity from other chromatographic methods (HPLC on RP-18 columns, for example) and can be used to separate natural products that are difficult to isolate with silica-based support materials.

Another major problem associated with the preparative separation of polyphenols such as flavonoids is their sparing solubility in solvents employed in chromatography. Moreover, the flavonoids become less soluble as their purification proceeds. Poor solubility in the mobile phase used for a chromatographic separation can induce precipitation at the head of the column, leading to poor resolution, decrease in solvent flow or even blockage of the column. Centrifugal countercurrent chromatography is less sensitive to the injection volume of the sample than other chromatographic methods and in cases of low solubility, the sample can be introduced in a larger amount of solvent, without affecting the resolution too much.

12.3.1 Preparative applications

12.3.1.1 Stilbene glycosides

Stilbenes are a class of relatively simple phenolic compounds, which includes resveratrol. Resveratrol possesses a wide range of pharmacological properties, including the inhibition of low-density lipoprotein oxidation, inhibition of smooth muscle cell proliferation and platelet aggregation, anti-inflammatory and anti-cancer activities (Pace-Asciak *et al.*, 1995). Glucosylated resveratrol analogues have comparable biological activities. After trans-epithelial passage, they can be hydrolysed into deglucosylated forms as resveratrol in the intestine (Henry-Vitrac *et al.*, 2006). In *in vitro* tests, the glucosylated analogues show even more powerful bioactivities. For example, resveratrol and piceid have similar antioxidant capacity, but piceid appears to be more efficacious than resveratrol as a consequence of the reaction of the latter with its radical form (Fabris *et al.*, 2008). Resveratrol–glycoside

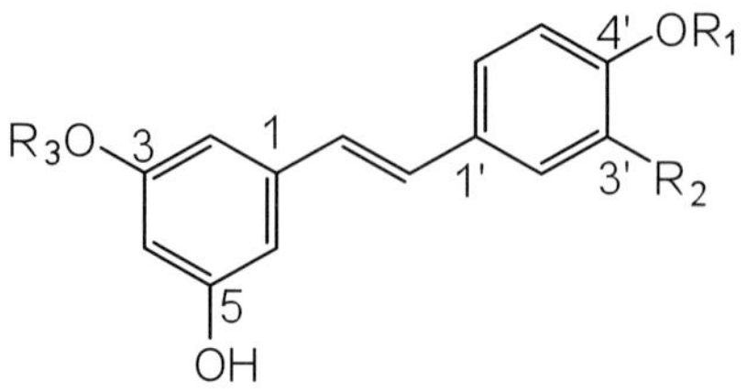

	R_1	R_2	R_3
1 Piceid	H	H	Glc
(3,5,4′-trihydroxystilbene 3-O-β-D-glucopyranoside)			
2 Resveratroloside	Glc	H	H
(3,5,4′-trihydroxystilbene 4′-O-β-D-glucopyranoside)			
3 Piceatannol glucoside	Glc	OH	H
(3,5,3′,4′-tetrahydroxystilbene 4′-O-β-D-glucopyranoside)			

Fig. 12.2   Stilbene glucosides found in the invasive plant *Polygonum cuspidatum.*

and 4′-methoxy-3,3,5′-stilbene-3-glucoside were more effective than resveratrol against hepatitis B virus (HBV), perhaps by inhibition of HBV copying and induction of apoptosis in HepG 2.2.15 cells. The important role of the hydroxyl groups was demonstrated by comparison with stilbene itself (Ashikawa *et al.*, 2002). Piceatannol, with one more hydroxyl group, was shown to have strong anti-inflammatory, immunomodulatory, anti-proliferative, anti-leishmanial, anti-leukemic, and protein–tyrosine kinase inhibitory activities (Duarte *et al.*, 2008). Moreover, it can block β-amyloid -induced accumulation of reactive oxygen species (ROS), thereby protecting PC12 cells from oxidative stress. Piceatannol 4′-O-β-D-glucopyranoside stimulates the proliferation of cultured human skin fibroblasts and normal human epidermal keratinocytes.

A good source of these glucosylated resveratrol derivatives, as shown by an LC/MS investigation (Fan *et al.*, 2009), is the invasive plant *Polygonum cuspidatum* (Polygonaceae) (Fig. 12.2). In previous work, the three compounds, piceid, resveratroloside and piceatannol glucoside, have been isolated in low yield by a time-consuming procedure involving silica gel column chromatography and repeated semi-preparative HPLC (Vastano *et al.*, 2000). Since these stilbenes have important beneficial effects for health, a rapid method for their separation and purification was sought, providing a means of benefitting from the biomass available in areas with an abundance of the neophytic plants. Semi-preparative high-speed countercurrent chromatography proved to be a suitable method for this purpose (Fan *et al.*, 2009).

Powdered air-dried root material of *P. cuspidatum* was extracted first with dichloromethane and then with methanol. Stilbene glucosides (Fig. 12.2) in the methanol extract were separated by HSCCC on a Tauto TBE-300B centrifuge instrument (Shanghai, China) equipped with three PTFE coils (internal diameter of tubes 1.6 mm; total volume 260 mL) and with a 20 mL sample loop. The two-phase solvent system was selected according to the partition coefficients (K) of each stilbene glycoside. The K value was determined by

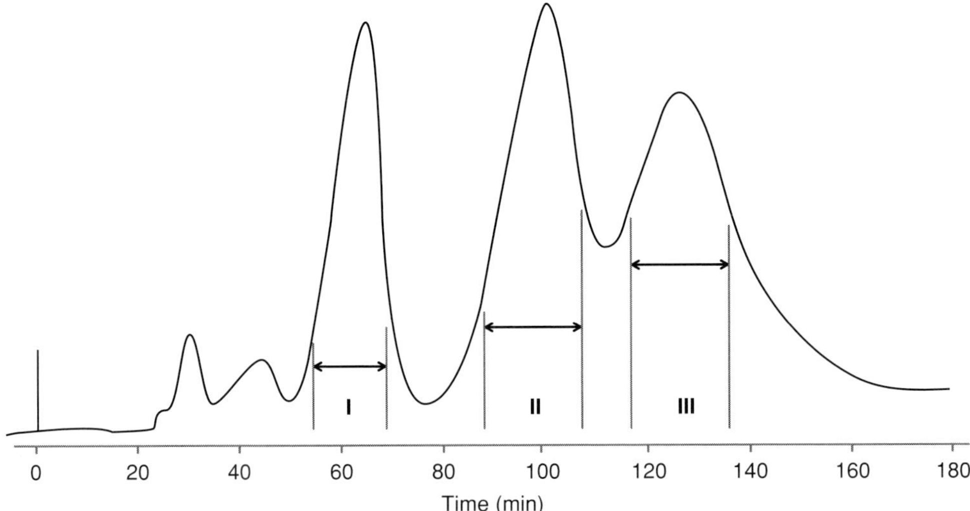

Fig. 12.3 Chromatogram of the HSCCC separation of crude MeOH extract from roots of *Polygonum cuspidatum* growing in Switzerland (solvent system: cyclohexane-ethyl acetate-methanol-water 1:5:1:5; detection: 290 nm; rotational speed: 900 rpm; flow rate: 3 mL/min; mobile phase: upper phase; sample: 500 mg; peak I: piceid; peak II: resveratroloside; peak III: piceatannol glucoside).

HPLC analysis. Approximately, 5 mg of crude MeOH extract was dissolved in 1 mL of each phase of the pre-equilibrated two-phase solvent system. After the test tube had been shaken vigorously for a few minutes and equilibrated again, the same volumes (10~20 µL) of upper phase and the lower phase were analysed by HPLC. The K value was expressed as the peak area of each stilbene glycoside in the lower phase divided by that in the upper phase. The chosen two-phase solvent system was cyclohexane–ethyl acetate–methanol–water (1:5:1:5, v/v/v/v). The upper phase was selected as mobile phase, as the partition coefficients of the compounds were 0.35, 0.30 and 0.50, respectively. The column was first filled with the same volume of upper and lower phases, (i.e. the instrument contains 50% of each phase) and then rotated at 900 rpm while the upper phase as mobile phase was pumped into the column at a flow rate of 3 mL/min. The sample (500 mg of methanol extract) dissolved in a 1:1 (v/v) mixture (10 mL) of each phase was injected. UV detection was performed at 290 nm, and 20 mL fractions were collected. Piceid (23 mg), resveratroloside (17 mg), piceatannol glucoside (15 mg) (Fig. 12.2) were isolated from 500 mg crude MeOH extract in one step. The total separation time was about 180 minutes (Fig. 12.3). Subsequent passage over a solid-phase extraction (SPE) column was used as a quick and efficient final purification to get rid of small quantities of anthraquinones and pigments. The HPLC profiles of the pure stilbene glycosides are shown in Fig. 12.4 (Fan *et al.*, 2009).

12.3.1.2 Flavonolignans

Large quantities of crude extract of *Silybum marianum* (Asteraceae) fruits have been separated on a Chinese-built HSCCC instrument. The countercurrent chromatograph consisted

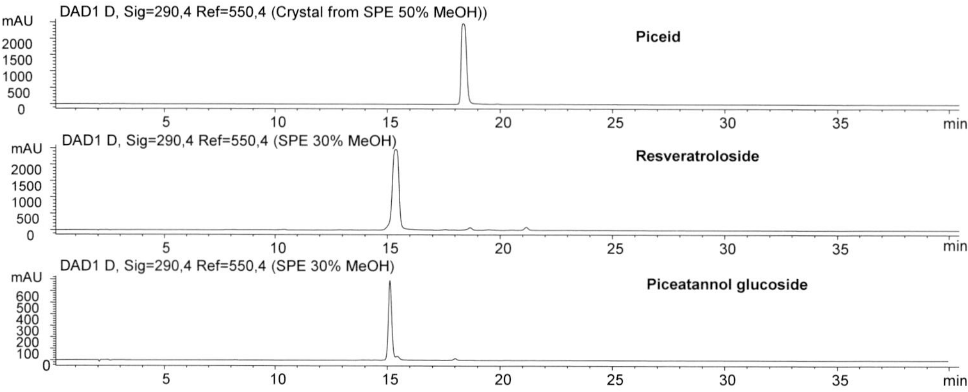

Fig. 12.4   HPLC analysis of purified stilbene glycosides. Conditions: Xterra[®] C_{18} column (5 μm, 3.5 × 150 mm; Waters, Milford, MA, USA); solvent 0.5% acetic acid in H_2O (A) and CH_3CN (B); gradient 10–40% B in 40 minutes, then to 100% B in 20 minutes; flow rate: 1 mL/min; detection 290 nm. (Reprinted with permission from Fan *et al.*, 2009.)

of three coils wound with 5 mm i.d. tubing, giving a total volume of 2460 mL. Dry fruits were extracted first with methanol and this extract was evaporated and washed with hexane and chloroform, before being extracted with ethyl acetate. A portion (12 g) of the latter was introduced into the instrument and eluted with the lower layer of the solvent system *n*-hexane–ethyl acetate–methanol–water (1:4:3:4) at a flow rate of 5 mL/min (Fig. 12.5).

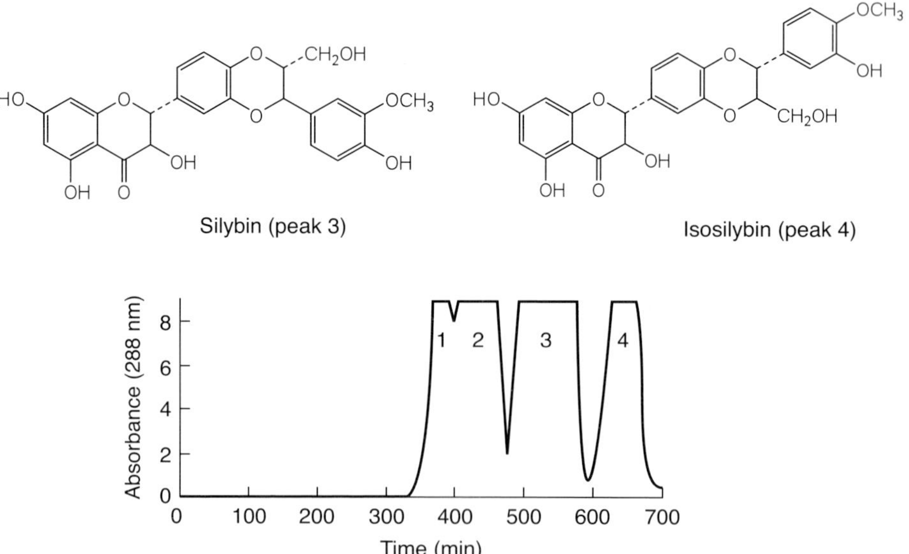

Fig. 12.5   Separation of a *Silybum marianum* extract on a 2460 mL HSCCC instrument (solvent system: *n*-hexane–ethyl acetate–methanol–water 1:4:3:4; detection: 288 nm; rotational speed: 650 rpm; flow rate: 5 mL/min; mobile phase: lower phase; sample: 12 g; peak 1: silychristin; peak 3: silybin; peak 4: isosilybin). (Reprinted with permission from Du *et al.*, 2002.)

The crude extract contains the anti-hepatotoxic substance silymarin, which itself comprises four components: silybin, isosilybin, silydianin and silychristin. These were all separated during the chromatographic run, although silydianin eluted with another component (peak 2); notable is the resolution of the isomers silybin (peak 3) and isosilybin (peak 4). Silybin (3.47 g) was obtained at a purity of 95.7% (Du *et al.*, 2002).

12.3.1.3 Flavonoids

There are numerous applications of the preparative separation of flavonoids by HSCCC (Marston *et al.*, 1990; Hostettmann *et al.*, 1998; Ito, 2005; Marston & Hostettmann, 2006; Di *et al.*, 2011). An example of the separation of flavonoid glycosides by HSCCC is shown in Fig. 12.6. The leaves of the African plant *Tephrosia vogelii* (Leguminosae) were first extracted with dichloromethane and then with methanol. Methanol extract (500 mg) was

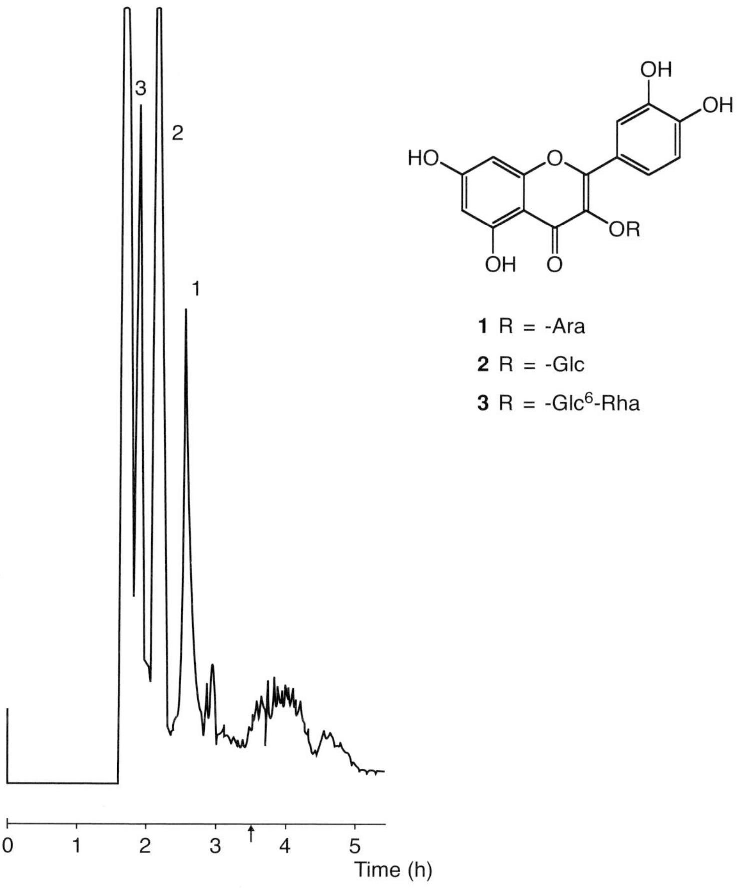

Fig. 12.6   Separation of flavonoid glycosides on a Quattro HSCCC instrument (solvent system: chloroform–ethanol–methanol–water 5:3:3:4; detection: 254 nm; rotational speed: 800 rpm; flow rate: 3 mL/min; mobile phase: upper phase; sample: 500 mg). (Reprinted with permission from Sutherland *et al.*, 1998.)

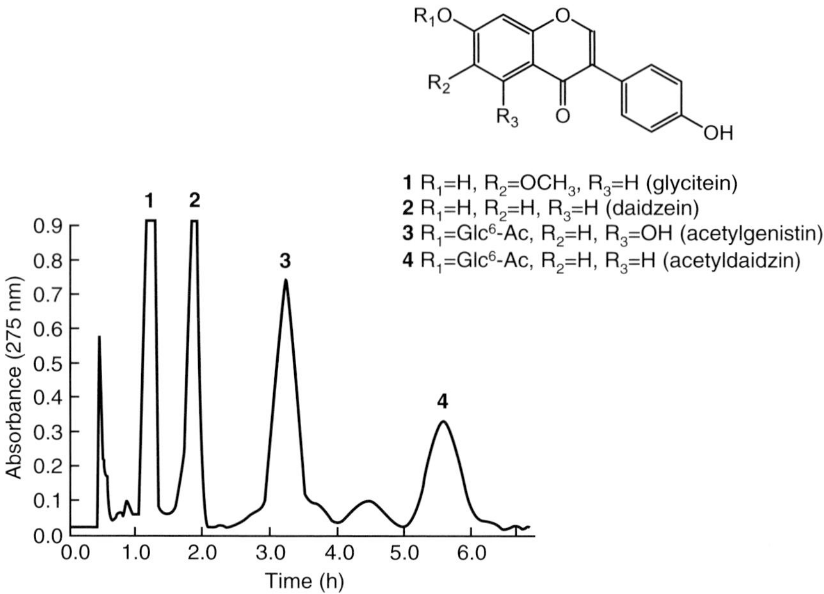

Fig. 12.7   Separation of a crude soybean extract by HSCCC (solvent system: chloroform–methanol–water 4:3:2; detection: 275 nm; rotational speed: 800 rpm; flow rate: 2 mL/min; mobile phase: lower phase; sample: 150 mg). (Reprinted with permission from Yang *et al.*, 2001.)

injected in a mixture of the two phases of the solvent system and elution of the three major glycosides was achieved within 3 hours (Sutherland *et al.*, 1998).

The technique of HSCCC was also employed as a key step in the purification of 26 phenolic compounds from the needles of Norway spruce (*Picea abies*, Pinaceae). An aqueous extract of needles (5.45 g) was separated with the solvent system $CHCl_3$–MeOH–*i*PrOH–H_2O 5:6:1:4, initially with the lower phase as mobile phase and then subsequently switching to the upper phase as mobile phase. Final purification of the constituent flavonol glycosides, stilbenes and catechins was by gel filtration and semi-preparative HPLC (Slimestad *et al.*, 1996).

There is great interest in soy isoflavones because of their reported bioactivities, which include estrogenic, anti-carcinogenic and anti-osteoporosis effects. Four pure isoflavones were obtained from a crude soybean extract after HSCCC with the solvent system $CHCl_3$–MeOH–H_2O 4:3:2 (Fig. 12.7). The isoflavones were isolated in 5–10 mg amounts after introduction of 150 mg of sample onto a P.C. Inc. 320 mL-capacity instrument (Yang *et al.*, 2001).

12.3.1.4 Anthocyanins

Anthocyanins provide a considerable separation challenge. The polarity and complexity of anthocyanin mixtures is often a barrier to the isolation of pure compounds. Countercurrent chromatography provides a solution to these problems. The solvent system *tert*-butyl methyl ether (TBME)–*n*-butanol–acetonitrile–water (2:2:1:5) acidified with trifluoroacetic acid can

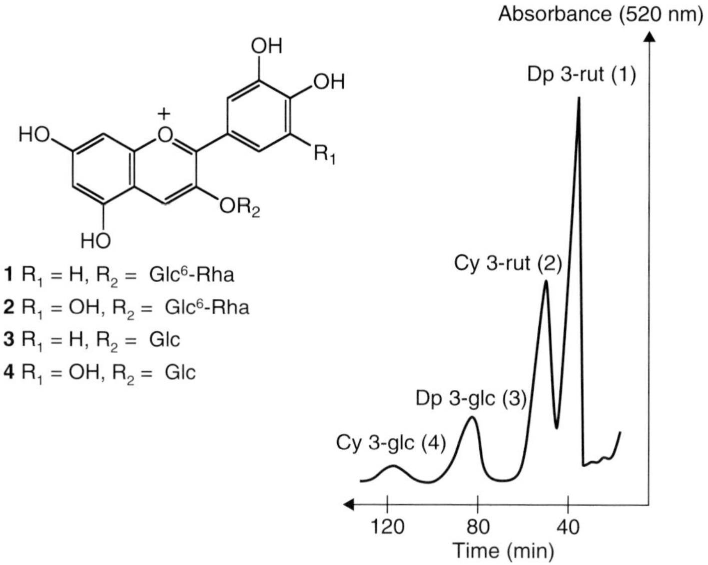

Fig. 12.8   Separation of anthocyanins from blackcurrant by HSCCC (**1**: delphinidin 3-rutinoside, **2**: cyanidin 3-rutinoside, **3**: delphinidin 3-glucoside, **4**: cyanidin 3-glucoside) (solvent system: TBME–*n*-butanol–acetonitrile–water 2:2:1:5; detection: 520 nm; rotational speed: 1000 rpm; flow rate: 5 mL/min; mobile phase: lower phase; sample: 430 mg). (Reprinted with permission from Degenhardt *et al.*, 2000a.)

be used, for example. When a blackcurrant (*Rubus nigrum*, Rosaceae) extract was introduced onto a 850 mL multilayer column (CCC-1000, Pharma-Tech), the solvent system provided a virtually baseline separation of four anthocyanins (Fig. 12.8). Anthocyanin pigments were also successfully fractionated from red cabbage, black chokeberry and roselle (Degenhardt *et al.*, 2000a).

12.3.1.5 Proanthocyanidins and tannins

Although hundreds of applications in the separation of low- to medium-polarity compounds exist in the literature (Conway, 1990; Hostettmann *et al.*, 1998; Marston & Hostettmann, 2006), the separation of highly polar compounds by HSCCC is still rare. In this context, centrifugal countercurrent chromatography has been successfully employed for the isolation of dimeric, trimeric and tetrameric proanthocyanidins from *Acacia mearnsii* (Leguminosae) (Fig. 12.9). A water extract of *A. mearnsii* bark was treated with ethanol and the precipitate was filtered off. The supernatant was extracted with ethyl acetate. HSCCC with a DE Spectrum-CCC instrument was performed on the ethyl acetate extract. The two-phase solvent system chosen was *t*-butyl methyl ether–acetonitrile–water with 0.1 % trifluoroacetic acid (TFA) at volume ratios of 2:2:3 (v/v/v). Crude extract (1 g) was then separated with the lower phase as mobile phase and detection at 254 nm. Final purification was achieved by gel filtration over Sephadex LH-20. LC–MS was used to determine the molecular mass of each proanthocyanin that was isolated. Structure elucidation of the isolated polyphenols was performed by LC–MS, LC–MS/MS) and NMR (1D and 2D spectroscopy). Samples

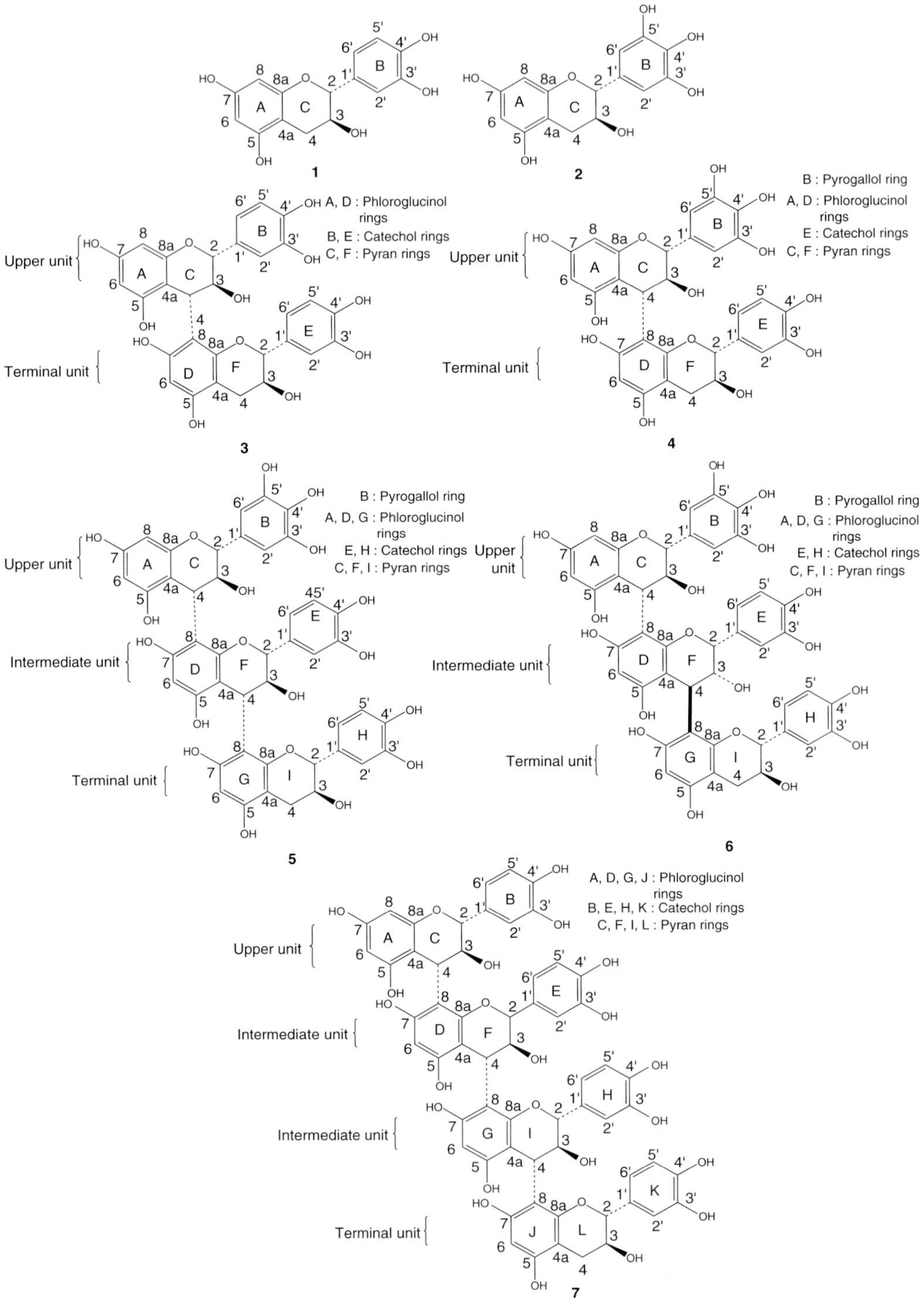

Fig. 12.9   Polyphenols isolated from a water extract of *Acacia mearnsii* bark (**1**: catechin; **2**: gallocatechin; **3**: procyanidin B; **4**: robinetinidol-(4β-8)-catechin; **5** and **6**: trimers; **7**: tetramer).

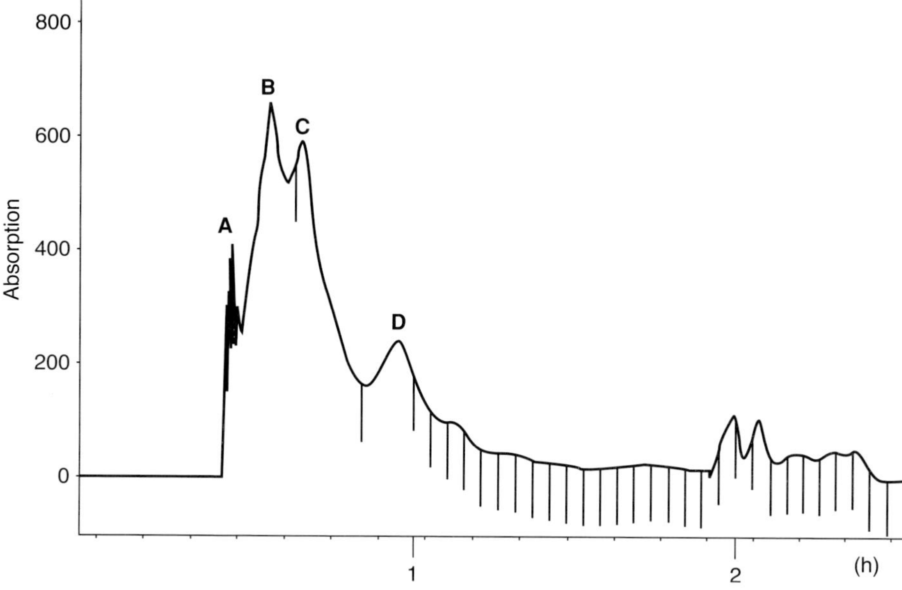

Fig. 12.10　HSCCC separation of tannins from an *Acacia mearnsii* stem bark water extract (solvent system: TBME–acetonitrile–water (containing 0.1% TFA) 2:2:3; detection: 280 nm; rotational speed: 1500 rpm; flow rate: 3 mL/min; mobile phase: lower phase; sample: 1 g; peak A: complex mixture; peak B: trimers **5** and **6**, tetramer **7**; peak C: dimers **3** and **4**; peak D: catechin **1** and gallocatechin **2**).

of proanthocyanidins differing slightly in the hydroxylation patterns of the aromatic rings either in the upper unit or the terminal unit of the proanthocyanidin molecule were separated in a single HSCCC run, followed by clean-up on Sephadex LH-20. Peak A in the HSCCC separation (Fig. 12.10) was a mixture and the components could not be separated. Peak B gave a tetramer (**7**, 5 mg) and two trimers (**5**, 14 mg, **6**, 20 mg). Peak C gave two dimers (**3**, 25 mg, **4**, 17 mg), while peak D gave catechin (**1**, 30 mg) and gallocatechin (**2**, 15 mg). In this manner, the characterisation of seven pure tannins from the stem bark of *Acacia mearnsii* was achieved.

A combination of gel filtration, HSCCC and semi-preparative HPLC was reported for the isolation of 8 dimeric proanthocyanidins of general structure shown in Fig. 12.11 from the stem bark of *Stryphnodendron adstringens* (Leguminosae). The HSCCC step involved separation with the upper layer of EtOAc–*n*PrOH–H$_2$O 35:2:2 as mobile phase (Palazzo de Mello *et al.*, 1996).

The strong affinity of tannins for proteins, polysaccharides and other high molecular weight compounds makes their purification difficult. They are usually separated by open or medium-pressure column chromatography over Sephadex LH-20, or over polymeric columns, such as Diaion HP-20, MCI-gel or TSK-gel, with aqueous alcohol as eluent. However, even with these methods, partial loss of polyphenolic compounds or tailing, due to adsorption and diffusion, is frequent (Yoshida & Hatano, 1997). Oligomeric hydrolysable tannins of high molecular weight are particularly difficult to purify by conventional methods. They are strongly adsorbed on chromatographic supports such as Toyopearl HW-40 and are

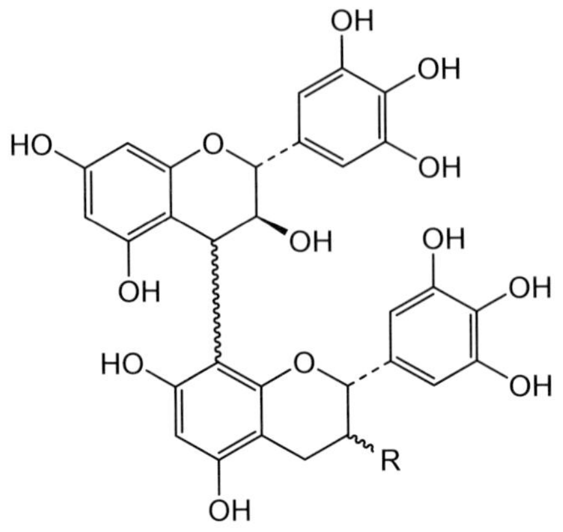

Fig. 12.11   General structure of dimeric proanthocyanidins isolated by HSCCC from the stem bark of *Stryphnodendron adstringens* (Leguminosae).

hardly separable from each other. These oligomeric tannins easily decompose during solid-phase chromatography. However, countercurrent chromatography has been successfully applied for the purification of the oligomeric fractions of several plant extracts, including *Heterocentron roseum* (Melastomataceae) leaves. By using, for example, *n*-butanol–*n*-propanol–water (4:1:5), the pure tetrameric tannin nobotanin K was isolated (Yoshida *et al.*, 1995).

Mango peels and kernels are by-products of the mango processing industry and contain flavonols, xanthones and gallotannins. Although the selective anti-microbial properties of hydrolysable tannins are well known, the anti-bacterial activities of purified gallotannins have only recently been studied. The reason for this is partly due to the limited number of pure gallotannins available. HSCCC provided an efficient way to obtain gallotannin standards. Countercurrent chromatography was performed on a TBE-300B instrument, with a volume of 305 mL. A prepurified extract (250 mg) was separated with the solvent system hexane–ethyl acetate–methanol–water 0.5:5:1:5 and the lower phase as the mobile phase. By this means, a sequential order of elution of the polyphenolics was achieved: mangiferin, tri-*O*-galloylglucose, tetra-*O*-galloylglucose, penta-*O*-galloylglucose, hexa-*O*-galloylglucose, hepta-*O*-galloylglucose, octa-*O*-galloylglucose, nona-*O*-galloylglucose, deca-*O*-galloylglucose. Their structures were elucidated by LC–MS (Engels *et al.*, 2010).

Tea (*Camellia sinensis*, Theaceae) contains a complex mixture of polyphenols. In order to separate the constituents, it is advisable to first prepare catechin-rich, flavonol glycoside-rich or proanthocyanidin-rich extracts. Pure compounds are then obtained by HSCCC of the relevant fractions. Catechins are normally separated by preparative HPLC or Sephadex gel chromatography but these procedures are rather time consuming and some degradation may occur. The countercurrent method is more rapid, employing *n*-hexane–ethyl acetate–methanol–water (3:10:3:10; lower phase as mobile phase), and can provide large quantities of epigallocatechin gallate, epicatechin gallate and epiafzelechin gallate. Flavonol glycosides were separated with ethyl acetate–water (1:1, lower phase as mobile phase). HSCCC of the proanthocyanidin-rich extract was performed with

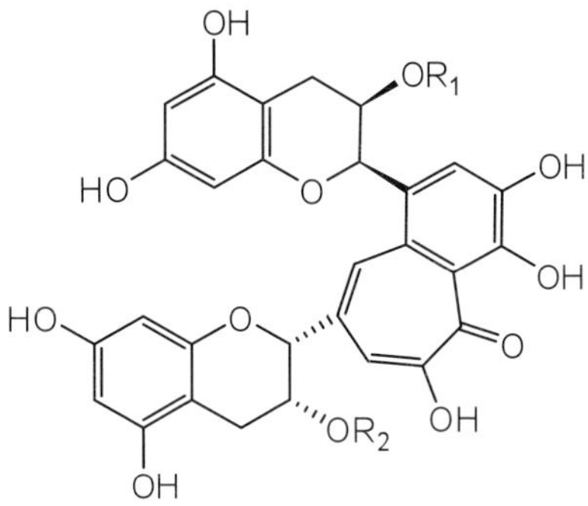

Fig. 12.12   Structure of theaflavins isolated from black tea.

n-hexane–ethyl acetate–methanol–water (1:5:1:5; lower phase as mobile phase). With ethyl acetate–ethanol–water (10:1:10; upper phase as mobile phase), epigallocatechin gallate-(4β→8)-epicatechin gallate could be obtained. Using the same solvent system, with the lower phase as mobile phase, the hydrolysable tannin strictinin (important in the formation of thearubigins) could be isolated. Bright orange-red theaflavins (Fig. 12.12) have also been isolated on a 850 mL capacity Pharma-Tech CCC-1000 instrument, using *n*-hexane–ethyl acetate–methanol–water (3:10:3:10; lower phase as mobile phase). Loading of 2–5 g was possible (Degenhardt *et al.*, 2000b).

Another study on tea catechins has been published by Yanagida *et al.* (2006), in which the solvent used was *tert*-butyl methyl ether-acetonitrile-0.1% aqueous trifluoroacetic acid 2:2:3, with the upper phase as the mobile phase.

12.3.2 Analytical applications

Although most HSCCC separations are on a preparative scale, analytical instruments do exist (Schaufelberger, 1991). However, these are mostly used to find suitable separation conditions for scale-up. A Pharma-Tech CCC-2000 instrument (column capacity 43 mL) was used to separate a crude flavonoid mixture (3 mg) into five components by rotating at 2000 rpm for 15 minutes. The solvent was chloroform–methanol–water 4:3:2 (lower phase as mobile phase) and the flow-rate was 5 mL/min (Zhang *et al.*, 1988).

An analytical-size instrument can also be employed for the isolation of pure compounds from small quantities of crude material. This was the case for the purification of acetyl-cholinesterase inhibitors from leaves of *Buddleja davidii* (Buddlejaceae). The leaves were extracted first with dichloromethane and then with methanol. A sample of the MeOH extract of the leaves (20 mg) was subjected to HSCCC on a Dynamic Extractions (Slough, UK) Mini centrifuge instrument (with a 17 mL coil) equipped with two HPLC pumps and a UV detector (254 nm), using chloroform–methanol–water 45:33:22 as the solvent system. The coil was first entirely filled with the two phases (1/1 v/v) and rotation was set to 2000 rpm. The lower phase was then pumped into the column at a flow rate of 1 mL/min using the head to tail mode (upper phase as stationary phase). After equilibrium was reached, the sample solution (20 mg in 0.5mL of each of the two phases (1/1 v/v)) was injected. Pure

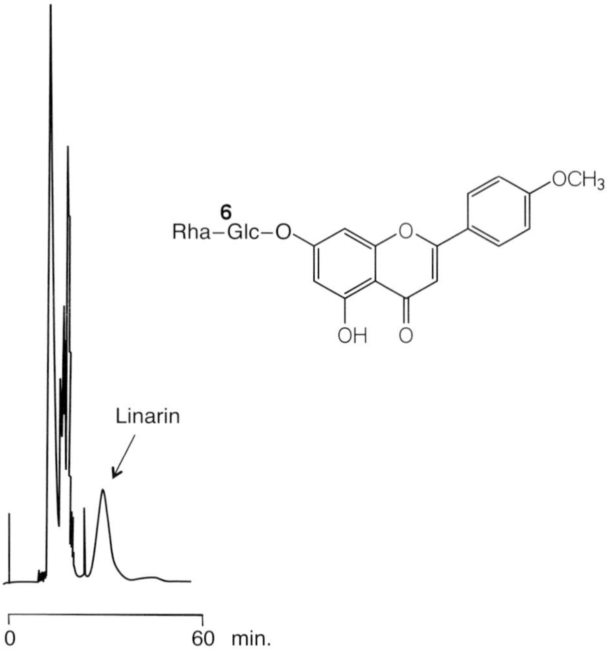

Fig. 12.13   HSCCC separation of the acetylcholinesterase inhibitor linarin from *Buddleja davidii* (Buddlejaceae) (solvent system: chloroform–methanol–water 45:33:22; detection: 254 nm; rotational speed: 2000 rpm; flow rate: 1 mL/min; mobile phase: lower phase)

linarin (3 mg of a white powder) eluted (Fig. 12.13) and after 60 minutes, the instrument was switched to reversed-phase operation and verbascoside eluted. Linarin inhibited the enzyme acetylcholinesterase at 0.01 µg in the bioassay (Fan *et al.*, 2008).

12.4 Extensions of the basic countercurrent chromatography method

Since the introduction of countercurrent chromatography and high-speed countercurrent chromatography by Ito, many developments and modifications have been made (Berthod, 2002). These modifications increase the versatility of the method. Some are applicable to the separation of polyphenols and some are not.

12.4.1 *Reversed-phase operation*

One of the strengths of countercurrent chromatography is that it can be run in the reversed-phase (RP) mode after normal-phase operation (or *vice versa*). This is normally achieved simply by switching the instrument from ascending mode to descending mode or *vice versa*. This mode of operation (also known as dual mode operation) allows the separation

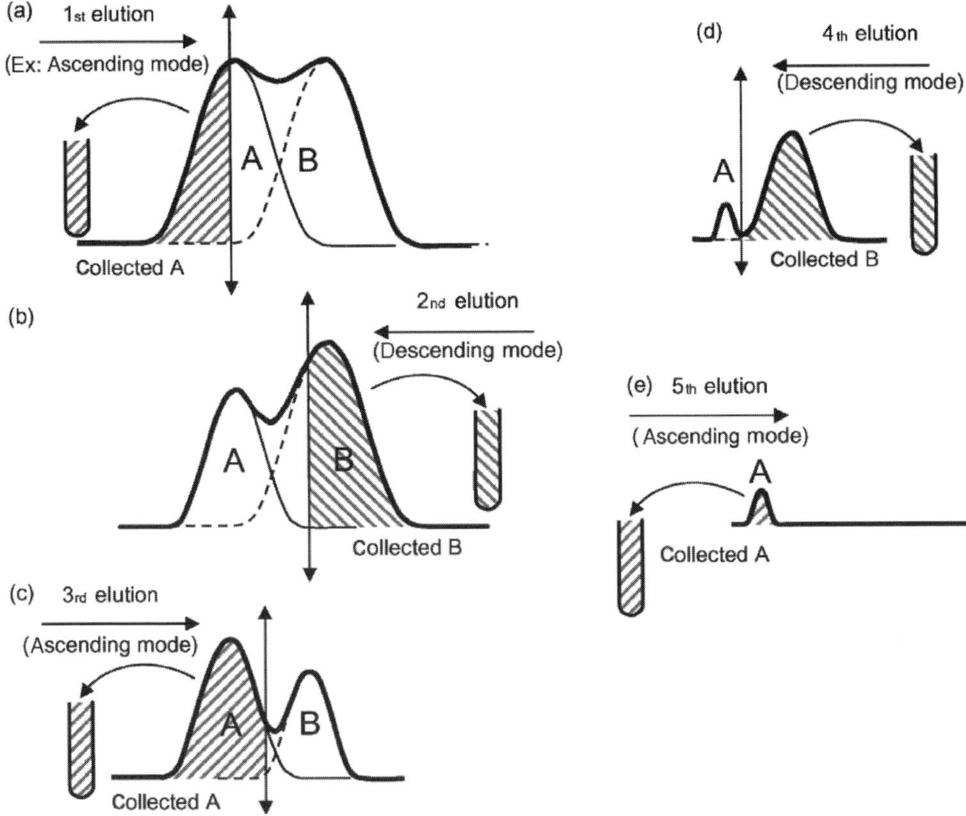

Fig. 12.14   The principle of MDM separation, as applied to a mixture of two poorly resolved compounds A and B. (Reprinted with permission from Delannay *et al.*, 2006.)

of compounds with very different polarities, as found, for example, in the crude extract of a plant.

12.4.2 *Multiple dual-mode operation*

Multiple dual-mode (MDM) operation involves a repeated switching between normal-phase and reversed-phase modes (Delannay *et al.*, 2006). It provides semi-continuous operation and results in more versatile and faster separations. An example of an MDM separation applied to a poorly resolved mixture A and B is shown in Fig. 12.14. Basically, there is peak-shaving to obtain components A and B, followed by re-injection of the remaining mixture and further collection of the pure components.

12.4.3 *Elution–extrusion*

The concept of elution–extrusion is shown in Fig. 12.15. In Fig. 12.15a, the two compounds 3 and 4 elute slowly when using the polar phase as the mobile phase. In the reversed-phase mode (Fig. 12.15b), compound 4 exits the instrument before compound 3. If the two phases

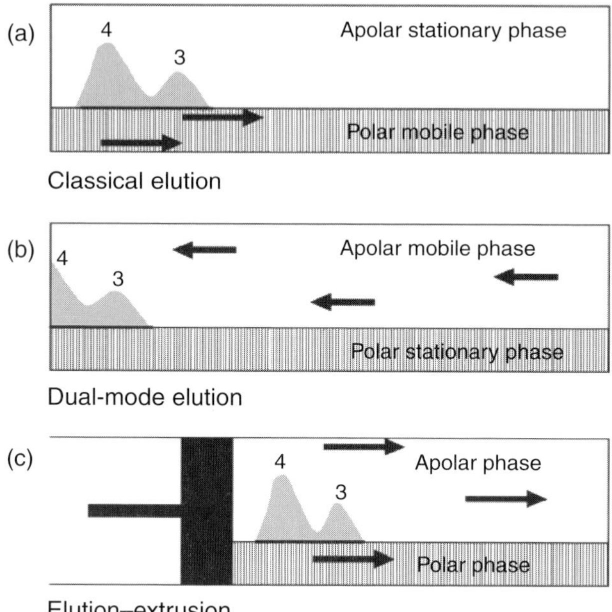

Fig. 12.15   Different elution modes in HSCCC. (a). Normal elution; (b). Reversed-phase elution; (c). Elution-extrusion, in which the content of the column is pushed out (extruded). (Reprinted with permission from Berthod *et al.*, 2003.)

(lower and upper) are pushed out simultaneously, either with nitrogen or with another solvent (methanol or methanol–water, for example), all the contents of the coil, including 3 and 4, are emptied out. Thus, a much lower volume of solvent is necessary to remove a sample from the coil than if normal elution is used (Berthod *et al.*, 2003). Elution–extrusion renders the HSCCC technique particularly suitable for the rapid separation of complex samples.

12.4.4 Gradient elution

A practical way to reduce run times is to introduce a solvent gradient. This also aids the separation of constituents of mixtures that differ widely in polarity. Although this concept may not seem obvious for CCC because of the risk of disturbing liquid–liquid eqilibria and causing elution of the stationary phase, gradients can indeed be used. Furthermore, in certain circumstances, the composition of one phase may be systematically varied while the other remains nearly constant. In order to check for these systems, the best way is to consult phase diagrams (see, e.g., Foucault, 1994).

A step-gradient elution has been employed to separate the coumarins xanthotoxin, isopimpinellin, bergapten, imperatorin and osthole from a *Cnidium monnieri* (Apiaceae) fruit extract. For the first 150 minutes of the chromatographic run on a rotating coil machine, the solvent system was light petroleum–ethyl acetate–methanol–water (5:5:5:5) (lower phase as mobile phase), modified to the proportions 5:5:6:4 for the next 100 minutes and finally 5:5:6.5:3.5 for the last 250 minutes (Liu *et al.*, 2004).

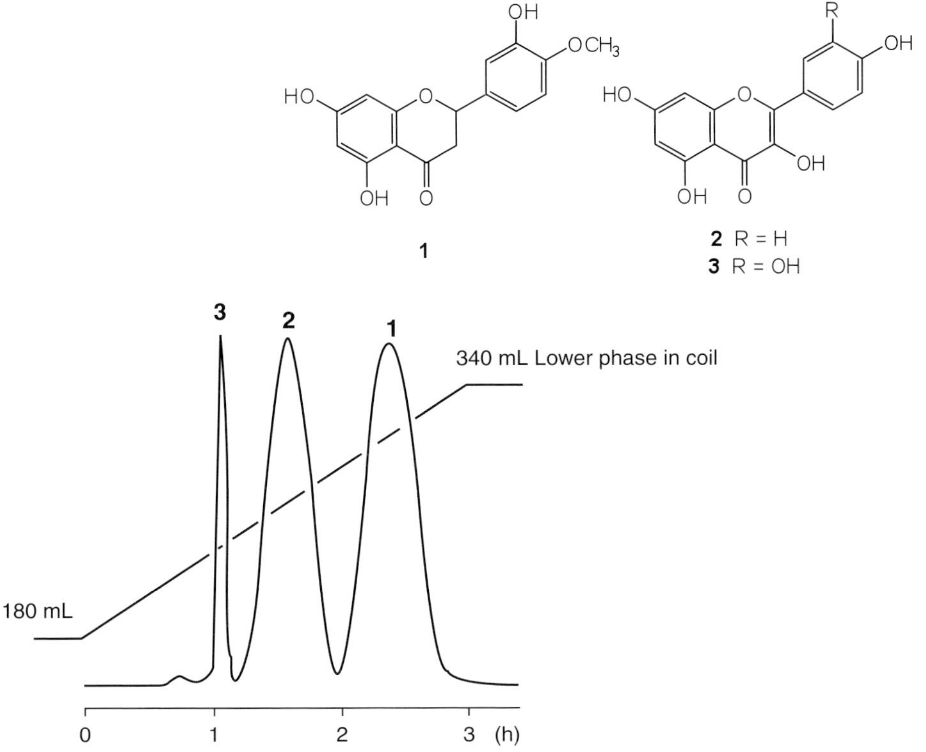

Fig. 12.16   Pseudo-gradient separation of hesperetin (**1**), kaempferol (**2**) and quercetin (**3**) on a P.C. Inc. Instrument (column capacity: 360 mL; solvent system: chloroform–methanol–water 33:40:27; detection: 254 nm; rotational speed: 700 rpm; flow rate upper phase: 4mL/min; flow rate lower phase: 1 mL/min; sample: 15 mg). (Reprinted with permission from Slacanin *et al.*, 1989.)

Flavonoids have been chromatographed with chloroform–methanol–water in the proportions 4:x:2, in which x was varied between 2.5 and 4.5 (Yuan *et al.*, 2002). Pseudo-gradient elution can be achieved by changing the proportions of phases in coils during a separation run. For the separation of three flavonoids (**1–3**), the HSCCC instrument was first filled with equivalent amounts of upper and lower phases of the solvent system chloroform–methanol–water 33:40:27. The flavonoid mixture was then injected. By pumping upper phase through the instrument at 4 mL/min with one pump and simultaneously pumping the lower phase at 1 mL/min with another pump, the content of lower phase in the coil increased from 180 mL to 340 mL over 3 hours. Separation of the three flavonoids was achieved in a time 5 hour shorter than for normal elution (Fig. 12.16) (Slacanin *et al.*, 1989).

12.4.5 HSCCC/MS

Coupling of HSCCC to mass spectrometry gives immediate structural information on the components being eluted. This has previously been achieved on the analytical scale and has now been performed in the preparative separation of a polyphenol sample (Gutzeit *et al.*, 2007). The latter method requires splitting of the eluent leaving the HSCCC instrument and

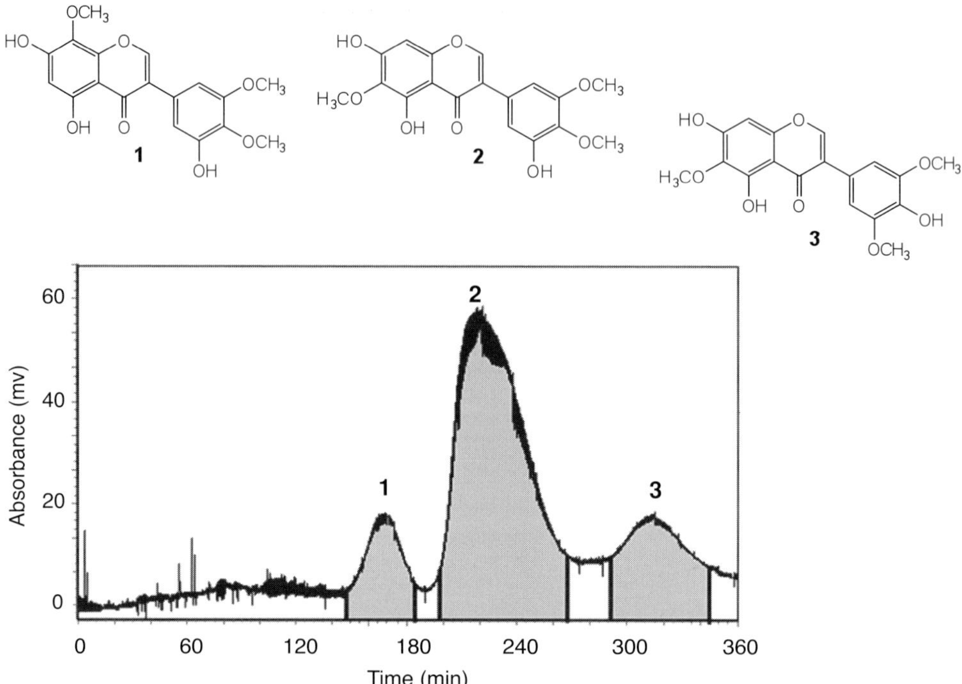

Fig. 12.17   Complexation HSCCC separation of isoflavones from rhizomes of *Belamcanda chinensis* (Iridaceae) on a TBE-300 instrument (solvent system: light petroleum–ethyl acetate–methanol–water 3:5:3:5 (containing 0.1 mol/L copper nitrate in the lower phase); detection: 254 nm; rotational speed: 850 rpm; flow rate: 1 mL/min; mobile phase: upper phase; sample: 100 mg). (Reprinted with permission from Liu *et al.*, 2011.)

necessitates very little carry-over of stationary phase, in order to obtain stable MS signals. HSCCC/ESI-MS-MS hyphenation has been employed for a target-guided preparative fractionation and isolation of polyphenols from sea buckthorn (*Hippophae rhamnoides* L. ssp. *rhamnoides*, Elaeagnaceae) fruit juice concentrate. Peak purity is accessible from complete ion traces and fractions can be evaporated directly for 1D and 2D NMR analysis. Flavonoid glycosides were identified by ESI-MS-MS in the negative ion mode (Gutzeit *et al.*, 2007).

The major xanthones of mangosteen (*Garcinia mangostana*, Guttiferae) have been screened and fractionated by a similar process. They were separated on a Kromaton instrument equipped with a 200 mL rotor, using heptane–ethyl acetate–methanol–water 2:1:2:1 as solvent (lower phase as mobile phase). The electrospray ionisation (ESI) spectra were run with the aid of a splitter (Destandau *et al.*, 2009).

12.4.6 Complexation HSCCC

It is well known that hydroxyflavonoids are able to chelate various metal ions, such as Cu(II), Fe(III) and Al(III). The main chelating sites are the 3-hydroxy-4-keto, the 5-hydroxy-4-keto and the *ortho*-dihydroxy groups. While these chelations have been extensively used for structure elucidation purposes (with UV spectra, for example), their application to separation has been little exploited. Recently, however, three flavonoid isomers were isolated by adding ions to the stationary phase of a HSCCC system. These isomers were difficult to separate by normal separation procedures. Copper nitrate (0.10 mol/L was added to the lower phase of a light petroleum–ethyl acetate–methanol–water (3:5:3:5) system and this was loaded into a TBE-300 instrument. After pumping upper phase into the rotating coil, the crude sample (100 mg) was added and separation was performed, yielding 9.2 mg of **1**, 46.4 mg of **2** and 1.3 mg of **3** (Fig. 12.17). In order to remove small amounts of copper salts that carried over with the mobile phase, the fractions were evaporated and partitioned between dichloromethane and acidified water.

12.4.7 High-performance CCC

The availability of a new generation of instruments has helped to stimulate the number of examples of separation by this high-speed countercurrent chromatography. For example, there are instruments which run at 1500–3000 rpm and which allow separation runs to be achieved in 1–2 hours or less, such as those produced by Dynamic Extractions (Slough, UK). These are known as high-performance CCC (HPCCC) instruments and they are capable of scale-up and separating quantities of sample which vary from milligrams to kilograms (Sutherland *et al.*, 2009). In work with natural products, one-step purifications of compounds are possible from crude plant extracts.

References

Abbott, T.P. & Kleiman, R. (1991) Solvent selection guide for countercurrent chromatography. *Journal of Chromatography*, **538**, 109–118.

Ashikawa, K., Majumdar, S., Banerjee, S., Bharti, A.C., Shishodia, S. & Aggarwal, B.B. (2002) Piceatannol inhibits TNF-induced NF-κB activation and NF-κB-mediated gene expression through suppression of IκBα kinase and p65 phosphorylation. *Journal of Immunology*, **169**, 6490–6497.

Berthod, A. (Ed.) (2002) *Countercurrent Chromatography – The Support-Free Liquid Stationary Phase*. Elsevier, Amsterdam.

Berthod, A., Ruiz-Angel, M.-J. & Carda-Broch, S. (2003) Elution–extrusion countercurrent chromatography. Use of the liquid nature of the stationary phase to extend the hydrophobicity window. *Analytical Chemistry*, **75**, 5886–5894.

Camacho-Frias, E. & Foucault, A. (1996) Les systemes de solvants en chromatographie de partage centrifuge. *Analusis*, **24**, 159–167.

Chen, F., Li, H.-B., Wong, R.N.S., Ji, B. & Jiang, Y. (2005) Isolation and purification of the bioactive carotenoid zeaxanthin from the microalga *Microcystis aeruginosa* by high-speed countercurrent chromatography. *Journal of Chromatography A*, **1064**, 183–186.

Conway, W.D. (1990) *Countercurrent Chromatography: Apparatus, Theory and Applications*. VCH, New York.

Degenhardt, A., Engelhardt, U.H., Wendt, A.-S. & Winterhalter, P. (2000b) Isolation of black tea pigments using high-speed countercurrent chromatography and studies on properties of black tea polymers. *Journal of Agricultural and Food Chemistry*, **48**, 5200–5205.

Degenhardt, A., Knapp, H. & Winterhalter, P. (2000a) Separation and purification of anthocyanins by high-speed countercurrent chromatography and screening for antioxidant activity. *Journal of Agricultural and Food Chemistry*, **48**, 338–343.

Delannay, E., Toribio, A., Boudesocque, L., *et al.* (2006) Multiple dual-mode centrifugal partition chromatography, a semi-continuous development mode for routine laboratory-scale purifications. *Journal of Chromatography A*, **1127**, 45–51.

Destandau, E., Toribio, A., Lafosse, M., Pecher, V., Lamy, C. & Andre, P. (2009) Centrifugal partition chromatography directly interfaced with mass spectrometry for the fast screening and fractionation of major xanthones. *Journal of Chromatography A*, **1216**, 1390–1394.

Di, D.-L., Zheng, Y.-Y., Chen, X.-F., Huang, X.-Y. & Feng, S.-L. (2011) Advances in application of high-speed countercurrent chromatography in separation and purification of flavonoids. *Chinese Journal of Analytical Chemistry*, **39**, 269–275.

Du, Q., Cai, W. & Ito, Y. (2002) Preparative separation of fruit extract of *Silybum marianum* using a high-speed countercurrent chromatograph with scale-up columns. *Journal of Liquid Chromatography and Related Technologies*, **25**, 2515–2520.

Duarte, N., Kayser, O., Abreu, P. & Ferreira, M.J.U. (2008) Antileishmanial activity of piceatannol isolated from *Euphorbia lagascae* seeds. *Phytotherapy Research*, **22**, 455–457.

Engels, C., Ganzle, M.O. & Schieber, A. (2010) Fractionation of gallotannins from mango (*Mangifera indica* L.) kernels by high-speed counter-current chromatography and determination of their antibacterial activity. *Journal of Agricultural and Food Chemistry*, **58**, 775–780.

Fabris, S., Momo, F., Ravagnan, G. & Stevanato, R. (2008) Antioxidant properties of resveratrol and piceid on lipid peroxidation in micelles and monolamellar liposomes. *Biophysical Chemistry*, **135**, 76–83.

Fan, P., Hay, A.-E., Marston, A. & Hostettmann, K. (2008) Acetylcholinesterase-inhibitory activity of linarin from *Buddleja davidii*, structure-activity relationships of related flavonoids, and chemical investigation of *Buddleja nitida*. *Pharmaceutical Biology*, **46**, 596–601.

Fan, P., Marston, A., Hay, A.-E. & Hostettmann, K. (2009) Rapid separation of three glucosylated resveratrol analogues from the invasive plant *Polygonum cuspidatum* by high-speed countercurrent chromatography. *Journal of Separation Science*, **32**, 2979–2984.

Foucault, A.P. (Ed.) (1994) *Centrifugal Partition Chromatography*. Marcel Dekker, New York.

Friesen, J.B. & Pauli, G.F. (2005) G.U.E.S.S. – a generally useful estimate of solvent systems for CCC. *Journal of Liquid Chromatography and Related Technologies*, **28**, 2777–2806.

Gutzeit, D., Winterhalter, P. & Jerz, G. (2007) Application of preparative high-speed counter-current chromatography/electrospray ionization mass spectrometry for a fast screening and fractionation of polyphenols. *Journal of Chromatography A*, **1172**, 40–46.

Henry-Vitrac, C., Desmouliere, A., Girard, D., Merillon, J.M. & Krisa, S. (2006) Transport, deglycosylation, and metabolism of *trans*-piceid by small intestinal epithelial cells. *European Journal of Nutrition*, **45**, 376–382.

Hostettmann, K., Marston, A. & Hostettmann, M. (1998) *Preparative Chromatography Techniques: Applications in Natural Product Isolation*. Springer, Berlin.

Ito, Y. (1981) Countercurrent chromatography. *Journal of Biochemical and Biophysical Methods*, **5**, 105–129.

Ito, Y. (2005) Golden rules and pitfalls in selecting optimum conditions for high-speed countercurrent chromatography. *Journal of Chromatography A*, **1065**, 145–168.

Liu, R., Feng, A., Sun, L. & Kong, J. (2004) Preparative isolation and purification of coumarins from *Cnidium monnieri* (L.) *Cusson* by high-speed countercurrent chromatography. *Journal of Chromatography A*, **1055**, 71–76.

Liu, W., Luo, J. & Kong, L. (2011) Application of complexation high-speed counter-current chromatography in the separation of 5-hydroxyisoflavone isomers from *Belamcanda chinensis* (L.) DC. *Journal of Chromatography A*, **1218**, 1842–1848.

Mandava, N.B. & Ito, Y. (1988) *Countercurrent Chromatography – Theory and Practice*. Marcel Dekker, New York.

Marston, A. & Hostettmann, K. (2006) Developments in the application of countercurrent chromatography to plant analysis. *Journal of Chromatography A*, **1112**, 181–194.

Marston, A., Slacanin, I. & Hostettmann, K. (1990) Centrifugal partition chromatography in the separation of natural products, *Phytochemical Analysis*, **1**, 3–17.

Oka, F., Oka, H. & Ito, Y. (1991) Systematic search for suitable solvent systems for high-speed countercurrent chromatography. *Journal of Chromatography*, **538**, 99–108.

Pace-Asciak, C.R., Hahn, S., Eleftherios, P.D., Soleas, G. & Goldberg, D.M. (1995) The red wine phenolics *trans*-resveratrol and quercetin block human platelet aggregation and eicosanoid synthesis: Implications for protection against coronary heart disease. *Clinica Chimica Acta*, **235**, 207–219.

Palazzo de Mello, J. , Petereit, F. & Nahrstedt, A. (1996) Prorobinetinidins from *Stryphnodendron adstringens*. *Phytochemistry*, **42**, 857–862.

Pauli, G.F., Pro, S.M. & Friesen, J.B. (2008) Countercurrent separation of natural products. *Journal of Natural Products*, **71**, 1489–1508.

Schaufelberger, D.E. (1991) Applications of analytical high-speed counter-current chromatography in natural products chemistry. *Journal of Chromatography*, **538**, 45–57.

Slacanin, I., Marston, A. & Hostettmann, K. (1989) Modifications to a high-speed countercurrent chromatograph for improved separation capability. *Journal of Chromatography*, **482**, 234–239.

Slimestad, R., Marston, A. & Hostettmann, K. (1996) Preparative separation of phenolic compounds from *Picea abies* by high-speed countercurrent chromatography. *Journal of Chromatography A*, **719**, 438–443.

Sutherland, I., Hewitson, P. & Ignatova, S. (2009) Scale-up of countercurrent chromatography: Demonstration of predictable isocratic and quasi-continuous operating modes from the test tube to pilot/process scale. *Journal of Chromatography A*, **1216**, 8787–8792.

Sutherland, I.A., Brown, L., Forbes, S., *et al.* (1998) Countercurrent chromatography (CCC) and its versatile application as an industrial purification and production process. *Journal of Liquid Chromatography and Related Technologies*, **21**, 279–298.

Vastano, B.C., Chen, Y., Zhu, N., Ho., C.-T., Zhou, Z. & Rosen, R.T. (2000) Isolation and identification of stilbenes in two varieties of *Polygonum cuspidatum*. *Journal of Agricultural and Food Chemistry*, **48**, 253–256.

Yanagida, A., Shoji, A., Shibusawa, Y., *et al.* (2006) Analytical separation of tea catechins and food-related polyphenols by high-speed counter-current chromatography. *Journal of Chromatography A*, **1112**, 195–201.

Yang, F., Ma, Y. & Ito, Y. (2001) Separation and purification of isoflavones from a crude soybean extract by high-speed countercurrent chromatography. *Journal of Chromatography A*, **928**, 163–170.

Yoshida, T., Haba, K., Arata, R., Nakata, F., Shingu, T. & Okuda, T. (1995) Tannins and related polyphenols of melastomataceous plants. VII. Nobotanins J and K, trimeric and tetrameric hydrolysable tannins from *Heterocentron roseum*. *Chemical and Pharmaceutical Bulletin*, **43**, 1101–1106.

Yoshida, T. & Hatano, T. (1997) Modern countercurrent chromatography. Application to preparative fractionation of polyphenols. *Analusis*, **25**, M20–M22.

Yuan, L.M., Ai, P., Chen, X.X., *et al.* (2002) Versatile two-phase solvent system for flavonoid prefractionation by high-speed countercurrent chromatography. *Journal of Liquid Chromatography and Related Technologies*, **25**, 889–897.

Zhang, T.-Y., Xiao, R., Xiao, Z.-Y., Pannell, L.K. & Ito., Y. (1988) Rapid separation of flavonoids by analytical high-speed countercurrent chromatography. *Journal of Chromatography*, **445**, 199–206.

Chapter 13
Strategies for the Controlled Synthesis of Oligomeric Polyphenols

Scott A. Snyder

Abstract: This chapter explores efforts to synthesize a very unique subset of complex polyphenols found in nature, namely, materials that are oligomeric. These compounds are produced by the merger of a simpler building block, such as catechin, resveratrol, or rosmarinic acid, with itself several different times and in several different ways to make more complex, bioactive materials that often serve as front-line chemical defenses against foreign pathogens such as fungi. Using several case studies related to these and other polyphenol-based materials, strategies are highlighted for how chemists can potentially achieve their controlled preparation in a laboratory setting with an aim toward further understanding their impressive chemical biology. The advantages and disadvantages of biomimetic approaches versus rational synthetic planning are also discussed.

Keywords: oligomer; natural product; resveratrol; ellagitannin; catechin; rosmaranic acid; polyphenol; total synthesis

13.1 Introduction

Although most natural products are produced with high specificity through enzyme-based synthesis, likely because of environmental pressure to generate a particular compound for a critical biochemical purpose, Nature sometimes abandons that control in service of producing compound collections (Fischbach & Clardy, 2007). This alternate, and arguably more rare form of synthesis, typically occurs when a producing organism encounters an unexpected environmental stress. In some of these cases, Nature uses enzymes to fashion a reactive secondary metabolite building block, one that may not only be able to combat that foreign pressure itself, but which also is capable of uniting with itself two, three, or even upward of over a dozen times to create new materials that are often more biochemically effective. The general goal is to access at least one compound that will enable organismal

Recent Advances in Polyphenol Research, Volume 3, First Edition. Edited by Véronique Cheynier, Pascale Sarni-Manchado and Stéphane Quideau.

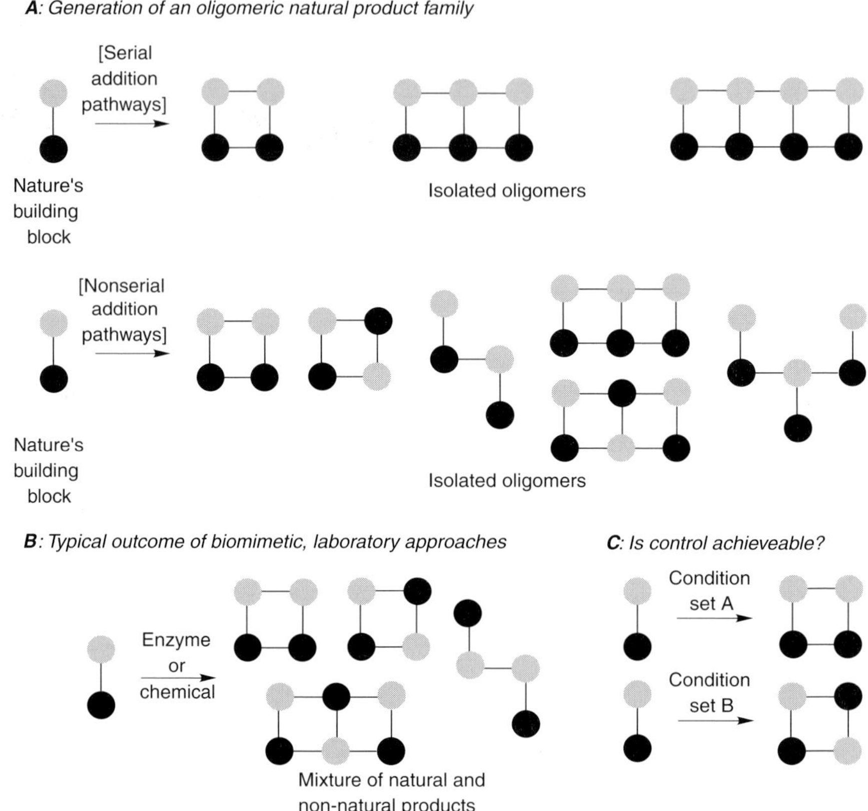

Fig. 13.1 Generalized picture of oligomeric natural product families, efforts to accomplish their laboratory synthesis, and the key synthetic challenge which they present. (Snyder *et al.*, 2011. Reproduced by permission of the Royal Society of Chemistry.)

survival. In this way, the produced natural products (often several dozen for a single producing species) can be likened to an immune system chemical response. It is worth noting that, although such compound collections are produced by a variety of organisms throughout the globe and cover most compounds classes (including terpenes and alkaloids), a large percentage are produced by plants using polyphenolic building blocks based on our analysis of relevant families (Snyder *et al.*, 2011).

As shown in Fig. 13.1, these oligomeric structures are usually produced in one of two modes. The first is the serial and predictable union of monomers into higher order structures; the polysaccharide amylose, composed of varying numbers of D-glucose molecules linked with α-stereochemistry at the anomeric position through the C-4 hydroxyl of each carbohydrate, would be one example of such an oligomer family (Cohen *et al.*, 2008). The second reflects less control in building block union, with far higher degrees of architectural diversity produced through the creation of unique bonds between different parts of the monomers, especially as the higher order structures are formed. Globally, both types of oligomer structures are fascinating, with their array of unique activity profiles and diversity of architectures offering considerable opportunity for biochemical explorations. The challenge, however, is obtaining the material supplies needed for such studies, as most oligomeric natural products are difficult to secure in pure form from natural sources in appreciable amounts, and chemical synthesis has generally proven unable to produce them cleanly and/or in high yield. As will be shown in the subsequent text, efforts over the course of the past quarter century have revealed that chemists can take Nature's building block and, through its exposure to a single enzyme or simple chemical, sometimes reproduce portions of Nature's oligomer collection by making several natural products concurrently; often, these materials are hard to separate, even through high-pressure liquid chromatography (HPLC) techniques, and include a large number of nonnatural products as well. In fact, when such efforts do produce a single or highly predominant material, it unfortunately tends to be a nonnatural structure.

This chapter seeks to explore the question of whether or not the controlled synthesis of such polyphenol-based oligomer families is possible, meaning can an entire collection of structures be produced at will, one at a time? As will be discussed through four specific groups of structures, two encompassing each type of oligomer synthesis (serial and diverse), recent endeavors have shown that such synthetic control is potentially possible. The keys appear to be inventive synthetic design that either tempers the reactivity of Nature's building block through the addition of extra atoms and/or carefully controlled reaction conditions in the case of serial oligomers, or involves unique starting points that enable the preparation of multiple, unique frameworks in a few, rationally designed, synthetic operations for the more structurally diverse oligomer collections.

13.2 Serial oligomer families

13.2.1 Overview

Similar to how other classes of serially produced molecules such as peptides and polysaccharides are synthesized controllably in the laboratory, the production of polyphenolic oligomers requires the addition of judiciously selected protecting groups onto Nature's building block coupled with suitable reaction conditions. The structural alterations are made to allow for proper tempering of the reactivity of various domains. Yet, while such ideas might seem obvious in light of how most complex molecules are synthesized in the laboratory (including nonoligomers), it is really only within the past decade that such concepts have been successfully introduced for the synthesis of oligomeric polyphenols. The reason for this apparent discrepancy is likely that these structures pose a number of additional reactivity concerns at various stages in addition to the key bond formations needed to join monomers to the growing chain, such as oxidation potential and/or general stability. In this section, we present brief overviews of the two main efforts toward such structures that have been accomplished to date, both of which target highly complex molecules of biochemical significance.

13.2.2 Catechin-derived oligomers

All oligomeric natural products of the catechin family derive, as the name implies, from the oligomerization of (+)-catechin (**1**, Fig. 13.2) or its epimer. The plants that produce them,

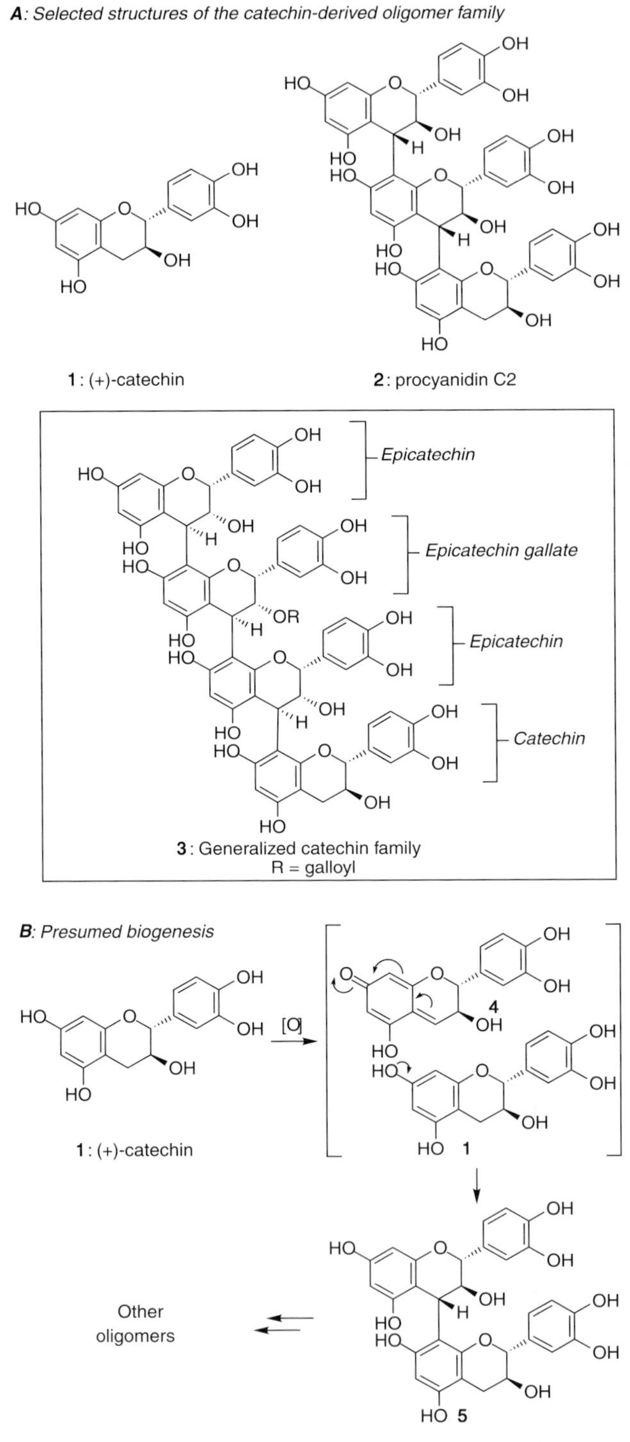

Fig. 13.2   The catechin-derived family of oligomeric natural products: structures and presumed biogenesis.

by large, are invasive weeds, which utilize both **1** as well as higher order structures such as procyanidin C2 (**2**) as herbicides to ward off competing plants from various nutritive sources (Ricardo da Silva *et al.*, 1991; Murray & Pizzorno, 1999). In some cases, they can even inhibit initial seed germination of competing plant species (Bais *et al.*, 2003).[1] Biosynthetically, structures such as **2** are believed to result from the addition process highlighted in Part B of Fig. 13.2, one which certainly can be deployed in a serial fashion to build up molecular complexity quickly. As indicated, an initial oxidation event converts **1** into **4**, setting the stage for a union with another molecule of (+)-catechin (**1**) to form **5** via a Friedel–Crafts reaction; this process can then repeat itself over, and over, again. The overall stereoselectivity in each new union should be governed by the preexisting chirality within the building blocks; thus, enzymatic participation outside of the initial formation of **1** and subsequent chemoselective oxidation process forming **4** is likely unnecessary in the initial dimerization.

While this biosynthesis appears relatively simple in comparison to those of some natural products and points to a clear synthetic blueprint for a potential laboratory synthesis, such an execution is likely quite challenging. In fact, it is hard to envision that simple reagents will be able to achieve the needed chemoselective oxidation in the presence of a second highly oxidation-prone catechol ring system. Moreover, even if that oxidation could be accomplished, it is then not clear how the chain length can be fully controlled, to say nothing of product separation from unwanted materials at the end of such an experiment given their likely polarity. Despite these challenges, in 2004, a team led by Suzuki developed a strategy that achieved the controlled preparation of a protected catechin trimer. As will be discussed, these researchers utilized a highly inventive synthetic design that addressed all the varied issues mentioned in the preceding text (Ohmori *et al.*, 2004).

Drawing from lessons taught in methods to achieve selective disaccharide formation via anomeric activation, the Suzuki team designed three unique catechin derivatives as their critical building blocks: compounds **6**, **7**, and **9** (Fig. 13.3). Their approach rested on the idea that the lone acetate within **6** could be expelled selectively to generate a reactive intermediate similar to that of **4** (see Fig. 13.2). If formed in the presence of **7**, it was then hoped that it could couple with that piece under reaction conditions that would leave its sulfide intact for a second round of coupling that would eventually enable its excision and the merger with compound **9** to form a trimer. In both events, the seemingly extraneous bromine atom within **6** would serve as a blocking group to ensure site-specific C–C bond formation. As shown in Fig. 13.3, when $BF_3 \bullet OEt_2$ was the activating Lewis acid in the first step (leading to material with high stereoselectivity: 93:7 d.r.), followed by the use of *N*-iodosuccinimide (NIS) in the second operation, this design was realized. It should be noted that any attempts to remove the phenolic protecting groups from the final material were not reported, likely because of the high aerial oxidation potential of the resultant product. Nevertheless, the developed

---

[1] One should note that the catechin family is related to a number of subtly different building blocks. These materials are collected within putative structure **3**, with their identity based on the stereochemical arrangement of groups on the central dihydropyran as well as the functionalization of its secondary hydroxyl group. Intriguingly, it has been shown that the orientation of these groups is related to activity (Chen *et al.*, 1996).

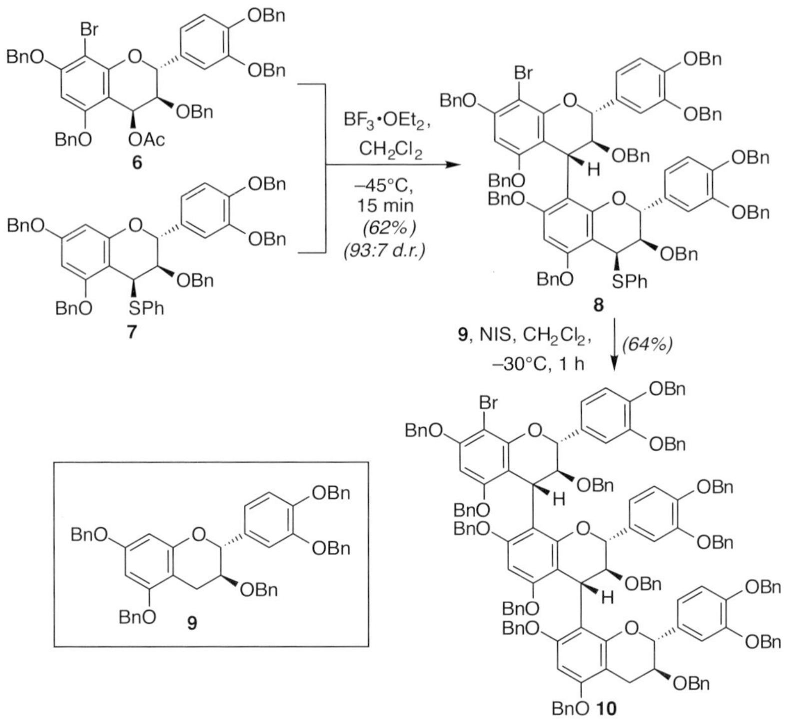

Fig. 13.3   Suzuki's elegant synthesis of a catechin-derived trimer (**10**) through careful substrate and reagent control.

strategy suggests a general approach toward other catechin-derived oligomers. Indeed, a recent report from the same team illustrates this very possibility through the synthesis of protected catechins containing up to 12 monomeric units (Ohmori *et al.*, 2011), again, in protected form.

13.2.3 *Ellagitannin-derived oligomers*

The ellagitannins are a diverse set of compounds, characterized as a group based on the union of the phenolic monomer gallic acid (3,4,5-trihydroxybenzoic acid) and/or its dimeric form ellagic acid with glucose molecules (Okuda *et al.*, 1995; Haslam, 1998; Feldman *et al.*, 1999). What is particularly interesting about these larger oligomers, beyond their sheer size, is the subsequent chemistry through which these pieces can be merged together through new bonds and rearrangements to afford unique structures, many of which constitute very ornate and highly complex architectures. One family that is illustrative of the class are the rubusuaviins, selected members of which are shown in Fig. 13.4 (Li *et al.*, 2007). Simple inspection reveals their dense functionalization, with all of their biaryl linkages existing as single atropisomers. What is unclear from a biosynthetic perspective are several elements, key of which are (1) when these biaryl bonds are formed, a question raised by the fact that one of these linkages is missing within rubusuaviin B (**12**) and (2) whether the resident

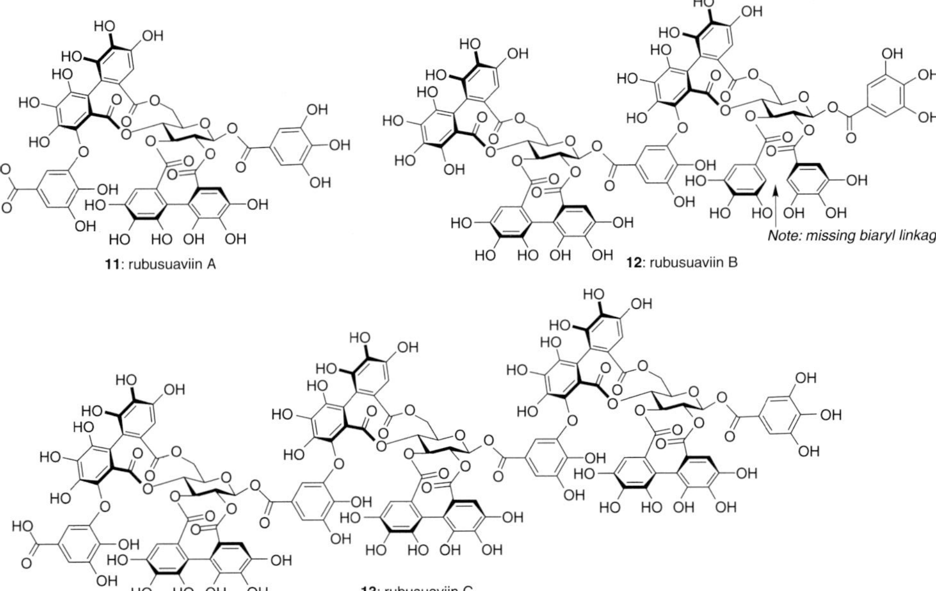

Note: Order of bond construction unknown; rubusuaviin A (11) could be a hydrolysis product of rubusuaviin C (13); rubusuaviin B could arise from the coupling of tellimagrandin II and cusuarictin (structures not shown)

Fig. 13.4 Selected structures of the rubusuaviin family of natural products (**11–13**). Synthetic approaches to oligomeric natural products. (Snyder *et al.*, 2011. Reproduced by permission of the Royal Society of Chemistry.)

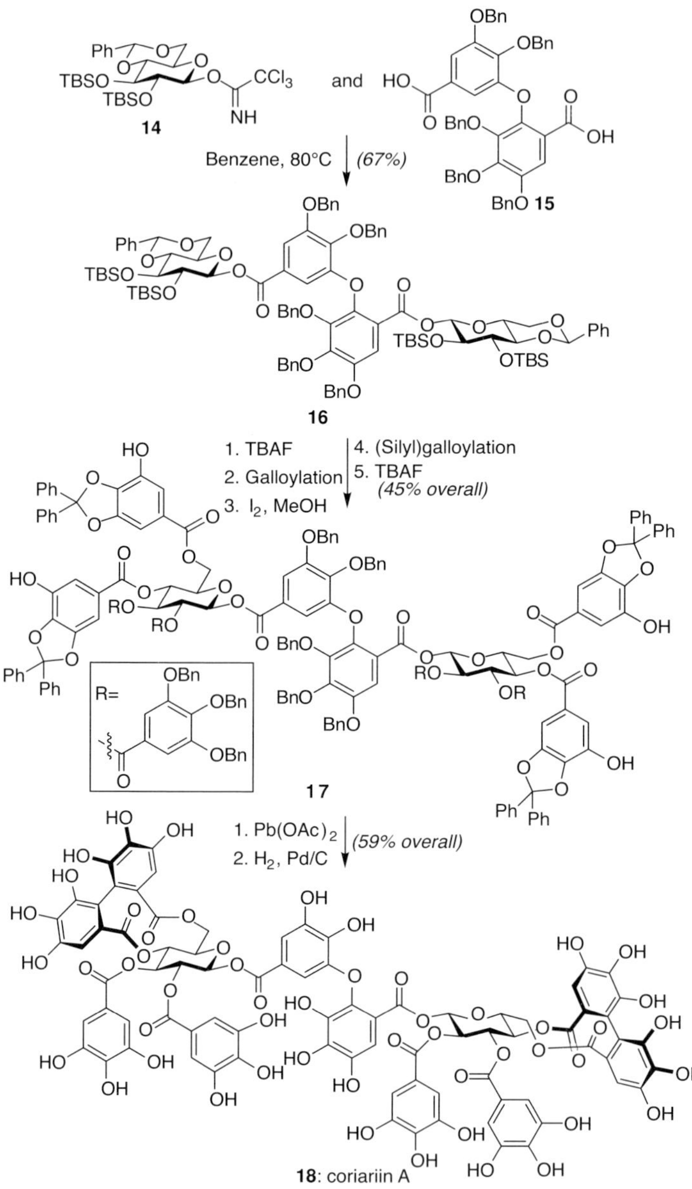

Fig. 13.5 Feldman's inventive total synthesis of coriariin A (**18**) using galloyl building blocks that have been carefully differentiated. Synthetic approaches to oligomeric natural products. (Snyder *et al.*, 2011. Reproduced by permission of the Royal Society of Chemistry.)

chirality, particularly that of the carbohydrate core, can control the generation of that biaryl stereochemistry on its own or are enzymes still necessary. From the standpoint of oligomers alone, it is also interesting to consider that if one were to synthetically "add" the one biaryl linkage missing within **12**, then this molecule is formally a dimer of **11**, while rubusuaviin C (**13**) is a trimer. Such homology provides a cascading level of serial oligomer construction beyond that possessed by most families, since it goes beyond just the ellagic acid portions. However, it should be noted that there is an alternate biogenetic theory that cannot be dismissed; namely, that rubusuaviin A (**11**) is a degradative by-product of rubusuaviin C (**13**) and that rubusuaviin B (**12**) could be the product of the merger of natural products tellimagrandin II and casuarictin, i.e., not a true oligomerization followed by subsequent functionalization (Pouységu *et al.*, 2011).

Many groups have performed valiant synthetic efforts to prepare such materials, and in 2000, the Feldman group (Feldman, 2005) accomplished the first preparation of a dimeric ellagitannin (Feldman & Lawlor, 2000). Their particular target was coriariin A (**18**, Fig. 13.5; Hatano *et al.*, 1986), a compound that has strong antitumor activity (Miyamoto *et al.*, 1987). As is readily discerned, the key structural difference for this material from the rubusuaviin family is an ethereally linked gallic acid core. It was this very core, in fact, which the Feldman team used as a key starting material in its protected form as diacid **15**. In the opening operations, both acids were coupled with a trichloroace-timidate activated D-glucose derivative (**14**) to forge the requisite β-glycoside linkages. The early incorporation of these sugar moieties was performed in expectation that their chirality might assist in later biaryl couplings from the standpoint of atroposelectivity. To test that hypothesis, a series of controlled deprotections and couplings of additional gallic acid building blocks with unique protecting group patterning and reactivity gave rise to the key test precursor (**17**) over 5 total steps. Pleasingly, when that key oxidative coupling (via a Wessely protocol) was attempted to form the two biaryl linkages, complete atroposelectivity was observed. As such, this synthesis suggests that atropoisomer formation can, potentially, follow core synthesis as part of Nature's preparation of the material. More generally, the success of the Wessely oxidation procedure (Bubb & Sternhell, 1970) also indicates that it can likely be deployed on other complex frameworks in light of the potential for the core ethereal linkage within **17** to cleave, an outcome that was not observed. The target molecule (coriariin A, **18**) was then completed through a standard global debenzylation (H$_2$, Pd/C). Clearly, this work points the path toward future endeavors, and it cannot be emphasized enough that this successful strategy was the product of many years of careful study of varied strategic alternatives. In fact, the approach was the third major generation of efforts within the Feldman laboratory toward these structures.

13.3 Oligomer families with diverse bond connections

13.3.1 Overview

Although the challenge of preparing any oligomeric natural product is high, families with diverse bond connections arguably provide a higher degree of synthetic difficulty than serially produced ones. The reason is that greater levels of chemoselectivity are needed to achieve control, since several different bonds can be formed from the key starting material. In this section, we will devote significant attention to the resveratrol family of oligomeric structures, highlighting those issues of control through various biomimetic studies. We will then discuss a potential solution for synthetic control that avoids the general problem of tempering the reactivity of Nature's building block by designing an alternate starting point for oligomer synthesis. Finally, we highlight the potential power of that solution to achieve controlled synthesis with a second family of structurally diverse polyphenolic oligomers: the helicterins, helisorins, and helisterculins.

13.3.2 *The resveratrol family of oligomeric natural products*

13.3.2.1 *Introduction*

Among all families of oligomeric structures, those which derive from resveratrol (**19**, Fig. 13.6) are perhaps the most architecturally diverse. Indeed, although resveratrol (Takaoka, 1940; Hillis & Inoue, 1967) itself is a relatively small molecule, its double bond,

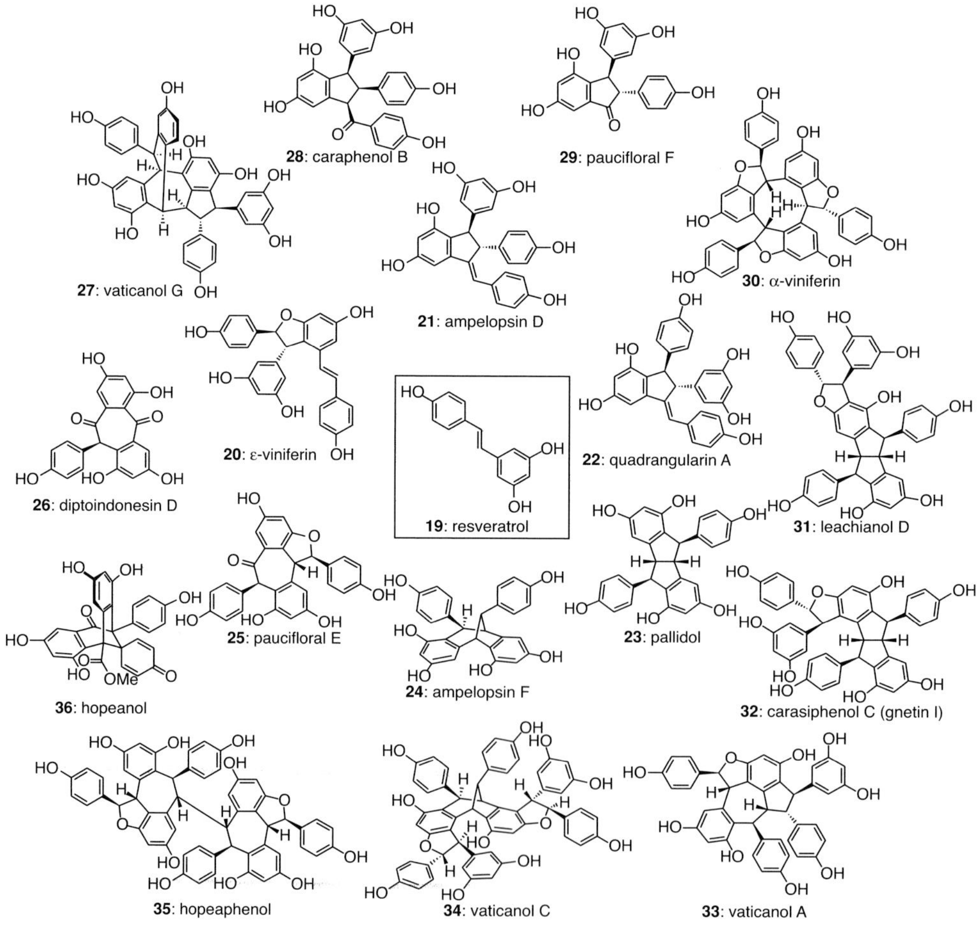

Fig. 13.6   Selected structures of the resveratrol family of oligomeric natural products.

two aromatic rings, and three phenol residues provide the reactive handles from which several hundred oligomeric natural products are constructed, including several with more than ten monomeric units. The structures within Fig. 13.6 (**20–36**) provide an overview of the type of complexity that can be generated, structures that include indane, dihydrofuran, seven-membered, eight-membered, and nine-membered ring systems in addition to various [3.2.1]- and [3.3.0]-bicycles; they also include glucosylated derivatives through both phenol and aromatic carbon atoms, though these compounds are not explicitly shown here (Diyasana *et al.*, 1985; Oshima *et al.*, 1990; Kurihara *et al.*, 1990; Tanaka *et al.*, 2000a, 2000b, 2001a, 2001b; Luo *et al.*, 2001; Takaya *et al.*, 2002a, 2002b; Ito *et al.*, 2003a, 2003b, 2003c, 2004, 2009a, 2009b; Supudompol *et al.*, 2004; Wang *et al.*, 2005; Guebailia *et al.*, 2006; Ge *et al.*, 2006a, 2006b, 2008; He *et al.*, 2009). In total, there are over six dozen species of plants throughout the world that fashion such materials, and their respective members can be found on nearly every continent. Most of the resveratrol-based oligomers are believed to be phytoalexins that are known to combat fungal infections (Landcake & Pryce, 1977). Grapevines, for instance, utilize resveratrol and its oligomers to counteract the *Botrytis* fungus, a disease to which some vintners actually expose their plants to produce grapes with intensified flavor. However, that process comes at the risk of total harvest loss if the plant cannot effectively manage the infection (van Baarlen *et al.*, 2004; Favaron *et al.*, 2009).

What is perhaps more interesting, however, are the biochemical properties of the family outside of their frontline antifungal capabilities. Indeed, resveratrol itself has demonstrated potent antitumor, antioxidant, antiaging, and cardiac protective properties in screens with mice (Jang *et al.*, 1997; Howitz *et al.*, 2003; Wood *et al.*, 2004; Szewczuk *et al.*, 2004, 2005; Chen *et al.*, 2005; Baur *et al.*, 2006; Milne *et al.*, 2007; Heiss *et al.*, 2007). Many of the higher order structures, however, have alternate activity in a variety of *in vitro* assays (Kitanaka *et al.*, 1990; Ohyama *et al.*, 1999; Huang *et al.*, 2001; Sahadin *et al.*, 2005). Vaticanol C (**34**), e.g., is a topoisomerase II inhibitor (Ito *et al.*, 2003a; Ohguchi *et al.*, 2005; Yamada *et al.*, 2006a, 2006b), while molecules related to hopeaphenol (**35**) have shown anti-HIV activity (Dai *et al.*, 1998). In fact, as the molecules get larger, they tend to become more specific and potent in their activity. Whether or not these and other activities extend to humans is currently unknown, though many have already begun to posit that resveratrol (**19**) itself deeply impacts human health through red wine consumption, the main source of the molecule in our diets. Though drinking such beverages is believed to be beneficial, no clinical data support this notion as of yet (Soleas *et al.*, 1997; Walle *et al.*, 2004). In any event, our goal in this chapter is not to evaluate the validity of such claims, as interesting as they may be, but rather to consider synthetic strategies by which the higher order structures can be accessed as well as whether or not their controlled synthesis is possible.

13.3.2.2 Biomimetic approaches

Although the biosynthesis of resveratrol itself is known, one which involves the sequential union of three molecules of malonyl-CoA to 4-coumaroyl-CoA with a terminating cyclization (Zhang *et al.*, 2006), little additional knowledge exists in terms of how the higher order structures are actually fashioned in Nature. Enzymatic participation in some form is often invoked in most biosynthetic proposals, largely because most plants produce single enantiomers of a given compound, but the mode of that intervention is unclear. Some possibilities that would make sense based on that for other families could be chemoselectively generating proper reactive intermediates and/or controlling the facial presentation of monomers. What does point to the participation of some chiral entity, as well as affording greater uniqueness to the family as a whole, is the observation that some plants prepare the antipode of a molecule produced by a different plant species (Tanaka *et al.*, 2000; He *et al.*, 2006).

Despite these questions, an examination of the structures within Fig. 13.6 does suggest several possibilities for their overall construction if an eye is directed toward the homologous components of the various members. For example, if one looks at the resveratrol dimers such as **20–25**, it becomes clear that these molecules could serve as the synthetic starting points for the higher order structures. Pallidol (**23**), e.g., clearly comprises the core for lechianol D (**31**) and carasiphenol C (**32**), just as ampelopsin F (**24**) is a likely precursor for vaticanol C (**34**). Similarly, dimerization of paucifloral E (**25**) could lead to structures like hopeaphenol (**35**), while oxidative cleavage of the double bond within ampelopsin D (**21**) could give rise to paucifloral F (**29**). Many more such connections exist, leading to the idea that if these dimeric starting points were formed with optical purity, then the higher order structures could potentially incorporate new resveratrol monomers through relative stereocontrol without needing further enzymatic participation.

The general pathway through which the main bond unions within the family are believed to be forged, or at least the key initial bonds, is through radical mechanisms (Sotheeswaran & Pasupathy, 1993). However, given the overall number and types of radicals that can be formed (both oxygen- and carbon-centered), these radical constructions may be deliberately uncontrolled to achieve architectural diversity. To provide a sense of some of these biosynthetic pathways, two representative examples are shown in Fig. 13.7. The molecule ε-viniferin (**20**), for instance, could arise through the merger of carbon-centered radicals **37** and **38** to initially form **39**, followed by a Friedel–Crafts cyclization. To produce optically active products, the first of these bond constructions (i.e., the formation of **39**) must be controlled by an exogenous chiral source. The *trans*-based dihydrofuran ring of **20**, a hallmark of nearly all such ring systems within the family, is likely the thermodynamically most stable isomer; as such, its formation may be the result of equilibrative pathways if an alternate dihydrofuran were initially formed (Kurosawa *et al.*, 2004; Lian & Hinkel, 2006). Similarly, pallidol (**23**) could arise through the union of two molecules of radical **37** and two subsequent Friedel–Crafts C–C bond constructions.

Once one moves beyond dimeric structures to higher order oligomers, several mechanistic pathways can usually be posited. For example, one could envision carasiphenol C (**32**) arising from the further union of radicals of type **37** with pallidol (**23**), assuming that positional control can be achieved in C–C bond formation versus the several other alternatives (an example of which would lead to natural product **31**, see Fig. 13.6), while the merger of *ent*-**20** (a known natural product) with the same radical could afford an alternate path to the same molecule (Sotheeswaran & Pasupathy, 1993). It is important to stress again that while these general pathways are all suggestive of high control in radical positioning, achieving

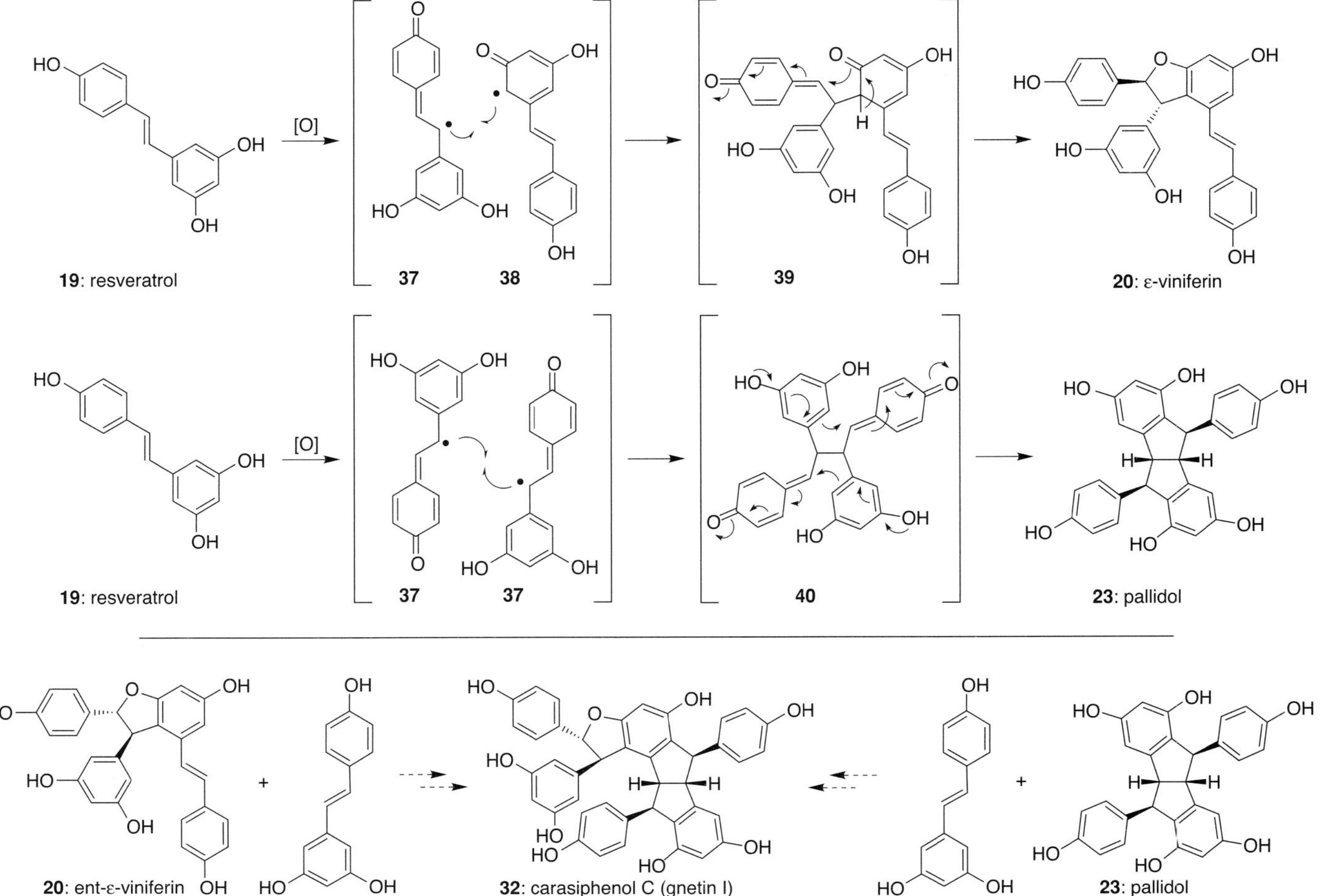

Fig. 13.7 Representative biosynthetic scheme showing the putative generation of two reveratrol dimers and possible pathways for a higher order structure.

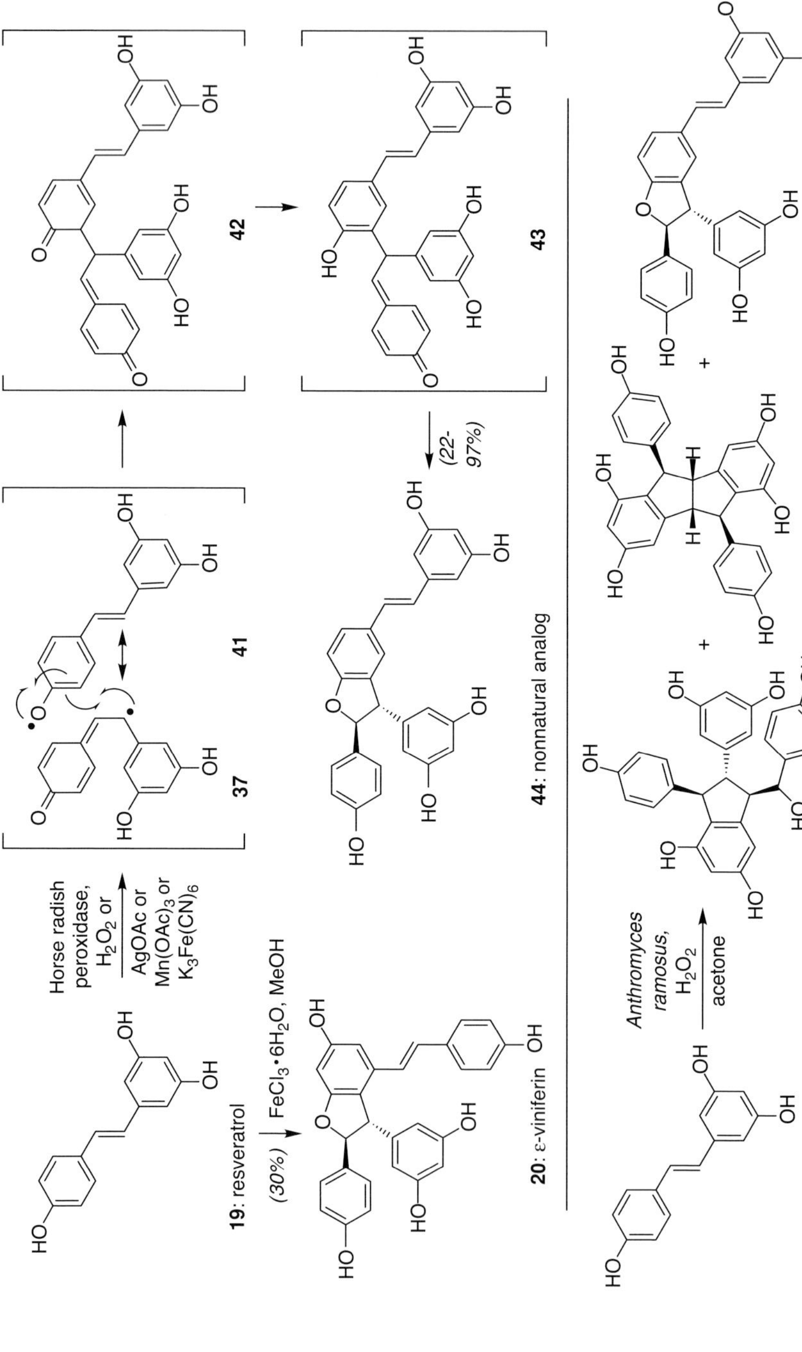

Fig. 13.8 Selected examples of efforts to effect direct resveratrol (**19**) dimerization; typically, nonnatural analogs of resveratrol dimers result.

that control in practice is likely far from simple. Not only can positional isomers of those radicals be formed, but there are many cases where a given radical has multiple resonance forms that can lead to different products as well. Of course, there are also different levels of reactivity to contend with between C- and O-centered radicals. As a result of these issues, most biomimetic studies attempting to execute such pathways in the laboratory encounter this challenge of controlled radical formation and reaction, suggesting that Nature may well avoid control as well in service of making many frameworks as posited above.

Figure 13.8 illustrates several representative studies along these lines. For example, if resveratrol (**19**) is exposed to simple horseradish peroxidase in the presence of H_2O_2, the predominant characterizable product is dihydrofuran **44**, a material likely generated through the indicated C–C and C–O bond formations and formed in 42% yield (Langcake & Pryce, 1977, 1977b). While reminiscent of ε-viniferin (**20**), the connectivities of **44** constitute a nonnatural analog. If the enzyme used for such a process is switched to *Anthromyces ramosus*, **44** is still formed as the predominant material, but compound **45** as well as pallidol (**23**) are also obtained in measurable quantities (Takaya *et al.*, 2005). Intriguingly, if simple single-electron transfer reagents are used instead, they also lead to the production of **44**, some even in quantitative yield (Sako *et al.*, 2004); however, a recent report does indicate that if $FeCl_3 \bullet 6H_2O$ is used specifically in methanol as solvent, then ε-viniferin (**20**) can be accessed in 30% yield (Yao *et al.*, 2004). Thus, the main conclusion that can be drawn from these results overall is that complete control is difficult to achieve from resveratrol (**19**) itself, and in those cases where high selectivity is observed (meaning yields at least above 40%), it is nonnatural products that are typically obtained. Only under very special conditions, ones that cannot be predicted *de novo*, are natural products formed.

In more recent efforts, several investigators have taken a slightly different approach in service of evaluating this reactivity challenge. Rather than exploring resveratrol itself, these endeavors have added extra atoms to its core in hopes of better controlling the radicals that are produced from it and thus shift the product distribution toward specific structures. These investigations may well make more sense from a "biomimetic" synthesis perspective in that Nature can certainly temporarily (or permanently) engage a given phenol in resveratrol with an acetate, carbohydrate, phosphate, or sulfonate moiety, thereby altering the radicals that could be generated and thus leading to different product distributions (Ito *et al.*, 2009b). Evolutionary pressure could be involved in this process, given that isolation efforts suggest that each plant synthesizes different arrays of resveratrol-based oligomers (in terms of structure, quantity, and overall number of compounds). It is important to note, however, that the presence of these extra groups may well be more masked than structural elucidation efforts may indicate, since natural product isolation procedures may excise them unintentionally.

Figure 13.9 illustrates some of the seminal studies along these lines. Arguably, the most successful ones have utilized *t*-butyl groups appended onto the *para*-phenol ring system of resveratrol. The goal of this design was to try to ensure that the radical formed from this ring system could only react via resonance form **47**. Pleasingly, these endeavors did in fact lead to the formation of indane **48** in 35% yield. Subsequent removal of these *t*-butyl groups through exposure to strong acid enabled the first total synthesis of quadrangularin A (**22**) to be achieved in the laboratory (Li *et al.*, 2006a). Since this initial report, these investigators

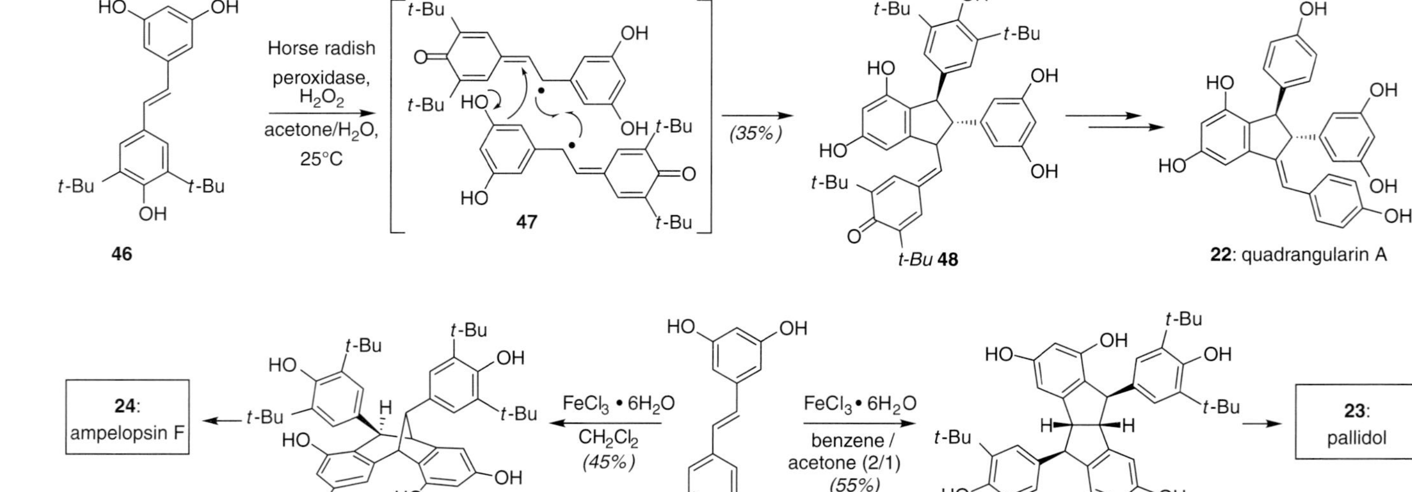

Fig. 13.9 Use of resveratrol derivative **46** to achieve total syntheses of several resveratrol dimers.

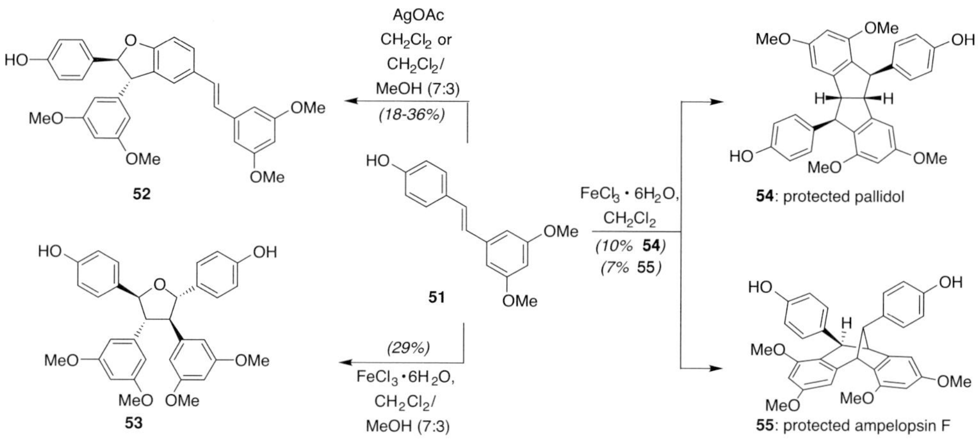

Fig. 13.10 Explorations into the effect of partial phenol protection of resveratrol on dimeric product formation using SET reagents.

have extended their findings with the discovery that product yields for this sequence could be enhanced slightly by using MnO_2 or Ag_2CO_3 as the oxidizing reagent. Even more interesting, if the oxidant is changed to $FeCl_3 \bullet 6H_2O$ and used in specific solvents, then both **49** or **50** could be obtained with relatively high selectivity as well (Li *et al.*, 2010a, 2010b); *t*-butyl group removal in these cases then allowed for the isolation of synthetic samples of both pallidol (**23**) and ampelopsin F (**24**).

Somewhat similar results are obtained if protective groups are placed on selected phenols rather than directly on the aryl ring. For instance, as shown in Fig. 13.10, dimethylated resveratrol derivative **51** could be converted into a number of different architectures in modest yield depending both on the oxidant and solvent used (Velu *et al.*, 2008); many of these frameworks are again nonnatural analogs, but under certain conditions both pallidol and ampelopsin F could be obtained in protected forms (**54** and **55**). Perhaps the greater message from these studies, as well as those presented earlier, is that engagement of resveratrol in alternate forms does allow access to natural products with higher frequency and apparent selectivity than with resveratrol alone. Yet, it is also interesting that so many additional nonnatural analogs can also be prepared from the same frameworks, thereby expanding the sense of what can be made from resveratrol's atoms and possible architectures that future isolation efforts may eventually reveal as also being structures produced in Nature.

Outside of efforts to make specific dimers, only a few studies have explored trimer or higher order oligomer synthesis through clear biomimetic synthesis. Figure 13.11 provides one of the most advanced examples, work in which ε-viniferin (**20**) was combined either with resveratrol itself (**19**) or its dimethylated form (**51**) and then exposed to horseradish peroxidase in the presence of H_2O_2. Both of these reactions ultimately provided a small but characterizable amount of davidiol A in either its native (**59**) or protected form (**58**), likely through the indicated pathway (He *et al.*, 2006). Whether such approaches can be achieved with higher selectivity and/or lead to other structures through the use of more specific chemical reagents or alternate enzymes remains to be seen, though such efforts would likely

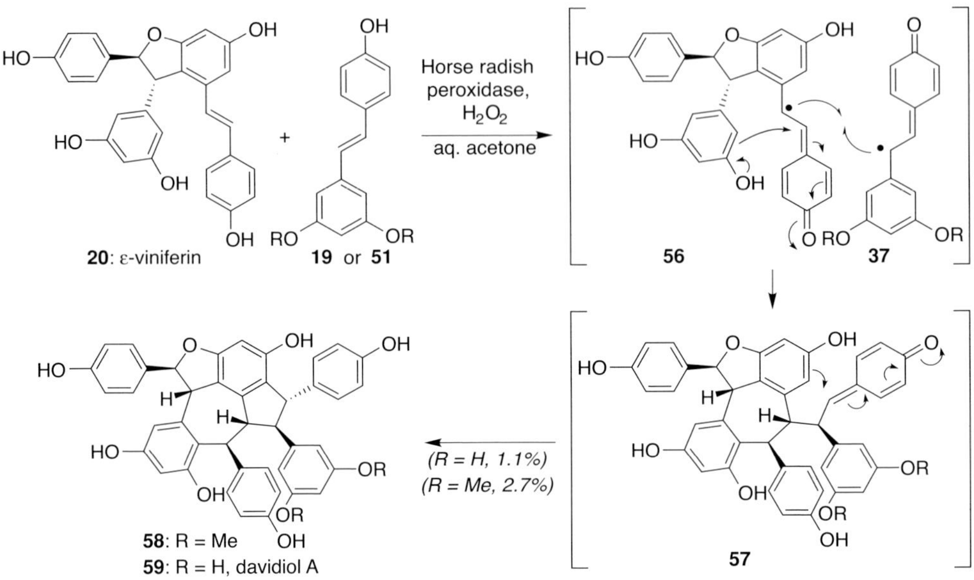

Fig. 13.11 Efforts to prepare the resveratrol trimer davidiol A (**59**) via couplings between ε-viniferin (**20**) and resveratrol (**19**). (From Snyder, 2011. Reproduced with permission.)

be best facilitated by access to an array of natural compounds to help guide HPLC-based separations and identification of the individual components within the product mixtures.

Finally, there is one additional approach for understanding the biological basis behind the formation of certain resveratrol-based oligomers that has shown a number of interesting results. Namely, instead of seeking to determine if a given molecule can be built from the ground up using resveratrol (**19**), this approach asks whether it could instead arise through the architectural rearrangement of another family member. One of the most impressive and insightful studies along these lines is shown in Fig. 13.12. Here, exposure of ε-viniferin (**20**) to various acid sources effected the generation of several natural products, including isoampelopsin D (**62**), ampelopsin D (**21**), and ampelopsin F (**24**). Product mixtures were obtained, largely due to stereochemical variety that result from the varied ring-opening and ring-reclosing processes proposed (i.e., access to either **61a** or **61b**). Nevertheless, this study suggests that ε-viniferin (**20**), not necessarily just resveratrol itself, could be a lynchpin to some of the higher order structures of the family (Takaya *et al.*, 2002; Niwa *et al.*, 2000). Again, controlled synthesis is not achieved, but that outcome is not likely to be Nature's specific goal if it seeks to combat a foreign pathogen. It is also reasonable to believe that structures even more complex than ε-viniferin (**20**) could lead to a number of other targets as well through such pathways; thus, there may be other privileged lynchpins in terms of accessing further structural diversity through uncontrolled rearrangement processes.

As a concluding comment, there are some additional biosynthetic approaches that have been posited in the literature for various domains within the family. Though none has progressed as far as the studies shown in the preceding text, the following citations are provided as sources of inspiration for additional ways to consider the family as a whole,

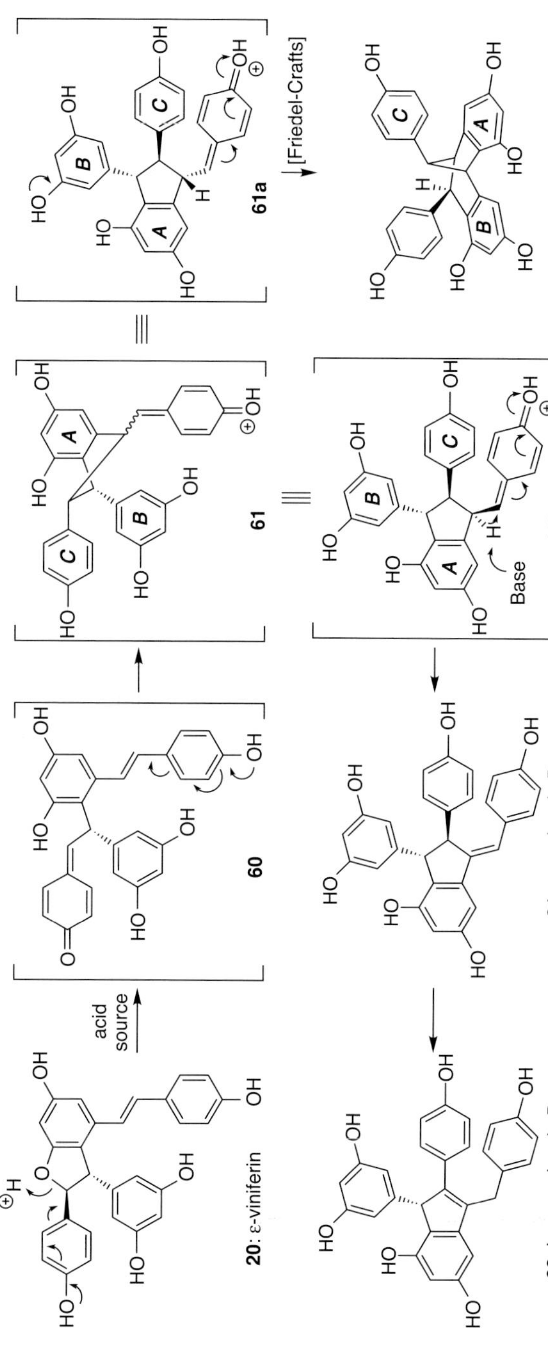

Fig. 13.12 Biogenetic explorations using ε-viniferin (**20**) instead of resveratrol as the starting material: an alternative paradigm for possible dimer formation. Note that the stereochemistry shown for both **21** and **62** is enantiomeric to that presented in Fig. 1.6. (From Snyder, 2011. Reproduced with permission.)

especially as some are based on nonradical bond constructions using resveratrol: Aguirre *et al.*, 1989; Engler *et al.*, 1994; 1995; Aguirre *et al.*, 1999; Thomas *et al.*, 2002a, 2002b; Li & Ferreira, 2003; Takaya *et al.*, 2005.

13.3.2.3 Stepwise synthesis approaches

Apart from the studies mentioned in the preceding text, some isolated resveratrol oligomers have been synthesized through fully rational chemical synthesis empowered by retrosynthetic analysis. We will not specifically delineate those routes here, though we note these works so that the unique ways that various domains have been tackled can be fully appreciated (Kim & Choi, 2009; Nicolaou *et al.*, 2009, 2010). What is essential to note about these particular approaches is that each has separately been able only to deliver a particular dimeric architecture, and nothing more complicated. As a result, it is unclear how any can be modified or adjusted to make other frameworks, meaning that none provides a family-level solution to the resveratrol synthesis problem (Snyder, 2011).

Given this state of affairs, we questioned whether it would be possible to utilize stepwise synthesis to prepare all the varied members through a single overarching strategy. Such an approach would need to use retrosynthetic analysis carefully and in a more global manner than typical (in service of also accessing diversity), and would seem to absolutely require a starting point other than resveratrol. However, what that starting point should be was not obvious, since we would need to use reagent-controlled cascades and chemoselective bond constructions to hopefully convert it into all of the varied structures, one at a time.

We began our efforts toward this goal in 2006, examining all the various members of the family looking for any critical design clues. Ultimately, molecules such as paucifloral F (**29**) and diptoindonesin D (**26**, see Fig. 13.6) proved to be key, as their three aromatic rings are inconsistent with direct resveratrol oligomerization. As mentioned earlier, these structures could result from oxidative cleavage of a standard resveratrol dimer. Nevertheless, it was the existence of these materials that led us to postulate that perhaps a building block of general structure **63** (Fig. 13.13) could lead to the architectural diversity of the entire

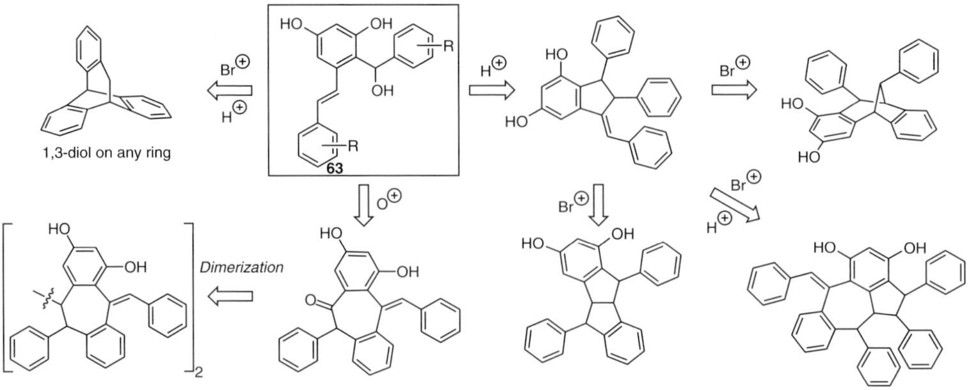

Fig. 13.13   Can intermediate **63** allow for the controlled production of all the carbogenic diversity of the resveratrol family of oligomers? (From Snyder, 2011. Reproduced with permission.)

family. We hoped that its array of functionality could enable access to all the carbogenic complexity of the resveratrol family following exposure to appropriate electrophiles. In general, this hypothesis has proven true, with nearly 20 natural products and a number of additional analogs synthesized with high control encompassing virtually every pathway defined within Fig. 13.13. In the remainder of this section, we present a portion of those studies (Snyder *et al.*, 2007, 2009).

We begin with our syntheses of indane-based structures. Following the preparation of building block **64**, we exposed it to a variety of acid sources in expectation of forming intermediate **66** (Fig. 13.14). Our hope was that the subsequent capture of its benzylic carbocation by various nucleophiles would lead to functionality to access multiple structures. Indeed, depending on the acid source used, we have been able to incorporate halogen, oxygen, and sulfur-based nucleophiles, with the latter two proving critical for the formation of both paucifloral F (**29**) and ampelopsin D (**21**) through the indicated sequences (Taylor, 1999).[2] Subsequent exposure of the latter of these targets (ampelopsin D, **21**) to HCl in MeOH at elevated temperature (80°C) afforded a third indane-based natural product, isoampelopsin D (**62**), with high control through an alkene isomerization. Application of the same chemistry to an alternate form of the key building block in form **69** led to the smooth synthesis of related natural products with their pendant aryl rings switched (i.e., **70** and **22**). This outcome is of import in that it was achieved despite the differential pushing of electron density within **69** in terms of resonance. In fact, a number of starting materials work in these processes, including one which led to even more highly oxidized compounds such as **71**.

With these achievements behind us, we next explored whether we could append additional ring systems onto the indane cores. If so, then we could potentially access the bicyclic frameworks of pallidol (**23**) and ampelopsin F (**24**). As indicated by the earlier work presented in Fig. 13.12 from the Niwa group, these compounds are formally one Friedel–Crafts reaction away from compounds of type **72** and **75** (Fig. 13.15). The question, however, is whether or not that reaction could be achieved in a controlled manner. Indeed, in that case the electrophile added to both faces of the alkene, leading to many products with differential stereochemistry. We anticipated that if an electrophile could be found that could add reversibly to the double bond and only be trapped through Friedel–Crafts chemistry, then the desired C–C bond construction could potentially be achieved selectively. Following several false starts, we ultimately tested electrophilic bromine for this purpose, and, as shown, that electrophile did, in fact, prove successful (Brown, 1997).

In the key transformation, we treated **72** and **75** separately with an electrophilic bromine source (either Br_2 or NBS). To our delight, we found that a smooth, regioselective, and quantitative bromination of the A-ring could be achieved. This operation was then followed by a second site-selective halogenation, this time of the pendant 3,5-dimethoxyphenyl ring

---

[2]It is worth noting that efforts to convert the ketone of molecules like paucifloral F (**29**) directly into the alkene of ampelosin D (**21**) have not proven successful despite numerous attempts. Only through the incorporation of the sulfide nucleophile could a fourth aromatic ring be appended; as such, the ability to so readily control product formation in the cascade-based cyclization was a critical component for the overall success of the approach.

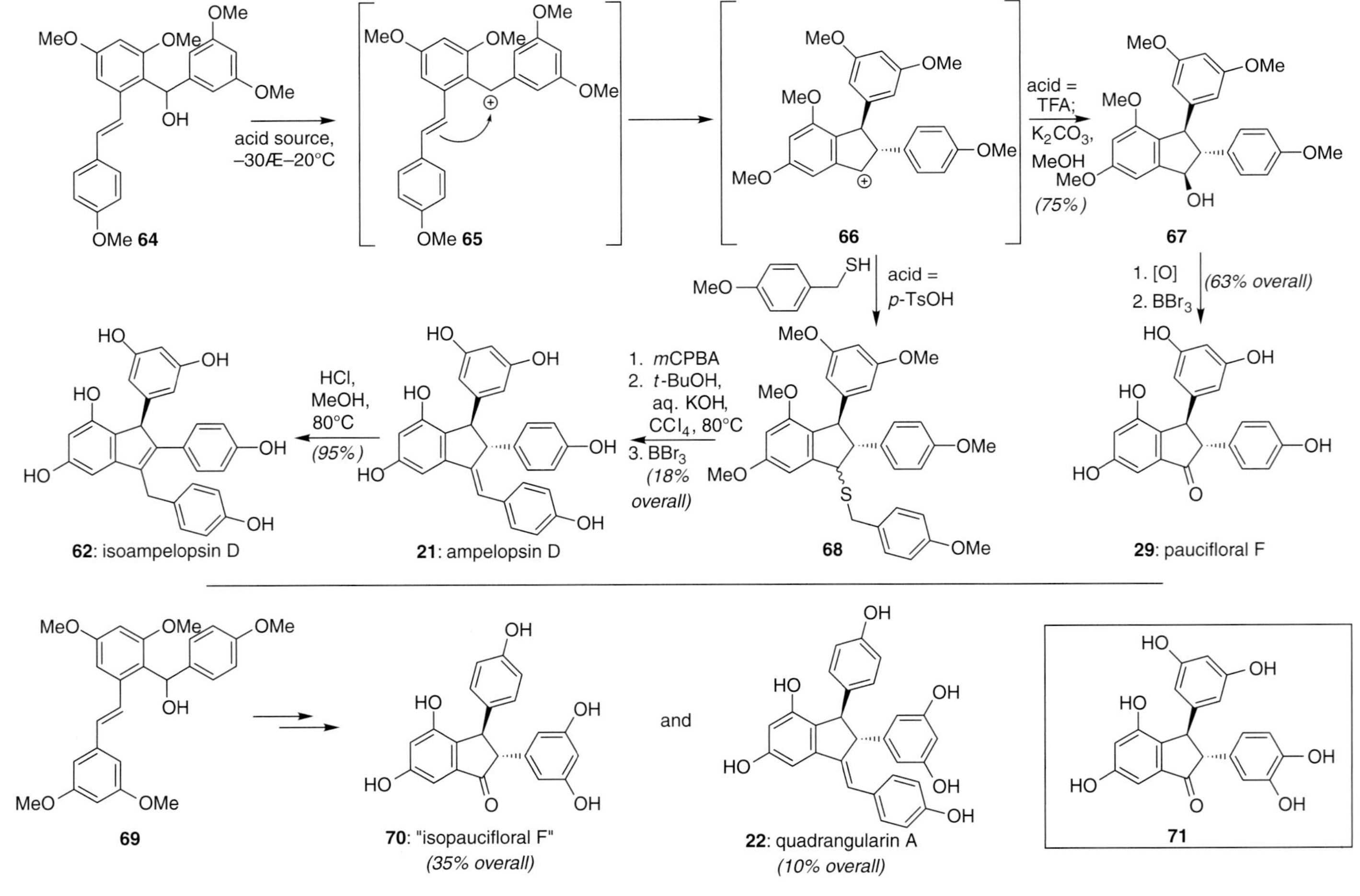

Fig. 13.14 Cascade-based approach to fashion indane-containing members of the resveratrol family of natural products from key triaryl intermediates.

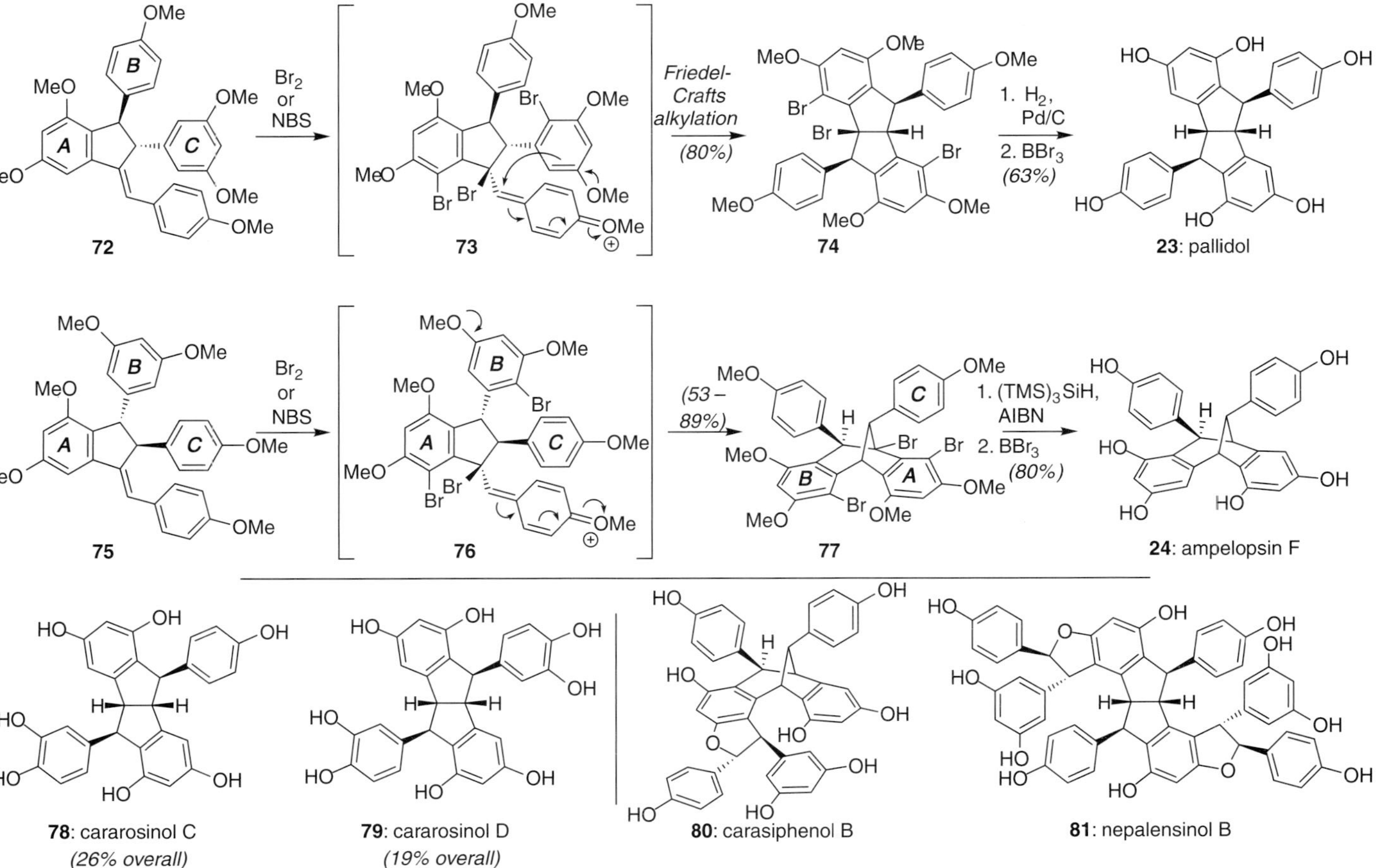

Fig. 13.15 Bromine-induced cyclization cascades to fashion pallidol (**23**) and ampelopsin F (**24**) from appropriate precursors. (Snyder *et al.*, 2011. Reproduced by permission of the Royal Society of Chemistry.)

system; this process was also quantitative. Finally, when more of the halogen source was added (or generated in situ)[3] the double bond was then engaged by the electrophile, and cyclization to either **74** or **77** proceeded smoothly to deliver the desired carbocyclic cores. We believe that the reversibility of bromonium formation was critical to the success of these sequences. Indeed, as shown in the second drawn sequence (i.e., **76→77**), product generation requires the addition of bromine onto the more hindered olefin face based on the positioning of its aryl C-ring neighbor. From a steric basis, this mode of addition must be less favored; however, only through the terminating cyclization could the reaction be driven forward as long as the bromine atom could add and eliminate with full reversibility. Gratifyingly, this chemistry could be used to access two members of the cararosinol family of natural products (i.e., **78** and **79**; Yang *et al.*, 2005), since it also succeeded in the presence of even with more highly electron-rich aromatic systems.

Current efforts are seeking to determine if the aryl bromines within **74** and **77** can be deployed to access structures such as carasiphenol B (**80**), nepalensinol B (**81**) and similar materials (Yamada *et al.*, 2006). Such investigations would provide a valuable dividend for the nonatom economic operation (Trost, 1991) of adding and then removing multiple bromines to prepare pallidol (**23**) and ampelopsin F (**24**). Moreover, we seek to determine if the resident stereochemistry can in fact direct dihydrofuran stereochemistry (a question of import since nepalensinol B (**81**) is of opposite configuration as carasiphenol C (**32**), see Fig. 13.6).

The remaining major elements of carbogenic complexity within the resveratrol family that have not yet been discussed are seven-membered rings. Natural products, such as paucifloral E (**25**), diptoindonesin D (**26**), vaticanols G (**27**) and A (**33**), hopeaphenol (**35**), among others, possess such domains. Our question was whether or not this system could also be accessed from our key precursors. The answer is yes, though one small structural adjustment is required: their alcohols must be converted to ketones. When these materials are exposed to electrophilic oxygen, then the seven-membered ring of protected hemsleyanol E (**82**) and diptoindonesin D (**83**) could be accessed (Fig. 13.16). Although this reaction looks simple, it is important to note that it proved exceedingly difficult to achieve in a flask, with only the use of *in situ* generated methyl(trifluoromethyl)dioxirane enabling the process to proceed in any yield. Given past efforts with natural products and acetate-protected starting materials (Niwa *et al.*, 2000; Takaya *et al.*, 2002), we hypothesize that methyl ether protection of the phenols may be the culprit for this unanticipated difficulty. As a side note, if **82** is treated with strong acid in a nonpolar solvent, natural product analog **84** is formed; to date, no such molecule has been isolated from Nature even though the free-phenol form of **82** is a known natural product. Thus, this phenonium-shift-based sequence affords access to additional structural diversity of potential biochemical value.

---

[3] We found that only two equivalents of molecular bromine were needed to effect the incorporation of three electrophilic halogens. Based on many control experiments, we believe that molecular oxygen, in combination with the substrate, provided the means for the conversion of bromide into bromine *in situ*. When NBS was used, a third equivalent of the reagent had to be added.

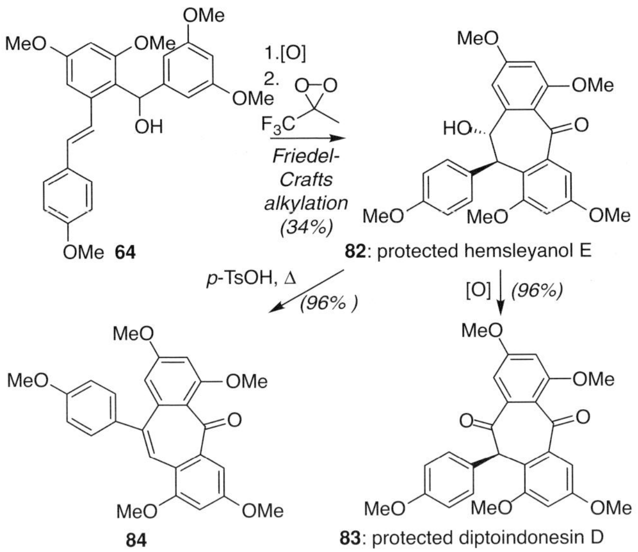

Fig. 13.16 Preparation of seven-membered rings for protected natural products such as diptoindonesin D (**83**) from key intermediate **64**. (From Snyder, 2011. Reproduced with permission.)

As a final comment on resveratrol-based structures, it is worth noting that other investigators (Kraus & Gupta, 2009; Chen *et al.*, 2010) have since used our key building block to prepare both shoreaphenol (**86**; Fig. 13.17; Saraswathym *et al.*, 1992) and amurensin H (**85**; Li *et al.*, 2006), while another (Jeffrey & Sarpong, 2009) has used its precursor

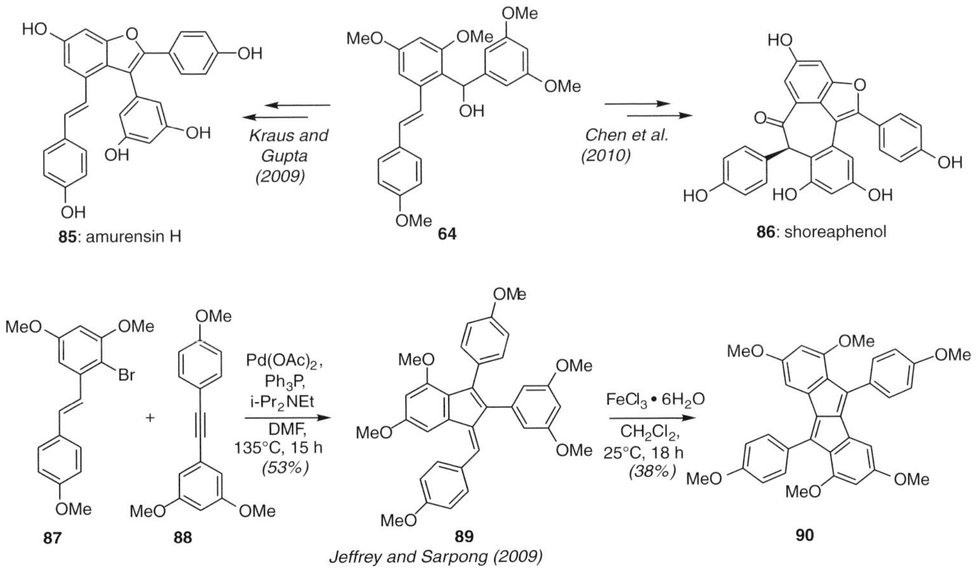

Fig. 13.17 Use of our key building blocks by other researchers to access resveratrol oligomers and analogs.

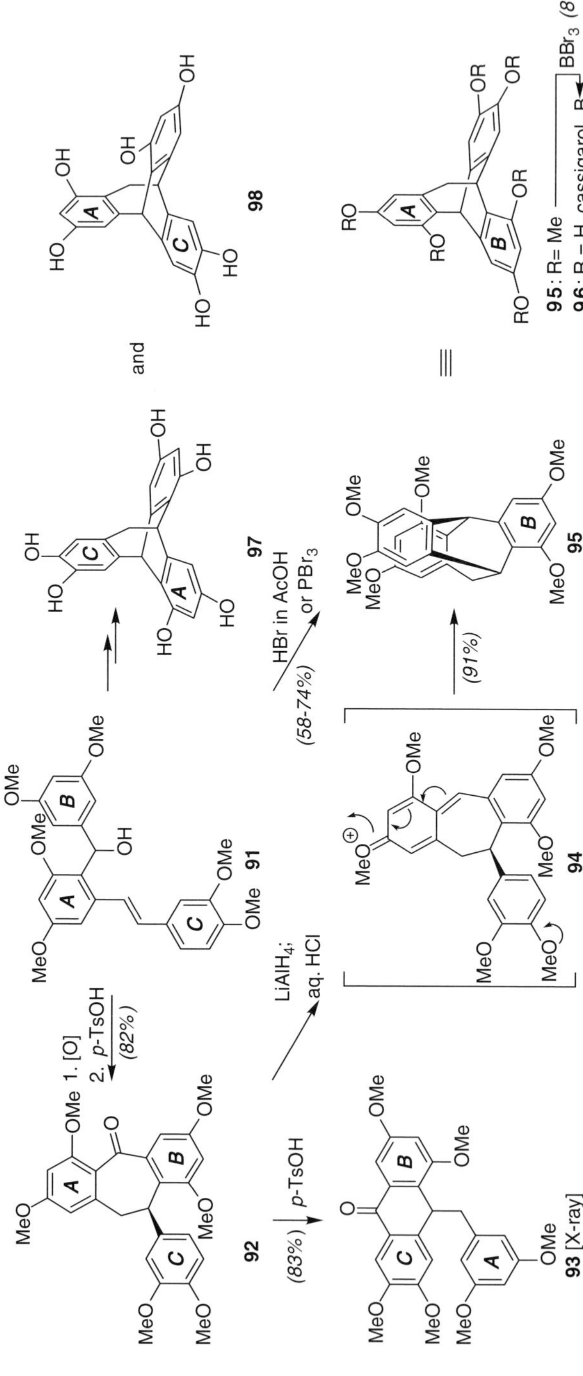

Fig. 13.18 Total synthesis of cassigarol B (**96**) and allied structures (**97** and **98**) from key intermediate **91**.

(**87**) in some innovative palladium-based cyclizations to access oxidized forms of several relevant frameworks. Time will tell if all of the frameworks of the family can indeed be derived from such pieces, though the structures prepared to date certainly seem to indicate that controlled synthesis of an entire oligomer family may well be possible if a different building block from Nature's starting material is employed.

Moreover, we believe that building blocks of type **63** (see Fig. 13.13) have value for accessing polyphenolic targets outside of the resveratrol class, and Figs. 13.18 and 13.19 present some of our more advanced work along these lines. In the first of these endeavors, we targeted the natural product cassigarol B (**96**; Baba *et al.*, 1988). As indicated, a series of Friedel–Crafts cyclizations provided the means to convert **91** into this material in a modicum of steps (just two to four operations); other chemoselective sequences (not shown) provided a way to access two other [3.2.2]-bicycles (**97** and **98**) in a controlled manner, noting that the aryl rings within each are rotated 120 degrees counterclockwise relative to the cassigarol B core. Careful control of reaction conditions, particularly in terms of temperature, proved critical for the selectivity of the various processes; for instance, the use of *p*-TsOH with the oxidized form of **91** afforded **92** in high yield when the reaction was maintained at low temperature, while use of higher temperatures induced **92** to convert into six-membered congener **93** through two proposed additional Friedel–Crafts-based processes (one being a retro Friedel–Crafts reaction; Snyder *et al.*, 2009). In the second of these efforts, two potent immunosuppressive natural products known as the dalesconols (**103** and **104**; Zhang *et al.*, 2008) were synthesized in several steps from key building block **100**, one which used naphthyl rings in place of the hydroxybenzenes of the resveratrol work (Snyder *et al.*, 2010). This synthesis was centered around a cyclization cascade that forged the entire polycyclic core from that critical starting material; in that key event, a sole carbon atom within the B-ring was used as both nucleophile and electrophile, ultimately becoming the quaternary center at the core junction of the molecules. While this event provided racemic product, it is noteworthy that these natural products were also isolated from Nature as a partial racemate (~30% e.e.). Moreover, it is worth noting that even minor alteration of reaction conditions and/or protecting groups with the same types of starting material in the key cyclization cascade afforded other unique structures, such as **105** and **107**, in moderate to good yield. As such, the starting materials used in these programs may globally be a "privileged" framework with which to access much architectural diversity (Welsch *et al.*, 2010).

13.3.3 *Rosmarinic-acid-derived oligomers*

Given the ability to achieve the controlled synthesis of the resveratrol family from an alternate starting point, we wondered whether or not that idea could be applied to another oligomer family. In that vein, we began explorations into preparing the natural products shown in Fig. 13.20: helisorin (**110**), helisterculin A (**111**), and the helicterins (**112** and **113**; Tezuka, *et al.*, 1999, 2000). These structures are formally dimeric and tetrameric assemblies of methylated forms of rosmarinic acid (**108**; Petersen and Simmonds, 2003) or a deprotected version of oresbiusin B (**109**; Huang *et al.*, 1996) if their unassigned chiral centers are of the same absolute configuration. Biosynthetically, there are a number of

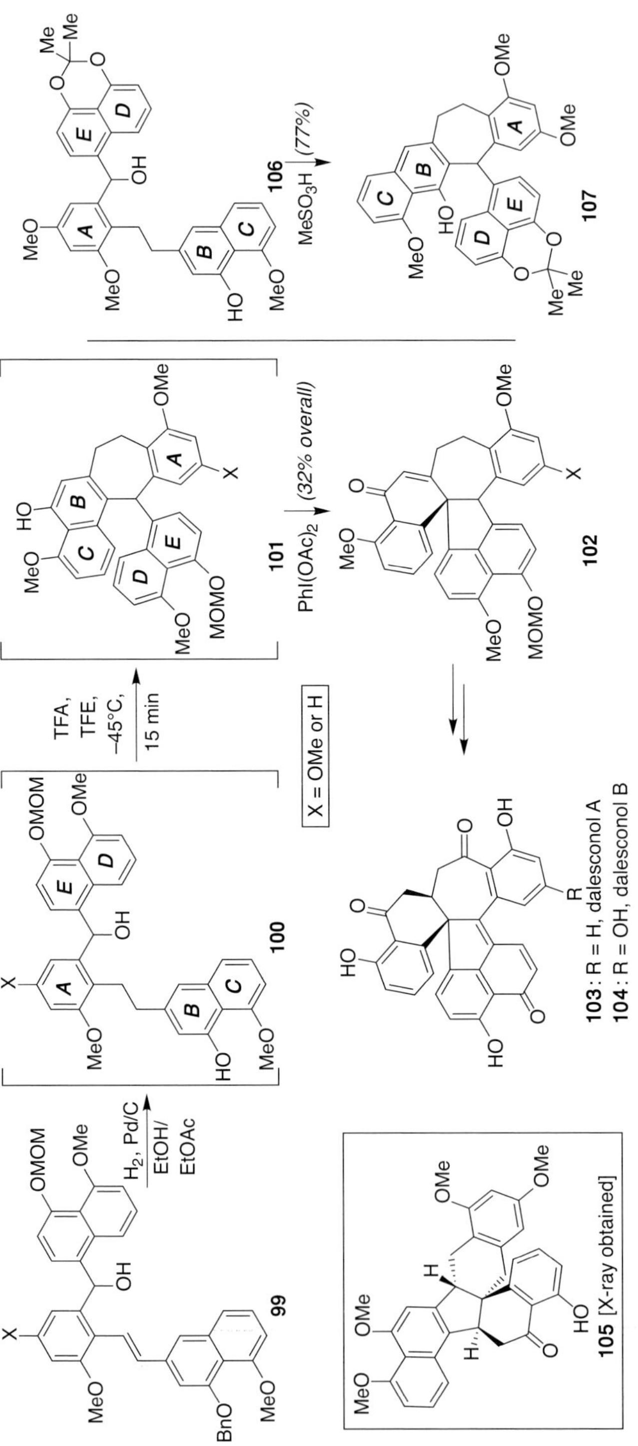

Fig. 13.19 Use of key precursor **99** to access the dalesconols (**103** and **104**) via a cascade-based cyclization sequence, as well as other unique nonnatural frameworks (such as **105** and **107**). (From Snyder, 2011. Reproduced with permission.)

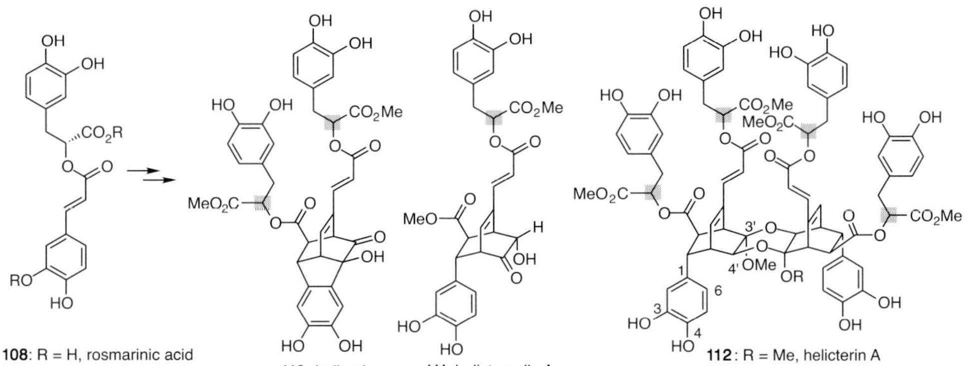

108: R = H, rosmarinic acid
109: R = Me, oresbiusin B

110: helisorin

111: helisterculin A

112: R = Me, helicterin A
113: R = H, helicterin B

Fig. 13.20   Structures of helicterin A and B (**112** and **113**) and related natural products (**110** and **111**); unassigned stereocenters are highlighted.

ways in which these types of compounds could arise from such precursors. As indicated in Fig. 13.21, the largest structures, the helicterins (represented here as **112**) are the likely product of a dimerization of hydroxyketones of type **114**; several dimerizations of [2.2.1]-based systems are known to produce such compounds (Creary & Rollin, 1977; Jauch *et al.*, 1991; Banks *et al.*, 1992). This material, in turn, could arise from diketone **115** if its more hindered carbonyl could be selectively reduced from its more hindered face. This compound could then be the product of rosmarinic acid monomer dimerization either through a Diels–Alder-based pathway, using **116** as dienophile and its oxidized form (**117**) as diene, or via controlled radical- and aldol-based bond constructions.

As was the case with the resveratrol-derived family, however, though such plans are easily written on paper, executing them in a reaction flask proved unachievable for us, at least in a series of model studies (Snyder & Kontes, 2009). For example, if model dienophile **120** was exposed, in great excess, to the oxidized form of **121**, the only Diels–Alder product observed (i.e., **122**) resulted from the reaction of the oxidized form of **121** with itself as both diene and dienophile.[4] This result was consistent with several variants of these materials as well. Only if we dramatically changed both the dienophile and the diene did we begin to achieve true intermolecular Diels–Alder events, but the resultant products (such as **124**) possessed the incorrect regiochemistry for the oligomer family in terms of the location of the electron-withdrawing group from the original dienophile (Liao *et al.*, 1999). Finally, radical-based reactions did lead to the controlled formation of a single product, but as indicated, the resultant structure was more of a resveratrol-family type (i.e. a dihydrofuran), not anything resembling the helicterin core.

Thus, based on these results, we felt that an alternate starting point was required to achieve success. Specifically, we wondered if a molecule such as **126** (Fig. 13.22), a fully functionalized variant of the previously synthesized (**122**) might constitute that type of

---

[4]The direct use of a diketone in this experiment was avoided because of the propensity of such materials to polymerize prior to reaction with any potential dienophile.

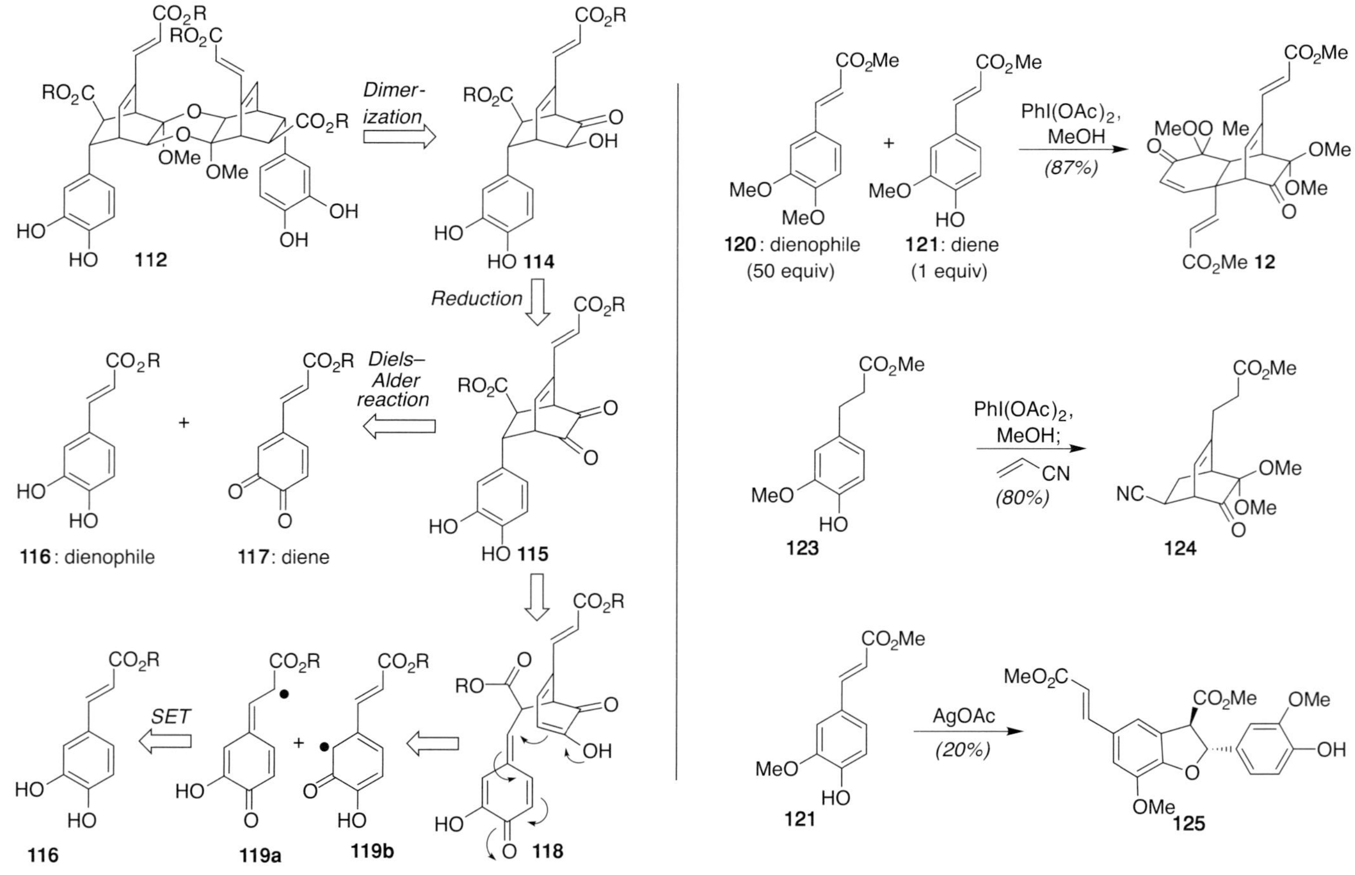

Fig. 13.21 Proposed biogenetic routes to the helicterins and selected model studies testing key elements of those proposals.

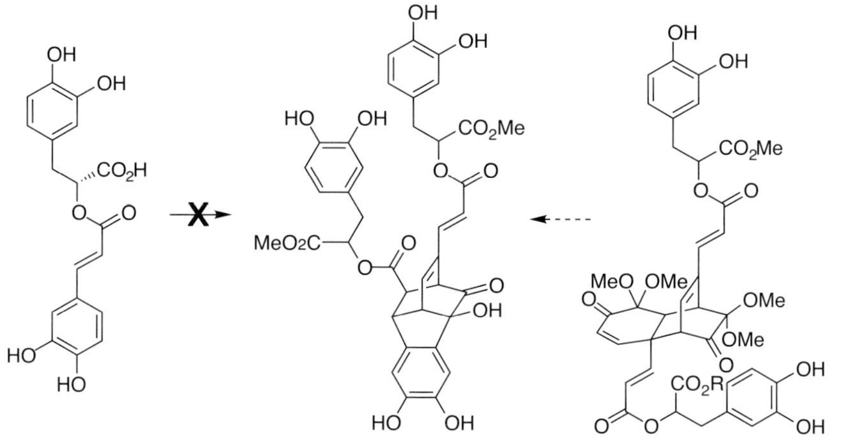

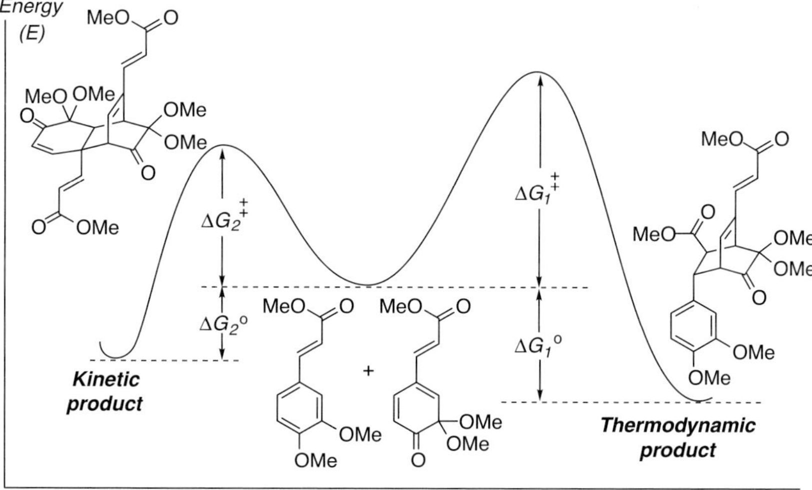

Fig. 13.22 Proposed alternate starting material to access helisorin (**110**) and related natural products through bond reorganization events.

material. The rationale for this selection was the idea that perhaps the Diels–Alder homo-dimeric product might constitute a kinetic outcome (Fig. 13.23). If heated to an appropriate temperature, perhaps it could revert to starting material and engage in an intermolecular Diels–Alder event with the desired dienophile (Singh & Samanta, 1999; Chittimalla *et al.*, 2002; Liao & Peddinti, 2002; Chittimalla *et al.*, 2006; Dong *et al.*, 2009). If true, then helicterin-like cores would have to constitute thermodynamic products (Fig. 13.23). As indicated in Fig. 13.24, exposure of compound **128** to just over five equivalents of **129** in mesitylene at 220°C for 30 minutes indeed provided compound **130** in 38% yield as a 1:1 mixture of diastereomers. Intriguingly, this result indicates that the remote chiral center

Fig. 13.23 Is the main Diels–Alder product a kinetic outcome?

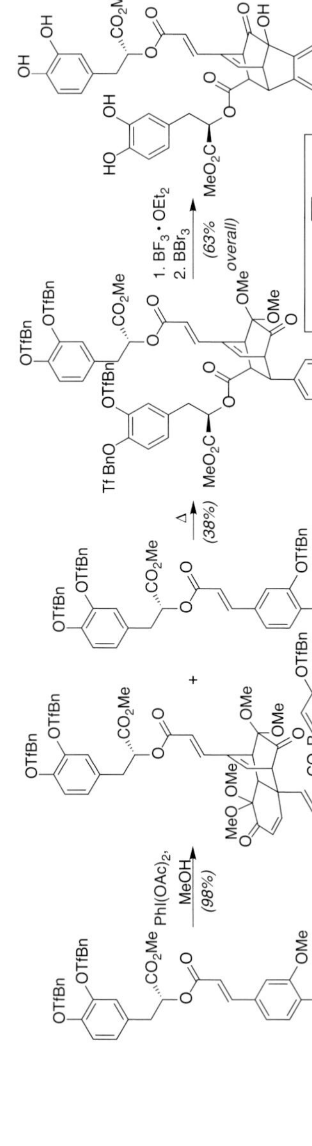

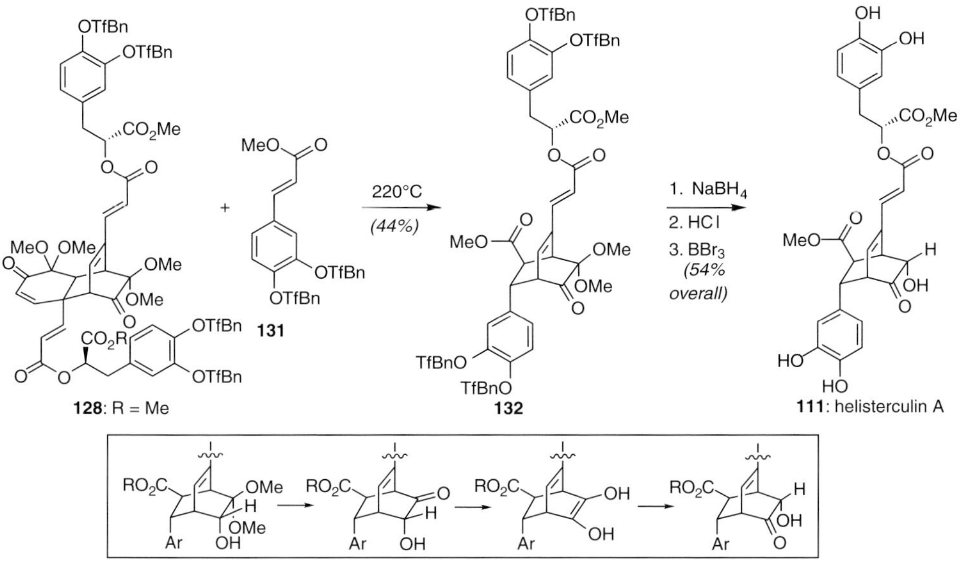

Fig. 13.25   Total synthesis of helisterculin A (**111**).

within the two reactants does not direct their facial presentation; thus, if this material is truly made through a Diels–Alder event in Nature, an enzyme must be involved to achieve diastereocontrol (Snyder & Kontes, 2011). From **130**, a Friedel–Crafts reaction and global deprotection of the 4-trifluoromethylbenzyl ethers, a protecting group judiciously chosen for its balance of stability versus lability in the presence of different Lewis acids, afforded the natural product helisorin (**110**) in 63% yield (Snyder & Kontes, 2009).

Application of the same general sequence using an alternate dienophile afforded the natural product helisterculin A (**111**, Fig. 13.25), with the key additional step here being an equilibration sequence to achieve the correct hydroxyketone stereochemistry; this process is noted mechanistically within the boxed structures when the reduced product formed from **132** was treated with HCl. Finally, use of **130** in similar equilibration chemistry, followed by a critical directed reduction (Evans & Chapman, 1986) and functional group manipulation, gave rise to hydroxyketal **134** as shown in Fig. 13.26. When this material was exposed to $BF_3 \bullet OEt_2$ followed by phenol protecting group removal with BBr_3, the large dimeric natural product helicterin B (**113**) was obtained in good yield if that final product was isolated quickly; indeed, with free catechol systems present, such compounds underwent aerial oxidization within minutes to afford polymeric materials. Of note, these sequences globally provided confirmation that all of the unassigned chiral centers do, in fact, correspond to rosmarinic acid. Equally important, they showed that **128** did indeed prove to be an intermediate that could lead to the entire family. Current work is seeking to determine if related strategies can afford access to the yunnaneic acid family of structures, compounds that look structurally like the helicterins but have slightly altered regiochemistry for key groups (Tanaka *et al.*, 1996). We are hopeful that these materials can be prepared, given that efforts to achieve the helicterin dimerization afforded compound **136** whose core is reminiscent of these types of architectures (as shown in Fig. 13.27).

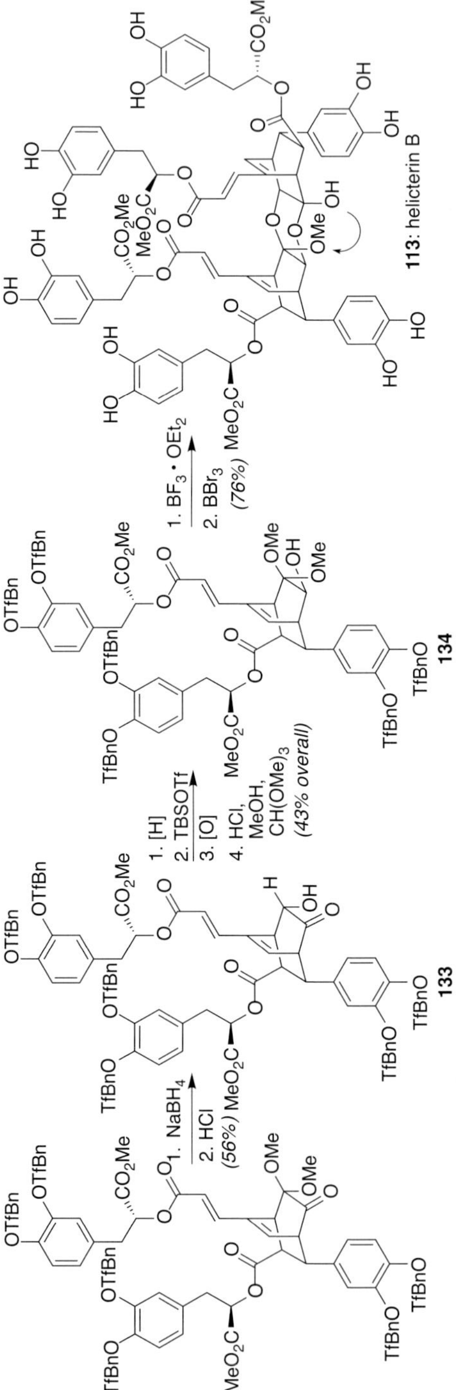

Fig. 13.26 Total synthesis of helicterin B (**113**).

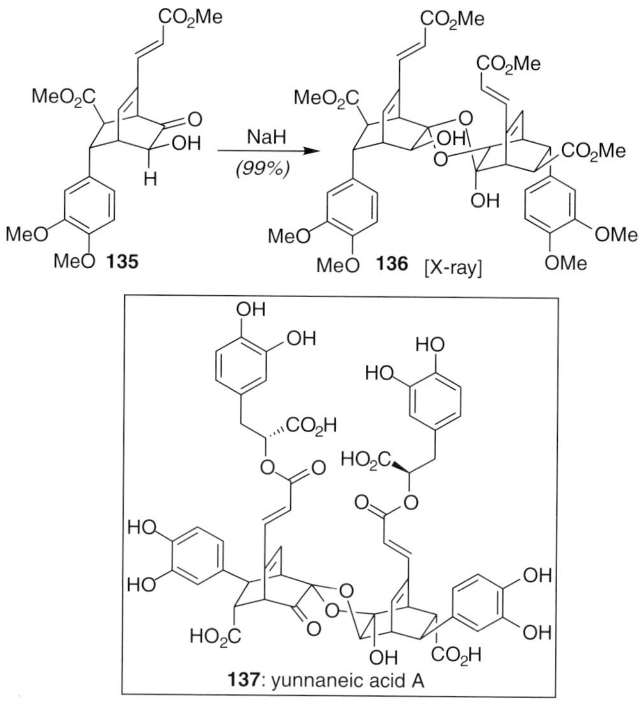

Fig. 13.27 An unexpected dimerization product may afford insight into the synthesis of yunnaneic acid A (**137**).

13.4 Conclusion

As revealed by the four main case studies within this chapter, the controlled synthesis of select members of oligomeric polyphenol families does appear possible, though it is clear in all cases that deliberate thought must be put into the overarching strategy to achieve success. These endeavors also hopefully demonstrate as well that polyphenolic natural products offer much opportunity for the development of new synthetic methods, particularly as more ornate and complex structures within these groups begin to be tackled. Finally, it is hoped that these endeavors globally will help to fuel additional biochemical studies, as the knowledge gained could be quite profound given what initial *in vitro* screens have suggested, as well as provide synthetic lessons applicable to oligomer families that are not polyphenol based.

Acknowledgments

The author would like to thank all of the undergraduate, graduate, and postdoctoral students within his group (past, present, and future) for their work on this exciting family of natural products and for their many contributions, both experimental and intellectual, which have made this project a true joy to pursue. Special thanks are noted to Ms. Alexandria Brucks,

Mr. Adel ElSohly, Mr. Daniel Griffith, Dr. Jan Marian van Hof, Dr. Ferenc Kontes, Mr. Daniel Treitler, and Mr. Daniel Wespe for their comments on the manuscript as well as assistance in its preparation. Financial support for our endeavors in the areas described from the National Institutes of Health (R01-GM84994), the Research Corporation for Science Advancement (Cottrell Scholar award to S.A.S.), the Alfred P. Sloan Foundation (Research Fellowship to S.A.S.), and Eli Lilly (Grantee award to S.A.S.) are gratefully acknowledged.

References

Aguirre, J.M., Aleso, E.N., Ibañez, A.F. Tombari, D.G. & Moltrasio Iglesias, G.Y. (1989) Reaction of 1,2-diarylethylamides with ethyl polyphosphate (EPP): correlation of the von Braun, Ritter, and Bischler-Napieralski reactions. *Journal of Heterocyclic Chemistry*, **26**, 25–27.

Aguirre, J.M., Aleso, E.N. & Moltrasio Iglesias, G.Y. (1999) Structure and stereochemistry of cyclodimers obtained by acid treatment of *trans*-stilbenes and of *N*-(1,2-diarylethyl) amides. *Journal of the Chemical Society, Perkin Transactions 1*, 1353–1358.

Baba, K., Maeda, K., Tabata, Y., Doi, M. & Kozawa, M. (1988) Chemical studies on the heartwood of *Cassia garrettiana* CRAIB. III.: structures of two new polyphenolic compounds. *Chemical and Pharmaceutical Bulletin*, **36**, 2977–2983.

Bais, H.P., Vepachedu, R., Gilroy, S., Callaway, R.M. & Vivanco, J.M. (2003) Allelopathy and exotic plant invasions: from molecules and genes to specific interactions. *Science*, **301**, 1377–1380.

Banks, M.R., Gosney, I., Grant, K.J., Reed, D. & Hodgson, P.K.G. (1992) Structure of Manasse's dimer from endo-2-hydroxyepicamphor by NOE difference spectroscopy. *Magnetic Resonance in Chemistry*, **30**, 996–999.

Baur, J.A., Pearson, K.J., Price, N.L., *et al.* (2006) Resveratrol improves health and survival of mice on a high-calorie diet. *Nature*, **444**, 337–342.

Brown, R.S. (1997) Investigation of the early steps in electrophilic bromination through the study of the reaction with sterically encumbered olefins. *Accounts of Chemical Research*, **30**, 131–137.

Bubb, W.A. & Sternhell, S. (1970) The Wessely acetoxylation. *Tetrahedron Letters*, **11**, 4499–4502.

Chen, D.Y.-K., Kang, Q. & Wu, R.T. (2010) Modular synthesis of polyphenolic benzofurans, and application in the total synthesis of malibatol A and shoreaphenol. *Molecules*, **15**, 5909–5927.

Chen, G., Shan, W., Ren, L., Dong, J. & Ji, Z. (2005) Synthesis and anti-inflammatory activity of resveratrol analogs. *Chemical and Pharmaceutical Bulletin*, **53**, 1587–1590.

Chen, Z.Y., Chan, P.T., Ho, K.Y., *et al.* (1996) Antioxidant activity of natural flavanoids is governed by number and location of their aromatic hydroxyl groups. *Chemistry and Physics of Lipids*, **79**, 157–163.

Chittimalla, S.K. & Liao, C.-C. (2002) The Diels–Alder reactions of 6,6-dimethoxycyclohexa-2,4-dienone generated by pyrolysis of its dimer. *Synlett*, 565–568.

Chittimalla, S.K., Shiao, H.-Y. & Liao, C.- C. (2006) Domino retro Diels–Alder/Diels–Alder reaction: an efficient protocol for the synthesis of highly functionalized bicyclo[2.2.2]octenones and bicyclo[2.2.2]octadienones. *Organic & Biomolecular Chemistry*, **4**, 2267–2277.

Cohen, R., Orlova, Y., Kovalev, M., Ungar, Y. & Shimon, E. (2008) Structural and functional properties of amylose complexes with genistein. *Journal of Agricultural and Food Chemistry*, **56**, 4212–4218.

Creary, X. & Rollin, A.J.J. (1977) Reaction of α-keto triflates with sodium methoxide. *The Journal of Organic Chemistry*, **42**, 4226-4230.

Dai, J.-R., Hallock, Y.F., Cardellina, J.H. & Boyd, M.R. (1998) HIV-inhibitory and cytotoxic oligostilbenes from the leaves of *Hopea malibato*. *Journal of Natural Products*, **61**, 351–353.

Diyasena, M.N.C., Sotheeswaran, S., Surendrakumar, S., Balasubramanian, S., Bokel, M., & Kraus, W. (1985) Balanocarpol, a new polyphenol from *Balanocarpus zeylanicus*(Trimen) and *Hopea jucunda*(Thw.)(Dipterocarpaceae). *Journal of the Chemical Society: Perkin Transactions 1*, 1807–1809.

Dong, S., Hamel, E., Bai, R., Covell, D.G., Beutler, J.A. and Porco, J.A. Jr. (2009) Enantioselective synthesis of (+)-chamaecypanone C: a novel microtubule inhibitor. *Angewandte Chemie International Edition*, **48**, 1494–1497.

Engler, T.A., Draney, B.W. & Gfesser, G.A. (1994) Evaluation of a synthetic route to ε-viniferin based on a new method for the stereoselective preparation of 2,3-diaryl-2,3-dihydrobenzofurans. *Tetrahedron Letters*, **35**, 1661–1664.

Engler, T.A., Gfesser, G.A. & Draney, B.W. (1995) Lewis acid-promoted reactions of unsymmetrically substituted stilbenes with 2-methoxy-1,4-benzoquinones: stereoselective synthesis of *trans*-2,3-diaryl-2,3-dihydrobenzofurans. *The Journal of Organic Chemistry*, **60**, 3700–3706.

Evans, D.A. & Chapman, K.T. (1986) The directed reduction of β-hydroxy ketones employing Me$_4$NHB(OAc)$_3$. *Tetrahedron Letters*, **27**, 5939–5942.

Favaron, F., Lucchetta, M., Odorizzi, S., Pais da Cunha, A.T. & Sella, L. (2009) The role of grape polyphenols on *trans*-resveratrol activity against *Botrytis cinerea* and of fungal laccase on the solubility of putative grape PR proteins. *Journal of Plant Pathology*, **91**, 579–588.

Feldman, K.S. (2005) Recent progress in ellagitannin chemistry. *Phytochemistry*, **66**, 1984–2000.

Feldman, K.S. & Lawlor, M.D. (2000) Ellagitannin chemistry. The first total synthesis of a dimeric ellagitannin, coriariin A. *Journal of the American Chemical Society*, **122**, 7396–7397.

Feldman, K.S., Sahasrabudhe, K., Quideau, S., Hunter, K.L. & Lawlor, M.D. (1999) Prospects and progress in ellagitannin synthesis. In: *Plant Polyphenols 2—Chemistry, Biology, Pharmacology, Ecology* (eds G.G. Gross, R.W. Hemingway & T. Yoshida), pp. 101–125. Kluwer Academic/Plenum Publishers, New York.

Fischbach, M.A. & Clardy, J. (2007) One pathway, many products. *Nature: Chemical Biology*, **3**, 353–355.

Ge, H.M., Huang, B., Tan, S.H., Shi, D.H., Song, Y.C. & Tan, R.X. (2006a) Bioactive oligostilbenoids from the stem bark of *Hopea exalata*. *Journal of Natural Products*, **69**, 1800–1802.

Ge, H.M., Xu, C., Wang, X.T., Huang, B. & Tan, R.X. (2006b) Hopeanol: a potent cytotoxin with a novel skeleton from *Hopea exalata*. *European Journal of Organic Chemistry*, 5551–5554.

Ge, H.M., Zhu, C.H., Shi, D.H., *et al.* (2008) Hopeahainol A: an acetylcholinesterase inhibitor from *Hopea hainanensis*. *Chemistry—A European Journal*, **14**, 376–381.

Guebailia, H.A., Chira, K., Richard, T., *et al.* (2006) Hopeaphenol: the first resveratrol tetramer in wines from North Africa. *Journal of Agriculture and Food Chemistry*, **54**, 9559–9564.

Haslam, E. (1998) *Practical Polyphenolics*. Cambridge University Press, Cambridge.

Hatano, T., Hattori, S. & Okuda, T. (1986) Tannins of *Coriaria japonica* A. GRAY. I. Coriariins A and B, new dimeric and monomeric hydrolyzable tannins. *Chemical and Pharmaceutical Bulletin*, **34**, 4092–4097.

He, S., Jiang, L., Wu, B., Li, C. & Pan, Y. (2009) Chunganenol: an unusual antioxidative resveratrol hexamer from *Vitis chunganensis*. *The Journal of Organic Chemistry*, **74**, 7966–7969.

He, Y.-H., Takaya, Y., Terashima, K. & Niwa, M. (2006) Determination of absolute structure of (+)-davidiol A. *Heterocycles*, **68**, 93–100.

Heiss, E.H., Schilder, Y.D.C. & Dirsch, V.M. (2007) Chronic treatment with resveratrol induces redox stress- and ataxia telangiectasia-mutated (ATM)-dependent senescence in p53-positive cancer cells. *The Journal of Biological Chemistry*, **282**, 26759–26766.

Hillis, W.E. & Inoue, T. (1967) Polyphenols of *Nothofagus* species. II. Heartwood of *Nothofagus fusca*. *Phytochemistry*, **6**, 59–67.

Howitz, K.T., Bitterman, K.J., Cohen, H.Y., *et al.* (2003) Small molecule activators of sirtuins extend *Saccharomyces cerevisiae* lifespan. *Nature*, **425**, 191–196.

Huang, H., Sun, H.-D., Wang, M.-S. & Zhou, S.-X. (1996) Phenolic compounds of *Isodon oresbius*. *Journal of Natural Products*, **59**, 1079–1080.

Huang, K.-S., Lin, M. & Cheng, G.-F. (2001) Anti-inflammatory tetramers of resveratrol from the roots of *Vitis amurensis* and the conformations of the seven-membered ring in some oligostilbenes. *Phytochemistry*, **58**, 357–362.

Ito, T., Abe, N., Oyama, M., *et al.* (2009a) Two novel resveratrol trimers from *Dipterocarpus grandiflorus*. *Helvetica Chimica Acta*, **92**, 1203–1216.

Ito, T., Abe, N., Oyama, M. & Iinuma, M. (2009b) Absolute structures of *C*-glucosides of resveratrol oligomers from *Shorea uliginosa*. *Tetrahedron Letters*, **50**, 2516–2520.

Ito, T., Akao, Y., Yi, H., *et al.* (2003a) Antitumor effect of resveratrol oligomers against human cancer cell lines and the molecular mechanism of apoptosis induced by vaticanol C. *Carcinogenesis*, **24**, 1489–1497.

Ito, T., Tanaka, T., Iinuma, M., *et al.* (2003b) New resveratrol oligomers in the stem bark of *Vatica pauciflora*. *Tetrahedron*, **59**, 5347–5363.

Ito, T., Tanaka, T., Iinuma, M., *et al.* (2003c) Two new oligostilbenes with dihydrobenzofuran, from the stem bark of *Vateria indica*. *Tetrahedron*, **59**, 1255–1264.

Ito, T., Tanaka, T., Iinuma, M., *et al.* (2004) Three new resveratrol oligomers from the stem bark of *Vatica pauciflora*. *Journal of Natural Products*, **67**, 932–937.

Jang, M., Cai, L., Udeani, G.O., *et al.* (1997) Cancer chemopreventive activity of resveratrol, a natural product derived from grapes. *Science*, **275**, 218–220.

Jauch, J., Schurig, V. & Walz, L. (1991) Structure of the dimer of racemic exo-3-hydroxybicyclo[2.2.1]heptan-2-one. *Zeitschrift für Kristallographie*, **196**, 255–260.

Jeffrey, J.L. & Sarpong, R. (2009) An approach to the synthesis of dimeric resveratrol natural products via a palladium-catalyzed domino reaction. *Tetrahedron Letters*, **50**, 1969–1972.

Kim, I. & Choi, J. (2009) A versatile approach to oligostilbenoid natural products—synthesis of permethylated analogues of viniferifuran, malibatol A, and shoreaphenol. *Organic and Biomolecular Chemistry*, **7**, 2788–2795.

Kitanaka, S., Ikezawa, T., Yasukawa, K., *et al.* (1990) (+)-α-viniferin, an anti-inflammatory compound from *Caragana chamlagu* root. *Chemical and Pharmaceutical Bulletin*, **38**, 432–435.

Kraus, G.A. & Gupta, V. (2009) A new synthetic strategy for the synthesis of bioactive stilbene dimers. A direct synthesis of amurensin H. *Tetrahedron Letters*, **50**, 7180–7183.

Kurihara, H., Kawabata, J., Ichikawa, S. & Mizutani, J. (1990) (–)-ε-viniferin and related oligostilbenes from *Carex pumila Thunb.* (*Cyperaceae*). *Agricultural and Biological Chemistry*, **54**, 1097–1099.

Kurosawa, W., Kobayashi, H., Kan, T. & Fukuyama, T. (2004) Total synthesis of (−)-ephedradine A: an efficient construction of optically active dihydrobenzofuran-ring via C–H insertion reaction. *Tetrahedron*, **60**, 9615–9628.

Langcake, P. & Pryce, R.J. (1977a) A new class of phytoalexins from grapevines. *Experientia*, **33**, 151–152.

Langcake, P. & Pryce, R.J. (1977b) Oxidative dimerisation of 4-hydroxystilbenes *in vitro*: production of a grapevine phytoalexin mimic. *Journal of the Chemical Society, Chemical Communications*, 208–210.

Li, H., Tanaka, T., Zhang, Y.-J., Yang, C.-R. & Kouno, I. (2007) Rubusuaviins A-F, monomeric and oligomeric ellagitannins from Chinese sweet tea and their α-amylase inhibitory activity. *Chemical and Pharmaceutical Bulletin*, **55**, 1325–1331.

Li, W., Li, H. & Hou, Z. (2006a) Total synthesis of (±)-quadrangularin A. *Angewandte Chemie International Edition*, **45**, 7609–7611.

Li, W., Li, H., Luo, Y., Yang, Y. & Wang, N. (2010a) Biosynthesis of resveratrol dimers by regioselective oxidative coupling. *Synlett*, 1247–1250.

Li, W., Li, H., Wang, A.X., Luo, Y. & Zang, P. (2010b) Unexpected formation of two novel resveratrol dimers. *Journal of Chemical Research (Synopses)*, **34**, 118–120.

Li, X.-C. & Ferreira, D. (2003) Stereoselective cyclization of stilbene derived carbocations. *Tetrahedron*, **59**, 1501–1507.

Li, Y., Yao, C., Bai, J., Lin, M. & Cheng, G. (2006b) Anti-inflammatory effect of amurensin H on asthma-like reaction induced by allergen in sensitized mice. *Acta Pharmacologica Sinica*, **27**, 735–740.

Liao, C.-C, Chu, C.-S., Lee, T.-H., *et al.* (1999) Generation, stability, dimerization, and Diels−Alder reactions of masked *o*-benzoquinones. Synthesis of substituted bicyclo[2.2.2]octenones from 2-methoxyphenols. *The Journal of Organic Chemistry*, **64**, 4102–4110.

Liao, C.-C. & Peddinti, R.K. (2002) Masked *o*-benzoquinones in organic synthesis. *Accounts of Chemical Research*, **35**, 856–866.

Lian, Y. & Hinkel, R.J. (2006) BiBr$_3$-initiated tandem addition/silyl-Prins reactions to 2,6-disubstituted dihydropyrans. *The Journal of Organic Chemistry*, **71**, 7071–7074.

Luo, H.-F., Zhang, L.-P. & Hu, C.-Q. (2001) Five novel oligostilbenes from the roots of *Caragana sinica. Tetrahedron*, **57**, 4849–4854.

Milne, J.C., Lambert, P.D., Schenk, S., *et al.* (2007) Small molecule activators of SIRT1 as therapeutics for the treatment of type 2 diabetes. *Nature*, **450**, 712–717.

Miyamoto, K., Kishi, N., Koshiura, R., Yoshida, T., Hatano, T. & Okuda, T. (1987) Relationship between the structures and the antitumor activities of tannins. *Chemical and Pharmaceutical Bulletin*, **35**, 814–822.

Murray, M. & Pizzorno, J. (1999) Procyanidolic oligomers. In: *The Textbook of Natural Medicine* (ed. M. Murray & J. Pizzorno), 2nd edn, pp. 899–902. Churchill, London.

Nicolaou, K.C., Kang, Q., Wu, T.R., Lim, C.S. & Chen, D.Y.-K. (2010) Total synthesis and biological evaluation of the resveratrol-derived polyphenol natural products hopeanol and hopeahainol A. *Journal of the American Chemical Society*, **132**, 7540–7548.

Nicolaou, K.C., Wu, T.R., Kang, Q. & Chen, D.Y.-K. (2009) Total synthesis of hopeahainol A and hopeanol. *Angewandte Chemie International Edition*, **48**, 3440–3443.

Niwa, M., Ito, J., Terashima, K., Koizumi, T., Takaya, Y., & Yan, K.-X. (2000) (–)-Ampelopsin D is different from (–)-quadrangularin A. *Heterocycles*, **53**, 1475–1478.

Ohguchi, K., Akao, Y., Matsumoto, K., *et al.* (2005) Vaticanol C-induced cell death is associated with inhibition of pro-survival signaling in HL60 human leukemia cell line. *Bioscience, Biotechnology, and Biochemistry*, **69**, 353–356.

Ohmori, K., Shono, N., Hatakoshi, Y., Yano, T. & Suzuki, K. (2011) Integrated synthetic strategy for higher catechin oligomers. *Angewandte Chemie International Edition*, **50**, 4862–4867.

Ohmori, K., Ushimaru, N. & Suzuki, K. (2004) Oligomeric catechins: an enabling synthetic strategy by orthogonal activation and C(8) protection. *Proceedings of the National Academy of Sciences, U.S.A.*, **101**, 12002–12007.

Ohyama, M., Tanaka, T., Ito, T., Iinuma, M., Bastow, K.F. & Lee, K.-H. (1999) Antitumor agents 200. Cytotoxicity of naturally occurring resveratrol oligomers and their acetate derivatives. *Bioorganic and Medicinal Chemistry Letters*, **9**, 3057–3060.

Okuda, T., Yoshida, T. & Hatano, T. (1995) Hydrolyzable tannins and related polyphenols. *Progress in the Chemistry of Organic Natural Products*, **66**, 1–117.

Oshima, Y., Ueno, Y., Hikino, H., Yang, L.L. & Yen, K.Y. (1990) Ampelopsins A, B and C, new oligostilbenes of *Ampelopsis brevipedunculata* var. *Hancei. Tetrahedron*, **46**, 5121–5126.

Petersen, M. & Simmonds, M.S.J. (2003) Rosmarinic acid. *Phytochemistry*, **62**, 121–125.

Pouységu, L., Deffieux, D., Malik, G., Natangelo, A. & Quideau, S. (2011) Synthesis of ellagitannin natural products. *Natural Product Reports*, **28**, 853–874.

Ricardo da Silva, J.M., Rigaud, J., Cheynier, V., *et al.* (1991) Procyanidin dimers and trimers from grape seed. *Phytochemistry*, **30**, 1259–1264.

Sahadin, Hakim, H.H., Juliawaty, L.D., *et al.* (2005) Cytotoxic properties of oligostilbenoids from the tree barks of *Hopea dryobalanoides. Zeitschrift fuer Naturforschung, C: Journal of Bioscience*, **60**, 723–727.

Sako, M., Hosokawa, H., Ito, T. & Iinuma, M. (2004) Regioselective oxidative coupling of 4-hydroxystilbenes: synthesis of resveratrol and ε-viniferin (*E*)-dehydrodimers. *The Journal of Organic Chemistry*, **69**, 2598–2600.

Saraswathym, A., Puroshothaman, K.K., Patra, A., Dey, A.K. & Kundu, A.B. (1992) Shoreaphenol, a polyphenol from *Shorea robusta. Phytochemistry*, **31**, 2561–2562.

Singh, V. & Samanta, B. (1999) Synthesis of tricyclo[7.2.2.0^{2,8}]tridecanes and photoreaction in the excited singlet state: a novel entry to the DCB carbon framework of phorbol. *Tetrahedron Letters*, **40**, 1807–1810.

Snyder, S.A. (2011) Synthetic approaches to the resveratrol-based family of oligomeric natural products. In: *Biomimetic Organic Synthesis* (eds. E. Poupon & B. Nay), pp. 695–721. Wiley-VCH, Weinheim.

Snyder, S.A., Breazzano, S.P., Ross, A.G., Lin, Y. & Zografos, A.L. (2009) Total synthesis of diverse carbogenic complexity within the resveratrol class from a common building block. *Journal of the American Chemical Society*, **131**, 1753–1765.

Snyder, S.A., ElSohly, A.M. & Kontes, F. (2011) Synthetic approaches to oligomeric natural products. *Natural Product Reports*, **28**, 897–924.

Snyder, S.A. & Kontes, F. (2009) Explorations into neolignan biosynthesis: concise total syntheses of helicterin B, helisorin, and helisterculin A from a common intermediate. *Journal of the American Chemical Society*, **131**, 1745–1752.

Snyder, S.A. & Kontes, F. (2011) Synthetic studies of biomimetic Diels–Alder processes toward the helicterin family of natural products. *Israel Journal of Chemistry*, **51**, 378–390.

Snyder, S.A., Sherwood, T.C. & Ross, A.G. (2010) Total syntheses of dalesconol A and B. *Angewandte Chemie International Edition*, **49**, 5146–5150.

Snyder, S.A., Zografos, A.L. & Lin, Y. (2007) Total synthesis of resveratrol-based natural products: a chemoselective approach. *Angewandte Chemie International Edition*, **46**, 8186–8191.

Soleas, G.J., Diamandis, E.P. & Goldberg, D.M. (1997) Resveratrol: a molecule whose time has come? and gone? *Clinical Biochemistry*, **30**, 91–113.

Sotheeswaran, S. & Pasuparthy, V. (1993) Distribution of resveratrol oligomers in plants. *Phytochemistry*, **32**, 1083–1092.

Supudompol, B., Likhitwitayawuid, K. & Houghton, P.J. (2004) Phloroglucinol derivatives from *Mallotus pallidus*. *Phytochemistry*, **65**, 2589–2594.

Szewczuk, L.M., Forti, L., Stivala, L.A. & Penning, T.M. (2004) Resveratrol is a peroxidase-mediated inactivator of COX-1 but not COX-2: a mechanistic approach to the design of COX-1 selective agents. *The Journal of Biological Chemistry*, **279**, 22727–22737.

Szewczuk, L.M., Lee, S.H., Blair, I.A. & Penning, T.M. (2005) Viniferin formation by COX-1: evidence for radical intermediates during co-oxidation of resveratrol. *Journal of Natural Products*, **68**, 36–42.

Takaoka, M. (1940) Phenolic substances of white hellebore (*Veratrum grandiflorum* loes. Fil.). synthesis of resveratrol and its derivatives. *Proceedings of the Imperial Academy of Tokyo*, **16**, 405–407.

Takaya, Y., Terashima, K., Ito, J., *et al.* (2005) Biomimic transformation of resveratrol. *Tetrahedron*, **61**, 10285–10290.

Takaya, Y., Yan, K.-X., Terashima, K., He, Y.-H. & Niwa, M. (2002a) Biogenetic reactions on stilbenetetramers from vitaceaeous plants. *Tetrahedron*, **58**, 9265–9271.

Takaya, Y., Yan, K.-X., Terashima, K., Ito, J. & Niwa, M. (2002b) Chemical determination of the absolute structures of resveratrol dimers, ampelopsins A, B, D and F. *Tetrahedron*, **58**, 7259–7265.

Tanaka, T., Iliya, I., Ito, T., *et al.* (2001a) Stilbenoids in lianas of *Gnetum parvifolium*. *Chemical and Pharmaceutical Bulletin*, **49**, 858–862.

Tanaka, T., Ito, T., Iinuma, M., Ohyama, M., Ichise, M. & Tateishi, Y. (2000a) Stilbene oligomers in roots of *Sophora davidii*. *Phytochemistry*, **53**, 1009–1014.

Tanaka, T., Ito, T., Nakaya, K., Iinuma, M. & Riswan, S. (2000b) Oligostilbenoids in stem bark of *Vatica rassak*. *Phytochemistry*, **54**, 63–69.

Tanaka, T., Ito, T., Nakaya, K., *et al.* (2001b) Six new heterocyclic stilbene oligomers from stem bark of *Shorea hemsleyana*. *Heterocycles*, **55**, 729–740.

Tanaka, T., Nishimura, A., Kouno, I., Nonaka, G. & Young, T.-J. (1996) Isolation and characterization of yunnaneic acids A–D, four novel caffeic acid metabolites from *Salvia yunnanensis*. *Journal of Natural Products*, **59**, 843–849.

Taylor, R.J.K. (1999) Recent developments in Ramberg–Bäcklund and episulfone chemistry. *Chemical Communications*, 217–227.

Tezuka, Y., Terazono, M., Kusumoto, T.I., *et al.* (1999) Helisterculins A and B, two new (7.5′,8.2′)-neolignans, and helisorin, the first (6.4′,7.5′,8.2′)-neolignan, from the Indonesian medicinal plant *Helicteres isora*. *Helvetica Chimica Acta*, **82**, 408–417.

Tezuka, Y., Terazono, M., Kusumoto, T.I., *et al.* (2000) Helicterins A–F, six new dimeric (7.5′,8.2′)-neolignans from the Indonesian medicinal plant *Helicteres isora*. *Helvetica Chimica Acta*, **83**, 2908–2919.

Thomas, N.F., Lee, K.C., Paraidathathu, T., *et al.* (2002a) Tandem pericyclic reactions in a new FeCl$_3$-promoted synthesis of catechol analogues of restrytisol C. *Tetrahedron*, **58**, 7201–7206.

Thomas, N.F., Lee, K.C., Paraidathathu, T., Weber, J.F.F. & Awang, K. (2002b) Manganese triacetate oxidative lactonization of electron-rich stilbenes possessing pyrocatechol and resorcinol substitution (resveratrol analogs). *Tetrahedron Letters*, **43**, 3151–3155.

Trost, B.M. (1991) The atom economy—a search for synthetic efficiency. *Science*, **254**, 1471.

van Baarlen, P., Legendre, L. & van Kan, J.A.L. (2004) Plant defense compounds against *Botrytis* infection. In: *Botrytis—Biology, Pathology and Control* (eds. Y. Elad, B. Williamson, P. Tudzynski & N. Delen), pp. 143–161. Kluwer Academic Publishers, the Netherlands.

Velu, S.S., Thomas, N.F., Buniyamin, I., *et al.* (2008) Regio- and stereoselective biomimetic synthesis of oligostilbenoid dimers from resveratrol analogues: influence of the solvent, oxidant, and substitution. *Chemistry—A European Journal*, **14**, 11376–11384.

Walle, T., Hsieh, F., DeLegge, M.H., Oatis, J.E. & Walle, U.K. (2004) High absorption but very low bioavailability of oral resveratrol in humans. *Drug Metabolism and Disposition*, **32**, 1377–1382.

Wang, S., Ma, D. & Hu, C. (2005) Three new compounds from the aerial parts of *Caragana sinica*. *Helvetica Chimica Acta*, **88**, 2315–2321.

Welsch, M., Snyder, S.A. & Stockwell, B.R. (2010) Privileged scaffolds for library design and drug discovery. *Current Opinion in Chemical Biology*, **14**, 347–361.

Wood, J.G., Rogina, B., Lavu, S., *et al.* (2004) Sirtuin activators mimic caloric restriction and delay aging in metazoans. *Nature*, **430**, 686–689.

Yamada, M., Hayashi, K., Hayashi, H., *et al.* (2006a) Stilbenoids of *Kobresia nepalensis* (Cyperaceae) exhibiting DNA topoisomerase II inhibition. *Phytochemistry*, **67**, 307–313.

Yamada, M., Hayashi, K., Ikeda, S., *et al.* (2006b) Inhibitory activity of plant stilbene oligomers against DNA topoisomerase II. *Biological and Pharmaceutical Bulletin*, **29**, 1504–1507.

Yang, G., Zhou, J., Li, Y. & Hu, C. (2005) Anti-HIV bioactive stilbene dimers of *Caragana rosea*. *Planta Medica*, **71**, 569–571.

Yao, C.-S., Lin, M. & Wang, Y.-H. (2004) Synthesis of the active stilbenoids by photooxidation reaction of *trans*-ε-viniferin. *Chinese Journal of Chemistry*, **22**, 1350–1355.

Zhang, Y., Li, S.-Z., Li, J., *et al.* (2006) Using unnatural protein fusions to engineer resveratrol biosynthesis in yeast and mammalian cells. *Journal of the American Chemical Society*, **128**, 13030–13031.

Zhang, Y.L., Ge, H.M., Zhao, W., *et al.* (2008) Unprecedented immunosuppressive polyketides from *Daldinia eschscholzii*, a mantis-associated fungus. *Angewandte Chemie International Edition*, **47**, 5823–5826.

Index